Quantum Mechanics

This original and innovative textbook takes the unique perspective of introducing and solving problems in quantum mechanics using linear algebra methods, to equip readers with a deeper and more practical understanding of this fundamental pillar of contemporary physics. Extensive motivation for the properties of quantum mechanics, Hilbert space, and the Schrödinger equation is provided through analysis of the derivative, while standard topics like the harmonic oscillator, rotations, and the hydrogen atom are covered from within the context of operator methods.

Advanced topics forming the basis of modern physics research are also included, such as the density matrix, entropy, and measures of entanglement. Written for an undergraduate audience, this book offers a unique and mathematically self-contained treatment of this hugely important topic. Students are guided gently through the text by the author's engaging writing style, with an extensive glossary provided for reference and numerous homework problems to expand and develop key concepts. Online resources for instructors include a fully worked solutions manual and lecture slides.

Andrew J. Larkoski is Acting Assistant Professor at SLAC National Accelerator Laboratory. He earned his Ph.D. from Stanford University and has held research appointments at Massachusetts Institute of Technology and Harvard University, and he was also a Visiting Assistant Professor at Reed College. A leading expert on particle physics, he has published over 50 articles, and won the LHC Theory Initiative Fellowship and the Wu-Ki Tung Award for Early-Career Research on QCD. His previous textbook with Cambridge, *Elementary Particle Physics: An Intuitive Introduction* (2019), offers a modern and accessible introduction to the guiding principles of particle physics.

Quantum Mechanics

A Mathematical Introduction

ANDREW J. LARKOSKI

SLAC National Accelerator Laboratory

Shaftesbury Road, Cambridge CB2 8EA, United Kingdom

One Liberty Plaza, 20th Floor, New York, NY 10006, USA

477 Williamstown Road, Port Melbourne, VIC 3207, Australia

314–321, 3rd Floor, Plot 3, Splendor Forum, Jasola District Centre, New Delhi – 110025, India

103 Penang Road, #05–06/07, Visioncrest Commercial, Singapore 238467

Cambridge University Press is part of Cambridge University Press & Assessment, a department of the University of Cambridge.

We share the University's mission to contribute to society through the pursuit of education, learning and research at the highest international levels of excellence.

www.cambridge.org
Information on this title: www.cambridge.org/highereducation/isbn/9781009100502
DOI: 10.1017/9781009118026

First published 2023

Printed in the United Kingdom by TJ Books Limited, Padstow, Cornwall

A catalogue record for this publication is available from the British Library.

ISBN 978-1-009-10050-2 Hardback

Additional resources for this publication at www.cambridge.org/larkoski-quantum.

Cambridge University Press & Assessment has no responsibility for the persistence or accuracy of URLs for external or third-party internet websites referred to in this publication and does not guarantee that any content on such websites is, or will remain, accurate or appropriate.

For Patricia, Henry, and Margaret

Contents

Preface

I joke with my students that I only know how to solve one differential equation: the linear, first-order, homogeneous equation whose solutions are, of course, just exponential functions. While a joke, there is some purpose behind it. One approach to quantum mechanics is just to identify the Schödinger equation as a second-order linear differential equation for the wavefunction, and then go to town finding its closed-form solutions, by hook or by crook. However, there's no physics in doing this and quantum mechanics fundamentally isn't a theory of differential equations. It is a theory in which linear operators and vector spaces and eigenvalues are central. So, while solving differential equations is very concrete and teaches students useful mathematical tools and tricks, it *isn't* quantum mechanics. In this book, I stick to the joke: no results require differential equations more complicated than the linear, first-order, homogeneous type and exploiting linear algebra is used to solve problems.

Approach

In this spirit, the motivation for a quantum mechanical description of Nature presented in this book is markedly different than other books. Instead of starting with the Schrödinger equation as an axiom or describing a two-state system with a state vector, I go back to linear algebra and work to rephrase results with which students are familiar in a more physical language. With this physical foundation front and center, I focus on a mathematical description of observables in quantum mechanics, which are, of course, Hermitian operators. However, from this starting point, it is abundantly clear that only an operator's eigenvalues could be the outcome of a measurement, as only the eigenvalues of an operator are basis-independent. There is no confusion about the states on the Hilbert space actually being unphysical because defining matrix elements of a Hermitian operator require a state and its conjugate-transpose. Focusing on observables clarifies and in some ways diminishes the role of the Hilbert space to just be the space of vectors that can be used to define matrix elements of operators.

Inevitably, through a course on quantum mechanics, students should have minor existential crises. Never before in a physics class have they been pushed so far out of their everyday experience and told to trust mathematics to guide them. Of course, the predictions of quantum mechanics have been borne out in innumerable experiments, but why there should be such a dramatic change to how the universe operates at small distances can be troubling. I don't think that there is any way to completely alleviate this struggle (and many seasoned physicists still get confused!), but I do think it can be minimized or constructively focused through a textbook application of the scientific method. Fundamentally, quantum mechanics says that the universe is described by linear operators on a complex-valued vector space. This, on its face, is perhaps not so troubling, and then from it the weirdness of quantum mechanics follows as a consequence. The scientific method provides no explanation for where its hypotheses come from, but these assumptions are much more likely to be palatable by a student than starting with the full weight of the Schrödinger equation. In this book, I work to follow this application of the scientific method as strictly as possible. Importantly, this is necessarily an ahistorical introduction to the subject.

This textbook was born out of an introductory course on quantum mechanics that I taught at Reed College in spring semester 2020. Normally, the date when I taught the course first would be irrelevant, but 2020 was no normal year. Halfway through that semester, the COVID-19 pandemic forced our campus, and essentially every institution of higher learning in the world, to close to in-person instruction. This dramatic and sudden move necessitated that I develop substantial lecture notes for students in the course, as my virtual lectures were irregular or non-existent, while attempting to personally manage the new world we found ourselves in. Since that time, we have all adapted to teaching and learning virtually, and I hope have become successful at doing so. Front and center to any format of instruction, in-person or online, is a readable and detailed textbook that shines a light when an instructor isn't accessible. My goal with this book is to be such a resource, in whatever the future of higher education brings.

Structure of This Book

This book is broadly organized into four parts. Chapters 1 through 4 slowly develop the mathematical formalism of quantum mechanics starting from basic principles of linear algebra. Especially for Chapter 2, it may seem like the pace is glacially slow and there would be substantial overlap with a dedicated course on linear algebra. However, an introductory linear algebra course taught by a mathematics department is often strictly devoted to understanding the properties of matrices and their action on real-valued vectors. This is indeed necessary background for quantum mechanics, but substantial generalization is needed. In particular, the linear derivative operator is used as a concrete and familiar example of a linear operator, and consequences of its linearity – like the Taylor expansion as exponentiation of it – gently introduce students to the power and breadth of linear algebra. Operators take center stage and the Hilbert space is introduced as the space of vectors that can be used to isolate individual matrix elements of operators, which physically correspond to observables. The momentum or spatial translation operator and the Hamiltonian or temporal translation operator are identified as special operators and their properties are developed through analogy with familiar classical momentum and energy. With the necessary linear algebra developed, Chapter 4 introduces the Dirac–von Neumann axioms of quantum mechanics and some of their most fundamental yet counterintuitive consequences, like the Heisenberg uncertainty principle.

Chapters 5 through 7 present several well-worn examples of one-dimensional quantum mechanical systems to illustrate the formalism developed in the early chapters. In particular, the infinite square well, harmonic oscillator, and free particle are introduced and analyzed in detail. Additionally, through them, other ideas in quantum mechanics are introduced within a concrete framework. For example, the correspondence principle is discussed with the infinite square well in which the quantum probability distributions transform to their classical counterparts at large energy. In the context of the harmonic oscillator, coherent states are introduced as eigenstates of the lowering operator and are explicitly shown to saturate the Heisenberg uncertainty principle. The S-matrix and its properties, such as the optical theorem and bound states as poles at complex-valued momentum, are introduced with the free particle. While these examples and other results presented here are general features of quantum mechanical systems, they are restricted to one spatial dimension for simplicity.

Chapters 8 and 9 move beyond one spatial dimension to describe more realistic systems in full three dimensions. Angular momentum is introduced as the operator that generates rotations. Additionally, Lie algebras and groups are discussed, along with the irreducible representations of the rotation group as indexed by half- or whole-integer values of spin. Spin-1/2 is observed to be a consequence of the simple-connectedness of three-dimensional rotations, as demonstrated by the plate trick. Angular momentum also provides a concrete framing for understanding quantum numbers and Noether's theorem in quantum mechanics. In Chapter 9, the hydrogen atom is presented and its Hamiltonian is diagonalized. Unlike standard treatments in other books, in which the hydrogen Hamiltonian is diagonalized in position space through solution of a differential equation, a different approach is taken. The conservation of the Laplace–Runge–Lenz vector is explicitly demonstrated classically, and then its Poisson brackets are upgraded to quantum commutators through canonical quantization. The hydrogen Hamiltonian is then diagonalized through analysis of the Lie algebra formed from the angular momentum and Laplace–Runge–Lenz operators.

Finally, Chapters 10 through 13 discuss more advanced topics and connect to ways that quantum mechanics is being studied contemporaneously. Chapter 10 introduces methods of approximation for determining eigenvalues of any operator, but focused on identification of the ground state of the Hamiltonian with an arbitrary potential. Perturbation theory, the variational method, the power method, and the WKB approximation are discussed and examples are presented to demonstrate their efficacy. Chapter 11 motivates and derives the path integral formulation of quantum mechanics, from the Schrödinger equation perspective used in most of the rest of the book. Conversely, it is demonstrated that the Schrödinger equation can be derived starting from the path integral and the harmonic oscillator path integral is explicitly calculated. Chapter 12 introduces the density matrix as the most general description of an ensemble of quantum particles. Numerous consequences of the density matrix are presented, including quantum entanglement, the von Neumann entropy, and Bell's inequalities. To close the book, Chapter 13 surveys some of the unresolved problems of quantum mechanics, including the measurement problem and decoherence, demonstrating that there's still a long way to go before we understand what quantum mechanics is.

Key Features

In addition to the unique motivation, topics, and analysis of problems in quantum mechanics, this book has a number of other features.

Boxed Fundamental Equations

The most important equations of quantum mechanics, and hence some of the most important equations in all of physics, presented in this book are boxed. Highlighted expressions include the Schrödinger equation, the Heisenberg uncertainty principle, the canonical commutation relation, and the path integral.

Supplementary Appendices

Several supplementary appendices are provided at the end of the text covering reviews of relevant mathematics of quantum mechanics (linear algebra, common integrals, probability) and Poisson brackets in classical mechanics. These topics are covered in the body of the text as well, but provide a quick, centralized reference.

Extensive In-Text Referencing

Foundational papers in quantum mechanics are referenced as footnotes throughout the textbook and are collected in the Bibliography at the end of the book for easy reference. Additionally, several textbooks and series of lecture notes at the undergraduate and graduate level are provided in Appendix C: Further Reading.

Glossary of Over 200 Quantum Mechanics Terms

With the formalism of quantum mechanics comes a whole new vocabulary for describing its properties and components. Throughout the text, over 200 technical terms or jargon are highlighted in boldface font and precise and concise definitions of all terms are provided in a the Glossary at the end of the book.

How to Teach With This Book

When I taught the quantum mechanics course in 2020, my lecture notes for the 13-week semester were the basis for this textbook. The notes covered material from every chapter in this book, but generally much less material was covered in the course than is presented here. As mentioned earlier, this was definitely less than ideal, but given the global circumstances still provided students with a broad survey of the topic. For an in-depth introduction of quantum mechanics, a course should go through the material significantly slower and possibly spend multiple weeks on individual chapters.

In its present form, the content of Chapters 1–9 of this book could be used for a solid one-semester (about 13 weeks) course. This would get up through the hydrogen atom, and particularly for Chapters 7, 8, and 9, multiple weeks would probably be necessary to cover each of them in detail. For a one-quarter course (about 10 weeks), perhaps only through Chapter 7 could be covered. However, as Chapters 5, 6, and 7 introduce features of quantum mechanics through explicit examples, some of the material in these chapters could be compressed or eliminated to have time to study later material, like angular momentum or hydrogen. I think that the entire book could be covered in detail in a two-quarter course (about 20 weeks). Especially for Chapters 11 and 12 on the path integral and density matrix, respectively, instructors might have additional topics they would supplement the course with to discuss more modern applications of these techniques, like quantum field theory or quantum computation. Textbooks or lecture notes that can be used to supplement this book appear in Appendix C: Further Reading.

Acknowledgments

First and foremost, I want to thank the Physics 342 Quantum Mechanics class for spring semester 2020. The lecture notes on which this textbook is based was developed for that course and those students were the first test subjects. More than just that though, this class endured the abrupt transition from in-person lectures to exclusively online learning. While we were all forced into the same situation outside of anyone's control, I can't thank the class enough for their patience, engagement, and dedication to the material throughout the entire semester.

My personal education in quantum mechanics has been influenced by essentially everyone with whom I have talked physics, and to everyone, thank you. However, there are a few people whose contribution to my understanding of quantum mechanics was so profound that I want to identify them individually.

My understanding of quantum mechanics was most profoundly influenced by a course that I *didn't* take. In my senior year as an undergraduate, I contacted Ann Nelson to ask if I could enroll in her legendary graduate-level quantum mechanics course. She said that I could as my grades were strong enough, but I would have plenty of time to take any physics class I wanted when I was in graduate school. That year was the last time I could take a course on art history, or learn Russian, or read the Beat poets, she said. So, I heeded her advice and took second-year French and just got the materials for her quantum course to study over the summer. (Apparently, this was good enough because I convinced Stanford University to waive the quantum mechanics requirement in graduate school!) This has been the single most profound piece of career advice I have ever received, that there's always time for physics later, and it is a story that I pass on to students when they ask for advice. For this and all your support, Ann, thank you.

As a graduate student, I was a teaching assistant for Pat Burchat's undergraduate course on quantum mechanics. An aspect of this course were recitation sections or worked problem sessions in which the teaching assistants would work with students in small groups on more conceptual quantum mechanics problems. Along with the other teaching assistant, Kiel Howe, we developed a worksheet that provided more motivation for the Schrödinger equation than that provided in the textbook (which was none, because the textbook for the course was that by David Griffiths). I thank both Pat and Kiel for working on that course and teaching quantum mechanics definitely gave me a new perspective on the subject.

Through the process of publishing two textbooks, from prospectus to external reviews to publication, I thank my editor at Cambridge University Press, Nick Gibbons, for his continued support. This book and my earlier book on particle physics have turned out better than I could have imagined, and Nick's guidance and oversight have made the publishing experience exceptional. For the highly structured layout of this book, I thank development editor Melissa Shivers for her comments and suggestions, ensuring that the final product is both readable and pedagogical.

While the perspective, on motivating quantum mechanics and introducing its parts and connections in this book are my own, the topics presented are well-established in numerous other resources. David Griffiths's book was my inspiration for the presentation here of the algebraic diagonalization of the harmonic oscillator and approximation methods (Chapters 6 and 10). J. J. Sakurai's unparalleled graduate-level book was much of the basis for my discussion of scattering in Chapter 7 and the density matrix in Chapter 12, though at a more elementary level. Other resources were extensively used as references for details throughout this book, such as Joseph Polchinski's *String Theory: Volume 1* textbook for the derivation of the path integral in Chapter 11; David Tong's lecture notes on the S-matrix in Chapter 7; Hitoshi Murayama's lecture notes for the calculation of the harmonic oscillator path integral in Chapter 11; and my textbook on particle physics for strange properties of rotations in three dimensions in Chapter 8. All of these are referenced in Appendix C: Further Reading.

Finally, I want to thank my family, especially my children, Henry and Margaret, and my wife, Patricia, for continuing to remind me that there's more to life than physics.

The valediction in this book is the same as in my particle physics book: any successes of this book are due to these people, and any failures are all my own.

Introduction

Welcome to the world of quantum mechanics! I'll be your guide through this fascinating, counterintuitive, and extremely rich subject. Quantum mechanics underlies nearly all of contemporary physics research; from the inner workings of atoms, to properties of materials, to the physics of neutron stars, to what happens at a black hole. Additionally, quantum mechanics and its consequences are exploited in modern technology and are beginning to produce major breakthroughs in computing. In this book, we will introduce the formalism, axioms, and a new way of thinking quantum mechanically that will require a re-interpretation of the classical physics you have learned.

1.1 Structure of This Book

I want to first motivate the structure of topics covered in this book, which is a bit different than other books on quantum mechanics. In your first course on (classical) physics, the organization of topics was likely the following:

(a) Introduce formalism and common language: kinematics; acceleration, velocity as concepts.
(b) Define relevant units and quantities: SI units; meters, seconds, kilograms, etc.
(c) Identify *the* fundamental equation: $\vec{F} = m\vec{a}$.
(d) *Many* examples of the fundamental equation: friction, blocks on ramps, springs, etc.
(e) Conservation laws: energy, momentum, angular momentum.
(f) Strange phenomena of rotations and oscillations: precession, etc.

For each broad topic, I have also provided explicit examples of each from introductory physics.

When first introduced to a topic, you need to learn the language in which you express that topic. In classical mechanics that is kinematics, and concepts like acceleration, while in quantum mechanics that will be **linear operators**, **states**, and something called the **Hilbert space**. This perspective requires agreed-upon measurement conventions (meters, etc.) and in quantum mechanics we will see that there is a new, fundamental quantity called $\hbar$ (read: "h-bar") that sets the scale for everything. With this language established, we then identify the fundamental equation. Classically (at least in introductory physics), this is Newton's second law, while quantum mechanically it will be the **Schrödinger equation**. The rest of the course is then a study of examples, or reformulation for special systems employing conservation laws: energy, momentum, and angular momentum conservation. These conservation laws will be absolutely central to quantum mechanics. Finally, the last week or two in an introductory course typically focuses on strange rotational phenomena, like precession, and we will see our share of (even stranger!) things near the end of the book. Thus, my goal is to treat quantum mechanics as if it were introductory physics, but for a vastly different, and more abstract, topic.

With that outline established, I want to motivate the goals of this book by asking what the goals of physics are. Physics is a science, and as such, knowledge increases through application of the scientific method. That is, we first make a hypothesis for how Nature works. Then, we test that hypothesis in experiment. If the hypothesis agrees with the outcome of experiment, then it lends evidence to the veracity of the hypothesis. Importantly, it does not "prove" the hypothesis; strict logical proof is not possible in empirical science. If it does not agree, it is discarded and a new hypothesis is put forth.

This is all well and good, and something you likely learned about in high school, but there was an important temporal phrasing that I used to describe the scientific method. We make a hypothesis for what happens in the *future* based on the data we have in the *present*. That is, the goal of science is to predict the dynamics or change in time of some natural system. If the prediction agrees with the experimental data, we claim that we "understand" it.

In physics, this can be made more mathematically precise. Making a physical prediction means that we determine how a system changes or evolves in time. Any quantity used to describe a system (energy, momentum, position, etc.) has a time derivative if it changes in time. Thus, to make predictions in physics, we want to determine the mechanism responsible for providing a quantity with a time derivative. In classical mechanics, this is very familiar: Newton's second law provides such a mechanism. Newton's second law is

$$\vec{F} = m\vec{a} = m\frac{d^2\vec{x}}{dt^2} = \frac{d\vec{p}}{dt}. \tag{1.1}$$

That is, forces are responsible for changing an object's momentum in time. Tell me all of the forces acting on an object, and Newton's second law tells me how to predict its momentum at any later time. Such a dynamical equation is also called an "equation of motion" and this understanding is why it is the fundamental equation of introductory physics.

So, our ultimate goal in physics, and in this book specifically, is to determine that fundamental equation in quantum mechanics and further how systems evolve under that equation. The veracity of the fundamental equation can be tested by comparison of its predictions to experimental data. Spoiler alert: the fundamental equation of quantum mechanics, the Schrödinger equation, has been tested for the past century and every experimental result agrees with its predictions. Which is of course why we teach it!

1.2 Fundamental Hypothesis of Quantum Mechanics

This entire framework rests on a very simple yet monumentally profound assumption of the universe: that its dynamics can be described through linear operators acting on a complex vector space. At one level, this is perhaps the simplest hypothesis that one could make about the universe. Linearity is the simplest action that a system can exhibit, and, among other things, leads to the Taylor expansion for approximating a function through a series of its derivatives. Linearity is also unique, while general non-linearity is essentially unconstrained. In science, we like constraining assumptions or hypotheses because from them almost everything else follows, with no additional assumptions necessary. We'll remind about linearity through matrices and other operations in the next chapter, and see what it imposes on the relationship of our mathematical description to actual experimental results.

Linearity of quantum mechanics is very special, and not something that we enforce for Newton's laws, for example. Why though is quantum mechanics described with a complex vector space? Complex numbers make no appearance in traditional formulations of classical mechanics, except for perhaps mathematical simplification, but in no sense are they fundamental to classical mechanics. We'll revisit the importance of complex numbers in quantum mechanics at the end of this book, but honestly no one knows why. Nevertheless, everything gets simpler with complex numbers. As some examples: polynomials over the real numbers have roots that are complex numbers; complex numbers are the minimal closed field extension of the real numbers; results of complex analysis are dramatically simpler than their counterparts in real analysis, such as Cauchy's theorem, the existence of derivatives of analytic functions, etc. So, by working with complex numbers, the system is extremely constrained, more so than working with real numbers. So, this is our hypothesis, guided by Occam's razor: the simplest thing that we can imagine for the universe is that it is fundamentally linear in a complex vector space. However, what in the world does that mean?

This hypothesis and the results we will present throughout this book will likely lead you to ask over and over "Why?" It is not the realm of science to answer why; the scientific method provides no deeper reason for why the hypotheses are what they are. Nevertheless, because of the stark distinction of quantum mechanics and our everyday experience, this seems to suggest a new worldview or philosophy that might be studied in its own right. Again, a larger philosophical interpretation lies outside of the scope of science and the scientific method, and as such any interpretation or meaning of quantum mechanics lies outside of this textbook. We will focus on making simple hypotheses and understanding their consequences, without much discussion of where those hypotheses come from. This approach can be called the "shut up and calculate"[1] interpretation of quantum mechanics, but really it is a strict scientific approach. While venturing out into unknown quantum mechanical adventures, the scientific method

---

[1] This pithy phrase is due to David Mermin; see N. D. Mermin, "What's wrong with this pillow?," *Phys. Today* **42**, 4–9 (1989). Because of its bluntness and pragmatism, it is often incorrectly attributed to Richard Feynman; see N. D. Mermin, "Could Feynman have said this?," *Phys. Today* **57**, 5–10 (2004).

will be our most trusted and reliable guide to making concrete progress and gaining real understanding.

The hypotheses, or axioms, of quantum mechanics are indeed strange and unfamiliar from what we are used to in classical mechanics, but much of that is a consequence of familiarity, rather than something more fundamental. We typically don't think about a grander philosophical interpretation about Newton's second law, for example, but it has philosophical implications, just as much as the "weirdness" of quantum mechanics that we will explore. Newton's laws, for instance, imply that our human experience can completely quantify and identify all physical phenomena of Nature. That seems quite restrictive and human-centric philosophically, but we typically don't concern ourselves with its consequence because it aligns with our mundane experience in the world. Perhaps the mathematical edifice of quantum mechanics is simply the most convenient description for Nature that we have invented up to now, and not a revelation of a deeper truth. However, when it keeps working so well for so long in so many different systems, it is human nature to think that there is more to it.

This book is designed to teach you how to make predictions in quantum mechanics, through following your nose as much as possible with the assumption of working with linear operations on a complex-valued vector space. What you choose to think beyond that is entirely up to you!

2

Linear Algebra

Quantum mechanics is ultimately and fundamentally a framework for understanding Nature through the formalism of linear algebra, vector spaces, and the like. In principle, your study of linear algebra from an introductory mathematics course would be necessary and sufficient background for studying quantum mechanics, but such a course is typically firmly rooted in the study of matrices and their properties. This is definitely relevant for quantum mechanics, but we will need a more general approach to linear algebra to be able to describe the dynamics of interesting physical systems and make predictions. Perhaps the most familiar and important linear operator, the derivative, was not even discussed within that context in a course on linear algebra. Because of this familiarity, studying properties of the derivative is an excellent place to begin to dip our toes into the shallow waters of the formalism of quantum mechanics.

2.1 Invitation: The Derivative Operator

Let's first consider a function of one variable, $f(x)$. Here, x is a position, for example, and $f(x)$ might be the amplitude of a jiggled rope at position x. Precisely what function $f(x)$ is, is not relevant for our current discussion. We can draw an example function as illustrated in Fig. 2.1. On this plot, I have identified the point x_0 at which the function takes a value $f(x_0)$. Now, from x_0, how do I move along in x to get to a new point a distance Δx to the right? The function value at this new point is, of course, $f(x_0 + \Delta x)$. However, to get there from the point x_0, that is, to establish the value $f(x_0 + \Delta x)$ exclusively from data at x_0, we can use the Taylor expansion

$$f(x_0 + \Delta x) = f(x_0) + \Delta x \left.\frac{df}{dx}\right|_{x=x_0} + \frac{\Delta x^2}{2} \left.\frac{d^2 f}{dx^2}\right|_{x=x_0} + \cdots . \qquad (2.1)$$

To move a distance Δx away from x_0, I need to know all of the derivatives of $f(x)$, evaluated at x_0. This seems very complicated and like we would actually need an infinite amount of data to proceed.

However, let's re-write this Taylor expansion in the compact form

$$f(x_0 + \Delta x) = \sum_{n=0}^{\infty} \frac{\Delta x^n}{n!} \left.\frac{d^n f}{dx^n}\right|_{x=x_0} . \qquad (2.2)$$

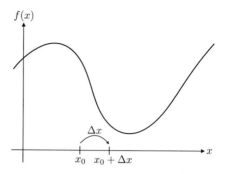

Fig. 2.1   Representation of the action of displacement from point x_0 to point $x_0 + \Delta x$ of a function $f(x)$.

Further, because the derivative is a linear operator, we can write

$$f(x_0 + \Delta x) = \left(\sum_{n=0}^{\infty} \frac{\Delta x^n}{n!} \frac{d^n}{dx^n} \right) f(x) \Big|_{x=x_0} . \tag{2.3}$$

Now, the sum in parentheses looks very familiar. If we just think of the multiple derivative d^n/dx^n as the multiplication of d/dx with itself n times, the sum has the form

$$\sum_{n=0}^{\infty} \frac{a^n}{n!} = e^a , \tag{2.4}$$

which is just the exponential function! Using this generous interpretation for now, the sum over derivatives is then

$$\sum_{n=0}^{\infty} \frac{\Delta x^n}{n!} \frac{d^n}{dx^n} = e^{\Delta x \frac{d}{dx}} . \tag{2.5}$$

So, the Taylor expansion of the function $f(x_0 + \Delta x)$ can be represented as

$$f(x_0 + \Delta x) = \left(\sum_{n=0}^{\infty} \frac{\Delta x^n}{n!} \frac{d^n}{dx^n} \right) f(x) \Big|_{x=x_0} = e^{\Delta x \frac{d}{dx}} f(x) \Big|_{x=x_0} . \tag{2.6}$$

This awkward "exponential of a derivative" is defined by its Taylor expansion. In this construction, we exploited the linearity of the derivative as an operator that acts on functions. In this chapter, we will remind about the properties of linear operators and then use that to provide a profoundly new interpretation of the derivative that we will exploit.

2.2 Linearity

You are most likely familiar with the property of linearity from a course on linear algebra (the similarity of names is not a coincidence). For a matrix $\mathbb{M}$ and vectors $\vec{u}$ and $\vec{v}$, the property of linearity means that both

(a) $\mathbb{M}(\vec{u} + \vec{v}) = \mathbb{M}\vec{u} + \mathbb{M}\vec{v}$ and
(b) $\mathbb{M}(a\vec{v}) = a\mathbb{M}\vec{v}$, where a is just a number (a **scalar**).

We will denote matrices throughout this book with "blackboard bold" font as $\mathbb{M}$, for example. We can generalize this definition of linearity to define a **linear operator** $\hat{O}$ as follows. Throughout this book, linear operators will be denoted by the caret ^ to distinguish them from mere functions. For two functions f and g, a linear operator satisfies both

(a) $\hat{O}(f+g) = \hat{O}f + \hat{O}g$ and
(b) $\hat{O}(af) = a\hat{O}f$, where a is a scalar number.

Many familiar objects are linear operators. For example, a simple function of position is a linear operator. If $\hat{O} \equiv \mathcal{O}(x)$ is just a function of x, then indeed it satisfies

(a) $\mathcal{O}(x)(f(x)+g(x)) = \mathcal{O}(x)f(x) + \mathcal{O}(x)g(x)$ and
(b) $\mathcal{O}(x)(af(x)) = a\mathcal{O}(x)f(x)$,

by the distributive law of multiplication and commutativity of multiplication of numbers. Additionally, the derivative $\hat{O} = d/dx$ is also a linear operator. That is, for two functions $f(x)$ and $g(x)$ and a scalar number a, we have

(a) $\frac{d}{dx}(f(x)+g(x)) = \frac{d}{dx}f(x) + \frac{d}{dx}g(x)$ and
(b) $\frac{d}{dx}(af(x)) = a\frac{d}{dx}f(x)$.

Crucially, linearity of the derivative depends on the existence of the Leibniz product rule. Specifically, the second requirement of linearity follows from the product rule

$$\frac{d}{dx}(af(x)) = a\frac{d}{dx}f(x) + f(x)\frac{da}{dx} = a\frac{d}{dx}f(x), \tag{2.7}$$

because a is assumed independent of x. Without the product rule, we would have no prescription for how multiplication by a constant is treated in differentiation. We will encounter many more linear operators throughout this book, but for now we will just focus on the derivative as our "canonical" linear operator and use it to gain intuition.

You likely have a good sense of and intuition for matrices: they act on vectors to rotate or scale them, they have a set of eigenvectors and eigenvalues that encode intrinsic information about the action and rank of the matrix, there are algorithmic procedures for calculating the determinant, etc. At its core, however, all a matrix is, is a linear operator that acts on vectors, and all of these properties of a matrix just mentioned simply follow from its linearity. So, if this is the case, I should be able to think about *any* linear operator as a matrix with concrete elements at a given row and column location. If this is the case, then the derivative, as a linear operator, somehow can be thought of as a matrix. However, the derivative acts on functions $f(x)$, not vectors, and there's no sense in which you can identify the element of the derivative in, say, the third row, fifth column. Or can you?

To provide a "matrix" interpretation of the derivative, we will think of our space in x as discrete. That is, we will put the x-coordinate on a grid as illustrated in Fig. 2.2. As the grid spacing $\Delta x \to 0$, we recover the continuous, real x line. On this grid, there is "nothing" between neighboring points, so we have to modify our definition of the derivative appropriately. First, for some function $f(x)$, on this grid, it is only evaluated

Fig. 2.2 Discretization of the continuous x coordinate onto a grid of points $x_0, x_1, x_2, \ldots$. The spacing between neighboring points is Δx.

at discrete x, so we denote $f(x_n) \equiv f_n$, the value of f at grid point n. Then, recalling the very first idea of a derivative from introductory calculus, the derivative of f at grid point n can be defined as

$$\frac{df_n}{dx} = \frac{f(x_{n+1}) - f(x_{n-1})}{2\Delta x} = \frac{f_{n+1} - f_{n-1}}{2\Delta x}. \tag{2.8}$$

As $\Delta x \to 0$, this of course returns the standard, continuous derivative we know and love. However, this form makes the matrix property of the derivative clear. Let's construct a vector of function values on the grid:

$$\vec{f} \equiv \begin{pmatrix} f_0 \\ f_1 \\ \vdots \\ f_n \\ \vdots \end{pmatrix}. \tag{2.9}$$

This vector may be infinite in length, but let's not worry about that. It is at least not continuous.

Now, the action of the derivative d/dx on this grid is simple, as a matrix $\mathbb{D}$ that multiplies $\vec{f}$:

$$\frac{df}{dx} \quad \to \quad \mathbb{D}\vec{f} = \begin{pmatrix} \ddots & \vdots & \vdots & \vdots & \vdots & \vdots & \cdots \\ \cdots & -\frac{1}{2\Delta x} & 0 & \frac{1}{2\Delta x} & 0 & 0 & \cdots \\ \cdots & 0 & -\frac{1}{2\Delta x} & 0 & \frac{1}{2\Delta x} & 0 & \cdots \\ \cdots & 0 & 0 & -\frac{1}{2\Delta x} & 0 & \frac{1}{2\Delta x} & \cdots \\ \cdots & \vdots & \vdots & \vdots & \vdots & \vdots & \ddots \end{pmatrix} \begin{pmatrix} \vdots \\ f_{n-1} \\ f_n \\ f_{n+1} \\ \vdots \end{pmatrix}. \tag{2.10}$$

This matrix has zeros on the diagonal, and the entries immediately off of the diagonal are $\pm 1/2\Delta x$, depending on whether the point is to the left or right of the point of interest. Clearly, this matrix ceases being sensible as $\Delta x \to 0$, as the off-diagonal elements diverge. However, this picture of a linear operator, any linear operator, as a matrix will be central to our language in quantum mechanics.

2.3 Matrix Elements

Another thing that is useful when thinking about matrices is the value of their individual elements. Consider a matrix $\mathbb{M}$, and let M_{ij} be the value (i.e., **matrix element**) at row i and column j. How do we identify such an element? Well, just given $\mathbb{M}$, the way we do it is to sandwich the matrix between two vectors that have non-zero entries at the ith and jth locations. That is, for vectors

$$
\vec{v}_i = \begin{pmatrix} 0 \\ \vdots \\ 0 \\ 1 \\ 0 \\ \vdots \\ 0 \end{pmatrix} \leftarrow i\text{th entry}, \qquad \vec{v}_j = \begin{pmatrix} 0 \\ \vdots \\ 0 \\ 1 \\ 0 \\ \vdots \\ 0 \end{pmatrix} \leftarrow j\text{th entry}, \tag{2.11}
$$

the matrix element M_{ij} is

$$
M_{ij} = \vec{v}_i^{\mathsf{T}} \mathbb{M} \vec{v}_j . \tag{2.12}
$$

For example, for the 2×2 matrix

$$
\mathbb{M} = \begin{pmatrix} M_{11} & M_{12} \\ M_{21} & M_{22} \end{pmatrix}, \tag{2.13}
$$

note that

$$
\vec{v}_1^{\mathsf{T}} \mathbb{M} \vec{v}_2 = (1 \ 0) \begin{pmatrix} M_{11} & M_{12} \\ M_{21} & M_{22} \end{pmatrix} \begin{pmatrix} 0 \\ 1 \end{pmatrix} = M_{12}, \tag{2.14}
$$

as promised.

Now, it is useful to be a bit more explicit with the matrix multiplication to illustrate what is going on. Note that we can express matrix multiplication as

$$
\vec{v}_i^{\mathsf{T}} \mathbb{M} \vec{v}_j = \sum_{a,b} (v_i)_a M_{ab} (v_j)_b = \sum_{a,b} \delta_{ia} M_{ab} \delta_{jb} = M_{ij} . \tag{2.15}
$$

There are a couple of things going on here. First, we denote the entries of the matrix and vector by the indices a and b. Matrix multiplication means to sum over a and b in the product of the elements of the vectors and the matrix. Next, we note that the vectors only have non-zero entries when $a = i$ (for $\vec{v}_i$) and $b = j$ (for $\vec{v}_j$). So, we can replace the vector elements with the **Kronecker δ** symbol, where

$$
\delta_{ab} = \begin{cases} 0, \text{if } a \neq b, \\ 1, \text{if } a = b. \end{cases} \tag{2.16}
$$

The Kronecker δs then pick out the single matrix element M_{ij} in the sums. Further, note the important properties that the length of the vectors is each 1 and distinct vectors are orthogonal:

$$
\vec{v}_i \cdot \vec{v}_i = \vec{v}_j \cdot \vec{v}_j = 1, \qquad\qquad \vec{v}_i \cdot \vec{v}_j = \delta_{ij} . \tag{2.17}
$$

This normalization of the vectors is necessary so that the product $\vec{v}_i^{\mathsf{T}} \mathbb{M} \vec{v}_j$ returns M_{ij}, and not M_{ij} scaled by some factor.

So, with this understanding, let's go back to the derivative as a matrix and extract its elements. The (i, j)th element of the derivative on a grid, $\mathbb{D}$, is

$$
\vec{f}_i^{\mathsf{T}} \mathbb{D} \vec{f}_j = D_{ij}, \tag{2.18}
$$

for some definition of basis vectors $\vec{f}_i$ and $\vec{f}_j$ whose properties we will determine. For this to hold, we must require that the length of the vectors $\vec{f}_i$ and $\vec{f}_j$ are both 1. Considering $\vec{f}_i$ first, this is

$$1 = \vec{f}_i \cdot \vec{f}_i = \vec{f}_i^{\mathsf{T}} \vec{f}_i = \sum_{a=1}^{N} (f_i^{\mathsf{T}})_a (f_i)_a, \tag{2.19}$$

for a grid of N total points. I've left the sum unevaluated, and also left the transpose on the f_i element in the left in the sum. Recall that this function vector is defined on a grid of spacing Δx. This normalization must be independent of the grid spacing: the vector squares to 1 for *any* Δx. The number of grid points in x is proportional to $1/\Delta x$; that is, if we consider the range $x \in [0,5]$, then the number of grid points is $N = 5/\Delta x$. To ensure that the sum is independent of Δx, this requires the transpose of the vector to be

$$\vec{f}_i^{\mathsf{T}} \vec{f}_i = \sum_{a=1}^{N} (f_i^{\mathsf{T}})_a (f_i)_a = \sum_{a=1}^{N} \Delta x \, (f_i)_a (f_i)_a = \sum_{a=1}^{N} \Delta x \, (f_i)_a^2. \tag{2.20}$$

Correspondingly, we have normalization and orthogonalization constraints that

$$1 = \sum_{a=1}^{N} \Delta x \, (f_i)_a^2 = \sum_{a=1}^{N} \Delta x \, (f_j)_a^2, \qquad \delta_{ij} = \sum_{a=1}^{N} \Delta x \, (f_i)_a (f_j)_a. \tag{2.21}$$

This identification in the discrete case then tells us how to interpret the continuous case. As $\Delta x \to 0$, these dot products just return the definition of the integral:

$$\lim_{\Delta x \to 0} \sum_{a=1}^{N} \Delta x \, (f_i)_a^2 \quad \to \quad \int_{x_0}^{x_1} dx \, f_i(x)^2 = 1, \tag{2.22}$$

for some function $f_i(x)$ on the domain $x \in [x_0, x_1]$. This requirement on a function is called an **L^2-norm** on the domain $x \in [x_0, x_1]$. It is the generalization of the Pythagorean/Euclidean norm to continuous functions. Similarly, orthogonality on this domain means that

$$\lim_{\Delta x \to 0} \sum_{a=1}^{N} \Delta x \, (f_i)_a (f_j)_a \quad \to \quad \int_{x_0}^{x_1} dx \, f_i(x) f_j(x) = \delta_{ij}. \tag{2.23}$$

Now, we have a prescription to determine the (i, j)th entry of the real, continuous, derivative. For two (appropriate) functions $f_i(x)$ and $f_j(x)$ that are L^2-normalized and orthogonal if $i \neq j$, the (i, j)th entry of the linear operator d/dx is

$$\left(\frac{d}{dx} \right)_{ij} = \int_{x_0}^{x_1} dx \, f_i(x) \frac{d}{dx} f_j(x), \tag{2.24}$$

on $x \in [x_0, x_1]$.

The derivative is just one example of a linear operator; we are able to generalize this matrix element procedure for any, arbitrary linear operator that acts on continuous functions. For some linear operator $\hat{\mathcal{O}}$ defined on the domain $x \in [x_0, x_1]$, its "(i, j)th" element is defined by the integral

$$\mathcal{O}_{ij} = \int_{x_0}^{x_1} dx \, f_i(x) \hat{\mathcal{O}} f_j(x), \tag{2.25}$$

and the functions $f_i(x)$, $f_j(x)$ are L^2-normalized:

$$1 = \int_{x_0}^{x_1} dx\, f_i(x)^2 = \int_{x_0}^{x_1} dx\, f_j(x)^2 \tag{2.26}$$

and orthogonal:

$$\delta_{ij} = \int_{x_0}^{x_1} dx\, f_i(x) f_j(x)\,. \tag{2.27}$$

Before continuing, we need to address one more aspect of this method for identification of matrix elements, $\mathcal{O}_{ij}$. To know everything about a linear operator, we need to know all of its matrix elements $\mathcal{O}_{ij}$, where i and j run over all possible values. For example, if we consider a 2×2 matrix $\mathbb{M}$, then i and j run over $1,2$: $i,j \in \{1,2\}$. For this matrix, we need to know four matrix elements: M_{11}, M_{12}, M_{21}, and M_{22} to completely specify the matrix. Requiring complete knowledge of a linear operator enforces properties on the set of vectors $\{\vec{v}_i\}$ that are used to define and isolate the matrix elements.

One constraint on the vectors $\{\vec{v}_i\}$ is simple. If $\mathbb{M}$ is an $N \times N$ matrix, then to identify all $N \times N$ matrix elements, there need to be (at least) N vectors in the set $\{\vec{v}_i\}$. We have already said that all vectors in the set are normalized, $\vec{v}_i \cdot \vec{v}_i = 1$, and orthogonal, $\vec{v}_i \cdot \vec{v}_j = 0$ if $i \neq j$. An $N \times N$ matrix acts on N-dimensional vectors; that is, it manipulates some N-dimensional space. If we have identified N **orthonormal** N-dimensional vectors $\{\vec{v}_i\}_{i=1}^N$, then these vectors necessarily span the whole space of N-dimensional vectors. That is, some N-dimensional vector $\vec{u}$ can be expressed as a linear combination of the $\vec{v}_i$s:

$$\vec{u} = \sum_{i=1}^N \alpha_i \vec{v}_i\,, \tag{2.28}$$

where the α_i are some numbers. If these properties hold, then we say that the set of vectors $\{\vec{v}_i\}_{i=1}^N$ are orthonormal and **complete**. Given a complete set of orthonormal vectors, we can find out anything we want about the matrix $\mathbb{M}$.

With a complete basis $\{\vec{v}_i\}$, the action of a matrix $\mathbb{M}$ on this basis (or any linear combination of them) returns another linear combination of the basis vectors. For example, for an $N \times N$ matrix $\mathbb{M}$ we choose for the basis vectors $\{\vec{v}_i\}_{i=1}^N$:

$$\vec{v}_i = \begin{pmatrix} 0 \\ \vdots \\ 0 \\ 1 \\ 0 \\ \vdots \\ 0 \end{pmatrix} \leftarrow i\text{th entry}\,. \tag{2.29}$$

The action of $\mathbb{M}$ on one such basis vector is

$$\mathbb{M}\vec{v}_i = \sum_{a,b=1}^N \vec{v}_a^{\mathsf{T}} \mathbb{M} \vec{v}_b\, (v_i)_b = \sum_{a,b=1}^N \vec{v}_a^{\mathsf{T}} \mathbb{M} \vec{v}_b\, \delta_{ib} = \sum_{a=1}^N \vec{v}_a^{\mathsf{T}} \mathbb{M} \vec{v}_i\,, \tag{2.30}$$

which is some other linear combination of the basis vectors $\vec{v}_a$. That is, under the action of a matrix $\mathbb{M}$, the space that is spanned by a complete basis $\{\vec{v}_i\}$ is unchanged. We say that the vector space is therefore **closed** under the operation of $\mathbb{M}$. The matrix $\mathbb{M}$ scrambles the space around, but can never leave the space.

2.4 Eigenvalues

However, as hinted at earlier, individual matrix elements are not universally well-defined. For example, let's take our 2×2 matrix $\mathbb{M}$. To identify its matrix elements M_{ij}, we need to specify a complete basis of orthonormal vectors. This specification is not unique. For example, let's take

$$\vec{v}_1 = \begin{pmatrix} 1 \\ 0 \end{pmatrix}, \qquad \vec{v}_2 = \begin{pmatrix} 0 \\ 1 \end{pmatrix}, \qquad (2.31)$$

and so $\vec{v}_1 \cdot \vec{v}_1 = \vec{v}_2 \cdot \vec{v}_2 = 1$ and $\vec{v}_1 \cdot \vec{v}_2 = 0$. Then we find, for example, that $M_{12} = \vec{v}_1^\mathsf{T} \mathbb{M} \vec{v}_2$. In this basis, we can express the matrix $\mathbb{M}$ as

$$\mathbb{M} = \begin{pmatrix} M_{11} & M_{12} \\ M_{21} & M_{22} \end{pmatrix}. \qquad (2.32)$$

Your friend, however, defines as their complete basis the vectors

$$\vec{u}_1 = \begin{pmatrix} \cos\theta \\ \sin\theta \end{pmatrix}, \qquad \vec{u}_2 = \begin{pmatrix} -\sin\theta \\ \cos\theta \end{pmatrix}, \qquad (2.33)$$

for some (or rather any) angle θ. These vectors are normalized:

$$\vec{u}_1 \cdot \vec{u}_1 = (\cos\theta \ \sin\theta) \begin{pmatrix} \cos\theta \\ \sin\theta \end{pmatrix} = \cos^2\theta + \sin^2\theta = 1, \qquad (2.34)$$

and similarly $\vec{u}_2 \cdot \vec{u}_2 = 1$. Also, you can show that $\vec{u}_1 \cdot \vec{u}_2 = 0$ and so we have two, orthonormal vectors $\vec{u}_1, \vec{u}_2$ which therefore define another complete basis in which to study $\mathbb{M}$. However, in this basis the (12) element of $\mathbb{M}$ is

$$\vec{u}_1^\mathsf{T} \mathbb{M} \vec{u}_2 = (\cos\theta \ \sin\theta) \begin{pmatrix} M_{11} & M_{12} \\ M_{21} & M_{22} \end{pmatrix} \begin{pmatrix} -\sin\theta \\ \cos\theta \end{pmatrix} \qquad (2.35)$$

$$= -M_{11}\cos\theta\sin\theta - M_{21}\sin^2\theta + M_{12}\cos^2\theta + M_{22}\cos\theta\sin\theta.$$

This is clearly very different than what we found using the $\vec{v}$ vector basis. Who is right and who is wrong?

Of course, neither you nor your friend are wrong: you have just expressed the elements of $\mathbb{M}$ in a different basis, and so find a different explicit representation of $\mathbb{M}$. However, the matrix $\mathbb{M}$ is independent of any particular basis, as the choice of basis was something that we did arbitrarily. $\mathbb{M}$ doesn't change with our whims; it is what it is.

As an example of something familiar that you've seen that manifests this property, consider analyzing the forces of a block on a ramp, which I have drawn in Fig. 2.3. The first step in such a problem, before even drawing a free-body diagram, is to determine the coordinate system you will use. Do you align the x-axis with the ramp, or do you

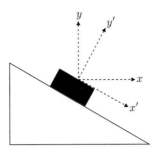

Fig. 2.3 A block of mass m placed on an inclined ramp. Two possible coordinate systems for analyzing the forces on the block are illustrated.

align it horizontally? Or, do you point it in some other direction entirely? You are allowed to orient your axes in any way you want: the physics is independent of your particular, idiosyncratic, choice. A pithy way to express this is that life (i.e., Nature) cannot imitate art (i.e., your coordinate system). Whatever coordinates we use, we must find the same acceleration of the block, for example.

Going back to our matrix example, what is the analogy of the coordinate basis-invariant physics? What properties of a matrix $\mathbb{M}$ are independent of the basis in which we choose to express its matrix elements? Well, if $\mathbb{M}$ is an $N \times N$ matrix, then clearly, and perhaps trivially, the dimension of the matrix N is basis-independent. Every orthonormal, complete basis for $\mathbb{M}$ must consist of N vectors.

Further, even though in general there are a continuous infinity of possible orthonormal, complete bases in which to express $\mathbb{M}$, there is only one for which $\mathbb{M}$ is expressed as a diagonal matrix, with only non-zero entries on the diagonal. Let's understand this basis a bit more. Let's assume the basis in which $\mathbb{M}$ is diagonal is $\{\vec{t}_i\}_{i=1}^{N}$. "Diagonal" means that the matrix elements M_{ij} are

$$M_{ij} = \lambda_i\, \delta_{ij} = \vec{t}_i^{\mathsf{T}} \mathbb{M} \vec{t}_j\,, \tag{2.36}$$

for some number λ_i. However, if $\mathbb{M}$ is diagonal in this basis, it looks like

$$\mathbb{M} = \begin{pmatrix} \lambda_1 & & & \\ & \lambda_2 & & \\ & & \ddots & \\ & & & \lambda_N \end{pmatrix}, \tag{2.37}$$

where all empty entries are 0. The matrix product is simply

$$\mathbb{M}\vec{t}_j = \lambda_j \vec{t}_j\,. \tag{2.38}$$

The matrix element from our expression earlier is

$$\vec{t}_i^{\mathsf{T}} \mathbb{M} \vec{t}_j = \lambda_j \vec{t}_i^{\mathsf{T}} \vec{t}_j = \lambda_j \vec{t}_i \cdot \vec{t}_j = \lambda_j\, \delta_{ij}\,. \tag{2.39}$$

On the right, we have used the orthonormality of the basis vectors. The collection of the λ_i numbers on the diagonal are of course nothing but the **eigenvalues** of the matrix $\mathbb{M}$, which can be found by solving the characteristic equation

$$\det(\mathbb{M} - \lambda \mathbb{I}) = 0, \tag{2.40}$$

where $\mathbb{I}$ is the $N \times N$ identity matrix. For such a matrix, the determinant of $\mathbb{M} - \lambda \mathbb{I}$ is in general an Nth-degree polynomial equation for λ, for which there are in general N solutions, corresponding to the $\lambda_1, \lambda_1, \ldots, \lambda_N$ values. Correspondingly, the collection of the $\vec{t}_i$ basis vectors which render $\mathbb{M}$ in this diagonal form are called its **eigenvectors**.

This "diagonal" basis for $\mathbb{M}$ is analogous to the proper frame in special relativity. Though energies, momenta, lengths, times, etc. are all different in every inertial reference frame, every frame can agree on what would be observed in the special frame which is at rest with respect to the object of interest. **Diagonalizing** a matrix $\mathbb{M}$ (that is, finding its eigenvalues) is like boosting to the "proper frame" of the matrix.

Thus, the only basis-independent quantities of a linear operator $\mathbb{M}$ are its dimension and collection of eigenvalues $\{\lambda_i\}$. This collection of eigenvalues of a matrix or linear operator is also called its **spectrum**. If you know the spectrum of a matrix, you know everything there is to know about the matrix.

Before studying more properties of the derivative operator, we pause for an example that illustrates the general properties of an orthonormal basis of functions.

Example 2.1 The **Legendre polynomials** provide a complete, orthonormal basis for all functions on $x \in [-1, 1]$. In this example we will study how they can be used to construct the derivative operator for functions on this domain. You will explore this more in Exercise 2.3.

There are an infinite number of Legendre polynomials, which is necessary because the number of possible functions on this domain is also infinite. However, in this problem we'll just use the first three Legendre polynomials to construct part of the matrix that represents the derivative operator d/dx on the domain $x \in [-1, 1]$. These polynomials are (up to normalization):

$$P_0(x) = \frac{1}{\sqrt{2}}, \tag{2.41}$$

$$P_1(x) = \sqrt{\frac{3}{2}} x,$$

$$P_2(x) = \sqrt{\frac{5}{8}} (3x^2 - 1).$$

These polynomials are plotted in Fig. 2.4.

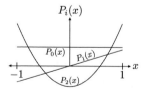

Fig. 2.4
A plot of the first three Legendre polynomials as defined in Eq. (2.41).

Solution

The first thing we will do is to verify that indeed these Legendre polynomials are orthonormal on $x \in [-1, 1]$. We will just present representative results here for conciseness. Let's explicitly verify that $P_2(x)$ is L^2-normalized on this domain and orthogonal to $P_1(x)$. Normalization requires that

$$1 = \int_{-1}^{1} dx \, P_2(x)^2 = \frac{5}{8} \int_{-1}^{1} dx \, (3x^2 - 1)^2 = \frac{5}{8} \int_{-1}^{1} dx \, (9x^4 - 6x^2 + 1) \tag{2.42}$$
$$= \frac{5}{8} \left(\frac{18}{5} - 4 + 2 \right),$$

which is indeed true. Orthogonality of $P_2(x)$ and $P_1(x)$ requires that

$$0 = \int_{-1}^{1} dx \, P_2(x) P_1(x) = \sqrt{\frac{15}{16}} \int_{-1}^{1} dx \, x(3x^2 - 1) = \sqrt{\frac{15}{16}} \int_{-1}^{1} dx \, (3x^3 - x), \tag{2.43}$$

which is also true because both x and x^3 are odd functions on $x \in [-1, 1]$. Other normalizations of the Legendre polynomials and orthogonality between distinct Legendre polynomials can be verified similarly.

With the orthonormality of this function basis established, we now move on to expressing the derivative operator d/dx in this basis. Because we have only identified three Legendre polynomials, we will only be able to construct a 3×3 matrix that represents the action of the derivative on a general quadratic polynomial. Recall that the element in the first row and second column of the d/dx matrix would be, for example:

$$\left(\frac{d}{dx} \right)_{12} = \int_{-1}^{1} dx \, P_0(x) \frac{d}{dx} P_1(x). \tag{2.44}$$

Note the different indexing of the Legendre polynomials and the label of a row or column. To determine the 3×3 derivative matrix, we can immediately establish that all entries in the first column are 0. An entry at row i of the first column takes the form

$$\left(\frac{d}{dx} \right)_{i1} = \int_{-1}^{1} dx \, P_{i-1}(x) \frac{d}{dx} P_0(x) = 0, \tag{2.45}$$

because the Legendre polynomial $P_0(x)$ is independent of x. Additionally, all entries in the third row of the matrix vanish, where the element in column j of the third row would be

$$\left(\frac{d}{dx} \right)_{3j} = \int_{-1}^{1} dx \, P_2(x) \frac{d}{dx} P_{j-1}(x) = 0. \tag{2.46}$$

This vanishes because the derivative operator reduces the degree of a polynomial by 1, and $P_2(x)$ is orthogonal to all polynomials of degree 0 and 1 on this domain.

For the remaining entries of the matrix, there are only two non-zero entries. The ij entry in this matrix can only be non-zero if the polynomials $P_{i-1}(x)$ and $\frac{d}{dx} P_{j-1}(x)$ are of the same degree. This can only happen for the pairs $i = 1, j = 2$ and $i = 2, j = 3$. For the first pair, the entry in the matrix is

$$\left(\frac{d}{dx} \right)_{12} = \int_{-1}^{1} dx \, P_0(x) \frac{d}{dx} P_1(x) = \frac{\sqrt{3}}{2} \int_{-1}^{1} dx = \sqrt{3}. \tag{2.47}$$

For the second pair, the matrix element is

$$\left(\frac{d}{dx}\right)_{23} = \int_{-1}^{1} dx\, P_1(x) \frac{d}{dx} P_2(x) = \frac{3\sqrt{15}}{2} \int_{-1}^{1} dx\, x^2 = \sqrt{15}. \tag{2.48}$$

Then, this restricted 3×3 derivative matrix on the domain $x \in [-1, 1]$ is

$$\frac{d}{dx} = \begin{pmatrix} 0 & \sqrt{3} & 0 \\ 0 & 0 & \sqrt{15} \\ 0 & 0 & 0 \end{pmatrix}. \tag{2.49}$$

As a matrix, this Legendre polynomial representation of the derivative operator is rather trivial. All eigenvalues λ of this matrix are 0, as found by solving the characteristic equation

$$\det \begin{pmatrix} -\lambda & \sqrt{3} & 0 \\ 0 & -\lambda & \sqrt{15} \\ 0 & 0 & -\lambda \end{pmatrix} = -\lambda^3 = 0. \tag{2.50}$$

Further, this matrix has a single eigenvector that has a non-zero length, namely

$$\vec{t} = \begin{pmatrix} 0 \\ 0 \\ 1 \end{pmatrix}. \tag{2.51}$$

However, the existence of this eigenvector is a bit misleading. If more Legendre polynomials are included to describe more general functions on $x \in [-1, 1]$, the only eigenvalues will be 0. Further, there will only ever be one eigenvector of non-zero length, because for any finite collection of Legendre polynomials, the last row of the derivative matrix will always only consist of 0 entries. We will address the reason why the derivative operator has strange properties in this basis and what properties we desire of a "good" basis in the next chapter.

2.5 Properties of the Derivative as a Linear Operator

Our discussion in most of this section has been limited to familiar finite, discrete matrices, but the structure applies to any general linear operator $\hat{O}$. Let's consider its consequences for the derivative operator d/dx. We will first establish the spectrum of eigenvalues and the corresponding eigenfunctions of the derivative, and then use these results to provide a physical interpretation of how the derivative acts.

2.5.1 Eigensystem of the Derivative Operator

Extrapolating from our matrix analysis, the eigenvalue equation for the derivative would be

$$\frac{d}{dx} f_\lambda(x) = \lambda\, f_\lambda(x), \tag{2.52}$$

for some eigenvalue λ and "eigenfunction" $f_\lambda(x)$. This eigenvalue equation, for some λ, is a simple differential equation for which the solution is just an exponential function:

$$f_\lambda(x) = e^{\lambda x}. \tag{2.53}$$

In our analogy, these $f_\lambda(x)$ functions should themselves be a complete, orthonormal basis. However, this form may suggest some problems with these requirements. First, the position x can range over all real values: $x \in (-\infty, \infty)$, and for real λ, the function $e^{\lambda x}$ is poorly behaved as $x \to \pm\infty$. Orthogonality of these functions would further mean, for two eigenvalues λ_1, λ_2 such that $\lambda_1 \neq \lambda_2$:

$$0 = \int_{-\infty}^{\infty} dx \, f_{\lambda_1}(x) f_{\lambda_2}(x) = \int_{-\infty}^{\infty} dx \, e^{(\lambda_1 + \lambda_2)x}, \tag{2.54}$$

but this is clearly crazy. The assumption we made that was too restricting was that the eigenvalues λ are real: if λ is imaginary, $\lambda = ik$, for a real number k, then these eigenfunctions are very nice, if complex, functions. With this modification, the eigenfunctions are now

$$f_k(x) = e^{ikx}, \tag{2.55}$$

which has absolute value or magnitude 1 for all x; it never diverges. This imaginary exponential is familiar from your study of **Fourier transforms**, for which you know that

$$\int_{-\infty}^{\infty} dx \, e^{i(k_1 - k_2)x} = 0, \tag{2.56}$$

if $k_1 \neq k_2$; a manifestation of orthogonality. This all seems good so far, other than the strange fact that we had to introduce the imaginary number to do it.

With this insight, let's re-write the eigenvalue equation for the derivative operator:

$$\frac{d}{dx} f_k(x) = ik f_k(x). \tag{2.57}$$

This is all well and good, but the "ik" on the right is a bit aesthetically displeasing, so let's move the i to the left:

$$\left(-i\frac{d}{dx}\right) f_k(x) = k f_k(x). \tag{2.58}$$

Recall that k is a real number. This is the eigenvalue equation for the modified derivative operator

$$\hat{O} = -i\frac{d}{dx}, \tag{2.59}$$

which, apparently, has exclusively real eigenvalues, in spite of (or rather, because of) the explicit factor of i in its expression.

From analysis of Fourier transforms, you know that a general function of x, say $g(x)$, can be expressed as a linear combination of the resulting imaginary exponential eigenfunction of the derivative operator:

$$g(x) = \sum_k \alpha_k f_k(x) \xrightarrow{\text{continuous } k} \int_{-\infty}^{\infty} dk \, \alpha(k) e^{ikx}, \tag{2.60}$$

where $\alpha(k)$ is some function of k, for all real $k \in (-\infty, \infty)$. Of course, these expressions say nothing more than that $\alpha(k)$ is the Fourier transform of $g(x)$ (and vice-versa).

2.5.2 Physical Interpretation of the Derivative Operator

These mathematical truisms are fascinating, but we need to breathe more physical life into them. In particular, we want to provide physical interpretations for everything that we've developed for linear operators in general, as this will set us up for taking some very large, and generous, steps beyond where the math stops leading us. In particular, once we've built up more of the structure of this linear edifice, we will really go overboard and take this entire framework as a hypothesis for how the universe works. For now, however, let's do the slightly more mundane task of just giving the derivative operator a physical intuition.

When we introduced the derivative operator, we had said that its purpose in life was to translate a function $f(x)$. That is, if we want to translate the argument of $f(x)$ by an amount a, we take the Taylor expansion about the point x:

$$f(x+a) = f(x) + a\frac{df}{dx} + \frac{a^2}{2}\frac{d^2f}{dx^2} + \cdots \qquad (2.61)$$

$$= \sum_{n=0}^{\infty} \frac{a^n}{n!}\frac{d^n}{dx^n}f(x) = e^{a\frac{d}{dx}}f(x).$$

Let's express this in another way. Let's just assume that we translate $f(x)$ by $\Delta x \ll 1$, so that we can safely ignore quadratic and higher terms in Δx. Then

$$f(x+\Delta x) \approx f(x) + \Delta x\frac{d}{dx}f(x) = \left(1 + \Delta x\frac{d}{dx}\right)f(x). \qquad (2.62)$$

Now, we can move another step of size Δx by applying the derivative operator again:

$$f(x+2\Delta x) \approx f(x) + 2\Delta x\frac{d}{dx}f(x) \approx \left(1 + \Delta x\frac{d}{dx}\right)^2 f(x), \qquad (2.63)$$

ignoring terms at quadratic order in Δx. We can continue this process, taking N steps of size Δx:

$$f(x+N\Delta x) \approx \left(1 + \Delta x\frac{d}{dx}\right)^N f(x). \qquad (2.64)$$

If we now identify $a \equiv N\Delta x$, we have

$$f(x+a) \approx \left(1 + \frac{a}{N}\frac{d}{dx}\right)^N f(x), \qquad (2.65)$$

so we have broken up moving a distance a from x into N individual steps as shown in Fig. 2.5. Now, as $N \to \infty$, we have that the action of these N steps exponentiates:

$$\lim_{N\to\infty}\left(1 + \frac{a}{N}\frac{d}{dx}\right)^N = e^{a\frac{d}{dx}}, \qquad (2.66)$$

Fig. 2.5
Representation of dividing up the interval from x to $x + a$ into N steps, each of length $\Delta x = a/N$.

Fig. 2.6
An object represented as a collection of point masses m_i at positions x_i.

just as we found with the Taylor expansion. So, the derivative d/dx translates us by an infinitesimal amount. Applying the derivative operator an infinite number of times enables us to move a finite distance. This general procedure of composition of infinitesimal actions to produce a finite action is referred to as **exponentiation**.

Okay, now let's consider some object, which can be visualized as a collection of point masses as illustrated in Fig. 2.6. The ith point mass has position x_i and mass m_i. This collection of points has a center-of-mass x_{cm} located at the point

$$x_{\mathrm{cm}} = \frac{m_1 x_1 + m_2 x_2 + \cdots + m_n x_n}{m_1 + m_2 + \cdots + m_n} = \frac{\sum_i m_i x_i}{\sum_i m_i}. \tag{2.67}$$

(In multiple spatial dimensions, this should be a vector equation, but for simplicity, let's stick to one dimension for now.) Now, let's act the derivative operator on this collection of points. That is, let's perform

$$\left(1 + \Delta x \frac{d}{dx_i}\right) x_i = x_i + \Delta x, \tag{2.68}$$

so that

$$\left(1 + \Delta x \sum_i \frac{d}{dx_i}\right) \frac{m_1 x_1 + m_2 x_2 + \cdots + m_n x_n}{m_1 + m_2 + \cdots + m_n} = \frac{\sum_i m_i (x_i + \Delta x)}{\sum_i m_i} \tag{2.69}$$

$$= \left(1 + \Delta x \frac{d}{dx_{\mathrm{cm}}}\right) x_{\mathrm{cm}} = x_{\mathrm{cm}} + \Delta x.$$

So, for a physical object with some well-defined center-of-mass, the derivative operator implements an infinitesimal translation of its center-of-mass. Let's keep going: if the center-of-mass of an object translates, then what quantity must that object also carry? Translation of the center-of-mass means that there is a velocity of the center-of-mass; or, that the object carries momentum.

Because the derivative translates the center-of-mass of an object and the center-of-mass translates if the object carries momentum, we therefore call the derivative operator the momentum operator. However, we have to be a bit careful with units as the object d/dx doesn't have the correct units of momentum. That can be corrected by multiplying by some constant c_p, which carries the necessary units. Further, we have established that the operator

$$\hat{O} = -i\frac{d}{dx} \tag{2.70}$$

has exclusively real eigenvalues. This is very nice because the momentum that we know and love is always, indeed, a real number. So, with this physical interpretation and these mathematical properties, we call the following linear, differential operator the **momentum operator** $\hat{p}$:

$$\hat{p} = -ic_p\frac{d}{dx}. \tag{2.71}$$

If this is momentum, then its units in SI are

$$[p] = [mv] = MLT^{-1}, \tag{2.72}$$

where m is mass, v is velocity, and M, L, and T are placeholders for the mass, length, and time units, respectively. The derivative alone has units of inverse length:

$$\left[\frac{d}{dx}\right] = L^{-1}, \tag{2.73}$$

so the units of c_p must be

$$[c_p] = MLT^{-1} \cdot L = ML^2T^{-1} = [Et], \tag{2.74}$$

which is equivalent to Joule-seconds, in SI.

Then, the eigenvalue equation for the momentum operator $\hat{p}$ is

$$\hat{p}\,h_p(x) = -ic_p\frac{d}{dx}h_p(x) = ph_p(x), \tag{2.75}$$

which has solution

$$h_p(x) = e^{i\frac{px}{c_p}}. \tag{2.76}$$

Now, p (with no caret) is just the eigenvalue of the momentum operator, which is simply a real-valued momentum.

2.6 Orthonormality of Complex-Valued Vectors

With these eigenfunctions for momentum, we are now in a position to understand the orthonormality of these complex basis functions. Going back to just considering properties of matrices, let's say that given a matrix $\mathbb{M}$, it has complex-valued eigenvectors $\{\vec{v}_i\}$ and real eigenvalues $\{\lambda_i\}$. The eigenvalue equation is then

$$\mathbb{M}\vec{v}_i = \lambda_i\vec{v}_i. \tag{2.77}$$

By orthonormality of the basis $\{\vec{v}_i\}$, we can further dot this expression by $\vec{v}_i$ appropriately to isolate the eigenvalue λ_i. However, note that for complex-valued $\vec{v}_i$, its dot product with itself is in general not real. The complex conjugate of the dot product of $\vec{v}_i$ with itself is

$$(\vec{v}_i \cdot \vec{v}_i)^* = \vec{v}_i^* \cdot \vec{v}_i^* \neq \vec{v}_i \cdot \vec{v}_i. \tag{2.78}$$

So we can't just dot Eq. (2.77) by $\vec{v}_i$ and expect to return the eigenvalue because a complex vector can't be normalized by a constraint on $\vec{v}_i \cdot \vec{v}_i$ alone. Instead, dotting the complex conjugate of the vector with itself is real-valued:

$$(\vec{v}_i^* \cdot \vec{v}_i)^* = \vec{v}_i \cdot \vec{v}_i^* = \vec{v}_i^* \cdot \vec{v}_i . \tag{2.79}$$

This can be used to enforce normalization and orthogonality of complex vectors $\vec{v}_i$ and $\vec{v}_j$ as

$$\vec{v}_i^* \cdot \vec{v}_j = \delta_{ij} . \tag{2.80}$$

The eigenvalue λ_i then can be isolated as

$$\vec{v}_i^{*\mathsf{T}} \mathbf{M} \vec{v}_i = \lambda_i , \tag{2.81}$$

where we both complex conjugate and transpose the vector multiplied on the left.

For the derivative operator's eigenfunctions, orthogonality can be expressed in a similar manner. Because $\exp[i\frac{px}{c_p}]$ is complex, to "dot" it with another eigenfunction, we must complex conjugate appropriately. That is, for $p_1 \neq p_2$, we have the orthogonality relation

$$0 = \int_{-\infty}^{\infty} dx \left(e^{i\frac{p_1 x}{c_p}} \right)^* e^{i\frac{p_2 x}{c_p}} = \int_{-\infty}^{\infty} dx\, e^{-i\frac{p_1 x}{c_p}} e^{i\frac{p_2 x}{c_p}} = \int_{-\infty}^{\infty} dx\, e^{i\frac{(p_2-p_1)x}{c_p}} . \tag{2.82}$$

From a physical perspective, this equation means that systems of different momenta are orthogonal in this space. We test orthogonality by seeing if one momentum plus the opposite of another momentum is at rest; i.e., at 0 momentum. In general, for any p_1, p_2, this integral is

$$\int_{-\infty}^{\infty} dx\, e^{i\frac{(p_2-p_1)x}{c_p}} = 2\pi c_p\, \delta(p_2 - p_1) , \tag{2.83}$$

where $\delta(x)$ is the **Dirac δ-function**, a continuous generalization of the Kronecker δ. $\delta(x) = 0$ if $x \neq 0$ and integrates to unity:

$$\int_a^b dx\, \delta(x) = 1 , \tag{2.84}$$

assuming $a < 0 < b$.

The expansion of a function into eigenstates of momentum enables a simple representation of the action of the momentum operator. Let's consider a function $g(x)$ that can be expressed as

$$g(x) = \int_{-\infty}^{\infty} dp\, \alpha(p)\, e^{i\frac{px}{c_p}} , \tag{2.85}$$

for some function of momentum $\alpha(p)$. Written in this form, we can explicitly demonstrate that the action of the momentum operator $\hat{p}$ is to move this function $g(x)$ around in the space of functions; that is, to modify its Fourier transform. Acting $\hat{p}$ on $g(x)$, we have

$$\hat{p} g(x) = -i c_p \frac{d}{dx} \int_{-\infty}^{\infty} dp\, \alpha(p)\, e^{i\frac{px}{c_p}} = \int_{-\infty}^{\infty} dp\, \alpha(p) \left(-i c_p \frac{d}{dx} \right) e^{i\frac{px}{c_p}} \tag{2.86}$$

$$= \int_{-\infty}^{\infty} dp\, \alpha(p)(-ic_p)\left(i\frac{p}{c_p}\right)e^{i\frac{px}{c_p}} = \int_{-\infty}^{\infty} dp\, \alpha(p)\, p\, e^{i\frac{px}{c_p}}$$

$$= \int_{-\infty}^{\infty} dp\, \beta(p)\, e^{i\frac{px}{c_p}}\,.$$

In the final line, $\beta(p) = p\,\alpha(p)$, which is just some other function of p. In addition to thinking of a linear operator as implementing motion throughout the space of vectors or functions, we also say that the linear operator transforms the vector or function to a new vector or function in the space. This space clearly has some rich properties from our study thus far, and in the next chapter we will enumerate them and give this space a name.

Exercises

2.1   In this chapter, we introduced the matrix $\mathbb{D}$ defined as the derivative acting on a grid with spacing Δx. This matrix has the form

$$\mathbb{D} = \begin{pmatrix} \ddots & \vdots & \vdots & \vdots & \cdots \\ \cdots & 0 & \frac{1}{2\Delta x} & 0 & \cdots \\ \cdots & -\frac{1}{2\Delta x} & 0 & \frac{1}{2\Delta x} & \cdots \\ \cdots & 0 & -\frac{1}{2\Delta x} & 0 & \cdots \\ & \vdots & \vdots & \vdots & \vdots & \ddots \end{pmatrix}, \qquad (2.87)$$

with only non-zero entries immediately above and below the diagonal. In this problem, we will just study the 2×2 and 3×3 discrete derivative matrices.

(a)   Explicitly construct 2×2 and 3×3 discrete derivative matrices, according to the convention above.

(b)   Now, calculate the eigenvalues of both of these matrices. Note that there should be two eigenvalues for the 2×2 matrix and three eigenvalues for the 3×3 matrix. Are any of the eigenvalues 0? For those that are non-zero, are the eigenvalues real, imaginary, or general complex numbers?

(c)   Determine the eigenvectors of this discrete derivative matrix, for each eigenvalue. Make sure to normalize the eigenvectors. Are eigenvectors corresponding to distinct eigenvalues orthogonal?

(d)   Now, consider exponentiating the discrete derivative matrix to move a distance Δx. Call the resulting matrix $\mathbb{M}$:

$$\mathbb{M} = e^{\Delta x \mathbb{D}}\,. \qquad (2.88)$$

What is the result of acting this exponentiated matrix on each of the eigenvectors that you found in part (c)?
Hint: Consider the Taylor expansion of this exponential. How does $\mathbb{D}^n$ act on an eigenvector of $\mathbb{D}$?

(e)   Now, just evaluate the exponentiated derivative matrices that you are studying in this problem. That is, determine the closed form of the 2×2 and 3×3 matrices $\mathbb{M}$, where

$$\mathbb{M} = e^{\Delta x \mathbb{D}} . \tag{2.89}$$

Hint: Can you write down a recursive formula for $\mathbb{D}^n$ that appears in the Taylor expansion of the exponential?

2.2 Why didn't we define the derivative matrix on a grid through the familiar "asymmetric" derivative:

$$\frac{df(x)}{dx} = \lim_{\Delta x \to 0} \frac{f(x + \Delta x) - f(x)}{\Delta x} ? \tag{2.90}$$

What are its eigenvalues? What happens when it is exponentiated?

2.3 In this exercise, we extend the analysis of the Legendre polynomials presented in Example 2.1.

(a) With only the first three Legendre polynomials, $P_0(x)$, $P_1(x)$, and $P_2(x)$, this is only a complete, orthonormal basis for general quadratic polynomials on $x \in [-1, 1]$. For a polynomial expressed as:

$$p(x) = ax^2 + bx + c , \tag{2.91}$$

for some constants a, b, c, re-write it as a linear combination of the Legendre polynomials. Express the coefficients of this linear combination as a three-dimensional vector.

(b) Act on this three-dimensional vector with the derivative matrix that you constructed in Example 2.1. Remember, the result represents another linear combination of Legendre polynomials. Does the result agree with what you would find from just differentiating the polynomial in part (a)?

(c) Now, construct the matrix that corresponds to the second derivative operator on the Legendre polynomials through explicit calculation of matrix elements, as in Example 2.1. Compare your result to simply squaring the (first) derivative matrix.

(d) Now, determine the matrix that corresponds to exponentiation of the derivative matrix; that is, determine the 3×3 matrix

$$\mathbb{M} = e^{\Delta x \frac{d}{dx}} , \tag{2.92}$$

where d/dx is shorthand for the matrix you constructed in Example 2.1. Note that the Taylor series of the exponential should terminate at a finite order in Δx. Act this on a general quadratic function $p(x)$ as defined in part (a) and show that the argument x just translates as $p(x + \Delta x)$.

2.4 Another linear operation that you are familiar with is anti-differentiation, or indefinite integration. In this problem, we will study features of anti-differentiation and relate it to differentiation.

(a) First, demonstrate that anti-differentiation is a linear operator. For anti-differentiation to be linear, what value must any integration constant be set to?

(b) If anti-differentiation is a linear operator, then we expect it in general to be able to be expressed as a matrix. Let's call this matrix $\mathbb{A}$ and assume that it acts on function vectors defined on a spatial grid. For a derivative matrix $\mathbb{D}$ and function vectors $\vec{f}, \vec{g}$, we assume

$$\mathbb{D}\vec{f} = \vec{g}. \tag{2.93}$$

We can solve for $\vec{f}$ given $\vec{g}$ as

$$\vec{f} = \mathbb{D}^{-1}\vec{g} = \mathbb{A}\vec{g}, \tag{2.94}$$

because anti-differentiation is the inverse operation of integration. However, does this exist, in general? If it does not exist, what additional constraint must you impose for the anti-differentiation matrix $\mathbb{A}$ to be well-defined?

2.5 Differentiation isn't the only operation that can be performed on functions. In this problem, we'll consider the operator $\hat{S}$ that takes the form

$$\hat{S} = -ix\frac{d}{dx}, \tag{2.95}$$

which is proportional to the product of the position and momentum operators.

(a) Determine the eigenvalues and eigenfunctions of this operator. That is, determine the values λ and functions $f_\lambda(x)$ such that

$$\hat{S}f_\lambda(x) = \lambda f_\lambda(x), \tag{2.96}$$

where we assume that the position coordinate $x \in (-\infty, \infty)$. Are the values of λ real? Can the eigenfunctions be L^2-normalized on $x \in (-\infty, \infty)$?

(b) Consider two eigenfunctions $f_{\lambda_1}(x)$ and $f_{\lambda_2}(x)$ for eigenvalues $\lambda_1 \neq \lambda_2$. Are these two functions orthogonal; that is, does

$$\int_{-\infty}^{\infty} dx\, f_{\lambda_1}(x)^* f_{\lambda_2}(x) = 0\,? \tag{2.97}$$

(c) Now, consider exponentiating this operator, in a similar manner to how we can exponentiate the momentum operator $\hat{p}$:

$$\hat{U}_S(\alpha) = e^{i\alpha\hat{S}} = e^{\alpha x \frac{d}{dx}}, \tag{2.98}$$

for some real-valued parameter α. How does this exponentiated operator act on a general function of x, $g(x)$? Just restrict your consideration to functions $g(x)$ that have a well-defined Taylor expansion about $x = 0$.
Hint: How does this exponentiated operator act on the position to power n, x^n?

2.6 Consider a general 2×2 matrix $\mathbb{M}$ which we can express as

$$\mathbb{M} = \begin{pmatrix} a & b \\ c & d \end{pmatrix}, \tag{2.99}$$

for some, in general complex, numbers a, b, c, d.

(a) What is the characteristic equation for this matrix; that is, what is the polynomial whose roots are eigenvalues of this matrix?

(b) What are the constraints on the element values a, b, c, d such that all eigenvalues are real numbers?

(c) Now, let's further impose the constraints that the determinant of $\mathbb{M}$ is 1 and its trace (the sum of the elements on the diagonal) is 0. What are the possible values for the eigenvalues now? If the eigenvalues are real, what is a possible form for the matrix $\mathbb{M}$?

2.7 In this chapter, we considered how two different orthonormal and complete bases lead to distinct representations of a matrix $\mathbb{M}$. In this problem, we will show explicitly that this change of basis does not affect the eigenvalues of the matrix.

(a) Consider a 2×2 matrix $\mathbb{M}$ which takes the form

$$\mathbb{M} = \left(\begin{array}{cc} M_{11} & M_{12} \\ M_{21} & M_{22} \end{array} \right), \tag{2.100}$$

when expressed in the basis

$$\vec{v}_1 = \left(\begin{array}{c} 1 \\ 0 \end{array} \right), \qquad \vec{v}_2 = \left(\begin{array}{c} 0 \\ 1 \end{array} \right). \tag{2.101}$$

As in this chapter, your friend chooses the basis

$$\vec{u}_1 = \left(\begin{array}{c} \cos\theta \\ \sin\theta \end{array} \right), \qquad \vec{u}_2 = \left(\begin{array}{c} -\sin\theta \\ \cos\theta \end{array} \right), \tag{2.102}$$

for some angle θ. In the text, we had computed the (12) element of matrix $\mathbb{M}$ in this basis; now, construct the entire matrix $\mathbb{M}$ in this new $\vec{u}$-vector basis.

(b) Construct the characteristic equation for the eigenvalues λ of matrix $\mathbb{M}$ in the $\vec{v}$-vector basis. Show explicitly that the characteristic equation for matrix $\mathbb{M}$ in the $\vec{u}$-vector basis is identical. That is, the eigenvalues of matrix $\mathbb{M}$ are independent of basis.

2.8 An orthonormal and complete basis for functions $g(\theta)$ that are periodic in $\theta \in [0, 2\pi)$, where $g(\theta) = g(\theta + 2\pi)$, are the imaginary exponentials

$$f_n(\theta) = \frac{e^{in\theta}}{\sqrt{2\pi}}, \tag{2.103}$$

where $n \in \mathbb{Z}$, an integer. In this problem, we will study this space.

(a) First show that these basis functions are orthonormal on $\theta \in [0, 2\pi)$.

(b) Consider the derivative operator on this space:

$$\hat{D} = -i\frac{d}{d\theta}. \tag{2.104}$$

What are its eigenvalues and eigenvectors?

(c) An equivalent way to express this space is to use a different basis in which sines and cosines are the basis elements:

$$c_n(\theta) = \frac{1}{\sqrt{\pi}} \cos(n\theta), \qquad s_n(\theta) = \frac{1}{\sqrt{\pi}} \sin(n\theta), \qquad (2.105)$$

for $n \neq 0$. Show that these basis functions are also orthonormal on $\theta \in [0, 2\pi)$. What is the normalized basis element for $n = 0$?

(d) In this sine/cosine basis, determine the matrix elements of the derivative operator defined in part (b). Can you explicitly write down the form of the derivative operator as a matrix in this basis?

3

Hilbert Space

Our study of linear algebra in the previous chapter suggested some interesting features of a deeper structure. First, as a study of the derivative operator suggested, we will consider a complex vector space, spanned by some set of orthonormal vectors $\{\vec{v}_i\}$, where orthonormality is defined by

$$\vec{v}_i^{*\mathsf{T}}\vec{v}_j = \vec{v}_i^* \cdot \vec{v}_j = \vec{v}_j^* \cdot \vec{v}_i = \delta_{ij}. \tag{3.1}$$

With this complete basis, the matrix elements of a linear operator $\hat{O}$ are defined to be

$$\mathcal{O}_{ij} = \vec{v}_i^{*\mathsf{T}}\hat{O}\vec{v}_j. \tag{3.2}$$

In this chapter, we will explore the consequences of these properties of complex-valued vector spaces. In particular, we will be led essentially uniquely to a special vector space, called the Hilbert space, and transformations on it that preserve its properties. The simple assumptions from which we start then quickly snowball into implying deep physical consequences for how Nature works and from them we are able to formulate fundamental hypotheses that will be the basis of quantum mechanics explored in the rest of this book.

3.1 The Hilbert Space

Up until now, we have considered arbitrary linear combinations of the basis vectors $\{\vec{v}_i\}$ to define the resulting space. However, there is something special about the requirement that the basis is both orthogonal and each element of the basis has length 1. The latter requirement, for instance, enables us to say for certain that the matrix element $\mathcal{O}_{ij}$ is exactly as defined in Eq. (3.2), and not just proportional to. Typically, for some general vector space, we do not require its basis elements to satisfy any sort of normalization condition. In the case at hand, however, there is something special about the matrix elements of $\hat{O}$ and, in particular, its eigenvalues. We had seen this with the derivative operator for which, with some physical intuition, we identified the eigenvalues of $\hat{p}$ to be real (and physical) momentum. To maintain and enforce this physical interpretation, we must enforce a normalization condition on every element in the space spanned by this orthonormal basis $\{\vec{v}_i\}$. That is, in addition to the normalization of the $\vec{v}_i$ basis elements, $\vec{v}_i^* \cdot \vec{v}_i = 1$, we require that the space under consideration only consists of those linear combinations of the $\vec{v}_i$ that are also normalized.

For concreteness, consider a vector $\vec{b}$ written as a linear combination of the $\vec{v}_i$s:

$$\vec{b} = \sum_i \alpha_i \vec{v}_i, \tag{3.3}$$

for some complex numbers α_i. Demanding that $\vec{b}$ is normalized, we have

$$\vec{b}^* \cdot \vec{b} = 1 = \left(\sum_i \alpha_i^* \vec{v}_i^* \right) \cdot \left(\sum_j \alpha_j \vec{v}_j \right) = \sum_{i,j} \alpha_i^* \alpha_j \vec{v}_i^* \cdot \vec{v}_j = \sum_{i,j} \alpha_i^* \alpha_j \delta_{ij} \tag{3.4}$$
$$= \sum_i |\alpha_i|^2.$$

That is, the square magnitude of the coefficients in the linear combination expansion must sum to 1.

This vector space is called the **Hilbert space** and denoted as $\mathcal{H}$,[1] and has the following properties:

(a) Completely spanned by some orthonormal set of complex-valued vectors $\{\vec{v}_i\}$.
(b) Only those complex linear combinations of the basis vectors $\{\vec{v}_i\}$ which themselves are unit normalized live in the Hilbert space.

Again, this second requirement ensures that matrix elements and eigenvalues of an operator $\hat{\mathcal{O}}$ are not rescaled by the action on any vector in the Hilbert space. While the Hilbert space is a complex vector space, we will often refer to its elements as **states** as well as vectors, as the term "state" suggests a more general structure and will connect with a physical interpretation shortly. Further, though we need a particular basis to be able to talk concretely about the elements of the Hilbert space, it is of course independent of any basis. This feature is analogous to how we can use any arbitrary coordinate system to analyze the forces of a block on a ramp, and we must find the same net acceleration, independent of coordinate choice.

All of this can be restated for continuous functions that define a Hilbert space. A complete, orthonormal set of complex functions $\{f_i(x)\}$ satisfying

$$\delta_{ij} = \int dx\, f_i^*(x) f_j(x), \tag{3.5}$$

where the integral extends over the domain of x, defines a basis of a Hilbert space. Those functions $g(x)$ in the Hilbert space are themselves normalized and expressed as some linear combination of the $f_i(x)$ functions:

$$g(x) = \sum_i \alpha_i f_i(x) \qquad \text{with} \qquad 1 = \int dx\, g^*(x) g(x) = \sum_i |\alpha_i|^2. \tag{3.6}$$

Thus, the Hilbert space of complex functions consists of those functions that are L^2-normalized.

---

[1] John von Neumann was the first to axiomatically construct this space in generality (J. von Neumann, "Allgemeine Eigenwerttheorie Hermitescher Funktionaloperatoren," *Math. Ann.* **102**, 49–131 (1930)), but others including David Hilbert and Hermann Weyl significantly contributed to its development.

With this understanding of the Hilbert space, we can introduce compact notation for denoting it. A Hilbert space $\mathcal{H}$ composed of N-dimensional, complex-valued vectors can be expressed as

$$\mathcal{H} = \left\{ \vec{b} \in \mathbb{C}^N \,\middle|\, \vec{b}^{*\mathsf{T}}\vec{b} = 1 \right\}. \tag{3.7}$$

Here, $\mathbb{C}$ denotes the complex numbers. For a Hilbert space $\mathcal{H}$ of continuous functions over the real numbers $x \in \mathbb{R}$, we can instead denote it as

$$\mathcal{H} = \left\{ g(x) \in \mathbb{C}, x \in \mathbb{R} \,\middle|\, \int_{-\infty}^{\infty} dx\, g^*(x)g(x) = 1 \right\}. \tag{3.8}$$

Note that these definitions of the Hilbert space are independent of any basis on the Hilbert space. Indeed, as we have argued, any basis is arbitrary and not fundamental, so these spaces necessarily must have definitions that are basis-independent if they are to have some physical interpretation.

3.2 Unitary Operators

A matrix $\mathbb{M}$ transforms or maps one vector $\vec{v}$ onto another vector $\vec{u}$:

$$\mathbb{M}\vec{v} = \vec{u}. \tag{3.9}$$

If both $\vec{v}$ and $\vec{u}$ exist in the same vector space V, then $\mathbb{M}$ is a map of the vector space to itself:

$$\mathbb{M} : V \to V. \tag{3.10}$$

A map of some space or set onto itself is referred to as an **automorphism**, and determining the set or properties of the automorphisms of a vector space enables a classification of all possible maps from one vector to another, while preserving the properties of the vector space of interest.

It's therefore useful and interesting to ask about the automorphisms of the Hilbert space. To do this, we will just use language familiar from finite-dimensional matrix algebra, but it will hold true for more general linear operators on any Hilbert space, with appropriate re-interpretation. Again, a general matrix $\mathbb{M}$ maps a vector $\vec{b}$ onto some other vector $\vec{c}$:

$$\mathbb{M}\vec{b} = \vec{c}. \tag{3.11}$$

Let's assume that $\vec{b}$ is a vector in the Hilbert space: $\vec{b} \in \mathcal{H}$. Then, it is normalized: $\vec{b}^* \cdot \vec{b} = 1$. What are the constraints on $\mathbb{M}$ such that $\vec{c}$ is also in the Hilbert space $\mathcal{H}$?

Let's just take the dot product of $\vec{c}$ with its complex conjugate:

$$\vec{c}^* \cdot \vec{c} = \vec{c}^{*\mathsf{T}}\vec{c} = \left(\mathbb{M}\vec{b}\right)^{*\mathsf{T}}\left(\mathbb{M}\vec{b}\right) = \left(\vec{b}^{\mathsf{T}}\mathbb{M}^{\mathsf{T}}\right)^* \mathbb{M}\vec{b} = \vec{b}^{*\mathsf{T}}\mathbb{M}^{*\mathsf{T}}\mathbb{M}\vec{b} = 1, \tag{3.12}$$

where we enforce that $\vec{c}^* \cdot \vec{c} = 1$ to be in the Hilbert space. To evaluate this dot product of $\vec{c}$ in terms of the matrix $\mathbb{M}$ and vector $\vec{b}$, we have to complex conjugate and transpose the matrix $\mathbb{M}$. This "transpose–conjugate" operation will come up over and over again, so we will give it a name. We call this transpose–conjugation **Hermitian conjugation** and denote it with a dagger $\dagger$:

$$\dagger = *\mathsf{T} \, . \tag{3.13}$$

With this notation, the dot product of $\vec{c}$ with its complex conjugate can be written as

$$\vec{c}^* \cdot \vec{c} = \vec{c}^\dagger \vec{c} = \vec{b}^\dagger \mathbb{M}^\dagger \mathbb{M} \vec{b} = 1 = \vec{b}^\dagger \vec{b} \, . \tag{3.14}$$

If this is to hold for an arbitrary normalized vector $\vec{b}$ and linear operator $\mathbb{M}$, then there is only one possibility. The product of the matrix and its Hermitian conjugate must be the identity matrix:

$$\mathbb{M}^\dagger \mathbb{M} = \mathbb{I} \, , \tag{3.15}$$

for which we refer to $\mathbb{M}$ as a **unitary matrix**. If we relax the restriction that $\mathbb{M}$ is a linear operator, there is one other possibility. $\mathbb{M}$ could first complex conjugate $\vec{b}$ and then act as a unitary matrix to maintain normalization. Such an operator is not linear, and is often referred to as "anti-unitary." Anti-unitary operators are not linear because they fail the second requirement of linearity. If $\hat{A}$ is an anti-unitary operator, then note that its action on a vector $\vec{v}$ multiplied by a complex number z is

$$\hat{A}(z\vec{v}) = z^* \hat{A}\vec{v} \, . \tag{3.16}$$

While anti-unitary operators have importance, especially with regards to the action of time reversal, we won't consider them further in this book. This result that only unitary or anti-unitary operators correspond to all possible automorphisms of the Hilbert space was first proved by Eugene Wigner in a result now known as **Wigner's theorem**.[2]

That is to say, unitary matrices or general unitary operators act on an element of the Hilbert space to produce another element of the Hilbert space. The Hilbert space is thus closed under the action of unitary, linear operators. It then behooves us to understand these unitary operators in detail.

First, what does "unitary" mean? "Unit" means "one," so somehow a unitary matrix represents "one," or so its name would suggest. To get a sense for this, let's just consider what a 1×1 unitary matrix would be. A 1×1 matrix is just a number, and as such is its own transpose. Let's call this number z, and so $z^\mathsf{T} = z$. Then, the unitary constraint on z is

$$z^\dagger z = z^* z = |z|^2 = 1 \, . \tag{3.17}$$

A complex number with unit length is just some number on the unit circle in the complex plane as illustrated in Fig. 3.1. z can therefore be expressed as

$$z = e^{i\phi} = \cos\phi + i\sin\phi \, . \tag{3.18}$$

Thus, we can interpret "unitary" to mean of unit length or magnitude.

---

[2] E. P. Wigner, *Gruppentheorie und ihre Anwendung auf die Quantenmechanik der Atomspektren*, Vieweg (1931).

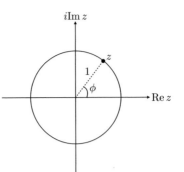

Fig. 3.1 Representation of a complex number z that lies on the unit circle in the complex plane an angle ϕ above the real axis.

Can we use this insight from regular complex numbers to understand general $N \times N$ unitary matrices? Let's take a hint from the exponential form of $z = e^{i\phi}$ and attempt to write a unitary matrix as

$$\mathbb{M} = e^{i\mathbb{T}}, \tag{3.19}$$

for some other matrix $\mathbb{T}$. As always with the exponential, we really think of it through its Taylor expansion. That is, exponentiating a matrix $\mathbb{T}$ means

$$e^{i\mathbb{T}} = \mathbb{I} + i\mathbb{T} - \frac{\mathbb{T}^2}{2} - i\frac{\mathbb{T}^3}{6} + \cdots = \sum_{n=0}^{\infty} \frac{i^n}{n!}\mathbb{T}^n. \tag{3.20}$$

In this way, every term of the Taylor expansion is just found by matrix multiplication of $\mathbb{T}$ with itself some number of times. The first term in the Taylor expansion is now the identity matrix $\mathbb{I}$, the generalization of the number 1 to $N \times N$ matrices.

With this form of the unitary matrix $\mathbb{M}$, let's see what the constraint on the matrix $\mathbb{T}$ is. Recall that, for $z = e^{i\phi}$, ϕ must be real for z to have unit magnitude. What is the generalization of a real number to matrices? It is not that every element of the matrix is real. Recall that individual matrix elements are basis-dependent and are not intrinsic properties of the matrix.

So, let's just Hermitian conjugate $\mathbb{M} = e^{i\mathbb{T}}$ and see what we find. We have

$$\mathbb{M}^\dagger = \left(e^{i\mathbb{T}}\right)^\dagger = \left(\mathbb{I} + i\mathbb{T} - \frac{\mathbb{T}^2}{2} - i\frac{\mathbb{T}^3}{6} + \cdots\right)^\dagger = \sum_{n=0}^{\infty} \left(\frac{i^n}{n!}\mathbb{T}^n\right)^\dagger \tag{3.21}$$

$$= \sum_{n=0}^{\infty} \frac{(-i)^n}{n!}\left(\mathbb{T}^\dagger\right)^n = e^{-i\mathbb{T}^\dagger}.$$

Note that both complex conjugation and transposition can be applied term-by-term in the Taylor expansion. So, we multiply

$$\mathbb{M}^\dagger \mathbb{M} = e^{-i\mathbb{T}^\dagger} e^{i\mathbb{T}} = \mathbb{I}. \tag{3.22}$$

If the matrix $\mathbb{T}$ was just a number, then multiplication in this exponential form is as simple as addition of the exponents. However, matrices do not in general commute with one another so we have to be careful with how to perform this product. We will see how to multiply exponentiated matrices much later in this book, while for now we will make a restriction that is consistent with our construction thus far.

We have stated that the only basis-independent data of a matrix are its eigenvalues and that there exists a basis in which a matrix can be expressed in diagonal form. A change of basis can be implemented by the action of other matrices, and so a matrix $\mathbb{M}$ can be expressed as

$$\mathbb{M} = \mathbb{V}\mathbb{D}\mathbb{U}, \tag{3.23}$$

where $\mathbb{D}$ is a diagonal matrix and $\mathbb{U}$ and $\mathbb{V}$ are two other matrices. If this change of basis of $\mathbb{M}$ is to still be an automorphism of the Hilbert space, then $\mathbb{U}$ and $\mathbb{V}$ must be unitary matrices. Further, we must have that $\mathbb{V} = \mathbb{U}^{\dagger}$, the Hermitian conjugate of $\mathbb{U}$, so that composition of $\mathbb{M}$ with itself is well-defined. Thus, for matrices $\mathbb{M}$ that act on states or vectors on Hilbert space, there must exist a unitary matrix $\mathbb{U}$ that diagonalizes $\mathbb{M}$:

$$\mathbb{M} = \mathbb{U}^{\dagger}\mathbb{D}\mathbb{U}. \tag{3.24}$$

A matrix or linear operator that can be expressed in this form necessarily commutes with its Hermitian conjugate:

$$\mathbb{M}\mathbb{M}^{\dagger} = \mathbb{U}^{\dagger}\mathbb{D}\mathbb{D}^{\dagger}\mathbb{U} = \mathbb{U}^{\dagger}\mathbb{D}^{\dagger}\mathbb{D}\mathbb{U} = \mathbb{M}^{\dagger}\mathbb{M}, \tag{3.25}$$

because diagonal matrices always commute. Such a matrix that commutes with its Hermitian conjugate is called **normal**, and normal matrices can always be diagonalized by a unitary matrix, by a result called the **spectral theorem**.[3] All unitary matrices are normal, so this implies that the exponentiated matrix $\mathbb{T}$ is also normal.

With this restriction to normal matrices, we are now in a position to simplify our exponentiated form of a unitary matrix. Because we assume that matrices $\mathbb{T}$ and $\mathbb{T}^{\dagger}$ commute, we have

$$\mathbb{M}^{\dagger}\mathbb{M} = e^{-i\mathbb{T}^{\dagger}}e^{i\mathbb{T}} = e^{i(\mathbb{T}-\mathbb{T}^{\dagger})} = \mathbb{I}. \tag{3.26}$$

By the Taylor expansion of the exponential, we must have

$$\mathbb{I} = e^{i(\mathbb{T}-\mathbb{T}^{\dagger})} = \mathbb{I} + i(\mathbb{T}-\mathbb{T}^{\dagger}) - \frac{(\mathbb{T}-\mathbb{T}^{\dagger})^2}{2} + \cdots. \tag{3.27}$$

The only way that this equality can hold is if the matrix $\mathbb{T}$ is equal to its Hermitian conjugate:

$$\mathbb{T} = \mathbb{T}^{\dagger}. \tag{3.28}$$

Such a matrix is called a **Hermitian matrix** and, apparently, is the generalization of a real number to matrices.

---

[3] The proof of the spectral theorem is provided in many modern textbooks in advanced matrix algebra; for example, see R. A. Horn and C. R. Johnson, *Matrix Analysis*, 2nd ed., Cambridge University Press (2013).

3.3 Hermitian Operators

Individual matrix elements of a matrix $\mathbb{T}$ are not intrinsic to the matrix, only its eigenvalues are. So, what are the eigenvalues of a Hermitian matrix? Let's assume that $\vec{v}$ is an eigenvector of a Hermitian matrix $\mathbb{T}$ and it is normalized: $\vec{v}^\dagger \vec{v} = 1$. Then, its eigenvalue equation is

$$\mathbb{T}\vec{v} = \lambda\vec{v}. \tag{3.29}$$

The eigenvalue λ can be isolated by multiplication with $\vec{v}^\dagger$ from the left:

$$\vec{v}^\dagger \mathbb{T}\vec{v} = \lambda\vec{v}^\dagger\vec{v} = \lambda. \tag{3.30}$$

Now, let's Hermitian conjugate both sides of this equation:

$$\left(\vec{v}^\dagger \mathbb{T}\vec{v}\right)^\dagger = \vec{v}^\dagger \mathbb{T}^\dagger\vec{v} = \lambda^*, \tag{3.31}$$

from properties of transposition and the fact that λ is just a number. Now, as $\mathbb{T}$ is Hermitian, this final result is identical to where we started:

$$\vec{v}^\dagger \mathbb{T}^\dagger\vec{v} = \lambda^* = \vec{v}^\dagger \mathbb{T}\vec{v} = \lambda. \tag{3.32}$$

Thus, $\lambda = \lambda^*$, and so eigenvalues of Hermitian matrices are exclusively real numbers. A real (i.e., Hermitian) matrix has exclusively real eigenvalues. As the only intrinsic properties of a matrix are its eigenvalues, this defines a Hermitian matrix (along with the property of being normal), given any basis in which it is expressed.

Before continuing, it's helpful to consider some simple examples of Hermitian matrices to gain some intuition.

Example 3.1   Let's consider the 2×2 matrix

$$\mathbb{T} = \begin{pmatrix} 1 & 0 \\ 0 & -1 \end{pmatrix}. \tag{3.33}$$

This matrix is indeed Hermitian: as all of its elements are real, complex conjugation does nothing. Further, it is a symmetric matrix (equal to its transpose), and so is Hermitian. The eigenvalues can be immediately read off to be $\lambda = \pm 1$.

Example 3.2   Now, consider the matrix

$$\mathbb{T} = \begin{pmatrix} 0 & -i \\ i & 0 \end{pmatrix}. \tag{3.34}$$

This is a matrix with only imaginary elements, for which its transpose is negative of itself: $\mathbb{T}^\mathsf{T} = -\mathbb{T}$. However, complex conjugation transforms $i \to -i$ and so this matrix is indeed Hermitian:

$$\mathbb{T}^\dagger = \begin{pmatrix} 0 & i \\ -i & 0 \end{pmatrix}^* = \begin{pmatrix} 0 & -i \\ i & 0 \end{pmatrix} = \mathbb{T}. \tag{3.35}$$

To calculate the eigenvalues of this matrix, we will use a trick that works for 2×2 matrices. The determinant of a matrix is its product of eigenvalues:

$$\det \mathbb{T} = \lambda_1 \lambda_2 = 0 \cdot 0 - i(-i) = -1. \tag{3.36}$$

The trace of a matrix, the sum of its diagonal elements, is also the sum of its eigenvalues:

$$\operatorname{tr} \mathbb{T} = \lambda_1 + \lambda_2 = 0 + 0 = 0. \tag{3.37}$$

Using these two identities, we immediately find that the eigenvalues are opposites $\lambda = \pm 1$, which are indeed real numbers.

3.3.1 The Momentum Operator

As an example coming from another direction, we have encountered operators with exclusively real eigenvalues before. The momentum operator $\hat{p}$ had real eigenvalues:

$$\hat{p} h_p(x) = \left(-ic_p \frac{d}{dx} \right) h_p(x) = p h_p(x), \tag{3.38}$$

where $p \in \mathbb{R}$ and $h_p(x)$ is an eigenfunction of momentum. As all eigenvalues of the momentum operator are real, this would seem to suggest that it is a Hermitian operator. While all Hermitian operators have exclusively real eigenvalues, the converse isn't true: every linear operator with real eigenvalues is not necessarily Hermitian because it may not also be a normal matrix. To determine if momentum is a Hermitian operator, we will just explicitly calculate its matrix elements in some basis. The quality of an operator to be Hermitian is basis-independent, so we can work in a specific basis and then generalize.

The momentum operator $\hat{p}$ is Hermitian if its matrix elements $(\hat{p})_{ij}$ and $(\hat{p})_{ji}$ are related by complex conjugation:

$$(\hat{p})_{ji} = (\hat{p})_{ij}^*, \tag{3.39}$$

where i and j denote rows and columns of $\hat{p}$. We can define these matrix elements with a complete, orthonormal basis of functions $\{f_i(x)\}$ on $x \in [x_0, x_1]$, where

$$(\hat{p})_{ji} = \int_{x_0}^{x_1} dx\, f_j^*(x) \hat{p} f_i(x) = \int_{x_0}^{x_1} dx\, f_j^*(x) \left(-ic_p \frac{d}{dx} \right) f_i(x) \tag{3.40}$$

$$\overset{\text{I.B.P.}}{=} -ic_p \left[\int_{x_0}^{x_1} dx\, \frac{d}{dx} \left(f_j^*(x) f_i(x) \right) - \int_{x_0}^{x_1} dx\, f_i(x) \frac{d}{dx} f_j^*(x) \right]$$

$$= (\hat{p})_{ij}^* - ic_p \int_{x_0}^{x_1} dx\, \frac{d}{dx} \left(f_j^*(x) f_i(x) \right).$$

On the second line, we used integration by parts (I.B.P.) to move the derivative from $f_i(x)$ onto $f_j^*(x)$. In the final line, we almost have what we want to demonstrate that the momentum is indeed Hermitian; however, there's a remaining total derivative term that integrates to the value of the functions evaluated on the boundary of the domain in x:

$$\int_{x_0}^{x_1} dx\, \frac{d}{dx}\left(f_j^*(x)f_i(x)\right) = f_j^*(x_1)f_i(x_1) - f_j^*(x_0)f_i(x_0)\,. \tag{3.41}$$

In general, this is non-zero, but we will see later that in examples of Hilbert spaces for physical systems, this indeed vanishes. A sufficient condition for this boundary-dependent term to vanish is to assume that the Hilbert space has no states that have support (i.e., take non-zero value) at its boundaries. With this assumption, this boundary term vanishes, and we have indeed shown that

$$(\hat{p})_{ji} = (\hat{p})_{ij}^*\,, \tag{3.42}$$

and so $\hat{p}^\dagger = \hat{p}$, and the momentum operator is Hermitian.

Because the momentum operator $\hat{p}$ is Hermitian, we can exponentiate it to construct a unitary spatial translation operator. We will denote this operator as $\hat{U}_p(\Delta x)$, for some displacement Δx, and it is expressed as

$$\hat{U}_p(\Delta x) = e^{i\frac{\Delta x \hat{p}}{c_p}} = e^{\Delta x \frac{d}{dx}}\,. \tag{3.43}$$

In the first equation, we note that $\hat{p}$ has units of momentum and for dimensional consistency the argument of the exponential must be unitless. To ensure unitlessness, we multiply $\hat{p}$ by the ratio $\Delta x/c_p$. Then, on the right, we have just inserted the explicit expression for $\hat{p}$ as a derivative operator. Thus, the unitary spatial translation operator is nothing more than the Taylor expansion operator we encountered early in this book.

3.3.2 Two Hypotheses

With all of these pieces, we then have a profound interpretation for Hermitian operators, which we will develop in most of the rest of this book. Unitary operators map the Hilbert space to itself, and unitary operators can be expressed as exponentiation of Hermitian operators. The only basis-independent information of an operator is its spectrum of eigenvalues and a Hermitian operator has exclusively real eigenvalues. The outcomes of physical experiment must both be real numbers and independent of any arbitrary basis, and our study of the momentum operator suggests a deep connection. Our first hypothesis is that

> 1. The eigenvalues of a Hermitian operator on a Hilbert space correspond to the results of physical measurement.

We thus also refer to these Hermitian matrices as, simply, **observables**. Further, we are able to make a profound interpretation of the vectors or functions that live in the Hilbert space. We make a second hypothesis that

> 2. The elements of a Hilbert space correspond to the possible physical states that a system can inhabit.

Thus, we say that the Hilbert space is also the **state space**. We will see what consequences these hypotheses have for making predictions of dynamics of physical systems shortly.

3.4 Dirac Notation

Up until now, we have used vector or matrix notation to represent operators or elements of the Hilbert space. This is perfectly sufficient, but there is a universally adopted, and in many ways more versatile, notation introduced by Paul Dirac, one of the founders of quantum mechanics. This is now referred to as **Dirac notation**,[4] and we'll introduce it in this section. For concreteness, let's consider a discrete, finite-dimensional Hilbert space, but we will generalize later to arbitrary Hilbert spaces. In that case, the Hilbert space $\mathcal{H}$ is

$$\mathcal{H} = \left\{ \vec{b} \in \mathbb{C}^N \,\middle|\, \vec{b}^\dagger \vec{b} = 1 \right\}, \tag{3.44}$$

where N is the dimension of the Hilbert space. A vector $\vec{b}$ in the Hilbert space $\mathcal{H}$ can be expressed as a **ket** in Dirac notation:

$$\vec{b} \Leftrightarrow |b\rangle. \tag{3.45}$$

A ket is analogous to a column vector in usual vector notation. The Hermitian conjugate of the vector $\vec{b}$, $\vec{b}^\dagger$, can be expressed as a **bra** in Dirac notation:

$$\vec{b}^\dagger \Leftrightarrow \langle b|. \tag{3.46}$$

The bra is analogous to the row vector in usual vector notation.

3.4.1 The Inner Product

Dirac notation is also sometimes referred to as "bra-ket" notation because of these names. As for why "bra" and "ket," well, putting them together you get "bra(c)ket," and smashing a bra with a ket indeed forms a bracket. In fact, consider two vectors $\vec{a}, \vec{b} \in \mathcal{H}$. Their dot product, or **inner product**, can be expressed as

$$\vec{a}^\dagger \vec{b} = \langle a|b\rangle, \qquad\qquad \vec{b}^\dagger \vec{a} = \langle b|a\rangle. \tag{3.47}$$

Note that the order of the bra-ket inner product is important, as opposite ordering corresponds to the complex conjugate:

$$\langle a|b\rangle = \langle b|a\rangle^*. \tag{3.48}$$

---

[4] P. A. M. Dirac, "A new notation for quantum mechanics," *Math. Proc. Cam. Phil. Soc.* **35**(3), 416–418 (1939). Very similar notation was introduced by Hermann Grassmann, a German polymath high school teacher, nearly a century earlier in H. Grassmann, *Die Ausdehnungslehre: Vollständig und in strenger Form begründet*, T.C.F. Enslin (1862).

Then, the normalization constraint on a vector $\vec{b} \in \mathcal{H}$ can be expressed as

$$\langle b|b \rangle = 1. \tag{3.49}$$

Dirac notation nicely encapsulates how vectors act on one another: dot products are taken by putting the flat sides of the bras and kets together. This also illustrates the use of the term "inner product": in a dot product, the vector nature of the objects is eliminated through multiplication "internal" to the two vectors, just returning a number. In Dirac notation, an inner product only features angle brackets facing outward, immediately telling you that an object of the form $\langle|\rangle$ is just a complex number.

Now, let's consider some linear operator that acts on the states in the Hilbert space; call it $\mathbb{M}$. Then, the (i, j)th entry of this matrix can be written in Dirac notation as

$$M_{ij} = \vec{v}_i^\dagger \mathbb{M} \vec{v}_j = \langle v_i | \mathbb{M} | v_j \rangle, \tag{3.50}$$

where $\{\vec{v}_i\} = \{|v_i\rangle\}$ is some complete, orthonormal basis of the Hilbert space. With an understanding of dot products in Dirac notation, orthonormality of the basis is easy to express:

$$\langle v_i | v_j \rangle = \delta_{ij}. \tag{3.51}$$

Is there a simple way to express completeness in Dirac notation?

3.4.2 The Outer Product and Completeness

We had said that "completeness" meant that the vector basis $\{\vec{v}_i\}$ can be used to find any and every element of an arbitrary linear operator $\mathbb{M}$ on the Hilbert space. To understand completeness, we need to introduce the notion of an **outer product** of vectors. For two vectors $\vec{a}, \vec{b} \in \mathcal{H}$, their outer product is

$$\vec{a}\vec{b}^\dagger = |a\rangle\langle b|, \qquad\qquad \vec{b}\vec{a}^\dagger = |b\rangle\langle a|. \tag{3.52}$$

Recall that the inner product of vectors $\vec{a}^\dagger \vec{b} = \langle a|b \rangle$ is just some complex number. On the other hand, their outer product $|a\rangle\langle b|$ is a matrix.

If you haven't seen outer products before, it can be a bit confusing, so before discussing general features, let's see how it works for two-dimensional vectors. Let's take the vectors

$$\vec{a} = \begin{pmatrix} a_1 \\ a_2 \end{pmatrix}, \qquad\qquad \vec{b} = \begin{pmatrix} b_1 \\ b_2 \end{pmatrix}. \tag{3.53}$$

Their inner product, $\vec{a}^\dagger \vec{b}$, is

$$\vec{a}^\dagger \vec{b} = (a_1^* \; a_2^*) \begin{pmatrix} b_1 \\ b_2 \end{pmatrix} = a_1^* b_1 + a_2^* b_2. \tag{3.54}$$

On the other hand, the outer product of the vectors, $\vec{a}\vec{b}^\dagger$, is

$$\vec{a}\vec{b}^\dagger = \begin{pmatrix} a_1 \\ a_2 \end{pmatrix} (b_1^* \; b_2^*) = \begin{pmatrix} a_1 b_1^* & a_1 b_2^* \\ a_2 b_1^* & a_2 b_2^* \end{pmatrix}. \tag{3.55}$$

To evaluate the outer product, we still do the usual row times column matrix multiplication. The elements at the resulting rows and columns of the outer product matrix are determined by the product of elements at that row in $\vec{a}$ and that column in $\vec{b}^\dagger$.

With this understanding, let's consider the orthonormal basis of vectors

$$\vec{v}_1 = \begin{pmatrix} 1 \\ 0 \end{pmatrix}, \qquad\qquad \vec{v}_2 = \begin{pmatrix} 0 \\ 1 \end{pmatrix}. \tag{3.56}$$

Their outer product is, for example:

$$\vec{v}_1 \vec{v}_2^\dagger = \begin{pmatrix} 1 \\ 0 \end{pmatrix} (0 \ 1) = \begin{pmatrix} 0 & 1 \\ 0 & 0 \end{pmatrix}; \tag{3.57}$$

that is, $\vec{v}_i \vec{v}_j^\dagger$ is the matrix with all 0 entries except a 1 in the ith row, jth column element. For the matrix $\mathbb{M}$, we know what element should be there:

$$M_{ij} = \vec{v}_i^\dagger \mathbb{M} \vec{v}_j = \langle v_i | \mathbb{M} | v_j \rangle. \tag{3.58}$$

So, we can completely reconstruct the matrix $\mathbb{M}$ by summing over a bunch of matrices that each only have one non-zero entry:

$$\mathbb{M} = \sum_{i,j} M_{ij} \vec{v}_i \vec{v}_j^\dagger = \sum_{i,j} \vec{v}_i \vec{v}_i^\dagger \mathbb{M} \vec{v}_j \vec{v}_j^\dagger = \sum_{i,j} |v_i\rangle \langle v_i | \mathbb{M} | v_j \rangle \langle v_j |. \tag{3.59}$$

For this identity to hold for an arbitrary orthonormal basis $\{\vec{v}_i\}$ and matrix $\mathbb{M}$, we must have that

$$\sum_i |v_i\rangle\langle v_i| = \mathbb{I}, \tag{3.60}$$

the identity matrix. This identity is called the **completeness relation**. An orthonormal basis $\{|v_i\rangle\}$ is complete if and only if

$$\sum_i |v_i\rangle\langle v_i| = \sum_i \vec{v}_i \vec{v}_i^\dagger = \mathbb{I}. \tag{3.61}$$

Let's see how this works for our two-dimensional vectors.

The outer products necessary for the completeness relation are

$$|v_1\rangle\langle v_1| = \begin{pmatrix} 1 \\ 0 \end{pmatrix}(1 \ 0) = \begin{pmatrix} 1 & 0 \\ 0 & 0 \end{pmatrix}, \quad |v_2\rangle\langle v_2| = \begin{pmatrix} 0 \\ 1 \end{pmatrix}(0 \ 1) = \begin{pmatrix} 0 & 0 \\ 0 & 1 \end{pmatrix}. \tag{3.62}$$

We then indeed find that

$$|v_1\rangle\langle v_1| + |v_2\rangle\langle v_2| = \begin{pmatrix} 1 & 0 \\ 0 & 1 \end{pmatrix} = \mathbb{I}. \tag{3.63}$$

We could have chosen a different orthonormal basis; for example, one we have studied earlier is

$$\vec{u}_1 = \begin{pmatrix} \cos\theta \\ \sin\theta \end{pmatrix}, \qquad\qquad \vec{u}_2 = \begin{pmatrix} -\sin\theta \\ \cos\theta \end{pmatrix}. \tag{3.64}$$

We have already shown that these vectors are orthonormal; are they also complete? The outer products are

$$\vec{u}_1\vec{u}_1^\dagger = \begin{pmatrix} \cos\theta \\ \sin\theta \end{pmatrix} (\cos\theta \;\; \sin\theta) = \begin{pmatrix} \cos^2\theta & \cos\theta\sin\theta \\ \cos\theta\sin\theta & \sin^2\theta \end{pmatrix}, \tag{3.65}$$

$$\vec{u}_2\vec{u}_2^\dagger = \begin{pmatrix} -\sin\theta \\ \cos\theta \end{pmatrix} (-\sin\theta \;\; \cos\theta) = \begin{pmatrix} \sin^2\theta & -\cos\theta\sin\theta \\ -\cos\theta\sin\theta & \cos^2\theta \end{pmatrix}.$$

This basis is also complete as

$$\vec{u}_1\vec{u}_1^\dagger + \vec{u}_2\vec{u}_2^\dagger = \begin{pmatrix} 1 & 0 \\ 0 & 1 \end{pmatrix} = \mathbb{I}. \tag{3.66}$$

In the following example, we introduce the Pauli matrices, and study more properties of changes of basis and completeness.

Example 3.3   The **Pauli matrices** σ_1, σ_2, σ_3, along with the identity matrix, form a complete basis for all 2×2 Hermitian matrices. The Pauli matrices are

$$\sigma_1 = \begin{pmatrix} 0 & 1 \\ 1 & 0 \end{pmatrix}, \qquad \sigma_2 = \begin{pmatrix} 0 & -i \\ i & 0 \end{pmatrix}, \qquad \sigma_3 = \begin{pmatrix} 1 & 0 \\ 0 & -1 \end{pmatrix}. \tag{3.67}$$

In Examples 3.1 and 3.2, we had studied some properties of σ_2 and σ_3. This representation of the Pauli matrices is in the basis of the eigenvectors of σ_3. Because σ_3 is already a diagonal matrix, its eigenvectors $\vec{v}_1$ and $\vec{v}_2$ are

$$\vec{v}_1 = \begin{pmatrix} 1 \\ 0 \end{pmatrix}, \qquad\qquad \vec{v}_2 = \begin{pmatrix} 0 \\ 1 \end{pmatrix}. \tag{3.68}$$

Can we express the Pauli matrices in a different basis?

Solution

What we will do here is to express the Pauli matrices in the basis of the eigenvectors of σ_1. Because the trace of σ_1 is 0 and its determinant is -1, the eigenvalues of σ_1 are $\lambda = \pm 1$. The eigenvectors $\vec{u}_1, \vec{u}_1$ of σ_1 then satisfy the linear equations

$$\sigma_1\vec{u}_1 = \vec{u}_1, \qquad\qquad \sigma_1\vec{u}_2 = -\vec{u}_2. \tag{3.69}$$

We can solve for the components of the vector $\vec{u}_1$ by explicitly writing

$$\vec{u}_1 = \begin{pmatrix} a \\ b \end{pmatrix}, \tag{3.70}$$

for some numbers a, b. Then, the linear equation that $\vec{u}_1$ satisfies is

$$\begin{pmatrix} 0 & 1 \\ 1 & 0 \end{pmatrix}\begin{pmatrix} a \\ b \end{pmatrix} = \begin{pmatrix} b \\ a \end{pmatrix} = \begin{pmatrix} a \\ b \end{pmatrix}, \tag{3.71}$$

or the components $a = b$. As discussed earlier, we must enforce that the length of $\vec{u}_1$ is unity, and so we can write

$$\vec{u}_1 = \frac{1}{\sqrt{2}} \begin{pmatrix} 1 \\ 1 \end{pmatrix}. \tag{3.72}$$

The same procedure can be followed to derive the components of $\vec{u}_2$; we find

$$\vec{u}_2 = \frac{1}{\sqrt{2}} \begin{pmatrix} 1 \\ -1 \end{pmatrix}. \tag{3.73}$$

Note that $\vec{u}_1$ and $\vec{u}_2$ are two-dimensional vectors, normalized, and mutually orthogonal; therefore, they span two-dimensional space.

Now, in this basis, we can express the Pauli matrices. By construction, the matrix σ_1 in the basis of its eigenvectors is diagonal, but the form of σ_2 and σ_3 will change. Let's just see how σ_3 transforms. Again, to determine the matrix elements in this basis, we sandwich σ_3 between the eigenvectors $\vec{u}_1$ and $\vec{u}_2$. We find

$$\begin{aligned} \vec{u}_1^\dagger \sigma_3 \vec{u}_1 &= 0, & \vec{u}_1^\dagger \sigma_3 \vec{u}_2 &= 1, \\ \vec{u}_2^\dagger \sigma_3 \vec{u}_1 &= 1, & \vec{u}_2^\dagger \sigma_3 \vec{u}_2 &= 0. \end{aligned} \tag{3.74}$$

Therefore, σ_3 in the basis of eigenvectors of σ_1 takes the form

$$\sigma_3 = \begin{pmatrix} 0 & 1 \\ 1 & 0 \end{pmatrix}, \tag{3.75}$$

which was the form of σ_1 in the basis of eigenvectors of σ_3. While we won't do it here, you can show that the form of σ_2 is just negated in transforming to the basis of eigenvectors of σ_1.

3.5 Position Basis and Continuous-Dimensional Hilbert Spaces

Our discussion thus far regarding Hilbert spaces, unitary and Hermitian operators, states, etc. has been completely general and independent of any particular basis on the Hilbert space. As we have stressed, any basis on Hilbert space just enables us to talk concretely about states and matrix elements of operators, but all physical quantities must be basis-independent. Nevertheless, there are some bases on Hilbert space whose nice physical interpretation gives them preference and they are therefore used nearly ubiquitously. In this section, we will introduce perhaps the most widely used basis on Hilbert space, the **position basis** or position space.

Let's consider a state $|g\rangle$ that lives on the Hilbert space $\mathcal{H}$. Given a discrete orthonormal and complete basis $\{|v_i\rangle\}$, $|g\rangle$ can be expressed as a linear combination

$$|g\rangle = \sum_i \alpha_i |v_i\rangle, \tag{3.76}$$

for some set of complex coefficients $\{\alpha_i\}$. Knowing that we are describing the Hilbert space by the basis $\{|v_i\rangle\}$, the $\{\alpha_i\}$ completely and uniquely specify the state $|g\rangle$. Thus we can say the set of $\{\alpha_i\}$ is the state $|g\rangle$ expressed in the $\{|v_i\rangle\}$ basis. These coefficients can be isolated by exploiting the orthonormality of the basis. Taking the inner product of a basis element $|v_j\rangle$ and the state $|g\rangle$, we have

$$\langle v_j|g\rangle = \sum_i \alpha_i \langle v_j|v_i\rangle = \sum_i \alpha_i \delta_{ij} = \alpha_j. \tag{3.77}$$

That is, to express a state in the Hilbert space in a particular basis, we take the inner product of that state with elements of the basis.

With this in mind, let's now consider the basis for Hilbert space that consists of eigenstates of the Hermitian position operator $\hat{x}$. An eigenstate of the position operator $|x'\rangle$ just returns its unique position when acted on by the position operator:

$$\hat{x}|x'\rangle = x'|x'\rangle, \tag{3.78}$$

where $x' \in (-\infty, \infty)$ is the real-valued position. Importantly, note that the value of position is continuous, and so there are a continuous infinity of position eigenstates. In the position eigenstate basis, we can determine the matrix elements of $\hat{x}$ and correspondingly provide a concrete expression for the position eigenstate. For two eigenstates $|x'\rangle$ and $|x''\rangle$, the matrix element of the position operator is

$$\langle x''|\hat{x}|x'\rangle = x'\langle x''|x'\rangle = x''\langle x''|x'\rangle. \tag{3.79}$$

Because $\hat{x}$ is Hermitian, it can act to the right or to the left on the position eigenstates and return the corresponding eigenvalues. Additionally, if the position eigenstates are assumed to be orthogonal, $\langle x''|x'\rangle = 0$ if $x' \neq x''$, this implies that the matrix elements of the position operator are proportional to Dirac δ-functions:

$$\langle x''|\hat{x}|x'\rangle = x'\delta(x'-x'') = x''\delta(x'-x''). \tag{3.80}$$

That is, the inner product of two position eigenstates is

$$\langle x''|x'\rangle = \delta(x'-x''). \tag{3.81}$$

Now, with this understanding, let's expand the state $|g\rangle$ in the position basis. Because the position eigenstates are labeled by a continuous position, their linear combination is expressed as a continuous sum or integral over eigenstates, with a coefficient that depends on the value of position. We can expand state $|g\rangle$ as

$$|g\rangle = \int_{-\infty}^{\infty} dx'\, g(x')|x'\rangle, \tag{3.82}$$

where $g(x')$ is a complex-valued function of real position x' that is the coefficient for that position eigenstate. Just as with the discrete-dimensional basis for the Hilbert space, we can isolate this coefficient function by hitting the state $|g\rangle$ with a bra state $\langle x|$:

$$\langle x|g\rangle = \int_{-\infty}^{\infty} dx'\, g(x')\langle x|x'\rangle = \int_{-\infty}^{\infty} dx'\, g(x')\,\delta(x'-x) = g(x). \tag{3.83}$$

Additionally, for $|g\rangle$ to live on the Hilbert space, it must be normalized and this constrains the function $g(x)$:

$$\langle g|g\rangle = 1 = \int_{-\infty}^{\infty} dx \, \langle g|x\rangle\langle x|g\rangle = \int_{-\infty}^{\infty} dx \, \langle x|g\rangle^* \langle x|g\rangle = \int_{-\infty}^{\infty} dx \, g^*(x)g(x) \qquad (3.84)$$
$$= \int_{-\infty}^{\infty} dx \, |g(x)|^2 \,.$$

In the position basis, normalization of a state means that the integral of its squared magnitude over all positions is 1.

These observations can be generalized to express inner products of different states on the Hilbert space and to express matrix elements of operators in the position basis. For two states $|g\rangle, |h\rangle \in \mathcal{H}$, their inner product can be expressed in position space as

$$\langle h|g\rangle = \int_{-\infty}^{\infty} dx \, \langle h|x\rangle\langle x|g\rangle = \int_{-\infty}^{\infty} dx \, h^*(x)g(x)\,, \qquad (3.85)$$

which we have already seen before as a continuous generalization of a dot product on a complex vector space. For an operator $\hat{\mathcal{O}}$ that acts on states on the Hilbert space, its matrix elements expressed in a general orthonormal basis $\{|f_i\rangle\}$ are, of course:

$$\mathcal{O}_{ij} = \langle f_i|\hat{\mathcal{O}}|f_j\rangle\,. \qquad (3.86)$$

These matrix elements can also be expressed when the basis states $\{|f_i\rangle\}$ themselves are expressed in the position basis:

$$\mathcal{O}_{ij} = \langle f_i|\hat{\mathcal{O}}|f_j\rangle = \int_{-\infty}^{\infty} dx \int_{-\infty}^{\infty} dx' \, \langle f_i|x\rangle\langle x|\hat{\mathcal{O}}|x'\rangle\langle x'|f_j\rangle \qquad (3.87)$$
$$= \int_{-\infty}^{\infty} dx \int_{-\infty}^{\infty} dx' \, f_i^*(x)\langle x|\hat{\mathcal{O}}|x'\rangle f_j(x')\,.$$

On the second line, we have just expressed the inner products of the position eigenstate with a basis state as an appropriate function of position. To continue, we need to evaluate the matrix element of $\hat{\mathcal{O}}$ in the position basis, $\langle x|\hat{\mathcal{O}}|x'\rangle$. Throughout this book, we will assume that (almost) any operator $\hat{\mathcal{O}}$ is a **local operator** that is diagonal in position space. That is, the matrix element is only non-zero if x and x' are equal:

$$\langle x|\hat{\mathcal{O}}|x'\rangle = \mathcal{O}_x \delta(x-x')\,, \qquad (3.88)$$

where $\mathcal{O}_x$ is the matrix element or expression of the operator $\hat{\mathcal{O}}$ in position space. Another way to say this is that in position space, the operator $\hat{\mathcal{O}}$ can be expressed through functions of position and some collection of derivatives with respect to position.

We can concretely write the operator in position space with the assumption of locality in the following way. First, the generalization of the completeness relation that we introduced for a discrete basis on Hilbert space to a continuous basis is

$$\mathbb{I} = \int_{-\infty}^{\infty} dx \, |x\rangle\langle x|\,. \qquad (3.89)$$

Here, $\mathbb{I}$ is the identity operator in this continuous space. Note that the outer product of position eigenstates in the integral $|x\rangle\langle x|$ only produces a matrix with non-zero entries on its diagonal. As locality enforces that $\hat{O}$ only has elements on the diagonal in position space, we can just multiply by the appropriate matrix element of $\hat{O}$ at that diagonal position, and sum up over all elements. That is, we can write

$$\hat{O} = \int_{-\infty}^{\infty} dx\, \mathcal{O}_x |x\rangle\langle x|, \tag{3.90}$$

where again $\mathcal{O}_x$ is the matrix element at that value of x.

One thing to note here is that the fact that a local operator is diagonal in position space does not mean that all local operators commute with one another. Unlike for a discrete-dimensional basis, the position basis is continuous and so there are sensible infinitesimal limits that one can define. For example, the derivative operator is defined through a limiting procedure of neighboring positions and through the limit its action at a single point is well-defined. Nevertheless, two infinitesimal limits do not need to commute, and both integration and differentiation involve infinitesimals. As a concrete example, we can express the derivative operator in position space as the limit

$$\frac{d}{dx} = \lim_{\Delta x \to 0} \int_{-\infty}^{\infty} dx\, \frac{|x+\Delta x\rangle\langle x| - |x\rangle\langle x+\Delta x|}{2\Delta x}, \tag{3.91}$$

which is the continuous generalization of Eq. (2.10). The $\Delta x \to 0$ limit ensures that the derivative is local as we have defined, but the approach to $\Delta x = 0$ is what is responsible for ensuring that the derivative is an interesting and non-trivial operator.

With this local assumption, the matrix element in our general basis $\{|f_i\rangle\}$ can be expressed as

$$\begin{aligned}
\mathcal{O}_{ij} &= \int_{-\infty}^{\infty} dx \int_{-\infty}^{\infty} dx'\, f_i^*(x)\langle x|\hat{O}|x'\rangle f_j(x') \tag{3.92}\\
&= \int_{-\infty}^{\infty} dx \int_{-\infty}^{\infty} dx'\, f_i^*(x)\mathcal{O}_x\delta(x-x')f_j(x')\\
&= \int_{-\infty}^{\infty} dx\, f_i^*(x)\mathcal{O}_x f_j(x).
\end{aligned}$$

In practice, we will rarely use the notation $\mathcal{O}_x$ to denote the operator's matrix elements in position space, but instead just write $\hat{O}$, with the understanding that it is what it needs to be so that it can act on a function of position x, $f_j(x)$. We've seen this form for matrix elements before when we studied the derivative or momentum operator, but now we have a more general formal understanding of where that representation comes from.

Before continuing, there's one more point that is important to make at this stage. Eigenstates of the position operator are not themselves in the Hilbert space because they are not unit normalized: $\langle x'|x\rangle = \delta(x'-x)$, which is infinite and non-normalizable if $x = x'$. Nevertheless, linear combinations of position eigenstates can be appropriately L^2-normalized and so live on Hilbert space, as we have seen. We'll return to this disconnect between a basis for the Hilbert space for which the basis elements are themselves not in Hilbert space in Chap. 7.

3.6 The Time Derivative

We'll devote this section to discuss the action of the time derivative operator, d/dt. Much of the discussion will parallel that of the spatial derivative operator d/dx, but its interpretation will be profoundly different. As a reminder of what the derivative does, consider a function of time $f(t)$ where we displace its argument by Δt. Using the Taylor expansion, we have

$$f(t + \Delta t) = f(t) + \Delta t \frac{d}{dt} f(t) + \frac{\Delta t^2}{2} \frac{d^2}{dt^2} f(t) + \cdots \tag{3.93}$$

$$= \sum_{n=0}^{\infty} \frac{\Delta t^n}{n!} \frac{d^n}{dt^n} f(t) = e^{\Delta t \frac{d}{dt}} f(t).$$

We should be familiar with exponentiation of a derivative from our analysis of d/dx, however, there's an important distinction between the spatial translation induced by d/dx and the time translation induced by d/dt. Given a spatial origin, we can freely move to the left or right of that particular point. That is, our spatial step Δx can be positive or negative. Time is not like that: regardless of how we beat on, we are borne ceaselessly into the future. Given a temporal origin we can only go but one way: forward. This correspondingly enforces a fixed sign for the time step Δt.

To determine what this sign is, we'll consider an analogy with something you have already encountered in studying waves. For a wave which has displacement $d(x,t)$ at position x and time t, recall that we can express it as

$$d(x,t) = A\cos(\omega t - kx + \phi_0), \tag{3.94}$$

where A is the amplitude of the wave, ω is the **angular frequency**, k is the **wavenumber**, and ϕ_0 is the initial phase. Note the relative signs between the temporal and spatial arguments in the sinusoidal function. For $k > 0$, this particular expression corresponds to a right-moving wave: as t increases, a crest moves right, as illustrated in Fig. 3.2. Another way to say this is that if you were sitting at a fixed point x_0, it would feel as if you were moving left as the wave passed. Correspondingly, the wave moves right.

Now, we have already fixed the momentum operator to be

$$\hat{p} = -ic_p \frac{d}{dx}, \tag{3.95}$$

with the overall "$-$" sign choice. For this choice of the sign of the momentum operator, we must have the opposite sign for the time translation operator to ensure that waves moving right (positive momentum) forward in time ($\Delta t > 0$) do indeed do that. That is, we can express the sinusoidal function $\cos(\omega t - kx)$ as

$$\cos(\omega t - kx) = \text{Re}\left[e^{-i\omega t + ikx}\right] = \text{Re}\left[e^{-i(\omega t - kx)}\right]. \tag{3.96}$$

We have already identified the eigenfunctions of momentum to satisfy

$$\hat{p}h_p(x) = p h_p(x), \tag{3.97}$$

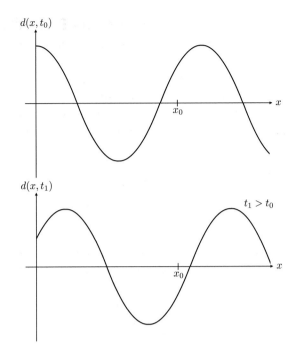

$d(x, t_0)$

x_0

x

$d(x, t_1)$

$t_1 > t_0$

x_0

x

Fig. 3.2   A snapshot of the displacement of a right-moving wave in space at two different times, where $t_0 < t_1$.

where the momentum value p is related to the wavenumber k as

$$k = \frac{p}{c_p} . \tag{3.98}$$

Then, the eigenstate $h_p(x)$ can be expressed as

$$h_p(x) = e^{i\frac{px}{c_p}} = e^{ikx} . \tag{3.99}$$

Note that with the "$+i$" in the exponent of the eigenfunction of momentum, $\hat{p}$ has a "$-i$" factor. Thus, if $e^{-i\omega t}$ represents an eigenfunction of the time derivative operator then, for all signs to work out, the Hermitian operator responsible for time translation must have a factor of "$+i$."

If the time translation operator in exponential form is

$$\hat{O}(\Delta t) = e^{\Delta t \frac{d}{dt}} , \tag{3.100}$$

for $\Delta t > 0$, we can write this in the unitary operator form

$$\hat{O} = e^{-i\hat{T}} , \tag{3.101}$$

if we identify the Hermitian operator $\hat{T}$ as

$$\hat{T} = \frac{\Delta t}{c_t} \cdot ic_t \frac{d}{dt} , \tag{3.102}$$

for some real constant c_t. Note the "$-i$" in the exponent of the unitary operator: this follows from the "$-i$" in Eq. (3.96) and our sign convention for the momentum operator. Further, this time derivative operator $\hat{T}$ is Hermitian for all the same reasons that the momentum operator is Hermitian.

What is the physical interpretation of this time derivative operator? We can think of it, again, in much the same way as we identified d/dx as the momentum operator. The time derivative operator d/dt generates a time translation of some object; that is, it forces the object to change in time. Conversely, if an object exhibits no change in time, then it just sits there, completely disengaged from every other object. When analyzing the dynamics of some system, if there is an object of that system that exhibits no change in time then we can consistently ignore it for our analysis. Or, it does not contribute to the system's **energy**: we can consistently set the energy of the unchanging object to 0. For an object or system to change in time, whatever it is, there must be a gradient of energy so that the system evolves to minimize its potential energy, for a given total energy. For example, a ball dropped from a bridge falls to the water below because the potential energy of the ball decreases as it falls. The same thing is true for, say, deformation of a squishy ball: the ball deforms to minimize its internal potential energy.

Thus, this physical picture suggests that non-trivial time dependence of some system ($d/dt \neq 0$) means that the system necessarily has non-trivial energy. So, somehow the time derivative corresponds to an energy operator. We will denote this energy operator as $\hat{H}$, as it will correspond to the **Hamiltonian**, familiar from classical mechanics. That is:

$$\hat{H} = ic_t \frac{d}{dt}. \tag{3.103}$$

The Hamiltonian is Hermitian: $\hat{H} = \hat{H}^\dagger$, ensuring that all of its eigenvalues are real. This is indeed a good thing as energy is always real.

If $\hat{H}$ has units of energy, we can correspondingly determine the units of c_t. We have

$$[\hat{H}] = [E] = [c_t] \left[\frac{d}{dt} \right] = [c_t]T^{-1}, \tag{3.104}$$

and so the units of c_t are

$$[c_t] = [E]T, \tag{3.105}$$

or Joule-seconds in SI. This is especially fascinating: recall that the unit of the c_p constant in the momentum operator is also Joule-seconds. Could it be that these two constants just happen to have the same units, but are otherwise unrelated? Sure, I guess it is possible, but a guiding principle of science is also **Occam's razor**: one should make the simplest assumptions in one's hypothesis. The simplest assumption, motivated by the fact that these constants have the same units, is that they are just the same: $c_t = c_p$. We will therefore make this hypothesis, which we will be able to test, making it science. Actually, we'll give these constants a new name:

$$c_p = c_t \equiv \hbar \qquad \text{(read: "h-bar")}, \tag{3.106}$$

which is called **Planck's reduced constant**,[5] or, more frequently in conversation, just h-bar. In SI units, the value of $\hbar$ is defined to be exactly

$$\hbar = 1.054571817\ldots \times 10^{-34}\ \text{J} \cdot \text{s}. \tag{3.107}$$

So, with this hypothesis, the momentum and energy operators are

$$\hat{p} = -i\hbar \frac{d}{dx}, \qquad\qquad \hat{H} = i\hbar \frac{d}{dt}. \tag{3.108}$$

Both operators are Hermitian (with appropriate assumptions) and so can be exponentiated to construct unitary spatial and temporal translation operators:

$$\hat{U}_p(x) = e^{i\frac{x\hat{p}}{\hbar}}, \qquad\qquad \hat{U}_H(t) = e^{-i\frac{t\hat{H}}{\hbar}}. \tag{3.109}$$

Remember the "$-$" sign for exponentiation of $\hat{H}$: this ensures that $t > 0$ translates you forward in time.

This formulation then enables states in the Hilbert space to have temporal dependence, in addition to spatial dependence. We can consider a state in position space $g(x,t) \in \mathcal{H}$ that encodes spatial and temporal dependence of some state of a system. For all of their similarities, time and space are different and this manifests itself in our definition of L^2-normalization for these time-dependent states. We really think of the function $g(x,t)$ as the complete spatial dependence of a state at time t. As such, the L^2-norm is only an integration over positions:

$$1 = \int_{-\infty}^{\infty} dx\, g^*(x,t) g(x,t), \tag{3.110}$$

for all t. The fact that states in the Hilbert space are normalized for all time means that the volume or number of states in the Hilbert space is constant in time. This means that one can choose a fixed complete, orthonormal basis in which to define states in the Hilbert space, and that basis works for all time. For example, let's say that $\{f_i(x)\}$ is some time-independent orthonormal basis expressed in position space, such that

$$\delta_{ij} = \langle f_i | f_j \rangle = \int_{-\infty}^{\infty} dx\, f_i^*(x) f_j(x). \tag{3.111}$$

Then a general state $g(x,t) \in \mathcal{H}$ can be expressed as

$$g(x,t) = \sum_i \alpha_i(t) f_i(x), \tag{3.112}$$

where the $\alpha_i(t)$ are some functions only of time, and all spatial dependence is carried by the $f_i(x)$ basis functions. For $g(x,t)$ to be normalized, we must have

$$1 = \int_{-\infty}^{\infty} dx\, g^*(x,t) g(x,t) = \sum_{i,j} \alpha_i^*(t) \alpha_j(t) \langle f_i | f_j \rangle = \sum_i |\alpha_i(t)|^2. \tag{3.113}$$

Normalization for all time means that the time-dependent coefficients' magnitude squared sum is 1, for all time.

---

[5] M. Planck, "Ueber irreversible Strahlungsvorgaenge," *Sitz.ber. Koenigl. Preuss. Akad. Wiss. Berlin* **5**, 440–480 (1899); *Ann. Phys.* **306**(1), 69–122 (1900).

With this time and space-dependent state in the Hilbert space, we give it a special name and symbol. We denote an element of the Hilbert space $\mathcal{H}$ expressed in position space as $\psi(x,t)$, where

$$\mathcal{H} = \left\{ \psi(x,t) \in \mathbb{C}; x,t \in \mathbb{R} \,\middle|\, \langle \psi|\psi \rangle = \int_{-\infty}^{\infty} dx\, \psi^*(x,t)\psi(x,t) = 1 \right\}, \qquad (3.114)$$

and call $\psi(x,t)$ the **wavefunction**. Essentially from here through the rest of this book, our goal will be to study the dynamics of wavefunctions of some system of interest.

Before continuing, we pause here to work through an example of time dependence in a two-state system.

Example 3.4   In this example, we will study the time dependence of a simple two-state system. Let's assume that $|1\rangle$ and $|2\rangle$ are energy eigenstates with energies E_1 and E_2, respectively. These states are orthonormal: $\langle i|j \rangle = \delta_{ij}$. We'll study the initial state at $t = 0$:

$$|\psi\rangle = \alpha_1|1\rangle + \alpha_2|2\rangle, \qquad (3.115)$$

where α_1, α_2 are some complex coefficients such that $|\psi\rangle$ is in the Hilbert space.

Solution

The first thing we will do is to give the state $|\psi\rangle$ time dependence through the action of the Hamiltonian $\hat{H}$. The unitary time-evolution operator is

$$\hat{U}_H(t) = e^{-i\frac{t\hat{H}}{\hbar}}. \qquad (3.116)$$

The state $|\psi\rangle$ at a general time t is therefore

$$|\psi(t)\rangle = \hat{U}_H(t)|\psi\rangle. \qquad (3.117)$$

So far, we don't have an explicit form for the Hamiltonian, but we do know its eigenvalues E_1, E_2, and we have an expression for $|\psi\rangle$ in terms of the eigenstates of the Hamiltonian. Therefore, we can act the time-evolution operator on the individual eigenstates to find

$$\hat{U}_H(t)|1\rangle = e^{-i\frac{tE_1}{\hbar}}|1\rangle, \qquad\qquad \hat{U}_H(t)|2\rangle = e^{-i\frac{tE_2}{\hbar}}|2\rangle. \qquad (3.118)$$

By linearity of the time-evolution operator, we can just tack on these exponential factors to the individual eigenstates in the expression for $|\psi\rangle$ to give it time dependence. That is:

$$\hat{U}_H(t)|\psi\rangle = |\psi(t)\rangle = \alpha_1 e^{-i\frac{tE_1}{\hbar}}|1\rangle + \alpha_2 e^{-i\frac{tE_2}{\hbar}}|2\rangle. \qquad (3.119)$$

With this time-dependent state, let's compute the outer product $|\psi(t)\rangle\langle\psi(t)|$ for generic t, in the basis of energy eigenstates. As we have already expressed $|\psi(t)\rangle$ in bra-ket notation, we can express this outer product in bra-ket notation too:

$$|\psi(t)\rangle\langle\psi(t)| = \left(\alpha_1 e^{-i\frac{tE_1}{\hbar}}|1\rangle + \alpha_2 e^{-i\frac{tE_2}{\hbar}}|2\rangle\right)\left(\alpha_1^* e^{i\frac{tE_1}{\hbar}}\langle 1| + \alpha_2^* e^{i\frac{tE_2}{\hbar}}\langle 2|\right) \qquad (3.120)$$

$$= |\alpha_1|^2|1\rangle\langle 1| + \alpha_1\alpha_2^* e^{-i\frac{t(E_1-E_2)}{\hbar}}|1\rangle\langle 2| + \alpha_1^*\alpha_2 e^{i\frac{t(E_1-E_2)}{\hbar}}|2\rangle\langle 1| + |\alpha_2|^2|2\rangle\langle 2|.$$

Note that we had to complex conjugate all numbers that appeared in the expression for $\langle\psi(t)|$ with respect to the ket $|\psi(t)\rangle$. As a 2×2 matrix, this can be expressed as

$$|\psi(t)\rangle\langle\psi(t)| = \begin{pmatrix} |\alpha_1|^2 & \alpha_1\alpha_2^* e^{-i\frac{t(E_1-E_2)}{\hbar}} \\ \alpha_1^*\alpha_2 e^{i\frac{t(E_1-E_2)}{\hbar}} & |\alpha_2|^2 \end{pmatrix}. \qquad (3.121)$$

The diagonal entries of this matrix are just the absolute squares of the coefficients in the expression for $|\psi\rangle$ at time $t = 0$. We will provide their physical interpretation in the next section.

Finally, we can explicitly construct the Hamiltonian operator $\hat{H}$ in its basis of eigenstates $|1\rangle, |2\rangle$. In its eigenstate basis, $\hat{H}$ is necessarily diagonal and the entries along the diagonal are just its eigenvalues:

$$\hat{H} = E_1|1\rangle\langle 1| + E_2|2\rangle\langle 2| = \begin{pmatrix} E_1 & 0 \\ 0 & E_2 \end{pmatrix}. \qquad (3.122)$$

For real-valued energies E_1, E_2, this is Hermitian.

3.7 The Born Rule

Let's now work to provide a physical interpretation of the wavefunction, and by extension, any state in the Hilbert space. To motivate this, recall the interpretation of the dot product. Consider two vectors (real-valued for now) $\vec{u}, \vec{v}$ that both have unit length:

$$\vec{u}\cdot\vec{u} = \vec{v}\cdot\vec{v} = 1. \qquad (3.123)$$

Now, their dot product is

$$\vec{u}\cdot\vec{v} = |\vec{u}||\vec{v}|\cos\theta = \cos\theta, \qquad (3.124)$$

because of their unit length and where θ is their relative angle. The dot product is a measure of how much of one vector lies along the direction of the other vector. A picture of this is shown in Fig. 3.3. If the dot product is large, then we can say that the vectors have a large overlap: knowing one vector provides significant information about the other vector. By contrast, if the dot product is zero, the vectors have no overlap and you know very little about $\vec{u}$, say, if you know $\vec{v}$. You only know that they are orthogonal.

Now, back to the wavefunction. Of course, the wavefunction overlaps completely with itself; this is just an expression of the normalization condition:

$$\int_{-\infty}^{\infty} dx\, \psi^*(x,t)\psi(x,t) = 1 = \langle\psi|\psi\rangle. \qquad (3.125)$$

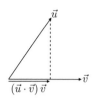

Fig. 3.3 Illustration of how the dot product of two vectors $\vec{u}$ and $\vec{v}$ isolates the amount of vector $\vec{u}$ along the direction of vector $\vec{v}$ (and vice-versa).

The wavefunction also overlaps completely with the entirety of the orthonormal basis $\{f_i(x)\}$:

$$\langle \psi | \psi \rangle = \sum_i |\alpha_i(t)|^2 = 1 . \qquad (3.126)$$

Note here that $\alpha_i(t)$ is the "dot product" of ψ with f_i:

$$\langle f_i | \psi \rangle = \int_{-\infty}^{\infty} dx\, f_i^*(x) \psi(x,t) = \sum_j \int_{-\infty}^{\infty} dx\, \alpha_j(t) f_i^*(x) f_j(x) = \alpha_i(t) . \qquad (3.127)$$

So $\alpha_i(t) = \langle f_i | \psi \rangle$ is the overlap of the wavefunction with the basis state $|f_i\rangle$. However, $\alpha_i(t)$ alone is complex-valued and unconstrained, until demanding L^2-normalization of the wavefunction. With the normalization

$$1 = \sum_i |\alpha_i(t)|^2 = \sum_i |\langle f_i | \psi \rangle|^2 = \sum_i \langle \psi | f_i \rangle \langle f_i | \psi \rangle , \qquad (3.128)$$

we can interpret $|\alpha_i(t)|^2$ as the actual, true fraction of the wavefunction ψ that lies along the direction of $f_i(x)$. As an absolute value, it is always non-negative and is restricted to lie between 0 and 1: $0 \leq |\alpha_i(t)|^2 \leq 1$.

These considerations lead us to hypothesize that these magnitude-squared coefficients represent the *probability* of ψ representing the basis element f_i. This extremely profound consequence is called the **Born rule**, after Max Born who postulated it.[6] Let's give this interpretation some more meat before we understand probability a bit more.

3.7.1 Measurements as Eigenvalues of Operators

Let's assume we have a Hilbert space $\mathcal{H}$ such that

$$\mathcal{H} = \{ |\psi\rangle \mid \langle \psi | \psi \rangle = 1 \} . \qquad (3.129)$$

On this Hilbert space, there is a unitary, linear operator $\hat{O}$ such that

$$\hat{O}^\dagger \hat{O} = \mathbb{I} , \qquad (3.130)$$

---

[6] M. Born, "Zur Quantenmechanik der Stoßvorgänge," *Z. Phys.* **37**(12), 863–867 (1926); reprinted and translated in J. A. Wheeler and W. H. Zurek (eds.), *Quantum Theory and Measurement*, Princeton University Press (1963), p. 52.

and such a unitary operator can be expressed as the exponentiation of some linear, Hermitian operator $\hat{T}$ such that $\hat{T} = \hat{T}^\dagger$ and

$$\hat{O} = e^{i\hat{T}} . \tag{3.131}$$

The only basis-independent data of the operator $\hat{T}$ are its eigenvalues and, as a Hermitian operator, all of the eigenvalues of $\hat{T}$ are real-valued. These two facts motivate the eigenvalues of $\hat{T}$ as possible results of a physical experiment. Further, we can provide an orthonormal basis on $\mathcal{H}$ with the eigenstates of $\hat{T}$; call them $\{|v_i\rangle\}$, such that

$$\hat{T}|v_i\rangle = \lambda_i|v_i\rangle . \tag{3.132}$$

For distinct eigenvalues, $\lambda_i \neq \lambda_j$, different basis elements are orthogonal and can be fixed to be normalized:

$$\langle v_i|v_j\rangle = \delta_{ij} , \tag{3.133}$$

and we will assume completeness:

$$\mathbb{I} = \sum_i |v_i\rangle\langle v_i| . \tag{3.134}$$

Thus, the physical interpretation of the eigenvalue equation

$$\hat{T}|v_i\rangle = \lambda_i|v_i\rangle \tag{3.135}$$

is that if your system is in the state $|v_i\rangle$, then the outcome of measuring the quantity that corresponds to the Hermitian operator $\hat{T}$ is *always* λ_i.

Now let's consider a general state $|\psi\rangle$ in this Hilbert space with this basis of eigenvectors of $\hat{T}$. We can write $|\psi\rangle$ expanded in this basis as

$$|\psi\rangle = \sum_i \alpha_i|v_i\rangle , \tag{3.136}$$

where we have suppressed explicit temporal dependence in the coefficients α_i. Given the state of the system $|\psi\rangle$, when we measure the quantity that corresponds to $\hat{T}$, we must get some value, but that value can be any eigenvalue of $\hat{T}$. This is just the statement of normalization of $|\psi\rangle$:

$$\langle\psi|\psi\rangle = \sum_i |\alpha_i|^2 = 1 . \tag{3.137}$$

The interpretation of the magnitude-squared terms $|\alpha_i|^2$ individually is that it is the *probability* of performing this measurement of the quantity represented by $\hat{T}$ for which the outcome is the eigenvalue λ_i for the state $|v_i\rangle$.

That is, given the state $|\psi\rangle$, if we perform the measurement corresponding to $\hat{T}$ over and over and over on this state, we expect that $|\alpha_i|^2$ is the fraction of the measurements that we perform for which the outcome is λ_i. We'll see many examples of this in action in the coming chapters. For the rest of this chapter, we'll review aspects of probability that we will need in going forward.

3.7.2 Terse Review of Probability

You are likely familiar with probability, if only colloquially.[7] There are three axioms of probability, which have already been stated but in a different language:

(a) The probability P of an event (= physical measurement) E is a non-negative real number:

$$P(E) \geq 0, \qquad P(E) \in \mathbb{R}. \tag{3.138}$$

(b) The total probability of any possible event occurring is unity:

$$\sum_E P(E) = 1. \tag{3.139}$$

(c) The probability P_{tot} of a collection of mutually exclusive events $\{E_i\}$ occurring is just the sum of their individual probabilities:

$$\sum_i P(E_i) = P_{\text{tot}}. \tag{3.140}$$

We had seen axiom 1 from the normalization of a state on Hilbert space: the magnitude-squared coefficients of $|\psi\rangle$ in the orthonormal basis $\{|v_i\rangle\}$ are non-negative, $|\alpha_i|^2 \geq 0$. Axiom 2 also follows from the normalization of a state on Hilbert space: $\langle\psi|\psi\rangle = 1$ (i.e., when we measure something on the state $|\psi\rangle$, there is always some result). Finally, axiom 3 follows from the orthogonality of the basis $\{|v_i\rangle\}$. Orthogonality of two states $|v_i\rangle, |v_j\rangle$ for $i \neq j$, means that their corresponding eigenvalues under action of some Hermitian operator $\hat{T}$ are distinct: $\lambda_i \neq \lambda_j$. Thus, outcomes of measuring the quantity that corresponds to $\hat{T}$ of λ_i and λ_j are mutually exclusive (i.e., they are distinct experimental outcomes).

To gain some familiarity with probability, let's consider a fair, six-sided die, with each side individually numbered with pips, shown in Fig. 3.4. A "fair die" means that the outcome of a roll of the die is equally likely to land on any of the sides. When we roll the die, some side must be up, so the total probability of any side being up must be 1. Different sides being up corresponds to mutually exclusive outcomes, so the axioms of probability imply that the probability for any one side of a fair die to be up is 1/6. That is, if you roll a fair die N times, as $N \to \infty$, you expect that $N/6$ rolls are 1, $N/6$ rolls are 2, etc.

Okay, but what if you roll the die just once? What value do you expect the die to return? One answer is simply that you expect each side with equal probability, so there is no well-defined answer. Indeed, but you could also interpret this question as what the average outcome of one roll would be. That is, roll the die $N \to \infty$ times, and ask what the mean outcome of the rolls is. To calculate this average, we simply sum up the values of each die for every one of the N rolls. If the outcome i has probability p_i, then

---

[7] We'll later discuss more aspects of probability relevant for quantum mechanics, as well. For a broader introduction to probability as a mathematical discipline and its uses, two good books are: E. T. Jaynes, *Probability Theory: The Logic of Science*, Cambridge University Press (2003) and S. Ross, *A First Course in Probability*, Pearson (2014).

Fig. 3.4 Illustration of a fair die whose sides are labeled with distinct numbers of pips.

you expect that $p_i N$ of the rolls return i, and the total sum of pips is $i p_i N$. To find the sum of all possible rolls, we then sum over i:

$$\text{Total} = \sum_{i=1}^{6} i p_i N. \tag{3.141}$$

The mean is the total divided by the number of rolls N. We call this mean the **expectation value** $E(\text{die})$:

$$E(\text{die}) = \sum_{i=1}^{6} i p_i. \tag{3.142}$$

With $p_i = 1/6$ for a fair die, this expectation value is

$$E(\text{die}) = \frac{1}{6} \sum_{i=1}^{6} i = \frac{1}{6} \frac{6(6+1)}{2} = \frac{7}{2} = 3.5. \tag{3.143}$$

This value is exactly halfway between the possible rolls of the die as shown in Fig. 3.5. Note that the center-of-mass is like an expectation value for the "mass" of an object. If we interpret $m_i / \sum_i m_i = p_i$, then the center-of-mass can be expressed as

$$x_{\text{cm}} = \frac{\sum_i x_i m_i}{\sum_i m_i} = \sum_i x_i p_i. \tag{3.144}$$

In the language of our Hilbert space, we've already seen this expectation value, though in a different context. Given a state $|\psi\rangle$, let's expand it in the eigenbasis $\{|v_i\rangle\}$ of a Hermitian operator $\hat{T}$ as

$$|\psi\rangle = \sum_i \alpha_i |v_i\rangle. \tag{3.145}$$

What is the expectation value of the outcome of a measurement of the quantity to which $\hat{T}$ corresponds? The outcomes are just its eigenvalues λ_i with probabilities $|\alpha_i|^2$, so the expectation value is

$$E(\hat{T}) = \sum_i \lambda_i |\alpha_i|^2 = \sum_i \alpha_i^* \alpha_j \langle v_i | \hat{T} | v_j \rangle = \langle \psi | \hat{T} | \psi \rangle. \tag{3.146}$$

That is, the expectation value of $\hat{T}$ on the state ψ is

$$E(\hat{T}) = \langle \psi | \hat{T} | \psi \rangle = \int_{-\infty}^{\infty} dx \, \psi^*(x,t) \hat{T} \psi(x,t), \tag{3.147}$$

where on the right we have written it in position space. This is also, equivalently, the matrix element of $\hat{T}$ in a basis that includes the state $|\psi\rangle$ that lies in the ψth row and

The expectation value for the outcomes of rolls of a fair die lies at the middle of the interval of possible rolls.

column on the diagonal of $\hat{T}$. That is, expectation values of a Hermitian operator just correspond to its matrix elements, in some basis.

Exercises

3.1   In this chapter, we introduced normal matrices as those matrices that commute with their Hermitian conjugate. This seems like a weak requirement, but it's very easy to construct matrices for which this is not true.

(a)   Consider a general, complex-valued 2×2 matrix $\mathbb{M}$ where

$$\mathbb{M} = \begin{pmatrix} a & b \\ c & d \end{pmatrix}, \tag{3.148}$$

for complex a,b,c,d. If $\mathbb{M}$ is normal, what constraints does that place on the values of a,b,c,d?

(b)   An upper-**triangular matrix** is a matrix that only has non-zero entries on the diagonal and above it; a lower-triangular matrix is defined analogously. For example, the 2×2 matrix

$$\begin{pmatrix} a & b \\ 0 & d \end{pmatrix} \tag{3.149}$$

is upper triangular. Show that a general upper-triangular matrix of arbitrary dimension cannot be normal if there are non-zero entries above the diagonal.

Hint: For a matrix $\mathbb{M}$, you just have to show that one element of $\mathbb{M}\mathbb{M}^\dagger$ does not equal the corresponding element of $\mathbb{M}^\dagger\mathbb{M}$.

3.2   In this exercise, we study the unitary matrix constructed from exponentiation of one of the Pauli matrices, introduced in Example 3.3. As Hermitian matrices, exponentiation should produce a unitary matrix, and we will verify that here.

(a)   As Hermitian matrices, the Pauli matrices can be exponentiated to construct a corresponding unitary matrix. Let's exponentiate σ_3 as defined in Eq. (3.67) to construct the matrix

$$\mathbb{A} = e^{i\phi\sigma_3}, \tag{3.150}$$

where ϕ is a real number. What is the resulting matrix $\mathbb{A}$? Write it in standard 2×2 form. Is it actually unitary? Remember that exponentiation of a matrix is defined by its Taylor expansion.

(b) Now, consider the Hermitian matrix constructed from the sum of σ_1 and σ_3:

$$\sigma_1 + \sigma_3 = \begin{pmatrix} 1 & 1 \\ 1 & -1 \end{pmatrix}. \tag{3.151}$$

Exponentiate this matrix to construct $\mathbb{B}$, where

$$\mathbb{B} = e^{i\phi(\sigma_1 + \sigma_3)}. \tag{3.152}$$

Can you express the result in the form of a 2×2 matrix?

Hint: Taylor expand the exponential function and sum up the terms with even powers of ϕ and odd powers of ϕ separately.

3.3 The unitary matrix that implements rotations on real two-dimensional vectors can be written as

$$\mathbb{M} = \begin{pmatrix} \cos\theta & \sin\theta \\ -\sin\theta & \cos\theta \end{pmatrix}, \tag{3.153}$$

where θ is the real-valued rotation angle.

(a) Verify that this matrix is indeed unitary.

(b) As a unitary matrix, it can be expressed as the exponential of a Pauli matrix in the form

$$\mathbb{M} = e^{i\theta\sigma_j}, \tag{3.154}$$

for some Pauli matrix σ_j defined in Eq. (3.67). Which Pauli matrix is it? Be sure to show your justification.

3.4 We introduced unitary operators as those that map the Hilbert space to itself (i.e., those that maintain the normalization constraint that we require of all vectors in the Hilbert space). However, we didn't show the converse, that there always exists a unitary operator that connects two vectors of the Hilbert space. For a Hilbert space $\mathcal{H}$ of two-dimensional vectors, consider the vectors $\vec{u}, \vec{v} \in \mathcal{H}$.

(a) We can express these vectors in a particular basis as

$$\vec{u} = \begin{pmatrix} e^{i\phi_1}\cos u \\ e^{i\phi_2}\sin u \end{pmatrix}, \qquad \vec{v} = \begin{pmatrix} e^{i\theta_1}\cos v \\ e^{i\theta_2}\sin v \end{pmatrix}, \tag{3.155}$$

for real parameters $u, v, \phi_1, \phi_2, \theta_1, \theta_2$. Show that these vectors are normalized to live in the Hilbert space.

(b) In this basis, determine the unitary matrix $\mathbb{U}$ such that

$$\vec{v} = \mathbb{U}\vec{u}. \tag{3.156}$$

Hint: Write the matrix $\mathbb{U}$ as

$$\mathbb{U} = \begin{pmatrix} a & b \\ c & d \end{pmatrix}, \tag{3.157}$$

and determine the constraints on a, b, c, d if $\mathbb{U}$ is unitary and maps $\vec{u}$ to $\vec{v}$.

(c) Can the matrix $\mathbb{U}$ be degenerate? That is, can its determinant ever be 0?

3.5   Consider two states on the Hilbert space $|u\rangle$ and $|v\rangle$. Construct the operator $\mathbb{U}_{uv}$ that maps one to another:

$$|v\rangle = \mathbb{U}_{uv}|u\rangle. \qquad (3.158)$$

Express $\mathbb{U}_{uv}$ as the outer product of Dirac bras and kets. Is the operator $\mathbb{U}_{uv}$ unitary? Can you make it unitary?

3.6   In this chapter, we introduced completeness as the requirement that an orthonormal basis $\{|v_i\rangle\}_i$ satisfies

$$\sum_i |v_i\rangle\langle v_i| = \mathbb{I}. \qquad (3.159)$$

In this problem, we will study different aspects of this completeness relation.

(a)   What if the basis is not orthogonal? Consider the vectors

$$\vec{v}_1 = \begin{pmatrix} 1 \\ 0 \end{pmatrix}, \qquad \vec{v}_2 = \begin{pmatrix} e^{i\phi_1}\sin\theta \\ e^{i\phi_2}\cos\theta \end{pmatrix}. \qquad (3.160)$$

What is their outer product sum $|v_1\rangle\langle v_1| + |v_2\rangle\langle v_2|$? For what value of θ does it satisfy the completeness relation? Does the result depend on ϕ_1 or ϕ_2?

(b)   In Example 2.1 and Exercise 2.3 of Chap. 2, we studied the first three Legendre polynomials as a complete basis for all quadratic functions on $x \in [-1,1]$. There we had demonstrated orthonormality:

$$\int_{-1}^{1} dx\, P_i(x)P_j(x) = \delta_{ij}. \qquad (3.161)$$

What about completeness? First, construct the "identity matrix" $\mathbb{I}$ formed from the outer product of the first three Legendre polynomials:

$$\mathbb{I} = P_0(x)P_0(y) + P_1(x)P_1(y) + P_2(x)P_2(y). \qquad (3.162)$$

Use the form of the Legendre polynomials presented in Eq. (2.41).

(c)   You should find something that does not look like an identity matrix. How can we test it? Provide an interpretation of this outer product identity matrix by integrating against an arbitrary quadratic function:

$$\int_{-1}^{1} dx\,(ax^2 + bx + c)\,\mathbb{I}, \qquad (3.163)$$

for some constants a, b, c. What should you find? What do you find?

3.7   In this exercise, we continue the study of the two-state system introduced in Example 3.4. Recall that $|1\rangle$ and $|2\rangle$ are energy eigenstates with energies E_1 and E_2, respectively, which are orthonormal: $\langle i|j\rangle = \delta_{ij}$. The initial state is

$$|\psi\rangle = \alpha_1|1\rangle + \alpha_2|2\rangle, \qquad (3.164)$$

where α_1, α_2 are some complex coefficients.

(a)   Calculate the expectation value of the Hamiltonian $\hat{H}$ from Eq. (3.122) in the time-dependent state $|\psi(t)\rangle$, $\langle\psi(t)|\hat{H}|\psi(t)\rangle$. Does it depend on time?

(b) Let's assume that there is some other Hermitian operator on this Hilbert space, called $\hat{O}$. We know its action on the energy eigenstates:

$$\hat{O}|1\rangle = |1\rangle - |2\rangle, \qquad\qquad \hat{O}|2\rangle = -|1\rangle + |2\rangle. \qquad (3.165)$$

Express $\hat{O}$ in both bra-ket notation and familiar matrix notation. Is it actually Hermitian?

(c) What is the expectation value of the unitary operator $\hat{O}$ from part (b), $\langle\psi(t)|\hat{O}|\psi(t)\rangle$? Does it depend on time? What does it simplify to if both α_1 and α_2 are real?

3.8 **Neutrinos** are very low mass, extremely weakly interacting particles that permeate the universe. About a quadrillion will pass through you while you read this problem. There are multiple types, or flavors, of neutrinos, and they can oscillate into one another as time passes. A model for the oscillations of neutrinos is the following. Consider two neutrinos that are also energy eigenstates, $|\nu_1\rangle$ and $|\nu_2\rangle$, with energies E_1 and E_2, respectively. Neutrinos are produced and detected not as energy eigenstates, but as eigenstates of a different Hermitian operator, called the flavor operator. There are two flavors of neutrino, called electron $|\nu_e\rangle$ and muon $|\nu_\mu\rangle$ neutrinos, and these can be expressed as a linear combination of $|\nu_1\rangle$ and $|\nu_2\rangle$:

$$|\nu_e\rangle = \cos\theta|\nu_1\rangle + \sin\theta|\nu_2\rangle, \qquad |\nu_\mu\rangle = -\sin\theta|\nu_1\rangle + \cos\theta|\nu_2\rangle, \qquad (3.166)$$

for some mixing angle θ. These flavor-basis neutrinos then travel for time T until they hit a detector where their flavor composition is measured.[8]

(a) Assume that the initial neutrino flavor is exclusively electron, $|\nu_e\rangle$. What is the electron-neutrino state after time T, $|\nu_e(T)\rangle$?

(b) After time T, what is the probability for the detector to measure an electron neutrino? What about a muon neutrino? The detector only measures those flavor eigenstates as defined by Eq. (3.166). You should find that the probability of measuring the muon neutrino is

$$P_\mu = \sin^2(2\theta)\sin^2\left(\frac{(E_1 - E_2)T}{2\hbar}\right). \qquad (3.167)$$

Hint: The flavor basis is orthogonal and complete, so it can be used to express the identity operator.

(c) Describe why this phenomenon is called neutrino *oscillations*.

---

[8] Why and how neutrinos oscillate is actually quite subtle, and requires quantum mechanical and special relativistic considerations. See chapter 12 of A. J. Larkoski, *Elementary Particle Physics: An Intuitive Introduction*, Cambridge University Press (2019).

4

Axioms of Quantum Mechanics and their Consequences

We've come a long way from the fundamental mathematical properties of linear operators to the profound physical interpretations of them. We have finally developed the necessary background for deriving the fundamental equation of quantum mechanics, the Schrödinger equation.

Before we do so, however, let's summarize our results thus far. Our summary will be important enough that we will refer to the following statements as the **axioms of quantum mechanics**:

Axiom 1 *Observables (outcomes of physical experiment) correspond to Hermitian operators on a Hilbert space $\mathcal{H}$.*

Axiom 2 *The state of a system is represented by a state vector $|\psi\rangle$ on the Hilbert space $\mathcal{H}$, such that $\langle\psi|\psi\rangle = 1$.*

Axiom 3 *The expectation value of an observable corresponding to the Hermitian operator $\hat{T}$ on the state $|\psi\rangle$ is $\langle\psi|\hat{T}|\psi\rangle$.*

These axioms are called the **Dirac–von Neumann axioms**,[1] and, as we have discussed, their motivation lies firmly in linear algebra and analogy with probability axioms. Our goal in this chapter will be to use these axioms and our earlier results to derive the fundamental equation of quantum mechanics, and its consequences.

4.1 The Schrödinger Equation

Recall again what the fundamental goal of all of physics, nay, all of science is. The scientific method is to first make a hypothesis, motivated by all experimental data available thus far. Then, we test that hypothesis with new experimental data. Depending on the outcome of the experiment, we either modify our hypothesis or gain confidence that Nature works as our hypothesis suggests. As mentioned early in this book, there is a time ordering to the scientific method: we make a hypothesis in the present and test it in the future. Thus our goal in physics is to predict how a system evolves forward in time, given data at the present.

---

[1] P. A. M. Dirac, *The Principles of Quantum Mechanics*, Oxford University Press (1981); J. von Neumann, *Mathematical Foundations of Quantum Mechanics: New Edition*, Princeton University Press (2018).

Indeed, the fundamental equation of classical mechanics, Newton's second law, describes how the momentum of a system changes in time:

$$\vec{F} = \frac{d\vec{p}}{dt}\,. \tag{4.1}$$

This is a differential equation in time and can be solved for all future times $t > 0$, given data of the momentum $\vec{p}$ at $t = 0$ and the force $\vec{F}$. If we know momentum (and position, etc.) at all times, we can definitively say that we have "solved" the system. Our goal in quantum mechanics will be the same: we want the equation that governs time dependence of a state vector $|\psi\rangle$ on Hilbert space, as it completely quantifies the physical state.

We have all of the pieces in place to do this. We want to figure out how the state $|\psi(t)\rangle$ at time t in the Hilbert space transforms at a slightly later time:

$$|\psi(t)\rangle \rightarrow |\psi(t+\Delta t)\rangle = |\psi(t)\rangle + \Delta t \frac{d}{dt}|\psi(t)\rangle + \cdots, \tag{4.2}$$

where we have Taylor expanded on the right. We had earlier motivated that the energy or Hamiltonian operator $\hat{H}$ translates forward in time. To translate an amount of time Δt, we act on $|\psi\rangle$ with the unitary operator

$$\hat{U}_H(\Delta t) = e^{-i\frac{\Delta t \hat{H}}{\hbar}}\,, \tag{4.3}$$

so that

$$|\psi(t+\Delta t)\rangle = e^{-i\frac{\Delta t \hat{H}}{\hbar}}|\psi(t)\rangle = \left(1 - i\frac{\Delta t \hat{H}}{\hbar} + \cdots\right)|\psi(t)\rangle\,, \tag{4.4}$$

where again we Taylor expand to linear order in the time Δt. If these two expressions for $|\psi(t+\Delta t)\rangle$ are to equal one another, we must have that

$$\frac{d}{dt}|\psi(t)\rangle = -i\frac{\hat{H}}{\hbar}|\psi(t)\rangle\,, \tag{4.5}$$

or, as it is usually written:

$$\boxed{i\hbar\frac{d}{dt}|\psi\rangle = \hat{H}|\psi\rangle\,,} \tag{4.6}$$

where time dependence is now implicit. This is called the **Schrödinger equation** and describes the time evolution of a state $|\psi\rangle$ for a given Hamiltonian $\hat{H}$ in the Hilbert space $\mathcal{H}$.[2]

---

[2] E. Schrödinger, "Quantisierung als Eigenwertproblem," *Ann. Phys.* **384**(4), 361–376(6), 489–527, **385**(13), 437–490, **386**(18), 109–139 (1926); "Der stetige Übergang von der Mikro- zur Makromechanik," *Naturwiss.* **14**, 664–666 (1926); E. Schrödinger, "An undulatory theory of the mechanics of atoms and molecules," *Phys. Rev.* **28**(6), 1049 (1926).

4.1.1 The Schrödinger Equation in Position Space

It's also useful to project this general expression for the Schrödinger equation into position space, as an equation for the time evolution for the wavefunction, $\psi(x,t)$. Recall that the wavefunction is the projection

$$\psi(x,t) = \langle x|\psi(t)\rangle. \tag{4.7}$$

So, by acting with the bra $\langle x|$ on the Schrödinger equation, we have

$$i\hbar \frac{d}{dt}\langle x|\psi(t)\rangle = \langle x|\hat{H}|\psi(t)\rangle = \int_{-\infty}^{\infty} dx' \, \langle x|\hat{H}|x'\rangle\langle x'|\psi(t)\rangle, \tag{4.8}$$

where we note that the position basis is time-independent and we have inserted a complete set of position eigenstates on the right. Assuming locality of the Hamiltonian, it is diagonal in position space:

$$\hat{H} = \int dx'' \, \hat{H} |x''\rangle\langle x''|, \tag{4.9}$$

where, as mentioned in the previous chapter, we have overloaded the definition of $\hat{H}$ to be the Hamiltonian that acts on the appropriate objects in the specific basis for the Hilbert space in which we are working. Inserting this into the projection of the Schrödinger equation to position space, we have

$$i\hbar \frac{d}{dt}\langle x|\psi(t)\rangle = \int_{-\infty}^{\infty} dx' \int_{-\infty}^{\infty} dx'' \, \hat{H}\langle x|x''\rangle\langle x''|x'\rangle\langle x'|\psi(t)\rangle \tag{4.10}$$
$$= \hat{H}\langle x|\psi(t)\rangle,$$

because $\langle x|x''\rangle = \delta(x - x'')$, for example. In position space, the Schrödinger equation is then

$$i\hbar \frac{d}{dt}\psi(x,t) = \hat{H}\psi(x,t). \tag{4.11}$$

Throughout this book, we will assume that the Hamiltonian is time-independent, for simplicity. Really, what this means is, in a basis-independent way, that the eigenvalues of $\hat{H}$ are independent of time. Many interesting physical systems exhibit such a property, and this will allow us to express the Schrödinger equation in a more useful way. Assuming that the eigenvalues of the Hamiltonian are time-independent, the eigenvalue equation for the Hamiltonian is

$$\hat{H}|\alpha_i(t)\rangle = E_i|\alpha_i(t)\rangle = i\hbar \frac{d}{dt}|\alpha_i(t)\rangle, \tag{4.12}$$

for an eigenstate $|\alpha_i(t)\rangle$ with corresponding energy eigenvalue E_i. This linear differential equation in time is simple enough to solve:

$$|\alpha_i(t)\rangle = e^{-i\frac{E_i t}{\hbar}}|\psi_i\rangle, \tag{4.13}$$

where $|\psi_i\rangle$ is some constant in time ket vector that is an eigenstate of $\hat{H}$.

These $|\psi_i\rangle$ eigenstates of the Hamiltonian provide a natural basis for the Hilbert space. For a time-independent Hamiltonian $\hat{H}$, the set of eigenstates $\{|\psi_i\rangle\}$ such that

$$\hat{H}|\psi_i\rangle = E_i|\psi_i\rangle \tag{4.14}$$

provides an orthonormal and assumed complete basis for states in the Hilbert space $\mathcal{H}$.[3] A general state $|\psi\rangle \in \mathcal{H}$ can then be written as

$$|\psi\rangle = \sum_i \beta_i |\alpha_i(t)\rangle = \sum_i \beta_i e^{-i\frac{E_i t}{\hbar}} |\psi_i\rangle \,, \tag{4.15}$$

where β_i is some constant in time, complex coefficient. Our problem of finding the state for all times t is then reduced to solving for the time-independent eigensystem $\{|\psi_i\rangle, E_i\}$ where

$$\hat{H}|\psi_i\rangle = E_i|\psi_i\rangle \,. \tag{4.16}$$

As an eigenvalue equation, we will typically drop the subscripts i. We will often also just consider the time-independent state, evaluated at $t = 0$:

$$|\psi(t=0)\rangle = \sum_i \beta_i |\psi_i\rangle \,, \tag{4.17}$$

and the β_i coefficients can be found through orthonormality of the $|\psi_i\rangle$:

$$\langle\psi_i|\psi(t=0)\rangle = \sum_j \beta_j \langle\psi_i|\psi_j\rangle = \sum_j \beta_j \delta_{ij} = \beta_i \,. \tag{4.18}$$

We claim victory in completely solving a quantum system if we know all of the eigenvalues of $\hat{H}$. We sometimes colloquially say that our goal is to "diagonalize the Hamiltonian," as when the Hamiltonian only has non-zero elements on the diagonal, then elements on the diagonal are just the eigenvalues.

Working in the position basis, we can provide more physical meaning to the Schrödinger equation. In classical mechanics, the energy of a particle consists of two components: kinetic energy and potential energy. For a particle of mass m, its kinetic energy K is

$$K = \frac{1}{2}mv^2 = \frac{p^2}{2m} \,, \tag{4.19}$$

where p is the momentum of the particle. The potential energy U of a particle can depend on the particle's position x. We denote this as

$$U = V(x) \,, \tag{4.20}$$

and we will often just refer to $V(x)$ as the "potential" (dropping the word "energy"). Thus the total energy classically is

$$E = K + U = \frac{p^2}{2m} + V(x) \,. \tag{4.21}$$

---

[3] Here we are also assuming that all energy eigenvalues E_i are distinct. In Chap. 9 we will discuss what happens if there are distinct eigenstates $|\psi_i\rangle$ and $|\psi_j\rangle$ that have the same energy eigenvalue.

Quantum mechanically, energy is something that is measured as the outcome of an experiment, so we need to interpret this "kinetic + potential" as some Hermitian operator with appropriate energy eigenvalues. All we have to do in this case is to re-interpret the classical momentum p, a coordinate or function on phase space, as a quantum operator $\hat{p}$, and similarly for x to $\hat{x}$:

$$\hat{H} = \hat{K} + \hat{U} = \frac{\hat{p}^2}{2m} + V(\hat{x}). \tag{4.22}$$

Now, in the position basis, the position operator $\hat{x}$ just becomes the position eigenvalue x and the momentum operator is the derivative:

$$\hat{p} \rightarrow -i\hbar \frac{d}{dx}. \tag{4.23}$$

Then, the Hamiltonian in the position basis is

$$\hat{H} = -\frac{\hbar^2}{2m} \frac{d^2}{dx^2} + V(x). \tag{4.24}$$

The Schrödinger equation expressed as an eigenvalue problem for the Hamiltonian in the position basis can then be expressed as a second-order differential equation:

$$\left(-\frac{\hbar^2}{2m} \frac{d^2}{dx^2} + V(x) \right) \psi(x) = E\psi(x). \tag{4.25}$$

In this book, we will go back and forth between the abstract linear operator form of the Schrödinger equation and its specific representation in the position basis. Either way, the physical system of interest is completely specified by the form of the potential $V(x)$. We will concretely study several example systems with different potentials: the free particle $V(x) = 0$, the harmonic oscillator $V(x) = \frac{1}{2}kx^2$, and the hydrogen atom $V(x) = -k/x$. These different potentials will produce different eigenstates and energy eigenvalues. We'll consider other potentials, as well, and the example of the hydrogen atom is more complicated than presented thus far because it is a three-dimensional system, which adds its own interesting challenges.

4.1.2 An Exact Solution of the Schrödinger Equation

One point to re-emphasize is that because the Hamiltonian is Hermitian, normalization of the wavefunction $\psi(x,t)$ is preserved by the Schrödinger equation for all time. This was an assumption/axiom we used to derive the Schrödinger equation, and simply follows because the general time evolution is implemented by a unitary operator:

$$\hat{U}_H(t) = e^{-i\frac{t\hat{H}}{\hbar}}. \tag{4.26}$$

The Schrödinger equation is nothing more than the infinitesimal form of the action of $\hat{U}_H(t)$ on the wavefunction $\psi(x,t)$. Indeed, consider the wavefunction at time t:

$$\psi(x,t) = \hat{U}_H(t)\psi(x,t=0) = e^{-i\frac{t\hat{H}}{\hbar}} \psi(x,t=0), \tag{4.27}$$

which assumes that the Hamiltonian is time-independent. We can just explicitly check that this solves the Schrödinger equation:

$$i\hbar\frac{d}{dt}\psi(x,t) = i\hbar\frac{d}{dt}e^{-i\frac{t\hat{H}}{\hbar}}\psi(x,t=0) = i\hbar\left(-i\frac{\hat{H}}{\hbar}\right)e^{-i\frac{t\hat{H}}{\hbar}}\psi(x,t=0) \qquad (4.28)$$
$$= \hat{H}\psi(x,t),$$

which is indeed the Schrödinger equation. Thus, given an initial wavefunction $\psi(x,t=0)$, we can find the wavefunction at any later time by acting on it by $\hat{U}_H(t)$:

$$\psi(x,t) = e^{-i\frac{t\hat{H}}{\hbar}}\psi(x,t=0). \qquad (4.29)$$

For now, this exact solution of the Schrödinger equation will be a novelty, but we'll come back to it later in the book.

4.2 Time Evolution of Expectation Values

In the previous chapter, we had introduced the notion of an expectation value of a Hermitian operator $\hat{T} = \hat{T}^{\dagger}$, given a quantum state $|\psi\rangle$. The expectation value is nothing more than a matrix element of $\hat{T}$:

$$E_T = \langle\psi|\hat{T}|\psi\rangle. \qquad (4.30)$$

Unlike the normalization, $\langle\psi|\psi\rangle = 1$, for a general Hermitian operator, this expectation value is not constant in time. Can we determine how it changes in time?

To do this, we will write the expectation value as a function of time:

$$E_T(t) = \langle\psi|\hat{T}|\psi\rangle(t). \qquad (4.31)$$

If the time is evolved forward by a small amount Δt, we can Taylor expand:

$$E_T(t+\Delta t) = E_T(t) + \Delta t\frac{d}{dt}E_T(t) + \cdots. \qquad (4.32)$$

Note the total derivative here: we have encapsulated all time dependence of the expectation value, explicit or implicit. Equivalently, we can individually evolve the states and operator in the expectation value.[4] To evolve the ket state $|\psi(t)\rangle$, note that

$$|\psi(t+\Delta t)\rangle = e^{-i\frac{\Delta t\hat{H}}{\hbar}}|\psi(t)\rangle = |\psi(t)\rangle - i\frac{\Delta t\hat{H}}{\hbar}|\psi(t)\rangle + \cdots, \qquad (4.33)$$

---

[4] While we choose to endow the state $|\psi\rangle$ with time dependence here, we could have instead assumed that the state is time-independent and the operator $\hat{T}$ is evolved in time with the Hamiltonian. The former approach that we take is called the Schrödinger picture, while constant states but time-evolved operators is called the Heisenberg picture of quantum mechanics. Either picture produces identical physical predictions because the only observable quantities in quantum mechanics are expectation values of Hermitian operators.

where we have Taylor expanded the exponential. The time evolution of the bra state is just the Hermitian conjugate:

$$\langle\psi(t+\Delta t)| = \langle\psi(t)| + \langle\psi(t)|i\frac{\Delta t\hat{H}}{\hbar} + \cdots. \tag{4.34}$$

Note that we keep the bra $\langle\psi(t)|$ to the left of the Hamiltonian $\hat{H}$ as we think of the bra as a row vector. Further, the Hermitian operator $\hat{T}$ in general can have time dependence itself. In that case, its time evolution is then

$$\hat{T}(t+\Delta t) = \hat{T}(t) + \Delta t\frac{\partial\hat{T}(t)}{\partial t} + \cdots. \tag{4.35}$$

Note that now this Taylor expansion involves the partial derivative with respect to time t; that is, it is only sensitive to the explicit time dependence of the operator.

Therefore, to linear order in the small time step Δt, the expectation value evolves to

$$\langle\psi|\hat{T}|\psi\rangle(t+\Delta t) = \langle\psi|\left(1+i\frac{\Delta t\hat{H}}{\hbar}\right)\left(\hat{T}+\Delta t\frac{\partial\hat{T}}{\partial t}\right)\left(1-i\frac{\Delta t\hat{H}}{\hbar}\right)|\psi\rangle + \cdots \tag{4.36}$$

$$= \langle\psi|\hat{T}|\psi\rangle + \Delta t\frac{i}{\hbar}\langle\psi|\hat{H}\hat{T}-\hat{T}\hat{H}|\psi\rangle + \Delta t\langle\psi|\frac{\partial\hat{T}}{\partial t}|\psi\rangle + \cdots.$$

In this expression, we have suppressed dependence on the time t everywhere for compactness. Now, setting the total time evolution of the expectation value equal to this expansion in terms of the time evolution of all of its components, we find

$$\boxed{\frac{d}{dt}\langle\psi|\hat{T}|\psi\rangle = \frac{i}{\hbar}\langle\psi|\hat{H}\hat{T}-\hat{T}\hat{H}|\psi\rangle + \langle\psi|\frac{\partial\hat{T}}{\partial t}|\psi\rangle.} \tag{4.37}$$

In this book, we will explicitly only consider operators $\hat{T}$ that have no time dependence for which $\partial\hat{T}/\partial t = 0$. In that case, the time evolution of the expectation value simplifies:

$$\frac{d}{dt}\langle\psi|\hat{T}|\psi\rangle = \frac{i}{\hbar}\langle\psi|\hat{H}\hat{T}-\hat{T}\hat{H}|\psi\rangle. \tag{4.38}$$

The most intriguing part of this equation governing the time dependence of the expectation value is the difference of the product of operators $\hat{H}$ and $\hat{T}$. We call this product difference the **commutator** of $\hat{H}$ and $\hat{T}$ and denote it as

$$\hat{H}\hat{T}-\hat{T}\hat{H} \equiv [\hat{H},\hat{T}]. \tag{4.39}$$

We say that the expectation value is independent of time if the Hamiltonian $\hat{H}$ and the operator $\hat{T}$ **commute**; that is, if $[\hat{H},\hat{T}] = 0$.

The existence of commutators should be very unfamiliar to your experience in classical mechanics. There, everything is just a function of real numbers on phase space, and multiplication of real numbers commutes:

$$ab = ba, \tag{4.40}$$

for $a, b \in \mathbb{R}$. However, as we had long motivated this topic, operators are like matrices and for two matrices $\mathbb{A}$ and $\mathbb{B}$, it need not be true that

$$\mathbb{A}\mathbb{B} = \mathbb{B}\mathbb{A}. \tag{4.41}$$

For a simple example of this, consider the 2×2 matrices

$$\mathbb{A} = \begin{pmatrix} 1 & 0 \\ 0 & -1 \end{pmatrix}, \qquad\qquad \mathbb{B} = \begin{pmatrix} 0 & 1 \\ 1 & 0 \end{pmatrix}. \qquad (4.42)$$

Also, note that both $\mathbb{A}$ and $\mathbb{B}$ are Hermitian matrices. The product $\mathbb{A}\mathbb{B}$ is

$$\mathbb{A}\mathbb{B} = \begin{pmatrix} 1 & 0 \\ 0 & -1 \end{pmatrix} \begin{pmatrix} 0 & 1 \\ 1 & 0 \end{pmatrix} = \begin{pmatrix} 0 & 1 \\ -1 & 0 \end{pmatrix}. \qquad (4.43)$$

On the other hand, the product $\mathbb{B}\mathbb{A}$ is

$$\mathbb{B}\mathbb{A} = \begin{pmatrix} 0 & 1 \\ 1 & 0 \end{pmatrix} \begin{pmatrix} 1 & 0 \\ 0 & -1 \end{pmatrix} = \begin{pmatrix} 0 & -1 \\ 1 & 0 \end{pmatrix}, \qquad (4.44)$$

and so their commutator is

$$[\mathbb{A}, \mathbb{B}] = \mathbb{A}\mathbb{B} - \mathbb{B}\mathbb{A} = \begin{pmatrix} 0 & 2 \\ -2 & 0 \end{pmatrix} \neq 0. \qquad (4.45)$$

It is natural, then, to expect that two Hermitian operators $\hat{H}$ and $\hat{T}$ on the Hilbert space do not commute, just like matrix multiplication isn't necessarily commutative.

4.2.1 An Example: Time Dependence of Expectation of Momentum

To make some sense out of this in the context of time evolution of expectation values, we'll end this section with a calculation of the time dependence of the expectation value of the momentum operator, $\hat{p}$. We will assume that $\hat{p}$ itself has no explicit time dependence, $\partial\hat{p}/\partial t = 0$, and so our evolution equation is

$$\frac{d}{dt}\langle\hat{p}\rangle = \frac{i}{\hbar}\langle[\hat{H}, \hat{p}]\rangle. \qquad (4.46)$$

In this expression, we have introduced the compact notation in which the explicit state $|\psi\rangle$ we evaluate the expectation value in is suppressed; that is

$$\langle\hat{p}\rangle \equiv \langle\psi|\hat{p}|\psi\rangle, \qquad\qquad \langle[\hat{H}, \hat{p}]\rangle \equiv \langle\psi|[\hat{H}, \hat{p}]|\psi\rangle. \qquad (4.47)$$

So what is the commutator of $\hat{H}$ and $\hat{p}$? First, recall the expansion of the Hamiltonian into the sum of kinetic and potential energies:

$$\hat{H} = \frac{\hat{p}^2}{2m} + V(\hat{x}). \qquad (4.48)$$

Focusing on the kinetic energy part, note that this commutes with momentum $\hat{p}$:

$$\hat{K}\hat{p} = \frac{\hat{p}^2}{2m}\hat{p} = \hat{p}\frac{\hat{p}^2}{2m} = \hat{p}\hat{K}, \qquad (4.49)$$

as $2m$ is just a scalar, constant number. The commutator of the Hamiltonian and momentum is then just the commutator of the potential $V(\hat{x})$ and $\hat{p}$:

$$[\hat{H}, \hat{p}] = [V(\hat{x}), \hat{p}]. \qquad (4.50)$$

We might think that this remaining commutator is also just 0, but remember that $\hat{p}$ in position space is a derivative:

$$\hat{p} = -i\hbar\frac{d}{dx}. \tag{4.51}$$

As a derivative, it must act on some function $f(x)$ to its right. Another way to say this is that the purpose in life of operators like $\hat{H}$ or $\hat{p}$ is to act on states or functions in some space. Our trick for evaluating the commutator when a derivative is involved is to put a function $f(x)$ to the right of the commutator, evaluate the commutator, and then remove $f(x)$. Such a function is called a **test function**, and its sole purpose is to keep us honest with manipulations of the derivative. Remember, the derivative is only a linear operator with the Leibniz product rule, and the test function reminds us to employ the rule.

So, working in position space, we compute

$$[V(x),\hat{p}]f(x) = V(x)\left(-i\hbar\frac{d}{dx}\right)f(x) - \left(-i\hbar\frac{d}{dx}\right)V(x)f(x) \tag{4.52}$$

$$= V(x)\left(-i\hbar\frac{d}{dx}\right)f(x) - \left(-i\hbar\frac{d}{dx}V(x)\right)f(x) - V(x)\left(-i\hbar\frac{d}{dx}\right)f(x),$$

where we used the product rule on the second line. Note that in the final expression, the first and third terms explicitly cancel, while the second term remains. We can then toss away the test function, and we find that the commutator of the potential and momentum is

$$[V(\hat{x}),\hat{p}] = i\hbar\frac{dV(\hat{x})}{d\hat{x}}. \tag{4.53}$$

Plugging this into our differential equation for the time dependence of $\langle\hat{p}\rangle$, we find

$$\frac{d\langle\hat{p}\rangle}{dt} = \left\langle\frac{i}{\hbar}\left(i\hbar\frac{dV(\hat{x})}{d\hat{x}}\right)\right\rangle = -\left\langle\frac{dV(\hat{x})}{d\hat{x}}\right\rangle. \tag{4.54}$$

Does this look familiar? For a conservative force F_{cons} in classical mechanics, we can express that force as a spatial derivative of a potential energy U:

$$F_{\text{cons}} = -\frac{dU}{dx}. \tag{4.55}$$

Therefore, Eq. (4.54) is just like Newton's second law! However, it is a statement of the time dependence of expectation values of the momentum operator $\hat{p}$, and not just the time dependence of a particle's functional momentum p. This whole structure is a manifestation of **Ehrenfest's theorem**, which is often colloquially stated as "expectation values of quantum mechanical operators satisfy the corresponding classical equations of motion."[5] While slightly misleading, it is correct in spirit.

---

[5] P. Ehrenfest, "Bemerkung über die angenäherte Gültigkeit der klassischen Mechanik innerhalb der Quantenmechanik," Z. Phys. **45**(7–8), 455–457 (1927).

4.2.2 The Canonical Commutation Relation

This whole procedure required properties of the commutation relation of momentum $\hat{p}$ with the position operator $\hat{x}$. In position space, this commutator can be evaluated in the same way as we did with $V(x)$ and $\hat{p}$. Working in position space and using a test function $f(x)$, we have

$$[x, \hat{p}] f(x) = x \left(-i\hbar \frac{d}{dx} \right) f(x) - \left(-i\hbar \frac{d}{dx} \right) x f(x) \tag{4.56}$$

$$= x \left(-i\hbar \frac{d}{dx} \right) f(x) - (-i\hbar) f(x) - x \left(-i\hbar \frac{d}{dx} \right) f(x)$$

$$= i\hbar f(x).$$

Thus, their commutator is

$$\boxed{[\hat{x}, \hat{p}] = i\hbar.} \tag{4.57}$$

This is a profoundly important relation in quantum mechanics, so central that it is called the **canonical commutation relation**.[6]

If you are familiar with **Poisson brackets** in the Hamiltonian formulation of classical mechanics, the quantum mechanical commutator is its spiritual descendent. For two functions $f(x, p)$ and $g(x, p)$ on phase space defined by position x and momentum p, their Poisson bracket is

$$\{f, g\} \equiv \frac{df}{dx} \frac{dg}{dp} - \frac{df}{dp} \frac{dg}{dx}. \tag{4.58}$$

The Poisson bracket of position and momentum individually is

$$\{x, p\} = \frac{dx}{dx} \frac{dp}{dp} - \frac{dx}{dp} \frac{dp}{dx} = 1. \tag{4.59}$$

Note that up to a factor of $i\hbar$ this is the same as the commutator of the quantum mechanical operators $\hat{x}$ and $\hat{p}$. In fact, a formal relationship between the quantum commutator and classical Poisson brackets can be defined via

$$\lim_{\hbar \to 0} \frac{[\cdot, \cdot]_{\text{quantum}}}{i\hbar} = \{\cdot, \cdot\}_{\text{classical}}, \tag{4.60}$$

where the $\cdot$ are placeholders for quantum operators or classical functions on phase space. This is a manifestation of the **correspondence principle**, the way in which classical mechanics can emerge from quantum mechanics. This particular relationship between Poisson brackets and the commutation relation was introduced by Paul Dirac in his PhD thesis.[7] Going the other direction, from Poisson brackets to commutators, is referred to as **canonical quantization**, and for a very long time was the standard method

---

[6] This was first guessed by Max Born and Pascual Jordan: M. Born and P. Jordan, "Zur Quantenmechanik," *Z. Phys.* **34**, 858–888 (1925), from initial observations of position and momentum matrix operators by Werner Heisenberg: W. Heisenberg, "A quantum-theoretical reinterpretation of kinematic and mechanical relations," *Z. Phys.* **33**, 879–893 (1925).

[7] P. A. M. Dirac, "Quantum mechanics," PhD thesis, University of Cambridge (1926). Dirac's hand-written thesis is available online at http://purl.flvc.org/fsu/lib/digcoll/dirac/dirac-papers/353070

for "quantizing" a classical theory. Indeed, the time evolution of a function $f(x,p)$ in classical mechanics can be represented as

$$\frac{df(x,p)}{dt} = \{f(x,p),H(x,p)\} + \frac{\partial f(x,p)}{\partial t}, \qquad (4.61)$$

where $H(x,p)$ is the classical Hamiltonian function on phase space. This is in exact analogy with the corresponding time evolution of expectation values in quantum mechanics, as per Ehrenfest's theorem. For a brief introduction or refresher on Poisson brackets, see Appendix B.

Getting back to the canonical commutation relation in particular, $[\hat{x},\hat{p}] = i\hbar$, what is particularly striking is that the commutator is non-zero. This has an interesting mathematical consequence we will discuss now, and the physical consequence will be discussed in the following section. Let's first assume that the eigenstates of $\hat{x}$ and $\hat{p}$ are identical. That is, for some set of states $\{|v_i\rangle\}$, we would have

$$\hat{x}|v_i\rangle = x_i|v_i\rangle \qquad \text{and} \qquad \hat{p}|v_i\rangle = p_i|v_i\rangle. \qquad (4.62)$$

If this were the case, then $\hat{x}$ and $\hat{p}$ would commute because their eigenvalues are just numbers:

$$\hat{x}\hat{p}|v_i\rangle = x_ip_i|v_i\rangle = p_ix_i|v_i\rangle = \hat{p}\hat{x}|v_i\rangle. \qquad (4.63)$$

However, we know that they do not commute, $[\hat{x},\hat{p}] \neq 0$, and so our assumption that their eigenstates are the same must be false. Another way to express this is that when thought of as matrices, the operators $\hat{x}$ and $\hat{p}$ cannot be **simultaneously diagonalized**. Two matrices that exclusively have non-zero elements on the diagonal commute, and so there necessarily cannot be a basis in which $\hat{x}$ and $\hat{p}$ are both diagonal. If we express the Hilbert space in the basis of eigenstates of momentum, for example, then necessarily eigenstates of position will be non-trivial linear combinations of multiple states with different momentum, and vice-versa.

4.3 The Uncertainty Principle

In this section, we'll provide more physical understanding of the non-commutativity of the position and momentum operators, and more generally, two arbitrary, non-commuting Hermitian operators. To start, we'll need to introduce the concept of **variance**, a robust measure of the spread of a probability distribution about its mean.

4.3.1 Some More Elementary Probability: The Variance

To motivate the variance, let's go back to our familiar example of rolling a die. Let's call p_i the probability of the ith side coming up on the roll. Then, the total probability is unity:

$$\sum_{i=1}^{6} p_i = 1. \qquad (4.64)$$

We had already introduced the expectation value as the probability-weighted mean. In this case, the expectation value is

$$E(\text{roll}) = \langle \text{roll} \rangle = \sum_{i=1}^{6} i \, p_i \,. \tag{4.65}$$

The variance is a measure of the deviation of the probability distribution of rolls about its mean. Specifically, the variance, denoted as σ_{roll}^2, is the mean square difference:

$$\sigma_{\text{roll}}^2 = \langle (\text{roll} - \langle \text{roll} \rangle)^2 \rangle = \sum_{i=1}^{6} (i - \langle \text{roll} \rangle)^2 \, p_i \,. \tag{4.66}$$

To understand the variance a bit more, let's expand out the square:

$$\sum_{i=1}^{6} (i - \langle \text{roll} \rangle)^2 \, p_i = \sum_{i=1}^{6} (i^2 - 2i \langle \text{roll} \rangle + \langle \text{roll} \rangle^2) \, p_i \,. \tag{4.67}$$

Recall that the expectation value $\langle \text{roll} \rangle$ is just a number; for a fair die, $\langle \text{roll} \rangle = 3.5$. Thus, once summed over i, the second term in the expanded square can just be expressed in terms of the expectation value again. Then

$$\sigma_{\text{roll}}^2 = \sum_{i=1}^{6} (i^2 - 2i \langle \text{roll} \rangle + \langle \text{roll} \rangle^2) \, p_i = \sum_{i=1}^{6} (i^2 - \langle \text{roll} \rangle^2) \, p_i \tag{4.68}$$

$$= \langle \text{roll}^2 \rangle - \langle \text{roll} \rangle^2 \,.$$

That is, the variance is the difference between the probability-weighted mean of the squared rolls and the squared expectation value.

Let's calculate this for a fair die to give it some meaning. We have

$$\sigma_{\text{roll}}^2 = \langle \text{roll}^2 \rangle - \langle \text{roll} \rangle^2 = \sum_{i=1}^{6} i^2 p_i - \left(\frac{7}{2} \right)^2 \tag{4.69}$$

$$= \frac{1 + 4 + 9 + 16 + 25 + 36}{6} - \frac{49}{4} = \frac{35}{12} \simeq 2.92 \,.$$

The square root of the variance is called the **standard deviation** σ_{roll} and we find for the fair die that

$$\sigma_{\text{roll}} \simeq 1.71 \,. \tag{4.70}$$

The interpretation of the standard deviation is that, on any given roll, you expect the outcome to be approximately within one σ_{roll} of the expectation value, $\langle \text{roll} \rangle$. Being within one σ_{roll} of the expectation value for the fair die means that your roll is in the range

$$\text{roll} \in [3.5 - 1.71, 3.5 + 1.71] = [1.79, 5.21] \,. \tag{4.71}$$

This is true for all possible rolls, except for 1 and 6. Thus, two-thirds of the time (for rolls of 2, 3, 4, or 5) a roll is within one standard deviation of the expectation value. This is indeed "most" of the time, as the probability of rolling one of these four numbers is greater than 50%.

Another interesting feature of the variance is that it is only zero if there is a unique possible outcome. That is, let's consider a very biased die for which $p_3 = 1$, and so $p_i = 0$, for $i = 1, 2, 4, 5, 6$. Then, the mean of such a biased die is

$$\langle \text{bias roll} \rangle = \sum_{i=1}^{6} i \, p_i = 3 \cdot 1 = 3 \,. \tag{4.72}$$

The variance of this biased die would then be

$$\sigma_{\text{bias roll}}^2 = \sum_{i=1}^{6} i^2 \, p_i - \langle \text{bias roll} \rangle^2 = 3^2 \cdot 1 - 3^2 = 0 \,, \tag{4.73}$$

as claimed. Now, let's consider a slightly less biased die, one for which $p_3 = 1 - \epsilon$ and $p_1 = \epsilon$, for $\epsilon \in [0, 1]$ and $p_i = 0$ for $i = 2, 4, 5, 6$. The mean now is

$$\langle \text{less bias roll} \rangle = \sum_{i=1}^{6} i \, p_i = 1 \cdot \epsilon + 3 \cdot (1 - \epsilon) = 3 - 2\epsilon \,. \tag{4.74}$$

The variance, on the other hand is

$$\sigma_{\text{less bias roll}}^2 = \sum_{i=1}^{6} i^2 \, p_i - \langle \text{less bias roll} \rangle^2 = 1^2 \cdot \epsilon + 3^2 \cdot (1 - \epsilon) - (3 - 2\epsilon)^2 \tag{4.75}$$
$$= 4\epsilon(1 - \epsilon) \,.$$

This clearly only vanishes if $\epsilon = 0$ or 1, which corresponds to the biased die having probability 1 of returning a single, unique number.

This property imbues the standard deviation or variance with a particularly interesting interpretation. We can think of the standard deviation as a measure of the "uncertainty" of outcomes, according to some probability distribution. If the variance is strictly 0, then every time the experiment is performed, you know exactly what the result will be: there is no uncertainty because the result is unique. However, if the variance is non-zero, then multiple outcomes will be likely, and in any given instance of the experiment you can't know what the result might be. As the variance gets larger and larger, the range of possible outcomes grows, and you become more and more uncertain as to what the outcome of any individual instance of the experiment might be.

The variance also exhibits a "translation invariance" by which all outcomes can be shifted by a constant, and the variance is unchanged. For example, let's consider a fair die whose six sides are the numbers $i + \lambda$, where $i = 1, 2, 3, 4, 5, 6$ and λ is a positive integer. That is, we have just painted λ more pips on each side of the die. The expectation value of a roll correspondingly increases by λ:

$$\langle \text{shifted roll} \rangle = \sum_{i=1}^{6} (i + \lambda) p_i = \frac{7}{2} + \lambda \,, \tag{4.76}$$

because the sum of probabilities is unity. This shift is exactly canceled in the variance, however. For this shifted die, its variance is

$$\sigma^2_{\text{shifted roll}} = \sum_{i=1}^{6}(i+\lambda)^2 p_i - \langle\text{shifted roll}\rangle^2 = \sum_{i=1}^{6}(i^2+2i\lambda+\lambda^2)p_i - (\langle\text{roll}\rangle+\lambda)^2$$

$$= \sum_{i=1}^{6} i^2 p_i + 2\lambda\langle\text{roll}\rangle + \lambda^2 - \langle\text{roll}\rangle^2 - 2\lambda\langle\text{roll}\rangle - \lambda^2$$

$$= \sigma^2_{\text{roll}}, \tag{4.77}$$

the same variance as our familiar fair die. Thus, the variance is really just a measure of the spread about the mean, with no sensitivity to the particular value of the mean.

In the following example, we will study statistical properties of a simple continuous probability distribution.

Example 4.1   The example of the fair die is a gentle introduction to statistical properties of probability distributions. However, most of the measurements we can make on a system do not return discrete values, like the die, but rather continuous values, like a position or momentum. So expectation values or variances of continuous distributions are defined by evaluating integrals, rather than finite sums. Let's study a very simple continuous probability distribution $p(x)$ of a random variable x. We will assume that the domain is $x \in [0,l]$, for some maximum length l and $p(x)$ is uniform, or constant, on that domain. What is the probability distribution, it expectation value, and its variance?

Solution

As a uniform probability distribution, $p(x)$ is independent of x on $x \in [0,l]$. Further, it must be normalized and integrate to 1, so we can easily determine its value:

$$1 = \int_0^l dx\, p(x) = p(x)\int_0^l dx = p(x)l, \tag{4.78}$$

or $p(x) = 1/l$. The expectation value of x, $\langle x \rangle$, is the integral of x over its domain, weighted by the probability distribution:

$$\langle x \rangle = \int_0^l dx\, x\, p(x) = \frac{1}{l}\int_0^l dx\, x = \frac{1}{l}\cdot\frac{l^2}{2} = \frac{l}{2}. \tag{4.79}$$

Not surprisingly, the expectation value of x for a uniform distribution is exactly in the middle. Another way to interpret this in a context that you have likely seen before, is of the center-of-mass of an extended object. For an approximately one-dimensional object with uniform mass distribution, like a ruler, you can balance it on one finger if you place your finger at the center of the ruler. At that point, your finger exerts a normal force to oppose gravity, and your finger exerts no torque.

To calculate the variance of this distribution, we need the second moment of the probability distribution, $\langle x^2 \rangle$. This is just the integral of x^2 over the domain, weighted by the probability distribution $p(x)$:

$$\langle x^2 \rangle = \int_0^l dx\, x^2\, p(x) = \frac{1}{l}\int_0^l dx\, x^2 = \frac{1}{l}\cdot\frac{l^3}{3} = \frac{l^2}{3}. \tag{4.80}$$

You might recognize this second moment of the uniform probability distribution from the moment of inertia of a uniform rod about one of its ends. The moment of inertia I is just this second moment multiplied by the mass of the rod. For the problem at hand, we want the variance σ^2, where

$$\sigma^2 = \langle x^2 \rangle - \langle x \rangle^2 = \frac{l^2}{3} - \frac{l^2}{4} = \frac{l^2}{12}. \tag{4.81}$$

You might recognize this expression from the moment of inertia for a rod about its center (again, up to a factor of the mass of the rod). These two expressions for the second moment and the variance of a probability distribution are related to one another by, effectively, the parallel axis theorem, with the amount of displacement given by the square of the expectation value, $\langle x \rangle^2$.

In Fig. 4.1, we have illustrated this probability distribution and the location of the expectation value of x. Further, we have shaded the region of the distribution that lies within one standard deviation σ of the expectation value, demonstrating again that most of the area under the probability distribution is in this region.

4.3.2 Uncertainty Principle from Commutation Relations

The interpretation of the variance as a measure of uncertainty is interesting, and intimately related to the impossibility of simultaneous diagonalization of two Hermitian operators $\hat{A}$ and $\hat{B}$ which do not commute: $[\hat{A},\hat{B}] \neq 0$. This lack of simultaneous diagonalization means that an eigenstate of $\hat{A}$ cannot be an eigenstate of $\hat{B}$, and vice-versa, so it is impossible to construct a state $|\psi\rangle$ for which the outcomes of measuring $\hat{A}$ and $\hat{B}$ are unique. This means that the variances of $\hat{A}$ and $\hat{B}$ on any state must be intimately related to one another. In particular, let's consider the product of variances, $\sigma_A^2\sigma_B^2$, and we will show that this product has a strict lower bound.

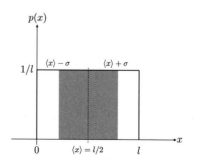

Fig. 4.1   Illustration of the uniform probability distribution on $x \in [0, l]$. The location of the expectation value is illustrated in the vertical dashed line, and the shaded region extends one standard deviation above and below the expectation value.

First, let's define these variances. The variance of Hermitian operator $\hat{A}$ on the state $|\psi\rangle$ is

$$\sigma_A^2 = \langle\psi|\hat{A}^2|\psi\rangle - \langle\psi|\hat{A}|\psi\rangle^2 = \langle\hat{A}^2\rangle - \langle\hat{A}\rangle^2\,, \tag{4.82}$$

where we suppress the particular state on the right as we want to establish a result for any state $|\psi\rangle$ on the Hilbert space. The variance of Hermitian operator $\hat{B}$ is defined similarly. Then, the product of variances can be expressed as

$$\sigma_A^2\sigma_B^2 = \left(\langle\hat{A}^2\rangle - \langle\hat{A}\rangle^2\right)\left(\langle\hat{B}^2\rangle - \langle\hat{B}\rangle^2\right) = \langle(\hat{A} - \langle\hat{A}\rangle)^2\rangle\langle(\hat{B} - \langle\hat{B}\rangle)^2\rangle\,. \tag{4.83}$$

In the form on the right, we can establish a lower bound on its value with the **Cauchy–Schwarz inequality**. First, note that the variance of $\hat{A}$ can be expressed as the inner product of some state with itself:

$$\langle(\hat{A} - \langle\hat{A}\rangle)^2\rangle = \langle\psi|(\hat{A} - \langle\hat{A}\rangle)(\hat{A} - \langle\hat{A}\rangle)|\psi\rangle = \langle\chi|\chi\rangle\,, \tag{4.84}$$

where

$$|\chi\rangle = (\hat{A} - \langle\hat{A}\rangle)|\psi\rangle\,. \tag{4.85}$$

Because $\hat{A}$ is Hermitian, $\langle\hat{A}\rangle$ is a real number and so the bra $\langle\chi|$ follows. We can similarly define

$$|\zeta\rangle = (\hat{B} - \langle\hat{B}\rangle)|\zeta\rangle\,, \tag{4.86}$$

so that the product of variances can be expressed as

$$\langle(\hat{A} - \langle\hat{A}\rangle)^2\rangle\langle(\hat{B} - \langle\hat{B}\rangle)^2\rangle = \langle\chi|\chi\rangle\langle\zeta|\zeta\rangle\,. \tag{4.87}$$

Now, we can use the Cauchy–Schwarz inequality. A detailed proof of the Cauchy–Schwarz inequality for any space with a bilinear inner product can be found in mathematics textbooks,[8] but here we will just motivate it. Note that the inner product of state $|\chi\rangle$ with itself is just its squared magnitude, and similarly for $|\zeta\rangle$. If these states were just familiar real vectors $\vec{u}, \vec{v} \in \mathbb{R}^N$, then note that their dot product would be

$$\vec{u} \cdot \vec{v} = |\vec{u}||\vec{v}|\cos\theta\,. \tag{4.88}$$

Squaring both sides of this dot product ensures that the cosine factor is non-negative:

$$(\vec{u} \cdot \vec{v})^2 = |\vec{u}|^2|\vec{v}|^2\cos^2\theta\,. \tag{4.89}$$

Then, because $0 \leq \cos^2\theta \leq 1$, we have the inequality that

$$(\vec{u} \cdot \vec{v})^2 \leq |\vec{u}|^2|\vec{v}|^2\,. \tag{4.90}$$

Applying this reasoning to the product of variances, this inequality can be expressed as

---

[8] Many books on real analysis will have a proof of Cauchy–Schwarz; two recommendations are T. Andreescu and B. Enescu, *Mathematical Olympiad Treasures*, Springer Science & Business Media (2011) and M. J. Steele, *The Cauchy–Schwarz Master Class: An Introduction to the Art of Mathematical Inequalities*, Cambridge University Press (2004).

$$\langle \chi | \chi \rangle \langle \zeta | \zeta \rangle \geq |\langle \chi | \zeta \rangle|^2 = |\langle (\hat{A} - \langle \hat{A} \rangle)(\hat{B} - \langle \hat{B} \rangle) \rangle|^2 \tag{4.91}$$
$$= |\langle (\hat{A}\hat{B} - \langle \hat{A} \rangle \langle \hat{B} \rangle) \rangle|^2.$$

In the second line, we have just expanded out the product.

Now, note that an expectation value of a general operator is just some complex number; let's call it z. In the case at hand, we have

$$z = \langle (\hat{A}\hat{B} - \langle \hat{A} \rangle \langle \hat{B} \rangle) \rangle, \tag{4.92}$$

and we have established that the lower bound of the product of variances is the square magnitude of z. As a complex number, z can be written as a real and imaginary part:

$$z = \text{Re}(z) + i\,\text{Im}(z), \tag{4.93}$$

where

$$\text{Re}(z) = \frac{z + z^*}{2}, \qquad\qquad \text{Im}(z) = \frac{z - z^*}{2i}, \tag{4.94}$$

and z^* is the complex conjugate of z. Note that both $\text{Re}(z)$ and $\text{Im}(z)$ are real numbers. The squared magnitude of z is

$$|z|^2 = zz^* = \text{Re}(z)^2 + \text{Im}(z)^2 \geq \text{Im}(z)^2, \tag{4.95}$$

because $\text{Re}(z)^2 \geq 0$. Now, for the complex number z at hand, note that its imaginary part is

$$\text{Im}(z) = \frac{z - z^*}{2i} = \frac{\langle (\hat{A}\hat{B} - \langle \hat{A} \rangle \langle \hat{B} \rangle) \rangle - \langle (\hat{A}\hat{B} - \langle \hat{A} \rangle \langle \hat{B} \rangle) \rangle^*}{2i} \tag{4.96}$$
$$= \frac{\langle \hat{A}\hat{B} - \hat{B}\hat{A} \rangle}{2i}.$$

This simple result follows from the Hermitivity of $\hat{A}$ and $\hat{B}$. The expectation value of any Hermitian matrix is a real number, and

$$\langle \hat{A}\hat{B} \rangle^* = \langle \hat{B}\hat{A} \rangle. \tag{4.97}$$

In this final form, we recognize $\hat{A}\hat{B} - \hat{B}\hat{A} = [\hat{A}, \hat{B}]$, the commutator of Hermitian operators $\hat{A}$ and $\hat{B}$. Inserting this inequality into Eq. (4.91), we then finally find

$$\sigma_A^2 \sigma_B^2 \geq |\langle (\hat{A}\hat{B} - \langle \hat{A} \rangle \langle \hat{B} \rangle) \rangle|^2 \geq \left(\frac{\langle [\hat{A}, \hat{B}] \rangle}{2i} \right)^2. \tag{4.98}$$

We can remove the squares and express this as a relationship between standard deviations of the Hermitian operators $\hat{A}$ and $\hat{B}$ as

$$\boxed{\sigma_A \sigma_B \geq \left| \frac{\langle [\hat{A}, \hat{B}] \rangle}{2i} \right|.} \tag{4.99}$$

This is called the **generalized uncertainty principle**. Recall that the variances σ_A^2 and σ_B^2 are measures of how close to an eigenstate of $\hat{A}$ or $\hat{B}$ the state $|\psi\rangle$ is. If the commutator of $\hat{A}$ and $\hat{B}$ is non-zero, then $|\psi\rangle$ cannot be an eigenstate of $\hat{A}$ and $\hat{B}$ simultaneously; that is, if $[\hat{A}, \hat{B}] \neq 0$, the σ_A and σ_B cannot both be 0.

For the particular identification of $\hat{A} = \hat{x}$ and $\hat{B} = \hat{p}$, we have

$$[\hat{x}, \hat{p}] = i\hbar, \tag{4.100}$$

and so

$$\boxed{\sigma_x \sigma_p \geq \frac{\hbar}{2}.} \tag{4.101}$$

This is called the **Heisenberg uncertainty principle**.[9] A generic state $|\psi\rangle$ cannot be an eigenstate of $\hat{x}$ and $\hat{p}$ simultaneously. This means that there is necessarily uncertainty (i.e., variance) of an object's position and momentum. Colloquially, we can say that if we know a particle's momentum, then it is in an eigenstate of $\hat{p}$, so that $\sigma_p = 0$. However, for the uncertainty relation to hold, we must have $\sigma_x = \infty$: the position of the particle is completely unknown! This is very unlike classical mechanics, in which we can know arbitrary data to arbitrary precision of any object. The consequences of this uncertainty relation will be explored throughout this book.

4.3.3 Uncertainty Principle from Fourier Conjugates

While the generalized uncertainty principle derived above is, as its name suggests, a general result for any two Hermitian operators that act on Hilbert space, there is an interesting restriction that sheds some more light on what this uncertainty relationship means. Recall that when we found the momentum eigenstates in position space, we noted that a general wavefunction in position space can be expressed as the Fourier transform of a wavefunction in momentum space:

$$\psi(x) = \int_{-\infty}^{\infty} \frac{dp}{\sqrt{2\pi\hbar}} g(p) e^{i\frac{px}{\hbar}}, \tag{4.102}$$

where $\psi(x)$ is the position-space wavefunction and $g(p)$ is the momentum-space wavefunction. We can go backwards as well, solving for $g(p)$:

$$g(p) = \int_{-\infty}^{\infty} \frac{dx}{\sqrt{2\pi\hbar}} \psi(x) e^{-i\frac{px}{\hbar}}. \tag{4.103}$$

The factors of $\sqrt{2\pi\hbar}$ ensure that $\psi(x)$ and $g(p)$ have the correct units and we assume that both are L^2-normalized. Because the position- and momentum-space representations of the wavefunction are related by Fourier transformation, we refer to position and momentum as **Fourier conjugate variables**.

Fourier conjugate variables satisfy an uncertainty relationship, which we will derive. Position and momentum are not the only possible Fourier conjugate variables, so this

---

[9] It was introduced by Werner Heisenberg in W. Heisenberg, "Uber den anschaulichen Inhalt der quantentheoretischen Kinematik und Mechanik," *Z. Phys.* **43**, 172–198 (1927), but actually proved shortly thereafter by Earle Hesse Kennard and Hermann Weyl: E. H. Kennard, "Zur Quantenmechanik einfacher Bewegungstypen," *Z. Phys.* **44**, 326–352 (1927); H. Weyl, *The Theory of Groups and Quantum Mechanics*, Courier Corporation (1950).

will be a general result, but we will use x and p in what follows. For simplicity, we will also assume that the expectation values of both position and momentum are zero:

$$\langle x \rangle = \int_{-\infty}^{\infty} dx\, x\, |\psi(x)|^2 = 0, \qquad \langle p \rangle = \int_{-\infty}^{\infty} dp\, p\, |g(p)|^2 = 0. \tag{4.104}$$

Again, this is merely for simplicity and compactness; we showed above that the variance is translation-invariant, so setting these expectation values to 0 just means we have made a convenient choice of coordinates and frame. With these assumptions, we start from

$$1 = \int_{-\infty}^{\infty} dx\, |\psi(x)|^2 = \int_{-\infty}^{\infty} dx\, \psi^*(x)\, \psi(x). \tag{4.105}$$

Now, we use integration by parts to massage the integral:

$$\int_{-\infty}^{\infty} dx\, \psi^*(x)\, \psi(x) = x|\psi(x)|^2 \big|_{-\infty}^{\infty} - \int_{-\infty}^{\infty} dx\, x \left(\psi(x) \frac{d\psi^*(x)}{dx} + \psi^*(x) \frac{d\psi(x)}{dx} \right)$$

$$= -\int_{-\infty}^{\infty} dx\, x \left(\psi(x) \frac{d\psi^*(x)}{dx} + \psi^*(x) \frac{d\psi(x)}{dx} \right). \tag{4.106}$$

As $\psi(x)$ is L^2-normalized on $x \in (-\infty, \infty)$, it must vanish significantly fast, so the first term vanishes. Next, we bound the integral by taking the absolute value of the integrand:

$$-\int_{-\infty}^{\infty} dx\, x \left(\psi(x) \frac{d\psi^*(x)}{dx} + \psi^*(x) \frac{d\psi(x)}{dx} \right) \leq 2 \int_{-\infty}^{\infty} dx \left| x\, \psi^*(x) \frac{d\psi(x)}{dx} \right|. \tag{4.107}$$

Thus, putting it together we have

$$1 \leq 2 \int_{-\infty}^{\infty} dx \left| x\, \psi^*(x) \frac{d\psi(x)}{dx} \right|. \tag{4.108}$$

To continue, we next use the Cauchy–Schwarz inequality, for which

$$1 \leq 2 \int_{-\infty}^{\infty} dx \left| x\, \psi^*(x) \frac{d\psi(x)}{dx} \right| \tag{4.109}$$

$$\leq 2 \left[\int_{-\infty}^{\infty} dx\, |x\, \psi(x)|^2 \right]^{1/2} \left[\int_{-\infty}^{\infty} dx \left| \frac{d\psi(x)}{dx} \right|^2 \right]^{1/2}.$$

Now, note that the first integral on the second line is just the variance of position, with the assumption of 0 expectation value:

$$\int_{-\infty}^{\infty} dx\, |x\, \psi(x)|^2 = \int_{-\infty}^{\infty} dx\, x^2 |\psi(x)|^2 = \sigma_x^2. \tag{4.110}$$

The other integral, involving a derivative of the wavefunction, is actually the variance of momentum. From the Fourier transform, note that

$$\frac{d\psi(x)}{dx} = \int_{-\infty}^{\infty} \frac{dp}{\sqrt{2\pi\hbar}}\, g(p) \frac{d}{dx} e^{i\frac{px}{\hbar}} = \frac{i}{\hbar} \int_{-\infty}^{\infty} \frac{dp}{\sqrt{2\pi\hbar}}\, p\, g(p)\, e^{i\frac{px}{\hbar}}. \tag{4.111}$$

Then, the integral of the squared magnitude of this derivative is

$$\int_{-\infty}^{\infty} dx \left| \frac{d\psi(x)}{dx} \right|^2 \tag{4.112}$$

$$= \int_{-\infty}^{\infty} dx \left[\frac{i}{\hbar} \int_{-\infty}^{\infty} \frac{dp}{\sqrt{2\pi\hbar}} \, p \, g(p) \, e^{i\frac{px}{\hbar}} \right] \left[-\frac{i}{\hbar} \int_{-\infty}^{\infty} \frac{dp'}{\sqrt{2\pi\hbar}} \, p' \, g^*(p') \, e^{-i\frac{p'x}{\hbar}} \right]$$

$$= \frac{1}{\hbar^2} \int_{-\infty}^{\infty} dp \int_{-\infty}^{\infty} dp' \, p \, p' \, g(p) \, g^*(p') \int_{-\infty}^{\infty} \frac{dx}{2\pi\hbar} \, e^{i\frac{x}{\hbar}(p-p')} \, .$$

The integral over x just returns a δ-function fixing $p' = p$, so we find

$$\int_{-\infty}^{\infty} dx \left| \frac{d\psi(x)}{dx} \right|^2 = \frac{1}{\hbar^2} \int_{-\infty}^{\infty} dp \, p^2 \, |g(p)|^2 = \frac{\sigma_p^2}{\hbar^2} \, , \tag{4.113}$$

the variance of the momentum divided by $\hbar^2$.

With this, we have the inequality

$$1 \le \frac{2}{\hbar} \sigma_x \sigma_p \, , \tag{4.114}$$

or

$$\sigma_x \sigma_p \ge \frac{\hbar}{2} \, , \tag{4.115}$$

exactly the Heisenberg uncertainty principle. In the following example, we will see this constraint on variances in action.

Example 4.2 To concretely illustrate the Heisenberg uncertainty principle, let's consider an explicit form for the position-space wavefunction $\psi(x)$ and then determine the momentum-space wavefunction by Fourier transformation. Let's assume that the position-space wavefunction is uniform on $x \in [0, l]$, just like we did for the probability distribution in Example 4.1. How is the Heisenberg uncertainty principle manifest for this distribution?

Solution

The first thing we need to do is to appropriately normalize the wavefunction on its domain. We require that its absolute square integrates to 1 on $x \in [0, l]$, and so

$$1 = \int_0^l dx \, |\psi(x)|^2 = |\psi(x)|^2 \int_0^l dx = l \, |\psi(x)|^2 \, . \tag{4.116}$$

Up to an arbitrary complex phase, we will set

$$\psi(x) = \frac{1}{\sqrt{l}} \, . \tag{4.117}$$

In Chap. 5, we will address the (ir)relevance of the overall phase of the wavefunction. Then, with this uniform wavefunction, its Fourier transform is

$$g(p) = \int_0^l \frac{dx}{\sqrt{2\pi\hbar}} \, \psi(x) \, e^{-i\frac{px}{\hbar}} = \frac{1}{\sqrt{2\pi\hbar l}} \int_0^l dx \, e^{-i\frac{px}{\hbar}} = \frac{1}{\sqrt{2\pi\hbar l}} \frac{\hbar}{p} \frac{1 - e^{-i\frac{pl}{\hbar}}}{i}$$

$$= e^{-i\frac{pl}{2\hbar}} \sqrt{\frac{2\hbar}{\pi l}} \frac{\sin \frac{pl}{2\hbar}}{p} \,. \tag{4.118}$$

This momentum-space wavefunction takes the form of the sinc function, the form of the ratio $\frac{\sin x}{x}$. In Fig. 4.2, we have plotted the probability distributions of position from the wavefunction of $\psi(x)$ and of momentum from its Fourier conjugate $g(p)$. Note that while the domain of the wavefunction is bounded on $x \in [0,l]$, the domain of $g(p)$ is unbounded: $p \in (-\infty, \infty)$. That is, contributions from every value of momentum are necessary to describe the wavefunction.

We had already calculated the variance of this position-space wavefunction from the analysis of Example 4.1. We found

$$\sigma_x^2 = \frac{l^2}{12} \,. \tag{4.119}$$

Now, on to calculating the variance of momentum. We first need the expectation value of momentum, $\langle p \rangle$, which is

$$\langle p \rangle = \int_{-\infty}^{\infty} dp \, p \, |g(p)|^2 = \frac{2\hbar}{\pi l} \int_{-\infty}^{\infty} dp \, \frac{\sin^2 \frac{pl}{2\hbar}}{p} = 0 \,, \tag{4.120}$$

because the integrand is an odd function of p. The second moment of the momentum is much stranger, as we find

$$\langle p^2 \rangle = \int_{-\infty}^{\infty} dp \, p^2 \, |g(p)|^2 = \frac{2\hbar}{\pi l} \int_{-\infty}^{\infty} dp \, \sin^2 \frac{pl}{2\hbar} \,. \tag{4.121}$$

This integral is divergent, and takes infinite value because $\sin^2 p$ takes values between 0 and 1 for all $p \in (-\infty, \infty)$. So we find that $\langle p^2 \rangle = \sigma_p^2 = \infty$. Thus, it is definitely true that the Heisenberg uncertainty principle is satisfied:

$$\sigma_x^2 \sigma_p^2 = \infty \geq \frac{\hbar^2}{4} \,. \tag{4.122}$$

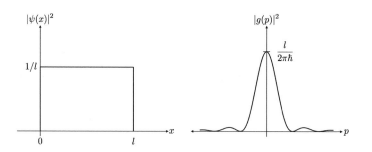

Fig. 4.2   Plots of the uniform position-space wavefunction from Eq. (4.117) (left) and its momentum-space Fourier conjugate of Eq. (4.118) (right). The distributions are the absolute squares of the wavefunctions, and the momentum-space wavefunction vanishes at momentum values $p = \pm(2n\pi\hbar)/l$, for $n = 1, 2, \ldots$.

One important feature that this example illustrates is the strangeness of the Fourier transform. Even though the position-space wavefunction is finite, bounded, and all of its moments exist, this does not guarantee that its momentum-space conjugate distribution satisfies these same properties. We will see many more Fourier conjugate distributions, and cases in which the Heisenberg uncertainty principle is non-trivial, however those cases are often highly particular and chosen carefully. In some sense this example illustrates that for general choice of a wavefunction, it is very challenging to identify a wavefunction and its Fourier transform that both have finite variances.

These two derivations of the Heisenberg uncertainty principle suggest an intimate relationship between the canonical commutation relations and Fourier conjugates. From the Fourier transformation perspective, we have established that a derivative in position space, d/dx, corresponds to the value of momentum p in momentum space. Conversely, a derivative in momentum space, d/dp, corresponds to the value of position x in position space. From the commutation relations, we had

$$[\hat{x}, \hat{p}] = i\hbar = \text{constant}. \tag{4.123}$$

Can we show that a constant commutator necessarily means that one operator can be expressed as the derivative of the other?

4.3.4 Constant Commutator as the Derivative Operator

To emphasize the generality of this result, let's assume that we have two operators, $\hat{A}$ and $\hat{B}$, that have a commutator that is a constant $c \in \mathbb{C}$:

$$[\hat{A}, \hat{B}] = c. \tag{4.124}$$

Unlike earlier, we won't even assume that $\hat{A}$ and $\hat{B}$ are Hermitian, but we'll mention what Hermitivity enforces later. Now, let's work in the space spanned by the eigenvectors of the $\hat{A}$ operator. For a normalized eigenstate $|v\rangle$ of $\hat{A}$, its eigenvalue equation is

$$\hat{A}|v\rangle = a|v\rangle, \tag{4.125}$$

where $a \in \mathbb{C}$ is an eigenvalue of $\hat{A}$. In the $\hat{A}$ basis, we assume that $\hat{A}$ is diagonal, so its action can be replaced by an eigenvalue a as eigenvalues on the diagonal are the only non-zero elements of $\hat{A}$. Then, the commutation relation in this $\hat{A}$ space can be expressed as

$$[\hat{A}, \hat{B}] = [a, \hat{B}] = c. \tag{4.126}$$

We will now show that, in this space, setting the operator

$$\hat{B} = -c\frac{d}{da}, \tag{4.127}$$

the commutation relations are satisfied.[10] As a derivative, we must enforce the Leibniz product rule to ensure the linearity of $\hat{B}$. As we have done before, we multiply the commutator on the right by a test function $f(a)$ to remind us to maintain linearity. Then, we have

$$[\hat{A},\hat{B}]f(a) = \left[a, -c\frac{d}{da}\right]f(a) = -ca\frac{df(a)}{da} + c\frac{d}{da}(af(a)) \qquad (4.128)$$

$$= -ca\frac{df(a)}{da} + cf(a) + ca\frac{df(a)}{da} = cf(a).$$

Now, dropping the test function $f(a)$, we have established that $[\hat{A},\hat{B}] = c$ implies that

$$\hat{B} = -c\frac{d}{da}, \qquad (4.129)$$

in the $\hat{A}$ operator space.

For general operators $\hat{A}$ and $\hat{B}$, their eigenvalues are not real, so a Fourier transform relating them to one another is strictly not well-defined. However, if $\hat{A}$ and $\hat{B}$ are both Hermitian, then their eigenvalues are real, and, like for position and momentum, they are indeed related by Fourier transformation. Further, if $\hat{A}$ and $\hat{B}$ are Hermitian, then their commutator c is necessarily imaginary. Taking the complex conjugate of the commutator, we have

$$[\hat{A},\hat{B}]^\dagger = c^* = [\hat{B},\hat{A}] = -c. \qquad (4.130)$$

Of course, this is what we find with position and momentum, for which we have $c = i\hbar$.

4.4 The Density Matrix: A First Look

We assumed, through the axioms of quantum mechanics, that probability is conserved, which had the consequence that linear operators on the Hilbert space of states must be unitary. Through identification of the Hamiltonian as the Hermitian operator that implements time evolution, we were led to the Schrödinger equation, Ehrenfest's theorem, and other results. The Schrödinger equation, for all of its virtues, has one glaring, awkward vice: it describes the time evolution of the wavefunction or a state in the Hilbert space, which is only a probability *amplitude*, and not an actual, observable, probability. Can we use the formalism we have developed to determine how honest probabilities evolve in time?

Expressed in position space, the probability density $\rho(x)$ of a quantum system is

$$\rho(x) = \psi^*(x)\psi(x), \qquad (4.131)$$

---

[10] Because the operator $\hat{A}$ necessarily commutes with itself, we can always add to $\hat{B}$ an arbitrary function of operator $\hat{A}$, and the commutation relations are still satisfied. Commutation relations of $\hat{B}$ with other operators could then be used to fix this function of $\hat{A}$, but in this book we will just focus on the simplest possible form for $\hat{B}$, given a constant commutation relation with $\hat{A}$.

where $\psi(x)$ is the wavefunction and time dependence has been suppressed, for now. This probability density is still a function of position x, and not just a single number. So, if we want to express this probability density in Dirac notation, we would *not* write it as

$$\rho(x) \neq \langle \psi | \psi \rangle = \int_{-\infty}^{\infty} dx\, \psi^*(x) \psi(x) = 1\,, \qquad (4.132)$$

which is not a function of position because we integrated over all positions. Okay, if the probability density is not the inner product of the state $|\psi\rangle$ with itself, what could it be? The only other operation we can perform with a bra $\langle \psi|$ and a ket $|\psi\rangle$ is the outer product. Therefore, the probability density must be the outer product

$$\rho = |\psi\rangle\langle\psi|\,. \qquad (4.133)$$

Recall that the outer product of bras and kets forms a *matrix*, and as this matrix describes the probability *density*, ρ is referred to as the **density matrix**. By the way, this argument is a good example of a "what else could it be?" proof in physics. If you have two options and one is eliminated, it must be the other option!

So how does this ρ evolve in time? That is, we would like to identify a differential equation in time for ρ. To do this, we do the usual Taylor expansion/time evolution trick. First, if ρ is a function of time and we translate it forward in time by Δt, we have

$$\rho(t + \Delta t) = \rho(t) + \Delta t \frac{d}{dt}\rho(t) + \mathcal{O}(\Delta t^2)\,. \qquad (4.134)$$

On the other hand, we can also just evolve the state $|\psi\rangle$ forward in time with the Hamiltonian. That is:

$$|\psi(t+\Delta t)\rangle\langle\psi(t+\Delta t)| = e^{-i\frac{\Delta t \hat{H}}{\hbar}}|\psi(t)\rangle\langle\psi(t)|e^{i\frac{\Delta t \hat{H}}{\hbar}} \qquad (4.135)$$

$$= \left(1 - i\frac{\Delta t \hat{H}}{\hbar} + \cdots\right)|\psi(t)\rangle\langle\psi(t)|\left(1 + i\frac{\Delta t \hat{H}}{\hbar} + \cdots\right)$$

$$= |\psi(t)\rangle\langle\psi(t)| - \Delta t\frac{i}{\hbar}[\hat{H}, |\psi(t)\rangle\langle\psi(t)|] + \cdots.$$

Setting the two sides of this equation equal to one another, we find that

$$\rho + \Delta t\frac{d\rho}{dt} = |\psi\rangle\langle\psi| - \Delta t\frac{i}{\hbar}[\hat{H}, |\psi\rangle\langle\psi|]\,, \qquad (4.136)$$

or

$$\frac{d\rho}{dt} = -\frac{i}{\hbar}[\hat{H}, \rho]\,. \qquad (4.137)$$

We'll revisit this equation near the end of this book, but it is called the **von Neumann equation**.[11] Intriguingly, note that the sign on the right side is different than what we found for time evolution of expectation values in Sec. 4.2.

This sign difference can be understood by making the connection with the time evolution of the density matrix and expectation values of a Hermitian operator $\hat{T}$. To do this, we first need to understand a bit more what the density matrix expresses and how

---

[11] J. von Neumann, *Mathematical Foundations of Quantum Mechanics: New Edition*, Princeton University Press (2018).

it can be used to calculate expectation values. To get a better understanding of the information that the density matrix contains, let's expand the state $|\psi\rangle$ in a complete, orthonormal basis on the Hilbert space. Let's call the basis states $\{|\psi_n\rangle\}$, and these could be, for example, the energy eigenstates we established earlier in this chapter. Then, we can expand $|\psi\rangle$ as

$$|\psi\rangle = \sum_n \beta_n |\psi_n\rangle, \tag{4.138}$$

where β_n is a complex probability amplitude. With this expansion, the density matrix for this state is

$$\rho = |\psi\rangle\langle\psi| = \sum_{m,n} \beta_n \beta_m^* |\psi_n\rangle\langle\psi_m|. \tag{4.139}$$

As we have assumed that the $\{|\psi_n\rangle\}$ basis is orthonormal and complete, the object $|\psi_n\rangle\langle\psi_m|$ corresponds to a matrix with a 1 in the nth row, mth column, and 0s everywhere else. Further, note that elements on the diagonal, for which $m = n$, are just the probabilities for the basis state n: $|\beta_n|^2$. The sum of these probabilities is necessarily 1, by the normalization of the wavefunction. Additionally, the sum of diagonal elements of a matrix is its trace and hence we establish that

$$\text{tr}\,\rho = \text{tr}\,|\psi\rangle\langle\psi| = \sum_n |\beta_n|^2 = 1. \tag{4.140}$$

Taking the trace of a matrix expressed as a sum over outer products of vectors is equivalent to taking the dot product of the vectors. That is:

$$\text{tr}\sum_{m,n} \beta_n \beta_m^* |\psi_n\rangle\langle\psi_m| = \sum_{m,n} \beta_n \beta_m^* \langle\psi_m|\psi_n\rangle = \sum_{m,n} \beta_n \beta_m^* \delta_{ij} = \sum_n |\beta_n|^2. \tag{4.141}$$

The trace turns an outer product into an inner product.

With this understanding of the trace, we can compactly express the expectation values of a Hermitian operator $\hat{T}$ with the density matrix. For a state $|\psi\rangle$, recall that the expectation value E_T is

$$E_T = \langle\psi|\hat{T}|\psi\rangle. \tag{4.142}$$

That is, the expectation value of $\hat{T}$ on state $|\psi\rangle$ is the inner product of the state $|\psi\rangle$ with $\hat{T}|\psi\rangle$. This dot product can be encapsulated with the trace of the outer product of these two states:

$$\langle\psi|\hat{T}|\psi\rangle = \text{tr}\left(|\psi\rangle\langle\psi|\hat{T}\right) = \text{tr}\,\rho\hat{T}, \tag{4.143}$$

where we identify the density matrix as $\rho = |\psi\rangle\langle\psi|$.

The trace has a number of interesting properties, three of which we will review here. First, the trace is *linear*. That is, the trace of a sum of two matrices is the sum of their individual traces. For matrices $\mathbb{M}$ and $\mathbb{N}$, the trace of their sum is

$$\text{tr}(\mathbb{M} + \mathbb{N}) = \sum_i (M_{ii} + N_{ii}) = \sum_i M_{ii} + \sum_i N_{ii} = \text{tr}\,\mathbb{M} + \text{tr}\,\mathbb{N}. \tag{4.144}$$

In the intermediate equalities, we have expressed the trace as the sum of diagonal elements of the matrices, denoted as M_{ii} for matrix $\mathbb{M}$, for example. Second, the trace is *cyclic*. For two matrices $\mathbb{M}$ and $\mathbb{N}$, the trace of their product is independent of the order of the product:

$$\text{tr}(\mathbb{M}\mathbb{N}) = \sum_{i,j} M_{ij} N_{ji} = \sum_{i,j} N_{ji} M_{ij} = \text{tr}(\mathbb{N}\mathbb{M}). \qquad (4.145)$$

In the intermediate equalities, we have expressed the matrix product through summing over appropriate rows and columns of the matrices, and the order of the product of individual matrix elements is irrelevant.

The third property of the trace is that the sum of the diagonal entries of a matrix $\mathbb{M}$ is also the sum of its eigenvalues. To demonstrate this property, we first consider the matrix $\mathbb{M}$ expressed in the basis in which it is diagonal. In that basis, the only non-zero entries are on the diagonal and they are the matrix's eigenvalues, λ_i:

$$\text{tr}\,\mathbb{M} = \sum_i M_{ii} = \sum_i \lambda_i . \qquad (4.146)$$

To express the matrix in any other basis, we can act on the matrix by a unitary matrix $\mathbb{U}$ on the left and right:

$$\mathbb{M} \to \mathbb{M}' = \mathbb{U}^\dagger \mathbb{M} \mathbb{U}. \qquad (4.147)$$

A single unitary matrix $\mathbb{U}$ transforms matrix $\mathbb{M}$ into another basis because, as mentioned near the beginning of this book, we assume that $\mathbb{M}$ is a normal matrix. Then, the trace of $\mathbb{M}'$ is the same as $\mathbb{M}$ by the cyclic property of the trace:

$$\text{tr}\,\mathbb{M}' = \text{tr}(\mathbb{U}^\dagger \mathbb{M} \mathbb{U}) = \text{tr}(\mathbb{U}\mathbb{U}^\dagger \mathbb{M}) = \text{tr}\,\mathbb{M}, \qquad (4.148)$$

and $\mathbb{U}\mathbb{U}^\dagger = \mathbb{I}$ because it is unitary.

With these properties established, we can then connect the time evolution of the density matrix to the time evolution of the expectation value of an operator $\hat{T}$. For simplicity we assume that the operator $\hat{T}$ has no explicit dependence on time. With this assumption, note that

$$\hat{T}\frac{d\rho}{dt} = \frac{d}{dt}(\rho\hat{T}) = -\frac{i}{\hbar}[\hat{H},\rho]\hat{T}. \qquad (4.149)$$

Now, we can take the trace of this equality and use the linearity of the trace:

$$\frac{d}{dt}\text{tr}(\rho\hat{T}) = \text{tr}\left(-\frac{i}{\hbar}[\hat{H},\rho]\hat{T}\right) = -\frac{i}{\hbar}\text{tr}(\hat{H}\rho\hat{T} - \rho\hat{H}\hat{T}) \qquad (4.150)$$

$$= -\frac{i}{\hbar}\left[\text{tr}(\hat{H}\rho\hat{T}) - \text{tr}(\rho\hat{H}\hat{T})\right].$$

Then, by the cyclicity of the trace, note that

$$\text{tr}(\hat{H}\rho\hat{T}) = \text{tr}(\rho\hat{T}\hat{H}) = \langle\psi|\hat{T}\hat{H}|\psi\rangle, \qquad (4.151)$$

and similarly for the other trace. Putting this all together, we then find that

$$\frac{d}{dt}\text{tr}(\rho\hat{T}) = -\frac{i}{\hbar}\left[\langle\psi|\hat{T}\hat{H}|\psi\rangle - \langle\psi|\hat{H}\hat{T}|\psi\rangle\right] = \frac{i}{\hbar}\langle\psi|[\hat{H},\hat{T}]|\psi\rangle, \qquad (4.152)$$

exactly as we had derived earlier for the time evolution of the expectation value.

We'll come back to the density matrix at the end of the book, and demonstrate that the density matrix describes a quantum system in ways that the wavefunction cannot.

Exercises

4.1   In our derivation of the generalized uncertainty principle, we needed to assume that the two operators that we considered were Hermitian. What if we relax this assumption? Consider two operators $\hat{C}$ and $\hat{D}$ that are not assumed to be Hermitian. Do they have an uncertainty principle?

4.2   In this chapter, we derived the canonical commutation relation between the position and momentum operators, but we had previously constructed the derivative operator $\mathbb{D}$ on a grid of positions in Chap. 2. In this problem, we will study the commutation relations of momentum and position on this grid and see if what we find is compatible with the continuous commutation relations.

     (a)   First, from the discrete derivative matrix $\mathbb{D}$ in Eq. (2.10), construct the Hermitian momentum matrix on the grid, $\mathbb{P}$. In this problem, we'll just explicitly consider 3×3 matrices.

     (b)   Now, in this same basis, construct the 3×3 Hermitian matrix that represents the position operator on this grid; call it $\mathbb{X}$.

     (c)   What is the commutator of $\mathbb{X}$ and $\mathbb{P}$, $[\mathbb{X}, \mathbb{P}]$? Is it what you would expect from the canonical commutation relation?

     (d)   Now, imagine taking Δx smaller and smaller, for the same interval in position. That is, consider larger and larger-dimensional momentum and position matrices. What do you expect the general form of the commutator $[\mathbb{X}, \mathbb{P}]$ is when $\mathbb{X}$ and $\mathbb{P}$ are $N \times N$ matrices? Does this have a sensible limit as $\Delta x \to 0$, or $N \to \infty$?

4.3   Consider the following operators that correspond to exponentiating the momentum and position operators, $\hat{x}$ and $\hat{p}$:

$$\hat{X} = e^{i\sqrt{\frac{2\pi c}{\hbar}}\hat{x}}, \qquad\qquad \hat{P} = e^{i\sqrt{\frac{2\pi}{c\hbar}}\hat{p}}. \qquad (4.153)$$

Here, c is some constant.

     (a)   What are the units of the constant c? From that, can you provide an interpretation of c?

     (b)   Calculate the commutator of $\hat{X}$ and $\hat{P}$, $[\hat{X}, \hat{P}]$. What does this mean?

     (c)   What is the range of eigenvalues of $\hat{x}$ and $\hat{p}$ for which the operators $\hat{X}$ and $\hat{P}$ are single-valued? Remember, $e^{i\pi} = e^{i3\pi}$, for example.

     (d)   Determine the eigenstates and eigenvalues of the exponentiated momentum operator, $\hat{P}$, in the position basis. That is, what values can λ take and what function $f_\lambda(x)$ satisfies

$$\hat{P} f_\lambda(x) = \lambda f_\lambda(x)? \qquad (4.154)$$

Do the eigenvalues λ have to be real valued? Why or why not?

(e)   Using part (b), what can you say about the eigenstates of $\hat{X}$?

(f)   What is the uncertainty principle for operators $\hat{X}$ and $\hat{P}$? Can you reconcile that with the Heisenberg uncertainty principle?

Hint: Think about the consequences of the answer to part (c).

4.4   Consider two Hermitian operators $\hat{A}$ and $\hat{B}$ and consider the unitary operators formed from exponentiating them:

$$\hat{U}_A = e^{i\hat{A}}, \qquad\qquad \hat{U}_B = e^{i\hat{B}}. \qquad (4.155)$$

If $\hat{A}$ and $\hat{B}$ were just numbers, then it would be easy to determine the unitary matrix formed from the product of $\hat{U}_A$ and $\hat{U}_B$. However, matrices do not in general commute, so it complicates this product. Using the Taylor expansion of the exponential, find the difference of unitary matrices

$$e^{i\hat{A}} e^{i\hat{B}} - e^{i(\hat{A}+\hat{B})}. \qquad (4.156)$$

Only consider terms in the Taylor expansion up through cubic order; that is, terms that contain at most the product of three matrices $\hat{A}$ and/or $\hat{B}$. Under what condition does this difference vanish?

4.5   An *instanton* is a quantum mechanical excitation that is localized in space like a particle. They are closely related to solitary waves or solitons that were first observed in the mid-nineteenth century as a traveling wave in a canal.[12] We'll study a model for instantons in this problem. Our simple model will be the following. We will consider a quantum system constrained on a circle, and we can define states on this circle by their winding number n, the number of times that the instanton wraps around the circle before it connects back to itself (think about winding a string around a cylinder and then tying it back together after n times around). The winding number n can be any integer, positive, negative, or zero, and the sign of the winding number encodes the direction in which it is wrapped.

States with different n are orthogonal, so we will consider the Hilbert space as spanned by the set of states $\{|n\rangle\}_{n=-\infty}^{\infty}$ which are orthonormal:

$$\langle m|n \rangle = \delta_{mn}, \qquad (4.157)$$

and we will assume they are complete.

(a)   On this Hilbert space, we can define a hopping operator $\hat{O}$ which is defined to act on the basis elements as

$$\hat{O}|n\rangle = |n+1\rangle. \qquad (4.158)$$

Show that this means that $\hat{O}$ is unitary.

(b)   Assume that the state $|\psi\rangle$ is an eigenstate of $\hat{O}$ with eigenvalue defined by an angle θ:

$$\hat{O}|\psi\rangle = e^{i\theta}|\psi\rangle. \qquad (4.159)$$

Express the state $|\psi\rangle$ as a linear combination of the winding states $|n\rangle$.

[12]   J. S. Russell, *Report on Waves*, Report of the 14th Meeting of the British Association for the Advancement of Science, York, September 1844 (1845).

(c) The Hamiltonian for this winding system $\hat{H}$ is defined to act as

$$\hat{H}|n\rangle = |n|E_0|n\rangle, \qquad (4.160)$$

where E_0 is a fixed energy and $|n|$ is the absolute value of the winding number n. Calculate the commutator of the hopping operator and the Hamiltonian, $[\hat{H}, \hat{O}]$.

(d) Determine the time dependence of the state $|\psi\rangle$; that is, evaluate

$$|\psi(t)\rangle = e^{-i\frac{\hat{H}t}{\hbar}}|\psi\rangle. \qquad (4.161)$$

4.6 The Heisenberg uncertainty principle establishes a robust lower bound on the product of variances of Hermitian operators, which depends on their commutator. We established this bound for any arbitrary, normalizable state $|\psi\rangle$; however, can we determine the state (or states) for which the Heisenberg uncertainty principle is saturated? That is, on what state do we have the equality

$$\sigma_x \sigma_p = \frac{\hbar}{2}? \qquad (4.162)$$

(a) In our derivation of the uncertainty principle, we exploited the Cauchy–Schwarz inequality at an early stage. In particular, Cauchy–Schwarz implies that

$$\langle(\hat{x}-\langle\hat{x}\rangle)^2\rangle\langle(\hat{p}-\langle\hat{p}\rangle)^2\rangle \geq |\langle(\hat{x}-\langle\hat{x}\rangle)(\hat{p}-\langle\hat{p}\rangle)\rangle|^2, \qquad (4.163)$$

where all expectation values are taken on some state $|\psi\rangle$. Argue that if this inequality is saturated and becomes an *equality*, then

$$(\hat{x}-\langle\hat{x}\rangle)|\psi\rangle = -\frac{2i\sigma_x^2}{\hbar}(\hat{p}-\langle\hat{p}\rangle)|\psi\rangle. \qquad (4.164)$$

Hint: Don't forget to use the saturated uncertainty principle, too.

(b) Now, express the relationship in Eq. (4.164) as a differential equation for the wavefunction $\psi(x) = \langle x|\psi\rangle$ in position space. Solve the differential equation for $\psi(x)$ for general $\langle\hat{x}\rangle$ and $\langle\hat{p}\rangle$. Don't worry about an overall normalization.

(c) Now, Fourier transform $\psi(x)$ to momentum space. How does the functional form of its Fourier transform compare to that of $\psi(x)$ in position space?

4.7 The Ehrenfest's theorem that we quoted in the text was that quantum expectation values satisfy the classical equations of motion. In Sec. 4.2.1, we had also stated that this isn't quite true. If the quantum expectation value of momentum satisfied the classical equations of motion, then its time derivative would be

$$\frac{d\langle\hat{p}\rangle}{dt} = -\frac{dV(\langle\hat{x}\rangle)}{d\langle\hat{x}\rangle}, \qquad (4.165)$$

but this is not what we had found for its time dependence in Eq. (4.54). In this equation, we'll study the differences between these expressions.

(a) For what potential $V(\hat{x})$ is the true time dependence of Eq. (4.54) equal to the simpler expression of Eq. (4.165)?

(b) Consider a power-law potential with

$$V(\hat{x}) = k\hat{x}^{2n},\tag{4.166}$$

where k is a constant that has the correct units to make the potential have units of energy and n is an integer greater than 1. For a general state $|\psi\rangle$, can you say how the time dependences of Eqs. (4.54) and (4.165) compare? Which expression produces a larger time derivative of the expectation value of momentum?

(c) For the form of the potential in part (b), what must the state $|\psi\rangle$ be for Eqs. (4.54) and (4.165) to be equal?

4.8 We've seen simple examples of expectation value and variance in analyzing the rolls of a fair die and the uniform distribution in Example 4.1. In this exercise, we will study these ideas on another very important continuous probability distribution. Consider the probability distribution $p(x)$ for $0 \le x < \infty$, where

$$p(x) = Ne^{-\lambda x},\tag{4.167}$$

where λ is a parameter and N is the normalization constant. This distribution is illustrated in Fig. 4.3.

(a) First, determine the value of N such that the probability distribution is normalized; that is, such that

$$1 = \int_0^\infty dx\, p(x).\tag{4.168}$$

(b) Now, calculate the mean or expected value of x on this normalized distribution; that is

$$\langle x \rangle = \int_0^\infty dx\, x\, p(x).\tag{4.169}$$

(c) Calculate the second moment of the distribution:

$$\langle x^2 \rangle = \int_0^\infty dx\, x^2\, p(x).\tag{4.170}$$

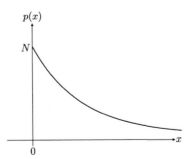

Fig. 4.3   An illustration of the exponential probability distribution of Eq. (4.167). The value of the distribution at $x = 0$ is the normalization factor, N.

With this result and part (b), what are the variance and standard deviations of this distribution? Is it indeed true that "most" of the distribution is within one standard deviation of the mean?

(d) The mean and variance can be generalized to define higher measures of the spread of the distribution about the mean. An nth spread for $n \geq 2$ is

$$\langle (x - \langle x \rangle)^n \rangle = \int_0^\infty dx \, (x - \langle x \rangle)^n \, p(x). \tag{4.171}$$

Note that for $n = 2$, this spread is just the variance. Calculate the value of the nth spread on this distribution.

(e) Another measure of the average of a distribution is the **median**, which is defined to be the point at which there is exactly half of the area of the distribution to the left and to the right of the point. What is the median for this exponential distribution? How does the value of the median compare to the value of the mean?

5

Quantum Mechanical Example: The Infinite Square Well

Starting in this chapter, and in the following two chapters, we will take all of the abstractness of quantum mechanics that we have developed and make it concrete in actual examples. Our goal for these examples will be to solve the time-independent eigenvalue problem for the Hamiltonian $\hat{H}$:

$$\hat{H}|\psi_n\rangle = E_n|\psi_n\rangle,\tag{5.1}$$

where $|\psi_n\rangle$ and E_n are the energy eigenstate and eigenvalue, respectively. Once we know the complete spectrum of the Hamiltonian, then we know the time evolution of an arbitrary quantum state $|\psi\rangle$ through application of the Schrödinger equation. The problem we will study in this chapter is the **infinite square well**, a very simple, introductory system in quantum mechanics. As the Hamiltonian encodes the energy of any system, we first need to express the energy of the infinite square well appropriately. Further, with the quantum dynamics of the system established, this provides a useful playground for probing general features of quantum mechanics within a simple, concrete example. Here, we compare the infinite square well to its corresponding classical system and demonstrate that there is a smooth transition from the quantum to the classical regime.

5.1 Hamiltonian of the Infinite Square Well

The system is defined by the potential in position space of

$$V(x) = \begin{cases} \infty, & x < 0, \\ 0, & 0 < x < a, \\ \infty, & x > a, \end{cases}\tag{5.2}$$

where a is the width of the potential well. The picture of the infinite square well's potential is illustrated in Fig. 5.1. Because of the infinitely high walls at $x = 0, a$, any object placed in the well stays in the well. That is, to leave the well would require infinite kinetic energy, which is not possible. We can thus think of the infinite square well problem as a particle in a box of width a with perfectly rigid walls. As the wavefunction $\psi(x,t)$ represents the probability amplitude for the particle to be at position x at time t, if the particle cannot leave the well then we must have that

$$\psi(x,t) = 0,\tag{5.3}$$

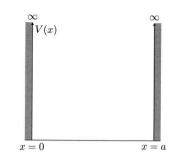

Fig. 5.1 An infinite square well potential, with perfectly rigid walls located at positions $x = 0, a$.

for $x < 0$ or $x > a$ for all time t. So, an eigenstate $|\psi_n\rangle$ of the Hamiltonian that satisfies

$$\hat{H}|\psi_n\rangle = \frac{\hat{p}^2}{2m}|\psi_n\rangle = E_n|\psi_n\rangle \tag{5.4}$$

only has support on the region where $0 < x < a$.

5.1.1 Energy Eigenvalues

To determine the energy eigenvalues of the infinite square well, let's focus on the intriguing form of the eigenvalue equation:

$$\frac{\hat{p}^2}{2m}|\psi_n\rangle = E_n|\psi_n\rangle, \tag{5.5}$$

for the nth-energy eigenvalue E_n. Now, the value of the energy E_n must be a real number, and so the momentum operator $\hat{p}$ must be Hermitian, as we have established. The Hermitivity of momentum is non-trivial for the infinite square well because we restrict the particle to lie in a finite domain. We'll come back to the consequence of the Hermitivity of momentum $\hat{p}$ for the structure of the Hilbert space in a second.

With the assumption that $\hat{p}$ is Hermitian, its eigenvalues are all real, so the energy eigenvalues must be all non-negative. This means that we can meaningfully take the square root of both sides of the eigenvalue equation to find

$$|\hat{p}||\psi_n\rangle = \sqrt{2mE_n}|\psi_n\rangle. \tag{5.6}$$

The absolute value of momentum $|\hat{p}|$ just means that we take the absolute value of its eigenvalues. Therefore, the energy eigenstates $|\psi_n\rangle$ of energy E_n correspond to fixed magnitude of momentum, where

$$|p_n| = \sqrt{2mE_n}, \tag{5.7}$$

where now p_n is an eigenvalue of $\hat{p}$. This makes sense from a physical understanding of the infinite square well. We had said that the infinite square well can be imagined as a particle in a box with perfectly rigid walls. If that particle has momentum, it will travel to one of the walls and bounce back, conserving its kinetic energy and thus just negating its momentum. It will just bounce back and forth with constant kinetic energy *ad infinitum*. We'll revisit this physical interpretation when we connect with the classical particle in a box later in this chapter.

We can always represent any state in the Hilbert space as a linear combination of momentum eigenstates. Recall that, for momentum p, the momentum eigenstate $|p\rangle$ in position space is

$$\langle x|p\rangle = e^{i\frac{px}{\hbar}}. \tag{5.8}$$

As an energy eigenstate of the infinite square well corresponds to a fixed magnitude of momentum, we just sum together the two momentum eigenstates with equal magnitude but opposite sign of momentum:

$$\langle x|\psi_n\rangle = \psi_n(x) = \alpha e^{i\frac{p_nx}{\hbar}} + \beta e^{-i\frac{p_nx}{\hbar}}, \tag{5.9}$$

where $p_n = \sqrt{2mE_n}$ and α and β are some complex coefficients.

To determine the value of these coefficients and momentum eigenvalue, we will enforce the Hermitivity of the momentum operator $\hat{p}$. As discussed in Sec. 3.3, for momentum to be Hermitian, states in the Hilbert space must have no support at the boundary of the region. As energy eigenstates span the Hilbert space, we must require that all energy eigenstates vanish at the boundaries of the well, $\psi_n(x) = 0$, for $x = 0, a$. This ensures that any linear combination of the energy eigenstates, that is, an arbitrary state in the Hilbert space, has no support at the boundaries. Enforcing this condition for the wavefunction established above, when $x = 0$ we find

$$\psi_n(x = 0) = \alpha + \beta = 0, \tag{5.10}$$

or $\beta = -\alpha$. With this constraint enforced, the wavefunction can then be written as

$$\psi_n(x) = \alpha\left(e^{i\frac{p_nx}{\hbar}} - e^{-i\frac{p_nx}{\hbar}}\right) = \gamma \sin\left(\frac{p_nx}{\hbar}\right), \tag{5.11}$$

for some other constant γ.

Next, let's enforce vanishing of this energy eigenstate wavefunction at the boundary $x = a$. As an element of the Hilbert space, this wavefunction must be L^2-normalized, so $\gamma \neq 0$. Thus, the only possible handle we can control to ensure that $\psi_n(x = a) = 0$ is fixing the eigenvalue of momentum $p_n \geq 0$. The factor $\sin\theta$ vanishes if its argument is an integer multiple of π, so we must enforce that

$$\frac{p_n a}{\hbar} = n\pi, \tag{5.12}$$

or

$$p_n = \frac{n\pi\hbar}{a}, \tag{5.13}$$

where n is a positive integer (for $n = 0$ the wavefunction would just be 0, and non-normalizable). This expression for momentum should be familiar: recall that the **de Broglie wavelength** λ_{dB} of a particle of momentum p is

$$\lambda_{\text{dB}} = \frac{h}{p} = \frac{2\pi\hbar}{p}, \tag{5.14}$$

where h is Planck's constant.[1] For an energy eigenstate of the particle in the infinite square well, we can express its de Broglie wavelength as

$$\lambda_{n,\mathrm{dB}} = \frac{2\pi\hbar}{p_n} = \frac{2a}{n},$$

(5.15)

which is the necessary size to exactly fit an integer n of half-wavelengths in the well of width a to ensure that the wavefunction vanishes at both boundaries. To illustrate this, the first three energy eigenstate wavefunctions are displayed in the well in Fig. 5.2.

With the momentum eigenvalues established, the nth-energy eigenvalue is just

$$E_n = \frac{p_n^2}{2m} = \frac{n^2\pi^2\hbar^2}{2ma^2},$$

(5.16)

where, as discussed earlier, $n \in \mathbb{N}$, the natural numbers. The energy eigenstate wavefunction is therefore

$$\psi_n(x) = \gamma \sin\left(\frac{p_n x}{\hbar}\right) = \gamma \sin\left(\frac{n\pi x}{a}\right),$$

(5.17)

where we still have an undetermined coefficient γ. As this energy eigenstate is an element of the Hilbert space, it must be L^2-normalized:

$$1 = \langle \psi_n | \psi_n \rangle = \int_0^a dx\, |\gamma|^2 \sin^2\left(\frac{n\pi x}{a}\right) = |\gamma|^2 \frac{a}{2}.$$

(5.18)

Thus, to ensure normalization of the wavefunction, we fix

$$|\gamma| = \sqrt{\frac{2}{a}},$$

(5.19)

and so the energy eigenstate wavefunctions are

$$\langle x | \psi_n \rangle = \psi_n(x) = \sqrt{\frac{2}{a}} \sin\left(\frac{n\pi x}{a}\right),$$

(5.20)

with energy eigenvalue

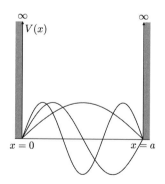

Fig. 5.2   Illustration of the first three energy eigenstate wavefunctions in the infinite square well. For the nth-energy eigenstate, n half-wavelengths fit in the well.

[1] L. V. P. R. de Broglie, "Recherches sur la théorie des quanta," *Ann. Phys.* **2**, 22–128 (1925).

$$E_n = \frac{n^2 \pi^2 \hbar^2}{2ma^2}.$$ (5.21)

These energy eigenstates completely span the Hilbert space for the infinite square well and ensure that momentum is a Hermitian operator.

In Fig. 5.3, we have illustrated the first six energy eigenvalues of the infinite square well, which show that the spacing of energies grows rapidly at higher levels. If the state of the particle in the well has a small expectation value of energy, then necessarily that state has very small contributions from high-energy eigenstates. Correspondingly, the wavefunction of a low-energy state in the infinite square well necessarily varies in space slowly, with all of its derivatives (proportional to momentum) relatively small. This latter point follows from the functional form of the energy eigenstate wavefunctions in Eq. (5.20).

5.1.2 A Re-analysis of the Hilbert Space

In Eq. (5.20), I performed a little sleight-of-hand. For normalization of the wavefunction, we established that the constant γ is required to satisfy

$$|\gamma| = \sqrt{\frac{2}{a}}.$$ (5.22)

However, all this would enforce is that γ can be written as

$$\gamma = \sqrt{\frac{2}{a}} e^{i\phi},$$ (5.23)

for some real-valued angle $\phi \in [0, 2\pi)$. When we wrote the expression for the energy eigenstate, we just ignored this exponential phase factor, and fixed γ to be a real number. Why can we do this?

A state $|\psi\rangle$ in the Hilbert space $\mathcal{H}$ is, by itself, unphysical. By "unphysical" I mean that there is no experiment we can perform to measure the state alone. The outcomes of any experiment are expectation values, or matrix elements, of Hermitian operators evaluated for the state of interest. Recall that the expectation value of a Hermitian operator $\hat{T}$ on state $|\psi\rangle$ is

$$E_T = \langle \psi | \hat{T} | \psi \rangle.$$ (5.24)

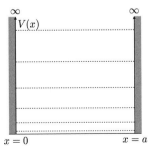

Fig. 5.3   Illustration of the spacing of the first few energy eigenvalues in the infinite square well potential, denoted by the dashed horizontal lines.

Note that such an expectation value is very general and includes, for example, the normalization of the state. If $\hat{T} = \mathbb{I}$, the identity operator, then

$$\langle \psi | \mathbb{I} | \psi \rangle = \langle \psi | \psi \rangle = 1 \,. \tag{5.25}$$

Now, consider a new state $|\psi'\rangle$ which is related to $|\psi\rangle$ by a complex phase factor:

$$|\psi'\rangle = e^{i\phi} |\psi\rangle \,. \tag{5.26}$$

All observable quantities (i.e., expectation values) are identical on the states $|\psi'\rangle$ and $|\psi\rangle$:

$$\langle \psi' | \hat{T} | \psi' \rangle = \langle \psi | e^{-i\phi} \hat{T} e^{i\phi} | \psi \rangle = \langle \psi | \hat{T} | \psi \rangle \,, \tag{5.27}$$

where we use the linearity of operator $\hat{T}$ to establish

$$e^{-i\phi} \hat{T} e^{i\phi} = \hat{T} \,. \tag{5.28}$$

Thus, if the overall phase of a state in the Hilbert space cannot affect any measurement that we might imagine performing, it has no physical consequence. Thus, we can, with impunity, set normalization constants however convenient, as we did with the energy eigenstates of the infinite square well.

Actually, this point places other constraints on the Hilbert space. All states in the Hilbert space are L^2-normalized, but further two states $|\psi\rangle$ and $|\psi'\rangle$ are physically identified if they only differ by an overall complex phase. If the Hilbert space $\mathcal{H}$ consists of all physically distinct states of our system, then we should not include both $|\psi\rangle$ and $|\psi'\rangle$ in the Hilbert space. That is, the Hilbert space is actually

$$\mathcal{H} = \left\{ |\psi\rangle \,|\, \langle \psi | \psi \rangle = 1 \,;\, |\psi'\rangle \simeq e^{i\phi} |\psi\rangle \,, \phi \in \mathbb{R} \right\} \,. \tag{5.29}$$

The symbol "$\simeq$" means "equivalent to" or "identified with" (not "equal to"). Mathematically, this defines a **conjugacy class** of states related by a phase. We only need one representative in the Hilbert space, and all other states that differ by a phase are mapped to it. Properly, the Hilbert space is thus not a vector space, but is actually a **ray space** because of this identification.

5.1.3 The Momentum Operator in the Energy Eigenbasis

With the eigenspace of the Hamiltonian of the infinite square well established, we have also identified the basis for the Hilbert space that is closed under time evolution. Additionally, this energy basis is convenient for expressing other operators on the well as explicit matrices. In this section, we'll study the momentum operator $\hat{p}$ in the energy eigenbasis, and you'll explore more in the exercises.

We would like to express the momentum operator as a matrix in the basis of energy eigenstates of the infinite square well. Equivalently, we would like to determine all matrix elements of momentum in this basis. So, to determine the element in the mth row and nth column of $\hat{p}$, we just sandwich the operator between the mth and nth-energy eigenstates:

$$(\hat{p})_{mn} = \langle \psi_m | \hat{p} | \psi_n \rangle = \int_0^a dx \sqrt{\frac{2}{a}} \sin\left(\frac{m\pi x}{a}\right) \left(-i\hbar \frac{d}{dx}\right) \sqrt{\frac{2}{a}} \sin\left(\frac{n\pi x}{a}\right) \qquad (5.30)$$

$$= -i\hbar \frac{2}{a} \frac{n\pi}{a} \int_0^a dx \sin\left(\frac{m\pi x}{a}\right) \cos\left(\frac{n\pi x}{a}\right) .$$

On the second line, we have just acted the derivative operator on the eigenstate in position space. To evaluate the integral of this mixed sine–cosine factor, we can re-express them as a sum or difference of complex exponentials. That is:

$$\sin\left(\frac{m\pi x}{a}\right) \cos\left(\frac{n\pi x}{a}\right) = \left(\frac{e^{i\frac{m\pi x}{a}} - e^{-i\frac{m\pi x}{a}}}{2i}\right) \left(\frac{e^{i\frac{n\pi x}{a}} + e^{-i\frac{n\pi x}{a}}}{2}\right) \qquad (5.31)$$

$$= \frac{1}{4i} \left(e^{i(m+n)\frac{\pi x}{a}} + e^{i(m-n)\frac{\pi x}{a}} - e^{-i(m-n)\frac{\pi x}{a}} - e^{-i(m+n)\frac{\pi x}{a}}\right)$$

$$= \frac{1}{2} \left[\sin\left((m+n)\frac{\pi x}{a}\right) + \sin\left((m-n)\frac{\pi x}{a}\right)\right] .$$

Of course, we could have used angle addition formulas for sine and cosine, but using the exponential form in the intermediate step eliminates the need to memorize those formulas.

So, our momentum matrix element is

$$(\hat{p})_{mn} = -i\frac{n\pi\hbar}{a^2} \int_0^a dx \left[\sin\left((m+n)\frac{\pi x}{a}\right) + \sin\left((m-n)\frac{\pi x}{a}\right)\right] \qquad (5.32)$$

$$= i\frac{n\hbar}{a} \left[\frac{1}{m+n} \cos\left((m+n)\frac{\pi x}{a}\right) + \frac{1}{m-n} \cos\left((m-n)\frac{\pi x}{a}\right)\right]_0^a$$

$$= i\frac{n\hbar}{a} \left[\frac{1}{m+n} \cos\left((m+n)\pi\right) + \frac{1}{m-n} \cos\left((m-n)\pi\right) - \frac{2m}{m^2 - n^2}\right] .$$

Now, we need to evaluate the remaining cosine factors. Note that as both m and n are integers, then so too are their sum and difference, so these terms are just cosine evaluated at multiples of π. We know that the result is just ± 1, depending on whether it is an even or odd multiple of π. Further, note that

$$\cos\left((m+n)\pi\right) = (-1)^{m+n} = (-1)^{m-n} = \cos\left((m-n)\pi\right) , \qquad (5.33)$$

so we finally have

$$(\hat{p})_{mn} = i\frac{n\hbar}{a} \left[\frac{1}{m+n} \cos\left((m+n)\pi\right) + \frac{1}{m-n} \cos\left((m-n)\pi\right) - \frac{2m}{m^2 - n^2}\right] \qquad (5.34)$$

$$= -2i\frac{\hbar}{a} \frac{mn}{m^2 - n^2} \left(1 - (-1)^{m+n}\right) .$$

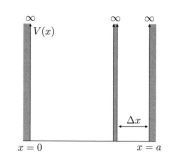

Fig. 5.4 A simple model of the Casimir effect with a thin hard wall placed a distance Δx from one side of the infinite square well potential.

With this expression for the matrix elements, we can explicitly write down the form of this matrix, which in the upper-left region is

$$\hat{p} = -2i\frac{\hbar}{a}\begin{pmatrix} 0 & -\frac{4}{3} & 0 & -\frac{8}{15} & 0 & -\frac{12}{35} & \cdots \\ \frac{4}{3} & 0 & -\frac{12}{5} & 0 & -\frac{20}{21} & 0 & \cdots \\ 0 & \frac{12}{5} & 0 & -\frac{24}{7} & 0 & -\frac{4}{3} & \cdots \\ \frac{8}{15} & 0 & \frac{24}{7} & 0 & -\frac{40}{9} & 0 & \cdots \\ 0 & \frac{20}{21} & 0 & \frac{40}{9} & 0 & -\frac{60}{11} & \cdots \\ \frac{12}{35} & 0 & \frac{4}{3} & 0 & \frac{60}{11} & 0 & \cdots \\ \vdots & \vdots & \vdots & \vdots & \vdots & \vdots & \ddots \end{pmatrix}. \tag{5.35}$$

Interestingly, the momentum operator is not diagonal in the energy basis, and in fact all diagonal entries, where $m = n$, are 0. Of course, we know that once we square momentum it is proportional to the Hamiltonian, and so the square of the matrix with these elements *is* diagonal. You will show this explicitly in the exercises.

Before moving on, we will work through an example of the force between two parallel plates as modeled by the infinite square well.

Example 5.1 The **Casimir effect** is the existence of an attractive or repulsive force between two parallel, electrically neutral plates held a very small distance apart.[2] A complete analysis and prediction of this effect requires the quantum theory of electromagnetism, well beyond the scope of this book, but we can nevertheless understand some features of this effect from a simple model.

The picture of this model is shown in Fig. 5.4. Our universe consists of an infinite square well of width a and there is a thin, infinite potential wall a distance Δx from the rightmost hard wall. The thin wall is free to move horizontally. We will also assume

---

[2] H. B. G. Casimir and D. Polder, "The influence of retardation on the London–van der Waals forces," *Phys. Rev.* **73**, 360 (1948); H. B. G. Casimir, "On the attraction between two perfectly conducting plates," *Indag. Math.* **10**, 261–263 (1948).

that there is a particle of mass m in each of the two parts of the larger well. In this example, we will determine the lowest energy of the system and the force due to quantum mechanical effects that is applied on the sliding, thin wall.

Solution

The two wells separated by the wall are separate systems, so their total energy E is simply the sum of the energy in each individually. From the figure, the width of the left well is $a - \Delta x$, and the width of the right well is Δx. Therefore, the sum of their lowest energies can be found from inserting the appropriate width into the expression of Eq. (5.16), with energy level $n = 1$. We have

$$E = E_{\text{left}} + E_{\text{right}} = \frac{\pi^2\hbar^2}{2m(a-\Delta x)^2} + \frac{\pi^2\hbar^2}{2m\Delta x^2}. \tag{5.36}$$

This expression for the energy is a function of the width Δx of one of the wells. As Δx varies, the energy varies, or, correspondingly, the two wells exert a force F on the thin barrier wall. This force can be found by differentiating the energy with respect to Δx, as is familiar from the relationship between a conservative force and potential energy. This relationship is

$$F = \frac{dE_{\text{gs}}}{d\Delta x} = \frac{d}{d\Delta x}\left(\frac{\pi^2\hbar^2}{2m(a-\Delta x)^2} + \frac{\pi^2\hbar^2}{2m\Delta x^2}\right) \tag{5.37}$$
$$= \frac{\pi^2\hbar^2}{m}\left(\frac{1}{(a-\Delta x)^3} - \frac{1}{\Delta x^3}\right).$$

Here, note that the sign in the relationship between force and energy is different than what you might be used to; typically, there is a minus sign in the relationship between conservative force and potential energy. However, in our set-up of the two infinite square wells, Δx was the width of the rightmost well, and so increasing Δx means that the barrier moves left, toward negative positions. This expression demonstrates that the barrier experiences no force if it is at the middle of the well, where $\Delta x = a/2$. Away from the middle, the force is a restoring force, pushing the barrier toward the middle. For a small displacement $\epsilon = a/2 - \Delta x$ from the center of the well, this restoring force has a Taylor expansion at lowest order of

$$F = -\frac{192\pi^2\hbar^2}{ma^4}\epsilon + \cdots, \tag{5.38}$$

where the ellipsis denotes terms at higher orders in ϵ. This takes the form of Hooke's law, and for small displacements about the center of the well, the barrier will harmonically oscillate.

Now, let's plug in some numbers to determine the scale of forces exerted by the well. Let's consider the particles in the wells to be electrons, with mass $m \simeq 10^{-30}$ kg, and set the overall width to be $a = 1$ m. Then, the prefactor of the force expression has the value

$$\frac{\pi^2\hbar^2}{m} \approx 10^{-36}\ \text{N·m}^3, \tag{5.39}$$

where we have used $\hbar = 10^{-34}$ J·s just to get an order of magnitude estimate. To even exert a magnitude of force of a nanonewton, $F = 10^{-9}$ N, the barrier would have to be extremely close to one of the outer walls. Specifically, the distance Δx would have to be approximately

$$\Delta x \simeq \left(\frac{\pi^2 \hbar^2}{mF} \right)^{1/3} \approx 10^{-9} \text{ m}, \tag{5.40}$$

or about a nanometer. To make this estimate, we assumed that we could ignore the $1/(a - \Delta x)^3$ contribution to the force as it will be orders and orders of magnitude smaller than $1/\Delta x^3$.

5.2 Correspondence with Classical Mechanics

One interesting thing to note about the energy eigenvalues of the infinite square well is that energies are all non-zero. Because the potential energy in the well is 0, this means that the particle must have a non-zero kinetic energy. Quantum mechanically, the particle must be moving, while classically, of course, it could just sit there at rest. This minimal energy that a quantum mechanical particle can have is often called the **zero-point energy** or **ground-state energy** and for the infinite square well this zero-point energy is

$$E_{\min} = E_1 = \frac{\pi^2 \hbar^2}{2ma^2}. \tag{5.41}$$

In this section, we will explore where this minimal energy comes from and work to connect this quantum infinite square well to the corresponding classical system.

5.2.1 Quantum Uncertainty in the Well

We'll focus our discussion around interpreting the energy eigenstates through the Heisenberg uncertainty principle, where the standard deviations for position and momentum are constrained by

$$\sigma_x \sigma_p \geq \frac{\hbar}{2}. \tag{5.42}$$

To calculate these standard deviations, we first need to calculate the expectation values of $\hat{x}$ and $\hat{p}$ operators on the energy eigenstates. This is actually trivial. For the expectation value of $\hat{x}$, we have

$$\langle \psi_n | \hat{x} | \psi_n \rangle = \int_0^a dx\, x\, \frac{2}{a} \sin^2 \frac{n\pi x}{a}. \tag{5.43}$$

Note that the function $\sin^2 \frac{n\pi x}{a}$ is symmetric in the well about $x = a/2$. That is, we can change the integration variable to $y = a - x$ and the integral is identical:

$$\langle \psi_n | \hat{x} | \psi_n \rangle = \int_0^a dx \, x \frac{2}{a} \sin^2 \frac{n\pi x}{a} = \int_0^a dy \, (a-y) \frac{2}{a} \sin^2 \frac{n\pi (a-y)}{a} \tag{5.44}$$

$$= \int_0^a dy \, (a-y) \frac{2}{a} \sin^2 \frac{n\pi y}{a} = a - \langle \psi_n | \hat{x} | \psi_n \rangle .$$

This then requires that for any energy eigenstate $|\psi_n\rangle$, $\langle \psi_n | \hat{x} | \psi_n \rangle = a/2$, the center of the well.

This result is natural when we consider the expectation value of momentum, $\langle \psi_n | \hat{p} | \psi_n \rangle$. In constructing the energy eigenvalues, we had noted that an energy eigenvalue corresponds to a fixed magnitude of momentum. There is no preference for moving left or right in the well, so the momentum eigenvalue p_n and its opposite $-p_n$ occurs with the same probability in an energy eigenstate. If a particle moves right just as much as it moves left, then its net motion is 0 and so $\langle \psi_n | \hat{p} | \psi_n \rangle = 0$ in an energy eigenstate. Note that this physical intuition is consistent with the direct calculation of the momentum operator in the energy eigenstate basis, Eq. (5.35), in which the momentum matrix has 0 diagonal entries.

The more non-trivial things to calculate are the second moments, $\langle \psi_n | \hat{x}^2 | \psi_n \rangle$ and $\langle \psi_n | \hat{p}^2 | \psi_n \rangle$. For the second moment of position, we must calculate the integral

$$\langle \psi_n | \hat{x}^2 | \psi_n \rangle = \int_0^a dx \, x^2 \frac{2}{a} \sin^2 \frac{n\pi x}{a} . \tag{5.45}$$

While this can be evaluated through applications of integration-by-parts, we won't go through the details here. The answer is

$$\langle \psi_n | \hat{x}^2 | \psi_n \rangle = \frac{a^2}{3} \left(1 - \frac{3}{2} \frac{1}{n^2 \pi^2} \right) . \tag{5.46}$$

For the second moment of momentum, $\langle \psi_n | \hat{p}^2 | \psi_n \rangle$, we can use a trick. To calculate this, note that the eigenvalue equation for the Hamiltonian states that

$$\hat{H} | \psi_n \rangle = \frac{\hat{p}^2}{2m} | \psi_n \rangle = E_n | \psi_n \rangle , \tag{5.47}$$

or

$$\hat{p}^2 | \psi_n \rangle = 2mE_n | \psi_n \rangle = \frac{n^2 \pi^2 \hbar^2}{a^2} | \psi_n \rangle . \tag{5.48}$$

Then, the second moment of $\hat{p}$ on an energy eigenstate is

$$\langle \psi_n | \hat{p}^2 | \psi_n \rangle = \langle \psi_n | \frac{n^2 \pi^2 \hbar^2}{a^2} | \psi_n \rangle = \frac{n^2 \pi^2 \hbar^2}{a^2} , \tag{5.49}$$

because $\langle \psi_n | \psi_n \rangle = 1$.

With these results, the variances of position and momentum in the nth-energy eigenstate are

$$\sigma_x^2 = \langle \psi_n | \hat{x}^2 | \psi_n \rangle - \langle \psi_n | \hat{x} | \psi_n \rangle^2 = \frac{a^2}{12} \left(1 - \frac{6}{n^2 \pi^2} \right) , \tag{5.50}$$

$$\sigma_p^2 = \langle \psi_n | \hat{p}^2 | \psi_n \rangle - \langle \psi_n | \hat{p} | \psi_n \rangle^2 = \frac{n^2 \pi^2 \hbar^2}{a^2} .$$

Now, the product of variances is

$$\sigma_x^2 \sigma_p^2 = \frac{n^2 \pi^2 \hbar^2}{12} - \frac{\hbar^2}{2} = \frac{\hbar^2}{4} \left(\frac{n^2 \pi^2}{3} - 2 \right). \tag{5.51}$$

Note that the factor in parentheses is greater than 1 for all $n \in \mathbb{N}$ and so we see that the Heisenberg uncertainty relation indeed holds:

$$\sigma_x^2 \sigma_p^2 \geq \frac{\hbar^2}{4}. \tag{5.52}$$

This parenthetical multiplicative factor is minimized on the ground state, when $n = 1$. In that case we find

$$\frac{\pi^2}{3} - 2 = 1.289868\ldots. \tag{5.53}$$

With this calculation, one way to interpret the ground state is that it is the "most quantum" of the states of the infinite square well: it is the state of the well that gets as close as possible to saturating the Heisenberg uncertainty principle.

The opposite limit is also interesting. As $n \to \infty$, the product of variances becomes

$$\sigma_x^2 \sigma_p^2 \to \frac{n^2 \pi^2 \hbar^2}{12} = \frac{ma^2}{6} E_n. \tag{5.54}$$

Recall that m and a are just some constants for a given particle and well; only E_n here is changing. The Heisenberg uncertainty relation in this $n \to \infty$ limit is then

$$\sigma_x^2 \sigma_p^2 = \frac{ma^2}{6} E_n \geq \frac{\hbar^2}{4}, \tag{5.55}$$

or

$$E_n \geq \frac{3}{2} \frac{\hbar^2}{ma^2}. \tag{5.56}$$

For large n, $E_n \to \infty$, so this relation is trivially satisfied. In fact, for large energies and small $\hbar$, this is nothing more than stating that $E_n \geq 0$, which must be true because the particle has only kinetic energy.

5.2.2 Classical Uncertainty in the Well

Let's belabor this point a bit and attempt to understand what this $E \to \infty$ ($\hbar \to 0$) limit means. To do this, let's figure out what the classical probability distributions for a particle in a box would be. Classically, an "infinite square well" just means a box with perfectly elastic walls: a ball hits the wall and only its direction of velocity is changed, not its magnitude. So, this perfect elasticity means that such a ball has a fixed magnitude of momentum, exactly as it did in the quantum case. So, if a ball has momentum magnitude $|p_0| > 0$ in the box, if you wait long enough, it will have bounced back and forth many times. Between each hit of the wall, it is traveling at constant speed, so its probability distribution of position is uniform over the box. That is, you are equally likely to find the ball anywhere in the box when you open it.

These data are sufficient to define the classical probability distributions of position and momentum, $p(x)$ and $p(p)$. Uniform in x means that $p(x)$ is independent of x and must integrate to 1 over the box:

$$1 = \int_0^a dx\, p(x) = p(x) \int_0^a dx = a, \tag{5.57}$$

so that

$$p(x) = \frac{1}{a}. \tag{5.58}$$

For the momentum distribution, there is only non-zero probability at p_0 or $-p_0$; every other value of momentum has 0 probability. Further, the probability of finding the ball going left or right is equal: leftness and rightness are not special. So, we can express the probability distribution of momentum with δ-functions:

$$p(p) = \frac{1}{2}\delta(p - p_0) + \frac{1}{2}\delta(p + p_0). \tag{5.59}$$

Recall that $\delta(x) = 0$ if $x \neq 0$, thus $p(p) = 0$ for $p \neq p_0, -p_0$. Additionally, note that the coefficients of the δ-functions are both $1/2$, reflecting their equal probability. Finally, this distribution is normalized:

$$\int_{-\infty}^{\infty} dp \left(\frac{1}{2}\delta(p - p_0) + \frac{1}{2}\delta(p + p_0) \right) = \frac{1}{2} + \frac{1}{2} = 1, \tag{5.60}$$

as

$$\int_{-\infty}^{\infty} dx\, \delta(x) = 1. \tag{5.61}$$

With these classical distributions, we can then calculate their variances. For position, its expectation value and second moment are

$$\langle x \rangle_{\text{classical}} = \int_0^a dx \frac{x}{a} = \frac{a}{2}, \tag{5.62}$$

$$\langle x^2 \rangle_{\text{classical}} = \int_0^a dx \frac{x^2}{a} = \frac{a^2}{3}.$$

Then, the classical variance of position is

$$\sigma_{x,\text{classical}}^2 = \langle x^2 \rangle_{\text{classical}} - \langle x \rangle_{\text{classical}}^2 = \frac{a^2}{3} - \frac{a^2}{4} = \frac{a^2}{12}. \tag{5.63}$$

We had found the same result in the uniform probability distribution of Example 4.1. The classical expectation values for momentum are

$$\langle p \rangle_{\text{classical}} = \int_{-\infty}^{\infty} dp \frac{p}{2} [\delta(p - p_0) + \delta(p + p_0)] = \frac{p_0 - p_0}{2} = 0, \tag{5.64}$$

$$\langle p^2 \rangle_{\text{classical}} = \int_{-\infty}^{\infty} dp \frac{p^2}{2} [\delta(p - p_0) + \delta(p + p_0)] = p_0^2.$$

Then, the classical variance of momentum is

$$\sigma_{p,\text{classical}}^2 = \langle p^2 \rangle_{\text{classical}} - \langle p \rangle_{\text{classical}}^2 = p_0^2. \tag{5.65}$$

Putting these results together, the product of these classical variances is

$$\sigma_{x,\text{classical}}^2 \sigma_{p,\text{classical}}^2 = \frac{a^2}{12} p_0^2 = \frac{a^2 m}{6} \frac{p_0^2}{2m} = \frac{ma^2}{6} E, \qquad (5.66)$$

where we identify the particle's energy as $E = p_0^2/2m$. If you recall, this is exactly the same relationship we found quantum mechanically as the energy level $n \to \infty$. Precisely, then, for large quantum mechanical energies, their properties *correspond* to classical states. This correspondence principle is one of the mysteries of quantum mechanics but must be true if the universe, all of it, is fundamentally quantum mechanical.

Another, graphical, way to think about this is as follows: let's draw the absolute square of a few energy eigenstate wavefunctions as in Fig. 5.5. As n increases, the probability oscillates more and more rapidly over smaller and smaller distances. For the nth state, there are n humps in the position space probability density $\psi_n^*(x)\psi_n(x)$ over the well. Over any small distance Δx, we can calculate the probability that the particle lies between x and $x + \Delta x$. We have

$$\int_x^{x+\Delta x} dx\, p_n(x) = \int_x^{x+\Delta x} dx'\, \frac{2}{a} \sin^2 \frac{n\pi x'}{a} \qquad (5.67)$$

$$= \frac{\Delta x}{a} + \frac{\sin \frac{2n\pi x}{a} - \sin \frac{2n\pi(x+\Delta x)}{a}}{2n\pi}.$$

Let's take the $n \to \infty$ limit now. The term with the sine factors vanishes, because it is suppressed by $1/n$. Thus, we have

$$\lim_{n\to\infty} \int_x^{x+\Delta x} dx\, p_n(x) = \frac{\Delta x}{a}. \qquad (5.68)$$

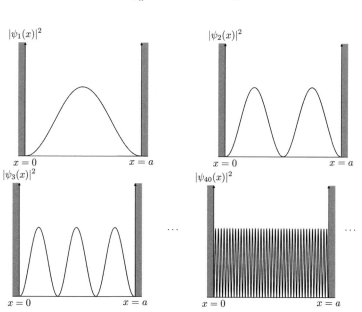

Fig. 5.5   Plots of the absolute square of a few of the energy eigenstate wavefunctions of the infinite square well.

This just corresponds to a probability distribution for position of $1/a$, exactly as we identified classically. That is, fast oscillations average out: as $n \to \infty$, we can just take the average of $\sin^2$:

$$\lim_{n\to\infty} p_n(x) = \lim_{n\to\infty} \frac{2}{a} \sin^2 \frac{n\pi x}{a} = \frac{1}{a}, \tag{5.69}$$

because $\sin^2 \frac{n\pi x}{a}$ oscillates increasingly rapidly between 0 and 1 as $n \to \infty$. Thus its average is $1/2$. So another way to state the correspondence principle is that at high energies, quantum fluctuations average out to classical behavior.

We will provide a more mathematically precise statement of the correspondence principle near the end of this book, but for now, a qualitative statement of the correspondence principle is:

> In the limit $\hbar \to 0$, the predictions of quantum mechanics reduce to those of classical mechanics.

Exercises

5.1   Consider a particle in a quantum state $|\psi\rangle$ of the infinite square well that is a linear combination of the two lowest-energy eigenstates:

$$|\psi\rangle \propto |\psi_1\rangle + 3|\psi_2\rangle. \tag{5.70}$$

In this problem, we will study features of this state.

(a)   In the definition of the state $|\psi\rangle$ presented above, I only wrote that it is proportional to this linear combination of energy eigenstates. By demanding that the state is normalized, $\langle\psi|\psi\rangle = 1$, determine the constant of proportionality to turn the relationship into an equality.

(b)   In the position basis, write the wavefunction $\psi(x) = \langle x|\psi\rangle$ for this normalized state. This is illustrated in the infinite square well in Fig. 5.6. What is the expectation value of position x on this state, $\langle\psi|\hat{x}|\psi\rangle$?

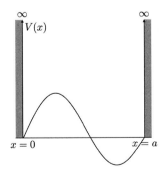

Fig. 5.6   Illustration of the wavefunction in the infinite square well defined from the state in Eq. (5.70).

(c)   What is the expectation value of the energy or Hamiltonian $\langle \psi | \hat{H} | \psi \rangle$ on this state?

(d)   Now, let's assume that the state $| \psi \rangle$ is the initial condition at time $t = 0$ for a state that evolves in time. Write the wavefunction at a general time t, $\psi(x,t)$. What is the expectation value of the position on this state at a general time? What about the expectation value of the energy or Hamiltonian? Do either of these expectation values change with time?

5.2   We discussed how the energy ground state of the infinite square well, $| \psi_1 \rangle$, was the state that minimized the uncertainty relation. That is, in this state, the product of the variance of the momentum and position, $\sigma_x^2 \sigma_p^2$, was minimized. Further, we demonstrated that for all energy eigenstates

$$\sigma_x^2 \sigma_p^2 \geq \frac{\hbar^2}{4}, \tag{5.71}$$

as required. However, consider the momentum eigenstate

$$\psi(x) = \frac{e^{i\frac{px}{\hbar}}}{\sqrt{a}}, \tag{5.72}$$

for some momentum p, in the infinite square well. Verify that it is indeed L^2-normalized on the well and calculate both variances σ_x^2 and σ_p^2. Is the uncertainty principle satisfied? Why or why not?

5.3   In this problem, we will work to understand the time evolution of wavefunctions that are localized in position in the infinite square well.

(a)   It will prove simplest later to re-write the infinite square well in a way that is symmetric for $x \to -x$. So, for an infinite square well in which the potential is 0 for $x \in [-\pi/2, \pi/2]$, find the energy eigenstates $\psi_n(x)$ and the corresponding energy eigenvalues E_n.

(b)   Now, let's consider the initial wavefunction $\psi(x)$ that is a uniform bump in the middle of the well:

$$\psi(x) = \begin{cases} \sqrt{\frac{2}{\pi}}, & |x| < \pi/4, \\ 0, & |x| > \pi/4. \end{cases} \tag{5.73}$$

From this initial wavefunction, determine the wavefunction at a general time t, $\psi(x,t)$.

(c)   Does this wavefunction "leak" into the regions where it was initially 0? Let's take the inner product of this time-dependent wavefunction with the wavefunction $\chi(x)$ that is uniform over the well:

$$\chi(x) = \frac{1}{\sqrt{\pi}}. \tag{5.74}$$

What is $\langle \chi | \psi \rangle$, as a function of time?

(d)   What is the first time derivative of the inner product at $t = 0$:

$$\frac{d\langle \chi | \psi \rangle}{dt} \bigg|_{t=0} ? \tag{5.75}$$

Can you think about what this means in the context of the Schrödinger equation?

(e) On the state $\psi(x,t)$, what are the expectation values of position and momentum for all time, $\langle \hat{x} \rangle$ and $\langle \hat{p} \rangle$?

5.4 A closely related system to the infinite square well is a particle on a ring. The Hamiltonian for a particle on a ring has 0 potential energy and the position on the ring can be represented by an angle $\phi \in [0, 2\pi)$. The wavefunction can then be expressed exclusively as a function of ϕ, $\psi(\phi)$, and it must be periodic: $\psi(\phi) = \psi(\phi + 2\pi)$.

(a) If the ring has radius a and the mass of the particle is m, write the Hamiltonian for this system in terms of the provided quantities. What is the momentum operator $\hat{p}$ on the ring and is it Hermitian?

(b) Determine the energy eigenstates of the Hamiltonian for the particle on the ring. What is the smallest allowed energy on the ring?

(c) Now, consider the uncertainty principle for the position and momentum operators on the ring. From our derivation in Chap. 4, we would expect that this uncertainty principle is

$$\sigma_\phi \sigma_p \geq \frac{\hbar}{2}.$$ (5.76)

Is this satisfied for every energy eigenstate on the ring? If not, can you identify what in our derivation of the uncertainty principle in Chap. 4 fails or does not apply for the particle on the ring?

5.5 We like to think of the $E \to \infty$ limit as the limit in which quantum mechanics "turns into" classical mechanics, but this clearly has limitations. The limitation that we will consider here is the fact that if the energy density of a particle's wavefunction is too high, then it will create a black hole. For a total energy E, a black hole is created if this is packed into a region smaller than its *Schwarzschild radius*, R_s.[3] For energy E, the Schwarzschild radius is

$$R_s = \frac{2G_N E}{c^4},$$ (5.77)

where G_N is Newton's constant and c is the speed of light.

(a) For a general energy eigenstate $|\psi_n\rangle$ of the infinite square well, determine its Schwarzschild radius. For what value of energy level n does the Schwarzschild radius equal the size a of the infinite square well? In this part, you can leave the answer in terms of the constants provided in the problem.

(b) What is the energy E_n for which the size of the well is the Schwarzschild radius? Evaluate this for a well that's the size of an atomic nucleus, $a = 10^{-15}$ m. Compare this energy to some "everyday" object's energy (something like the kinetic energy of a thrown ball, the energy of photons from the sun, etc.).

---

[3] K. Schwarzschild, "On the gravitational field of a mass point according to Einstein's theory," *Sitzungsber. Preuss. Akad. Wiss. Berlin (Math. Phys.)* **1916**, 189–196 (1916) [arXiv:physics/9905030 [physics]].

(c) Using part (a), what energy eigenstate level n does this correspond to? Take the mass m of the object to be that of the pion, a sub-atomic particle responsible for binding atomic nuclei. The mass of the pion m_π is

$$m_\pi = 2.4 \times 10^{-28} \text{ kg}. \qquad (5.78)$$

How does this compare to the energy level n where you would predict that the pion would be traveling at the speed of light, c? You can use the approximation that $\hbar = 10^{-34}$ J·s.

5.6 In Sec. 5.1.3, we explicitly constructed the momentum operator as a matrix in the energy eigenbasis. For the infinite square well, the square of the momentum operator is proportional to the Hamiltonian, so we should just be able to square this momentum matrix and immediately read off its eigenvalues as the energies of the infinite square well. This is also one of the few examples that is simple enough to explicitly demonstrate equivalence between the wave and matrix formulations of quantum mechanics.[4]

(a) Using the explicit form of the matrix element of the momentum operator in Eq. (5.34), express the matrix element of squared momentum operator $(\hat{p}^2)_{mn}$ as an infinite sum.
Solution:

$$(\hat{p}^2)_{mn} = -\frac{4\hbar^2 mn}{a^2} \sum_{l=1}^{\infty} \frac{l^2 \left(1 - (-1)^{m+l}\right)\left(1 - (-1)^{l+n}\right)}{(m^2 - l^2)(l^2 - n^2)}. \qquad (5.79)$$

(b) If the squared momentum matrix is to be proportional to the Hamiltonian, then in the energy eigenbasis it must be diagonal. Show that all off-diagonal matrix elements $(\hat{p}^2)_{mn}$ with $m \neq n$ are 0. For some values of row m and column n this is simple, given the expression in Eq. (5.79). What infinite sums must also vanish for $\hat{p}^2$ to be diagonal? Can you show this?

(c) Now, let's determine the energy eigenvalues. From these matrix elements, the nth diagonal entry of the Hamiltonian $\hat{H}$ should be

$$(\hat{H})_{nn} = -\frac{\hbar^2}{2m}(\hat{p}^2)_{nn}. \qquad (5.80)$$

Evaluating this directly for arbitrary values of n is a bit tricky, so instead we will satisfy ourselves with fixed values of n and then attempt to generalize. Let's consider just the first diagonal entry of the squared momentum operator, $(\hat{p}^2)_{11}$. What is the infinite sum that you must evaluate for this matrix element?

Hint: Can you partial fraction expand each term in the infinite series into a sum of simpler terms? You might also need the value of the infinite series

$$\sum_{k=1}^{\infty} \frac{1}{(2k-1)^2} = \frac{1}{1^2} + \frac{1}{3^2} + \frac{1}{5^2} + \cdots = \frac{\pi^2}{8}. \qquad (5.81)$$

---

[4] For more details about the history and mathematics of this problem, see J. Prentis and B. Ty, "Matrix mechanics of the infinite square well and the equivalence proofs of Schrödinger and von Neumann," *Am. J. Phys.* **82**, 583 (2014).

(d)   Can you evaluate the infinite sum that defines $(\hat{p}^2)_{nn}$ for arbitrary n? See the Prentis and Ty paper for details.

5.7   We had proved the canonical commutation relation $[\hat{x}, \hat{p}] = i\hbar$ in generality in the previous chapter. Of course, then, any matrix realization of the position and momentum operators must satisfy the commutation relation. In this chapter, we had constructed the momentum operator in the basis of energy eigenstates of the infinite square well, and we can verify the canonical commutation relation if we also have the position operator in the same basis.

(a)   Using a similar procedure as that for the momentum operator in Sec. 5.1.3, determine the matrix elements of the position operator $(\hat{x})_{mn}$ in the basis of energy eigenstates of the infinite square well.

(b)   The matrix form of the canonical commutation relation is $[\hat{x}, \hat{p}] = i\hbar\mathbb{I}$, where $\mathbb{I}$ is the appropriate identity matrix in the corresponding basis. Thus, the commutator has no non-zero elements off the diagonal. Simplify the expression for the off-diagonal elements of the commutator as much as possible. Using mathematical software, approximately evaluate the sum that you find for different values of row m and column n. Does the sum seem to vanish?

(c)   Now, simplify the expression for the diagonal elements of the commutator, $([\hat{x}, \hat{p}])_{nn}$, as much as possible. Again, using mathematical software, sum the first few (about 20) terms for a few values of n. Do you find that every diagonal entry of the commutator is indeed $i\hbar$?

(d)   Take the trace of the commutator as a matrix, $\mathrm{tr}[\hat{x}, \hat{p}]$; what do you find? How do you reconcile this with the result of part (c)?

5.8   The energy eigenstates of the infinite square well are of course not unique as a basis for the Hilbert space, which has no support at the boundaries of the well, where $x = 0$ and $x = a$. For example, the wavefunction

$$\zeta_1(x) = Nx(a-x) \tag{5.82}$$

vanishes at the boundaries of the well, but is clearly not the same as the sinusoidal ground-state wavefunction of the infinite square well. A comparison of this wavefunction with the wavefunction of the ground state of the infinite square well is shown in Fig. 5.7. In this problem, we will invert the usual problem, and identify the potential for which $\zeta_1(x)$ is the ground-state wavefunction.

(a)   First, normalize this wavefunction and determine the normalization constant N.

(b)   What is the overlap of $\zeta_1(x)$ with the ground-state wavefunction of the infinite square well? That is, what fraction of $\zeta_1(x)$ is described by the ground state of the infinite square well?

(c)   Now, let's assume that this wavefunction $\zeta_1(x)$ is an eigenstate of another Hamiltonian with potential $V(x)$:

$$\left(-\frac{\hbar^2}{2m}\frac{d^2}{dx^2} + V(x)\right)\zeta_1(x) = E_1\zeta_1(x), \tag{5.83}$$

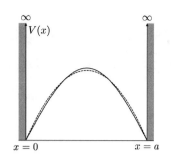

Fig. 5.7 Comparison between the ground-state wavefunction $\psi_1(x)$ of the infinite square well (solid) and the wavefunction $\zeta_1(x)$ (dashed) of Eq. (5.82).

for some ground-state energy E_1. Determine the potential $V(x)$ in terms of the energy E_1. How is this potential similar to and different from the infinite square well?

(d) Now, with this potential $V(x)$, can you guess the wavefunction of the first excited state? How much larger is its energy than the ground state? Can you guess higher excited states?

6

Quantum Mechanical Example: The Harmonic Oscillator

In this chapter, we will study the quantum **harmonic oscillator** which, just as in classical mechanics, will be our "canonical" quantum system and one which we can apply to a huge variety of systems. Unlike the infinite square well, we will find that the harmonic oscillator is *not* most naturally expressed in position space. This new formulation of the harmonic oscillator's Hamiltonian will demonstrate the power of exploiting commutation relations in quantum mechanics and will be a foundation for analyses in future chapters. Eigenstates of the harmonic oscillator, especially its lowest-energy state, have a deep connection to the Heisenberg uncertainty principle and open the door to studying a whole class of states that bridge the chasm between quantum wavefunctions and classical particles.

6.1 Representations of the Harmonic Oscillator Hamiltonian

Our first task is just to write the Hamiltonian down so as to be able to analyze the harmonic oscillator. From your introductory physics course, the harmonic oscillator's potential energy expressed as the position displacement from equilibrium is well known, so we will start there.

6.1.1 In Position Space

The potential energy of a spring or simple harmonic oscillator is

$$U(x) = \frac{1}{2}kx^2, \tag{6.1}$$

where k is the spring constant and x is the displacement from equilibrium. We can just directly "quantize" this to produce the corresponding quantum potential, but it will turn out to be convenient to re-write it slightly. Recall that the angular frequency ω of a spring is

$$\omega = \sqrt{\frac{k}{m}}, \tag{6.2}$$

or $k = m\omega^2$. We can then express the potential energy as

$$U(x) = \frac{1}{2}kx^2 = \frac{m\omega^2}{2}x^2. \tag{6.3}$$

This form is more general than that with the explicit spring constant because it represents a harmonic oscillator of any kind, and not just a spring. It is this form that we
will use in the Hamiltonian. The form of the parabolic harmonic oscillator potential
in position space is illustrated in Fig. 6.1.

Then, as an operator, the Hamiltonian of the harmonic oscillator is

$$\hat{H} = \frac{\hat{p}^2}{2m} + \frac{m\omega^2}{2}\hat{x}^2 . \tag{6.4}$$

Of course, we can represent $\hat{p}$ and $\hat{x}$ by their position space expressions, but that transforms the Hamiltonian eigenvalue problem into a differential equation that isn't so fun
to evaluate. Instead, we will stay in general operator land, only expressing a result in a
particular basis or representation at the last possible moment.

6.1.2 Factorization Into a Product Operator

With this strategy in mind, let's stare at this Hamiltonian and attempt to make sense
of it. The first thing we note about this Hamiltonian is that it is the sum of squares.
Whenever we see such a thing, we might first want to **factorize** it into a product of two
terms where each is linear in the two operators. If $\hat{p}$ and $\hat{x}$ were just simple numbers
(no hats), this factorization is of course easy:

$$\frac{p^2}{2m} + \frac{m\omega^2}{2}x^2 = \left(-\frac{ip}{\sqrt{2m}} + \sqrt{\frac{m\omega^2}{2}}x\right)\left(\frac{ip}{\sqrt{2m}} + \sqrt{\frac{m\omega^2}{2}}x\right) . \tag{6.5}$$

However, $\hat{x}$ and $\hat{p}$ do not commute as operators, so we have to be careful. Nevertheless,
let's just take this factorized product of the complex linear combination of $\hat{x}$ and $\hat{p}$ and
see what we find:

$$\left(-\frac{i\hat{p}}{\sqrt{2m}} + \sqrt{\frac{m\omega^2}{2}}\hat{x}\right)\left(\frac{i\hat{p}}{\sqrt{2m}} + \sqrt{\frac{m\omega^2}{2}}\hat{x}\right) = \frac{\hat{p}^2}{2m} + \frac{m\omega^2}{2}\hat{x}^2 + i\frac{\omega}{2}\hat{x}\hat{p} - i\frac{\omega}{2}\hat{p}\hat{x}$$

$$= \hat{H} + i\frac{\omega}{2}[\hat{x},\hat{p}] = \hat{H} + i\frac{\omega}{2}(i\hbar)$$

$$= \hat{H} - \frac{\hbar\omega}{2} . \tag{6.6}$$

Note the appearance of the commutator: if $\hat{x}$ and $\hat{p}$ were just numbers, the commutator
of course vanishes. So, we can factorize the Hamiltonian at the expense of adding one
remainder term:

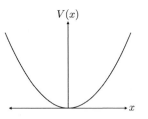

Fig. 6.1 An illustration of the harmonic oscillator potential.

$$\hat{H} = \frac{\hat{p}^2}{2m} + \frac{m\omega^2}{2}\hat{x}^2 = \left(-\frac{i\hat{p}}{\sqrt{2m}} + \sqrt{\frac{m\omega^2}{2}}\hat{x}\right)\left(\frac{i\hat{p}}{\sqrt{2m}} + \sqrt{\frac{m\omega^2}{2}}\hat{x}\right) + \frac{\hbar\omega}{2}. \qquad (6.7)$$

In what follows, it will be convenient to pull out the overall energy unit $\hbar\omega$ so that

$$\hat{H} = \hbar\omega\left[\left(-\frac{i\hat{p}}{\sqrt{2m\hbar\omega}} + \sqrt{\frac{m\omega}{2\hbar}}\hat{x}\right)\left(\frac{i\hat{p}}{\sqrt{2m\hbar\omega}} + \sqrt{\frac{m\omega}{2\hbar}}\hat{x}\right) + \frac{1}{2}\right]. \qquad (6.8)$$

Now everything in the square brackets is dimensionless, as all units are covered by $\hbar\omega$. Additionally, note that this indeed has units of energy because $\hbar$ has units of energy times time, and ω has units of (radians) per time.

In doing this factorization, we have isolated the operators $\hat{a}$ and $\hat{a}^\dagger$ that are the linear combinations of the momentum and position operators. We call

$$\hat{a} \equiv \frac{i\hat{p}}{\sqrt{2m\hbar\omega}} + \sqrt{\frac{m\omega}{2\hbar}}\hat{x}, \qquad\qquad \hat{a}^\dagger \equiv -\frac{i\hat{p}}{\sqrt{2m\hbar\omega}} + \sqrt{\frac{m\omega}{2\hbar}}\hat{x}. \qquad (6.9)$$

With these operators, the harmonic oscillator Hamiltonian is extremely compactly written as

$$\hat{H} = \hbar\omega\left(\hat{a}^\dagger\hat{a} + \frac{1}{2}\right). \qquad (6.10)$$

Note that $\hat{a}$ and $\hat{a}^\dagger$ are not Hermitian; indeed

$$(\hat{a})^\dagger = \hat{a}^\dagger, \qquad (6.11)$$

and vice-versa. Their product, however, is Hermitian:

$$(\hat{a}^\dagger\hat{a})^\dagger = (\hat{a})^\dagger(\hat{a}^\dagger)^\dagger = \hat{a}^\dagger\hat{a}, \qquad (6.12)$$

which of course is required as the Hamiltonian is Hermitian.

We can almost forget about the particular representation of $\hat{a}$ and $\hat{a}^\dagger$ in terms of $\hat{x}$ and $\hat{p}$, with one caveat: we need to know the commutation relation of $\hat{a}$ and $\hat{a}^\dagger$. Well, we can just evaluate it explicitly. Note that

$$\hat{a}^\dagger\hat{a} = \left(-\frac{i\hat{p}}{\sqrt{2m\hbar\omega}} + \sqrt{\frac{m\omega}{2\hbar}}\hat{x}\right)\left(\frac{i\hat{p}}{\sqrt{2m\hbar\omega}} + \sqrt{\frac{m\omega}{2\hbar}}\hat{x}\right) \qquad (6.13)$$

$$= \frac{\hat{p}^2}{2m\hbar\omega} + \frac{m\omega}{2\hbar}\hat{x}^2 + \frac{i}{2\hbar}[\hat{x},\hat{p}]$$

and

$$\hat{a}\hat{a}^\dagger = \left(\frac{i\hat{p}}{\sqrt{2m\hbar\omega}} + \sqrt{\frac{m\omega}{2\hbar}}\hat{x}\right)\left(-\frac{i\hat{p}}{\sqrt{2m\hbar\omega}} + \sqrt{\frac{m\omega}{2\hbar}}\hat{x}\right) \qquad (6.14)$$

$$= \frac{\hat{p}^2}{2m\hbar\omega} + \frac{m\omega}{2\hbar}\hat{x}^2 - \frac{i}{2\hbar}[\hat{x},\hat{p}].$$

Then, the commutation relation of $\hat{a}$ and $\hat{a}^\dagger$ is

$$[\hat{a}^\dagger,\hat{a}] = \hat{a}^\dagger\hat{a} - \hat{a}\hat{a}^\dagger = \frac{i}{\hbar}[\hat{x},\hat{p}] = \frac{i}{\hbar}(i\hbar) = -1, \qquad (6.15)$$

or

$$[\hat{a}, \hat{a}^\dagger] = 1 . \qquad (6.16)$$

In going forward, to find the energy eigenvalues of the harmonic oscillator Hamiltonian, we will just use

$$\hat{H} = \hbar\omega \left(\hat{a}^\dagger \hat{a} + \frac{1}{2} \right) , \quad \text{with} \quad (\hat{a})^\dagger = \hat{a}^\dagger \quad \text{and} \quad [\hat{a}, \hat{a}^\dagger] = 1 . \qquad (6.17)$$

This will demonstrate the power of the operator formalism of quantum mechanics.

6.2 Energy Eigenvalues of the Harmonic Oscillator

The first thing we note is that the eigenstates of the Hamiltonian are eigenstates of the operator $\hat{a}^\dagger \hat{a}$ individually. So let's see if we can understand this reduced operator's properties (i.e., its eigenvalues). As a Hermitian operator, the eigenvalues of $\hat{a}^\dagger \hat{a}$ are necessarily real. Can we learn anything about their sign?

6.2.1 What the Eigenvalues are Not

Let's first assume that $|\psi\rangle$ is an eigenstate of $\hat{a}^\dagger \hat{a}$ with eigenvalue λ. Further, let's assume that $\lambda < 0$. What are its consequences? As an eigenstate, $|\psi\rangle$ satisfies

$$\hat{a}^\dagger \hat{a} |\psi\rangle = \lambda |\psi\rangle . \qquad (6.18)$$

Let's now act with $\hat{a}$ on both sides of this equation. Then

$$\hat{a}\hat{a}^\dagger \hat{a} |\psi\rangle = \lambda \hat{a} |\psi\rangle = (\hat{a}\hat{a}^\dagger - \hat{a}^\dagger \hat{a} + \hat{a}^\dagger a)\hat{a} |\psi\rangle \qquad (6.19)$$
$$= (\hat{a}^\dagger \hat{a} + [\hat{a}, \hat{a}^\dagger])\hat{a} |\psi\rangle .$$

All we've done here is to add and subtract $\hat{a}^\dagger \hat{a}$ to the left side of the equality. Now, recall that the commutator is $[\hat{a}, \hat{a}^\dagger] = 1$. So we have

$$(\hat{a}^\dagger \hat{a} + 1)\hat{a} |\psi\rangle = \lambda \hat{a} |\psi\rangle , \qquad (6.20)$$

or, in a more suggestive way:

$$\hat{a}^\dagger \hat{a} (\hat{a} |\psi\rangle) = (\lambda - 1) (\hat{a} |\psi\rangle) . \qquad (6.21)$$

That is, if $|\psi\rangle$ is an eigenstate with eigenvalue $\lambda < 0$, then, necessarily, $\hat{a}^\dagger \hat{a}$ has another eigenstate $\hat{a} |\psi\rangle$ with eigenvalue $\lambda - 1 < \lambda < 0$. Now you might see the problem. By acting on $|\psi\rangle$ with $\hat{a}$, we can decrease its eigenvalue arbitrarily, unbounded from below. In particular, consider the state

$$\hat{a}^n |\psi\rangle = \underbrace{\hat{a}\hat{a} \cdots \hat{a}}_{n \text{ times}} |\psi\rangle . \qquad (6.22)$$

What is its eigenvalue?

We act with $\hat{a}^{\dagger}\hat{a}$ on it:

$$\hat{a}^{\dagger}\hat{a}\hat{a}^{n}|\psi\rangle = \hat{a}^{\dagger}\hat{a}^{n}\hat{a}|\psi\rangle. \tag{6.23}$$

Now, let's commute $\hat{a}^{\dagger}$ past $\hat{a}^{n}$. To evaluate $[\hat{a}^{\dagger},\hat{a}^{n}]$, we will use induction. Note that if $n = 1$, we had

$$[\hat{a}^{\dagger},\hat{a}] = -1. \tag{6.24}$$

Now consider $n = 2$:

$$[\hat{a}^{\dagger},\hat{a}^{2}] = \hat{a}^{\dagger}\hat{a}\hat{a} - \hat{a}\hat{a}\hat{a}^{\dagger} = \hat{a}\hat{a}^{\dagger}\hat{a} + [\hat{a}^{\dagger},\hat{a}]\hat{a} - \hat{a}\hat{a}\hat{a}^{\dagger} \tag{6.25}$$

$$= \hat{a}\hat{a}\hat{a}^{\dagger} + \hat{a}[\hat{a}^{\dagger},\hat{a}] + [\hat{a}^{\dagger},\hat{a}]\hat{a} - \hat{a}\hat{a}\hat{a}^{\dagger} = -2\hat{a}.$$

This suggests the general result

$$[\hat{a}^{\dagger},\hat{a}^{n}] = -n\hat{a}^{n-1}. \tag{6.26}$$

Let's prove it. Using induction, we note that it is true for $n = 1$. Assuming it is true for n, let's calculate $n + 1$:

$$[\hat{a}^{\dagger},\hat{a}^{n+1}] = \hat{a}^{\dagger}\hat{a}^{n+1} - \hat{a}^{n+1}\hat{a}^{\dagger} = \hat{a}^{\dagger}\hat{a}^{n}\hat{a} - \hat{a}^{n+1}\hat{a}^{\dagger} = \hat{a}^{n}\hat{a}^{\dagger}\hat{a} + [\hat{a}^{\dagger},\hat{a}^{n}]\hat{a} - \hat{a}^{n+1}\hat{a}^{\dagger}$$

$$= \hat{a}^{n}\hat{a}^{\dagger}\hat{a} - n\hat{a}^{n} - \hat{a}^{n+1}\hat{a}^{\dagger} = \hat{a}^{n+1}\hat{a}^{\dagger} + \hat{a}^{n}[\hat{a}^{\dagger},\hat{a}] - n\hat{a}^{n} - \hat{a}^{n+1}\hat{a}^{\dagger}$$

$$= -(n+1)\hat{a}^{n}, \tag{6.27}$$

exactly as assumed. Thus, connecting back to the eigenstates of $\hat{a}^{\dagger}\hat{a}$, we find

$$\hat{a}^{\dagger}\hat{a}\hat{a}^{n}|\psi\rangle = (\hat{a}^{n}\hat{a}^{\dagger}\hat{a} + [\hat{a}^{\dagger},\hat{a}^{n}]\hat{a})|\psi\rangle \tag{6.28}$$

$$= (\hat{a}^{n}\lambda - n\hat{a}^{n-1}\hat{a})|\psi\rangle = (\lambda - n)\hat{a}^{n}|\psi\rangle.$$

Thus we can decrease the eigenvalue as far as we want!

What is the problem with this? Well, the Hamiltonian encodes the energies of our system, and apparently if there is one eigenstate with negative energy, there are an infinite number of them! Further, these energies are unbounded from below; there is no smallest energy or no ground state. Thus, we must use some physics insight to proceed. A system with no minimum energy is very sick: it can lose an arbitrary amount of energy, and because it can and it is energetically favorable to do so, it does. To forbid this pathology, we must therefore enforce that there are no negative eigenvalues of $\hat{H}$ or $\hat{a}^{\dagger}\hat{a}$. This is very sensible. Recall that the Hamiltonian was the sum of squares of Hermitian operators:

$$\hat{H} = \frac{\hat{p}^{2}}{2m} + \frac{m\omega^{2}}{2}\hat{x}^{2}, \tag{6.29}$$

which would naturally seem to be exclusively non-negative, in a basis-independent sense.

So, with that constraint, we will look for eigenstates $|\psi\rangle$ with eigenvalue $\lambda \geq 0$ and

$$\hat{H}|\psi\rangle = \lambda|\psi\rangle = \hbar\omega\left(\hat{a}^{\dagger}\hat{a} + \frac{1}{2}\right)|\psi\rangle. \tag{6.30}$$

What are these eigenstates and eigenvalues?

6.2.2 What the Eigenvalues are

With the restriction that all eigenvalues of $\hat{a}^\dagger \hat{a}$ are non-negative, the smallest physical eigenvalue is 0; all other eigenvalues must be positive. So, let's just assume that $|\psi_0\rangle$ is the state for which its eigenvalue is 0 under action by $\hat{a}^\dagger \hat{a}$:

$$a^\dagger a |\psi_0\rangle = 0 |\psi_0\rangle = 0. \tag{6.31}$$

By the way, we say that a state with 0 eigenvalue of some operator is **annihilated** by that operator. "Annihilated" properly means taken out of the Hilbert space; in particular, 0 is not in the Hilbert space if for no other reason than it cannot have the correct, unitary normalization. Such a state is also an eigenstate of the Hamiltonian:

$$\hat{H}|\psi_0\rangle = \hbar\omega \left(\hat{a}^\dagger \hat{a} + \frac{1}{2} \right) |\psi_0\rangle = \frac{\hbar\omega}{2} |\psi_0\rangle. \tag{6.32}$$

As we have argued that $\hat{a}^\dagger \hat{a}$ can have no negative eigenvalues, this $|\psi_0\rangle$ corresponds to the smallest eigenvalue of $\hat{H}$. We therefore call it the ground state and the lowest energy of the Hamiltonian is

$$E_0 = \frac{\hbar\omega}{2}. \tag{6.33}$$

This is fascinating. Consider a classical harmonic oscillator; like a mass on a spring. It is definitely possible for the mass to be at rest (with 0 kinetic energy) at the spring's equilibrium point (with 0 potential energy). Thus such a classical system can have 0 energy. Quantum mechanically, however, this apparently can't be true: a quantum harmonic oscillator always has a non-zero energy, even in the ground state.

We had seen in the previous section how to go from one value of energy to another, using the action of the $\hat{a}$ and $\hat{a}^\dagger$ operators individually. Let's see if we can leverage this ground state into constructing states with larger eigenvalues. Let's consider the state formed by acting $\hat{a}^\dagger$ on $|\psi_0\rangle$, $\hat{a}^\dagger |\psi_0\rangle$. The action of the Hamiltonian on this state is

$$\hat{H}\hat{a}^\dagger |\psi_0\rangle = \hbar\omega \left(\hat{a}^\dagger \hat{a} + \frac{1}{2} \right) \hat{a}^\dagger |\psi_0\rangle = \hbar\omega \left(\hat{a}^\dagger \hat{a}^\dagger \hat{a} + \hat{a}^\dagger [\hat{a}, \hat{a}^\dagger] + \frac{1}{2}\hat{a}^\dagger \right) |\psi_0\rangle \tag{6.34}$$

$$= \hbar\omega \left(1 + \frac{1}{2} \right) \hat{a}^\dagger |\psi_0\rangle,$$

where we used that $\hat{a}|\psi_0\rangle = 0$. Then, apparently, the state $\hat{a}^\dagger |\psi_0\rangle$ is an eigenstate of the Hamiltonian with eigenvalue larger than the ground state by $\hbar\omega$.

This hints at a general construction of eigenvalues and eigenstates. Let's consider the state formed by acting on $|\psi_0\rangle$ with $\hat{a}^\dagger$ n times: $(\hat{a}^\dagger)^n |\psi_0\rangle$. The action of the Hamiltonian on such a state is

$$\hat{H}(\hat{a}^\dagger)^n |\psi_0\rangle = \hbar\omega \left(\hat{a}^\dagger \hat{a} + \frac{1}{2} \right) (\hat{a}^\dagger)^n |\psi_0\rangle \tag{6.35}$$

$$= \hbar\omega \left((\hat{a}^\dagger)^{n-1}\hat{a} + \hat{a}^\dagger [\hat{a}, (\hat{a}^\dagger)^n] + \frac{1}{2}(\hat{a}^\dagger)^n \right) |\psi_0\rangle$$

$$= \hbar\omega \left(\hat{a}^\dagger [\hat{a}, (\hat{a}^\dagger)^n] + \frac{1}{2}(\hat{a}^\dagger)^n \right) |\psi_0\rangle.$$

Last section, we had shown that the commutator of n $\hat{a}$s with one $\hat{a}^\dagger$ is

$$[\hat{a}^\dagger, \hat{a}^n] = -n\hat{a}^{n-1}.$$ (6.36)

Hermitian conjugating this, we have

$$[\hat{a}^\dagger, \hat{a}^n]^\dagger = -n(\hat{a}^\dagger)^{n-1} = [(\hat{a}^\dagger)^n, \hat{a}].$$ (6.37)

Note how the order of the commutator switches, as it is nothing more than the difference of the product of operators. So, we have

$$[\hat{a}, (\hat{a}^\dagger)^n] = n(\hat{a}^\dagger)^{n-1}.$$ (6.38)

Using this in our result above, we find

$$\hat{H}(\hat{a}^\dagger)^n|\psi_0\rangle = \hbar\omega\left((\hat{a}^\dagger)^n n + \frac{1}{2}(\hat{a}^\dagger)^n\right)|\psi_0\rangle = \hbar\omega\left(n + \frac{1}{2}\right)(\hat{a}^\dagger)^n|\psi_0\rangle.$$ (6.39)

So, through the action of $\hat{a}^\dagger$, we can construct eigenstates of the Hamiltonian that have eigenvalues

$$E_n = \hbar\omega\left(n + \frac{1}{2}\right),$$ (6.40)

for the nth application of $\hat{a}^\dagger$. We thus refer to the $\hat{a}$ and $\hat{a}^\dagger$ operators as **ladder operators**, as they can be used to go up and down the ladder of energy eigenstates of $\hat{H}$. The first five energy eigenvalues are illustrated as dashed lines in the harmonic oscillator potential in Fig. 6.2.

While we note that $(\hat{a}^\dagger)^n|\psi_0\rangle$ is an eigenstate of the Hamiltonian, it is only additionally in the Hilbert space if its norm is 1. Let's assume that $|\psi_0\rangle$ is normalized: $\langle\psi_0|\psi_0\rangle = 1$. What is the inner product of $(\hat{a}^\dagger)^n|\psi_0\rangle$ with itself? Note that its Hermitian conjugate is

$$\left((\hat{a}^\dagger)^n|\psi_0\rangle\right)^\dagger = \langle\psi_0|\hat{a}^n,$$ (6.41)

so the inner product is

$$\langle\psi_0|\hat{a}^n(\hat{a}^\dagger)^n|\psi_0\rangle = \langle\psi_0|\hat{a}^{n-1}(\hat{a}^\dagger)^n\hat{a} + \hat{a}^{n-1}[\hat{a}, (\hat{a}^\dagger)^n]|\psi_0\rangle$$ (6.42)
$$= \langle\psi_0|\hat{a}^{n-1}n(\hat{a}^\dagger)^{n-1}|\psi_0\rangle,$$

where we have used the commutator of $\hat{a}$ and $(\hat{a}^\dagger)^n$ as established earlier.

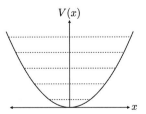

Fig. 6.2 Illustration of the spacing of the first few energy eigenvalues in the harmonic potential, denoted by the dashed horizontal lines.

We have the interesting recursion relation between the inner product of these states, where

$$\langle\psi_0|\hat{a}^n(\hat{a}^\dagger)^n|\psi_0\rangle = n\langle\psi_0|\hat{a}^{n-1}(\hat{a}^\dagger)^{n-1}|\psi_0\rangle \tag{6.43}$$

$$= n(n-1)\langle\psi_0|\hat{a}^{n-2}(\hat{a}^\dagger)^{n-2}|\psi_0\rangle$$

$$\vdots$$

$$= n!\langle\psi_0|\psi_0\rangle = n!\,.$$

So, the state $(\hat{a}^\dagger)^n|\psi_0\rangle$ is in general not in the Hilbert space. However, the fix is easy: we just need to include an appropriate normalization factor. We call

$$|\psi_n\rangle \equiv \frac{(\hat{a}^\dagger)^n}{\sqrt{n!}}|\psi_0\rangle \tag{6.44}$$

the nth-energy eigenstate of the harmonic oscillator Hamiltonian. This state is normalized:

$$\langle\psi_n|\psi_n\rangle = \frac{\langle\psi_0|\hat{a}^n(\hat{a}^\dagger)^n|\psi_0\rangle}{n!} = 1 \tag{6.45}$$

and satisfies the eigenvalue equation

$$\hat{H}|\psi_n\rangle = \hbar\omega\left(\hat{a}^\dagger\hat{a} + \frac{1}{2}\right)(\hat{a}^\dagger)^n|\psi_0\rangle = \left(n + \frac{1}{2}\right)\hbar\omega|\psi_n\rangle\,. \tag{6.46}$$

6.2.3 Orthonormality and Completeness of Energy Eigenstates

Simply from properties of the operators $\hat{a}$ and $\hat{a}^\dagger$, we have constructed an infinite tower of energy eigenstates and their eigenvalues. We can verify that these eigenstates of the Hamiltonian are also orthonormal and form a complete basis for the Hilbert space of the harmonic oscillator. For orthonormality, consider the two energy eigenstates $|\psi_n\rangle$ and $|\psi_m\rangle$, with $n \neq m$. Then, their inner product is

$$\langle\psi_n|\psi_m\rangle = \langle\psi_0|\frac{\hat{a}^n(\hat{a}^\dagger)^m}{\sqrt{n!m!}}|\psi_0\rangle\,. \tag{6.47}$$

To evaluate this, we'll use the property that we derived in Sec. 4.3.3 for the relationship between operators whose commutator is a constant. Because $[\hat{a},\hat{a}^\dagger] = 1$, in the $\hat{a}^\dagger$ basis we can express $\hat{a}$ as

$$\hat{a} = \frac{d}{d\hat{a}^\dagger}\,. \tag{6.48}$$

Thus, we can express the inner product of two distinct energy eigenstates as

$$\langle\psi_n|\psi_m\rangle = \frac{1}{\sqrt{n!m!}}\langle\psi_0|\frac{d^n}{d(\hat{a}^\dagger)^n}(\hat{a}^\dagger)^m|\psi_0\rangle\,. \tag{6.49}$$

Now, there are two possibilities if $n \neq m$: either $n > m$ or $n < m$. If $n > m$, there are more derivatives than powers of $\hat{a}^\dagger$ and so

$$\frac{d^n}{d(\hat{a}^\dagger)^n} (\hat{a}^\dagger)^m \bigg|_{n>m} = 0, \tag{6.50}$$

and thus the two energy eigenstates are orthogonal: $\langle \psi_n | \psi_m \rangle = 0$. On the other hand, if $n < m$ then

$$\frac{d^n}{d(\hat{a}^\dagger)^n} (\hat{a}^\dagger)^m \bigg|_{n<m} = n! (\hat{a}^\dagger)^{m-n}, \tag{6.51}$$

and so their inner product is

$$\langle \psi_n | \psi_m \rangle = \sqrt{\frac{n!}{m!}} \langle \psi_0 | (\hat{a}^\dagger)^{m-n} | \psi_0 \rangle. \tag{6.52}$$

Now, by construction, $\hat{a} | \psi_0 \rangle = 0$ and so its Hermitian conjugate is $\langle \psi_0 | \hat{a}^\dagger = 0$ and so the case in which $n < m$ also vanishes: $\langle \psi_n | \psi_m \rangle = 0$. Thus, we have constructed orthonormal eigenstates of the Hamiltonian.

Completeness of the energy eigenstates, as usual, is more subtle to analyze. Our standard definition of completeness is that the outer product of the energy eigenstates should reproduce the identity operator on the Hilbert space:

$$\sum_{n=0}^\infty |\psi_n\rangle \langle \psi_n| = \sum_{n=0}^\infty \frac{1}{n!} (\hat{a}^\dagger)^n |\psi_0\rangle \langle \psi_0| \hat{a}^n = \mathbb{I}. \tag{6.53}$$

Proving this relationship just using properties of the ladder operators is challenging, so we'll take a different route. Completeness is equivalently stated in the following way. Consider a general state $|\psi\rangle$ on the Hilbert space $\mathcal{H}$ of the harmonic oscillator. Completeness of the energy eigenstates means that *any* state in the Hilbert space can be expressed as a linear combination of them:

$$|\psi\rangle = \sum_{n=0}^\infty \beta_n |\psi_n\rangle = \sum_{n=0}^\infty \frac{\beta_n}{\sqrt{n!}} (\hat{a}^\dagger)^n |\psi_0\rangle, \tag{6.54}$$

for some complex coefficients β_n such that

$$\sum_{n=0}^\infty |\beta_n|^2 = 1. \tag{6.55}$$

Further, note that the state $|\psi_0\rangle$ is independent of $\hat{a}^\dagger$, and so the sum over n can be done independently of its action on the ground state:

$$|\psi\rangle = \left(\sum_{n=0}^\infty \frac{\beta_n}{\sqrt{n!}} (\hat{a}^\dagger)^n \right) |\psi_0\rangle = f(\hat{a}^\dagger) |\psi_0\rangle, \tag{6.56}$$

where $f(\hat{a}^\dagger)$ is some function of the operator $\hat{a}^\dagger$. $f(\hat{a}^\dagger)$ is therefore an operator itself, and it maps the energy ground state $|\psi_0\rangle$ to another state $|\psi\rangle$ in the Hilbert space. Just like the exponentiation of the derivative operator we studied at the very beginning of this book, this function is defined through its Taylor expansion:

$$f(\hat{a}^\dagger) = f(0) + \hat{a}^\dagger \left. \frac{df(\hat{a}^\dagger)}{d\hat{a}^\dagger} \right|_{\hat{a}^\dagger=0} + \frac{(\hat{a}^\dagger)^2}{2} \left. \frac{d^2 f(\hat{a}^\dagger)}{d(\hat{a}^\dagger)^2} \right|_{\hat{a}^\dagger=0} + \cdots \qquad (6.57)$$

$$= \sum_{n=0}^{\infty} \frac{(\hat{a}^\dagger)^n}{n!} \left. \frac{d^n}{d(\hat{a}^\dagger)^n} f(\hat{a}^\dagger) \right|_{\hat{a}^\dagger=0}.$$

The values of the derivatives of $f(\hat{a}^\dagger)$ are determined by the coefficients β_n by matching terms at the same order in $(\hat{a}^\dagger)^n$. This Taylor expansion assumes the existence of all derivatives of $f(\hat{a}^\dagger)$ near $\hat{a}^\dagger = 0$, and such a function is called **analytic**. Therefore, the Hilbert space of the harmonic oscillator consists of all normalizable states $|\psi\rangle$ that can be accessed from the ground state $|\psi_0\rangle$ by an operator $f(\hat{a}^\dagger)$ that is analytic in $\hat{a}^\dagger$ about $\hat{a}^\dagger = 0$.

Before constructing the energy eigenstate wavefunction in position space, we pause to consider an example, relating the raising and lowering operators to number and phase operators.

Example 6.1   At its core, the Hamiltonian of the harmonic oscillator involves the operator that is the product of the raising and lowering operators, $\hat{a}^\dagger \hat{a}$. We had shown that the physically sensible eigenvalues of this operator are just non-negative integers, n, and so this operator is also called the **number operator** as it counts the number of the energy eigenstate that is occupied. The number operator is often denoted as $\hat{N} = \hat{a}^\dagger \hat{a}$. Further, up to non-trivial commutation relations, $\hat{a}^\dagger \hat{a}$ is like the squared magnitude of the operator $\hat{a}$ and so this suggests another representation of the creation and annihilation operators in analogy with complex numbers. If we define a Hermitian phase operator $\hat{\Theta}$, then we can write

$$\hat{a}^\dagger = \sqrt{\hat{N}} e^{-i\hat{\Theta}}, \qquad\qquad \hat{a} = e^{i\hat{\Theta}} \sqrt{\hat{N}}. \qquad (6.58)$$

Note that the square root of the number operator is well-defined as all of its eigenvalues are non-negative. We'll study properties of this representation of the creation and annihilation operators here, and further in Exercise 6.6.

Solution

First, let's verify that indeed $\hat{N} = \hat{a}^\dagger \hat{a}$. To do this, let's just multiply the expressions for the raising and lowering operators together:

$$\hat{a}^\dagger \hat{a} = \left(\sqrt{\hat{N}} e^{-i\hat{\Theta}} \right) \left(e^{i\hat{\Theta}} \sqrt{\hat{N}} \right) = \sqrt{\hat{N}} \left(e^{-i\hat{\Theta}} e^{i\hat{\Theta}} \right) \sqrt{\hat{N}} = \hat{N}. \qquad (6.59)$$

In this series of products, we used the associativity of linear operators to multiply the exponential phase operators together. $\hat{\Theta}$ commutes with itself, so these exponential factors just cancel, leaving the number operator $\hat{N}$ alone.

Now, let's define the angle $\theta \in [0, 2\pi)$ to be an eigenvalue of the phase operator $\hat{\Theta}$ and $|\theta\rangle$ be its corresponding eigenstate. We want to express the state $|\theta\rangle$ in terms of energy eigenstates of the harmonic oscillator. Equivalently, we want to express the state $|\theta\rangle$ as a linear combination of the number operator's eigenstates $|n\rangle$, for $n = 0, 1, 2, \ldots$. Without loss of generality, we can write

$$|\theta\rangle = \sum_{n=0}^{\infty} \beta_n |n\rangle\,, \tag{6.60}$$

for some complex coefficients β_n. As an eigenstate of the phase operator $\hat{\Theta}$, it satisfies

$$\hat{\Theta}|\theta\rangle = \theta|\theta\rangle\,. \tag{6.61}$$

Further, note that this state is also an eigenstate of the exponentiated unitary operator

$$e^{i\hat{\Theta}}|\theta\rangle = e^{i\theta}|\theta\rangle\,. \tag{6.62}$$

So, if we had a way to construct the exponential unitary phase operator, then we could act it on the number operator eigenstate basis representation of the state $|\theta\rangle$.

To do this, we will use some trickery. Let's assume that the number operator $\hat{N}$ is actually invertible and say $\hat{N}^{-1}$ exists, where

$$\hat{N}\hat{N}^{-1} = \mathbb{I}\,, \tag{6.63}$$

the identity operator. Assuming this exists, we can relate the unitary phase operator to the number operator and the lowering operator as

$$e^{i\hat{\Theta}} = \hat{a}\,\frac{1}{\sqrt{\hat{N}}}\,. \tag{6.64}$$

This clearly makes no sense because the number operator $\hat{N}$ has a 0 eigenvalue. But, let's see how long we can ignore this subtlety and just formally keep going in the hopes that all terms of the form $1/0$ disappear by the end of our calculation.

From the analysis of the eigenstates of the harmonic oscillator, we had established that the action of the lowering operator $\hat{a}$ on the eigenstate $|n\rangle$ is as follows. From Eq. (6.44), we have the relationship

$$|n\rangle = \frac{(\hat{a}^{\dagger})^n}{\sqrt{n!}}|0\rangle\,. \tag{6.65}$$

The action of $\hat{a}$ on the state $|n\rangle$ is then

$$\hat{a}|n\rangle = \frac{\hat{a}(\hat{a}^{\dagger})^n}{\sqrt{n!}}|0\rangle = \frac{(\hat{a}^{\dagger})^n\hat{a} + [\hat{a}, (\hat{a}^{\dagger})^n]}{\sqrt{n!}}|0\rangle = \frac{n(\hat{a}^{\dagger})^{n-1}}{\sqrt{n!}}|0\rangle \tag{6.66}$$
$$= \sqrt{n}\,|n-1\rangle\,.$$

In this expression, we used the commutation relation of Eq. (6.38).

With this result, we can then act the unitary exponential phase operator on the eigenstate $|\theta\rangle$. First we have, by definition:

$$e^{i\hat{\Theta}}|\theta\rangle = e^{i\theta}|\theta\rangle = \sum_{n=0}^{\infty} \beta_n e^{i\theta}|n\rangle\,. \tag{6.67}$$

Next, we can use the lowering/number operator representation, where

$$e^{i\hat{\Theta}}|\theta\rangle = \sum_{n=0}^{\infty} \beta_n \hat{a}\,\frac{1}{\sqrt{\hat{N}}}|n\rangle = \sum_{n=0}^{\infty} \frac{\beta_n}{\sqrt{n}}\hat{a}|n\rangle = \sum_{n=0}^{\infty} \beta_n|n-1\rangle\,. \tag{6.68}$$

Then, by the orthogonality of distinct number states $\langle n|n'\rangle = \delta_{nn'}$, we just set the coefficients of each state equal, according to the two representations of the unitary phase operator. That is, we require

$$\beta_{n+1} = \beta_n e^{i\theta} \,. \tag{6.69}$$

For some initial coefficient β_0, the solution of this recursive equation is

$$\beta_n = e^{in\theta}\beta_0 \,. \tag{6.70}$$

Then, the eigenstate of the phase operator in the number operator basis is

$$|\theta\rangle = \beta_0 \sum_{n=0}^{\infty} e^{in\theta} |n\rangle \,. \tag{6.71}$$

We will leave the normalization β_0 undetermined.

6.2.4 The Ground State in Position Space

As every state in the Hilbert space is related to the ground state by repeated action of the raising operator $\hat{a}^\dagger$, it is worthwhile to identify what the ground state actually looks like in a convenient basis. The ground state $|\psi_0\rangle$ is annihilated by the lowering operator $\hat{a}$:

$$\hat{a}|\psi_0\rangle = \left(i\frac{\hat{p}}{\sqrt{2m\hbar\omega}} + \frac{m\omega}{2\hbar}\hat{x} \right)|\psi_0\rangle = 0 \,. \tag{6.72}$$

Acting with $\langle x|$ expresses this eigenvalue equation in the position basis as a differential equation for the ground state wavefunction:

$$\langle x| \left(i\frac{\hat{p}}{\sqrt{2m\hbar\omega}} + \frac{m\omega}{2\hbar}\hat{x} \right)|\psi_0\rangle = \left(\sqrt{\frac{\hbar}{2m\omega}}\frac{d}{dx} + \sqrt{\frac{m\omega}{2\hbar}}x \right)\psi_0(x) = 0 \,, \tag{6.73}$$

where $\psi_0(x) = \langle x|\psi_0\rangle$ is the ground-state wavefunction. Rearranging this differential equation, we have

$$\frac{d\psi_0(x)}{dx} = -\frac{m\omega}{\hbar}x\psi_0(x) \,, \tag{6.74}$$

which is an ordinary, linear, homogeneous differential equation that can easily be solved. We find

$$\psi_0(x) = Ne^{-\frac{m\omega}{2\hbar}x^2} \,, \tag{6.75}$$

where N is a normalization constant. This is fascinating: this function is the shape of a bell curve, normal distribution, or **Gaussian function**. A plot of this ground-state wavefunction is illustrated in Fig. 6.3. In the following section, we will see what makes this Gaussian so special and why the energy of a particle in the quantum harmonic oscillator is at least $\hbar\omega/2$. In the following example, we will construct the wavefunction for the first excited state, using the raising operator.

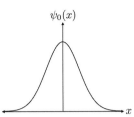

Fig. 6.3   Plot of the ground-state wavefunction of the harmonic oscillator potential, which takes the shape of a Gaussian function in position space.

Example 6.2   With the expression for the ground-state wavefunction, let's keep going and construct the wavefunction of the first excited state, $\psi_1(x)$, of the harmonic oscillator. Recall that the relationship between the ground state and the first excited state from Eq. (6.44) is

$$|\psi_1\rangle = \hat{a}^\dagger |\psi_0\rangle. \tag{6.76}$$

Solution

To determine the wavefunction, we just project this onto position space. The raising operator in position space is

$$\hat{a}^\dagger = -\sqrt{\frac{\hbar}{2m\omega}}\frac{d}{dx} + \sqrt{\frac{m\omega}{2\hbar}}x, \tag{6.77}$$

where we have just inserted the expression for the momentum operator in position space into Eq. (6.9). So, all we need to do to determine the first excited-state wavefunction is to take a derivative of the ground state. We then find

$$\psi_1(x) = \hat{a}^\dagger \psi_0(x) = \left(-\sqrt{\frac{\hbar}{2m\omega}}\frac{d}{dx} + \sqrt{\frac{m\omega}{2\hbar}}x\right) N e^{-\frac{m\omega}{2\hbar}x^2} \tag{6.78}$$

$$= N\sqrt{\frac{2m\omega}{\hbar}}x e^{-\frac{m\omega}{2\hbar}x^2}.$$

A plot of the functional form of this wavefunction is illustrated in Fig. 6.4. The presence of the normalization factor N from the ground state ensures that this excited state is also L^2-normalized. Further, this wavefunction is orthogonal to the ground-state wavefunction, where

$$0 = \int_{-\infty}^{\infty} dx\, \psi_1(x)\psi_0(x) = N^2 \sqrt{\frac{2m\omega}{\hbar}} \int_{-\infty}^{\infty} dx\, x\, e^{-\frac{m\omega}{\hbar}x^2}, \tag{6.79}$$

because the integrand is an odd function of x.

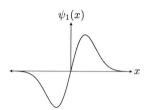

Fig. 6.4   Plot of the wavefunction for the first excited state of the harmonic oscillator, $\psi_1(x)$.

6.3 Uncertainty and the Ground State

As we observed in the infinite square well, the ground state of the harmonic oscillator has a non-zero energy. Why? Why is 0 energy not possible quantum mechanically while it is classically? Let's consider what $E = 0$ would mean. First, the harmonic oscillator potential is illustrated in Fig. 6.1. There is only one point on the potential where its energy is 0: at $x = 0$. All other points have greater energy, so if we say that the total energy is 0, we must require $x = 0$. Further, the kinetic energy is a square of momentum so it is necessarily non-negative: $K = p^2/2m \geq 0$. However, for $K = 0$, we must enforce that $p = 0$ as well. Thus, the only possible way for $E = 0$ is if both $x = 0$ and $p = 0$.

If this is true, then note that such a state has zero variance in both position and momentum: $\sigma_x^2 = \sigma_p^2 = 0$. However, now I think you see the problem. Such a state violates the very general bound we had established, the Heisenberg uncertainty principle. Because position and momentum do not commute, it is not possible for a state to have both $x = 0$ and $p = 0$ as then it would simultaneously be in an eigenstate of position and momentum. Thus, a zero energy state of the harmonic oscillator cannot exist.

Okay, but what is the ground state doing? Why is it what it is? Let's calculate the momentum and position variances in the ground state $|\psi_0\rangle$ and see what its corresponding product is. The first thing we need to do is to express operators $\hat{x}$ and $\hat{p}$ in terms of the ladder operators $\hat{a}$ and $\hat{a}^\dagger$. Inverting the relationships we have from Eq. (6.9), we find

$$\hat{x} = \sqrt{\frac{\hbar}{2m\omega}}(\hat{a} + \hat{a}^\dagger), \qquad\qquad \hat{p} = i\sqrt{\frac{m\hbar\omega}{2}}(\hat{a}^\dagger - \hat{a}). \qquad (6.80)$$

Starting with σ_x^2, we have

$$\sigma_x^2 = \langle\psi_0|\hat{x}^2|\psi_0\rangle - \langle\psi_0|\hat{x}|\psi_0\rangle^2. \qquad (6.81)$$

The expectation value of position is 0:

$$\langle\psi_0|\hat{x}|\psi_0\rangle = \sqrt{\frac{\hbar}{2m\omega}}\langle\psi_0|\hat{a} + \hat{a}^\dagger|\psi_0\rangle, \qquad (6.82)$$

because $\hat{a}|\psi_0\rangle = \langle\psi_0|\hat{a}^\dagger = 0$. This annihilation property can be used to simplify the expectation value of $\hat{x}^2$:

$$\langle \psi_0|\hat{x}^2|\psi_0\rangle = \frac{\hbar}{2m\omega}\langle \psi_0|(\hat{a}+\hat{a}^\dagger)^2|\psi_0\rangle \tag{6.83}$$

$$= \frac{\hbar}{2m\omega}\langle \psi_0|\hat{a}^2+(\hat{a}^\dagger)^2+\hat{a}^\dagger\hat{a}+\hat{a}\hat{a}^\dagger|\psi_0\rangle = \frac{\hbar}{2m\omega}\langle \psi_0|\hat{a}\hat{a}^\dagger|\psi_0\rangle$$

$$= \frac{\hbar}{2m\omega}\langle \psi_0|\hat{a}^\dagger\hat{a}+[\hat{a},\hat{a}^\dagger]|\psi_0\rangle = \frac{\hbar}{2m\omega},$$

using the commutation relation $[\hat{a},\hat{a}^\dagger]=1$ and the normalization of $|\psi_0\rangle$: $\langle \psi_0|\psi_0\rangle=1$.

Now let's do the same thing for momentum, $\hat{p}$. The expectation value of $\hat{p}$ alone vanishes for the same reason that the expectation value of $\hat{x}$ did. For $\hat{p}^2$, its expectation value is

$$\langle \psi_0|\hat{p}^2|\psi_0\rangle = -\frac{m\hbar\omega}{2}\langle \psi_0|(\hat{a}^\dagger-\hat{a})^2|\psi_0\rangle \tag{6.84}$$

$$= -\frac{m\hbar\omega}{2}\langle \psi_0|(\hat{a}^\dagger)^2+\hat{a}^2-\hat{a}^\dagger\hat{a}-\hat{a}\hat{a}^\dagger|\psi_0\rangle = \frac{m\hbar\omega}{2}\langle \psi_0|\hat{a}\hat{a}^\dagger|\psi_0\rangle$$

$$= \frac{m\hbar\omega}{2}.$$

With these results, we find that the product of variances of position and momentum is

$$\sigma_x^2\sigma_p^2 = \langle \psi_0|\hat{x}^2|\psi_0\rangle\langle \psi_0|\hat{p}^2|\psi_0\rangle = \frac{\hbar}{2m\omega}\frac{m\hbar\omega}{2} = \frac{\hbar^2}{4}. \tag{6.85}$$

Thus, the ground state of the harmonic oscillator saturates the Heisenberg uncertainty principle. One interpretation of why the ground-state energy is non-zero is to satisfy Heisenberg uncertainty. If the ground-state energy were lower, then the variance on momentum and position must decrease, but this is forbidden by the commutation relation of $\hat{x}$ and $\hat{p}$.

6.4 Coherent States

Let's understand this minimum uncertainty state a bit more. Recall that, in position space, the ground-state wavefunction $\psi_0(x)$ was a Gaussian:

$$\psi_0(x) \propto e^{-\frac{m\omega}{2\hbar}x^2}. \tag{6.86}$$

This Gaussian looks like a bump localized around $x=0$, as shown in Fig. 6.3. We can construct other minimum uncertainty states by recalling the translation-invariance property of the variance we discussed in Sec. 4.3.1. The variance is insensitive to the specific value of the mean of the distribution. In the case of the Gaussian, the variance doesn't care where its peak is located. That is, shifting the Gaussian with a peak located at $x=0$ to have its peak at $x=\Delta x$ does not affect its variance.

Using the raising and lowering operator expression for the position operator $\hat{x}$, we can see that a translation of $\hat{x}$ to $\hat{x}-\Delta x$ is equivalent to correspondingly translating a and $a^\dagger$:

$$\hat{x} = \sqrt{\frac{\hbar}{2m\omega}}(\hat{a}+\hat{a}^\dagger) \to \sqrt{\frac{\hbar}{2m\omega}}(\hat{a}+\hat{a}^\dagger) - \Delta x \tag{6.87}$$

$$= \sqrt{\frac{\hbar}{2m\omega}}\left((\hat{a}-\lambda)+(\hat{a}^\dagger-\lambda^*)\right).$$

Here, λ is a complex number and the simultaneous translation of $\hat{a}$ and $\hat{a}^\dagger$ ensures that relationship $(\hat{a}^\dagger)^\dagger = \hat{a}$ still holds after translation. Further, the spatial translation Δx is related to λ via

$$\Delta x = \sqrt{\frac{\hbar}{2m\omega}}(\lambda+\lambda^*) = \sqrt{\frac{2\hbar}{m\omega}}\,\text{Re}(\lambda), \tag{6.88}$$

where $\text{Re}(\lambda)$ is the real part of λ. Further, the ground state $|\psi_0\rangle$ was an eigenstate of the lowering operator $\hat{a}$ with 0 eigenvalue: $\hat{a}|\psi_0\rangle = 0|\psi_0\rangle$. If we consider a state $|\chi\rangle$ which is annihilated by the translated $\hat{a}$ such that

$$(\hat{a}-\lambda)|\chi\rangle = 0, \tag{6.89}$$

then $|\chi\rangle$ is just an eigenstate of $\hat{a}$ with eigenvalue equal to λ. Recall that $\hat{a}$ is not Hermitian, so λ is in general just some complex number. Does such an eigenstate of $\hat{a}$ correspond to a minimum uncertain state? Let's calculate and see.

First, we calculate the expectation value of $\hat{x}$ on this state:

$$\langle\chi|\hat{x}|\chi\rangle = \sqrt{\frac{\hbar}{2m\omega}}\langle\chi|\hat{a}+\hat{a}^\dagger|\chi\rangle = \sqrt{\frac{\hbar}{2m\omega}}(\lambda+\lambda^*). \tag{6.90}$$

Now, the expectation value of the square of $\hat{x}$:

$$\langle\chi|\hat{x}^2|\chi\rangle = \frac{\hbar}{2m\omega}\langle\chi|\hat{a}^2+(\hat{a}^\dagger)^2+\hat{a}\hat{a}^\dagger+\hat{a}^\dagger\hat{a}|\chi\rangle \tag{6.91}$$

$$= \frac{\hbar}{2m\omega}\left(\lambda^2+(\lambda^*)^2+\lambda^*\lambda+\langle\chi|\hat{a}\hat{a}^\dagger|\chi\rangle\right)$$

$$= \frac{\hbar}{2m\omega}\left(\lambda^2+(\lambda^*)^2+\lambda^*\lambda+\langle\chi|\hat{a}^\dagger\hat{a}+[\hat{a},\hat{a}^\dagger]|\chi\rangle\right)$$

$$= \frac{\hbar}{2m\omega}\left((\lambda+\lambda^*)^2+1\right).$$

So, the variance in position of this translated state $|\chi\rangle$ is

$$\sigma_x^2 = \langle\chi|\hat{x}^2|\chi\rangle - \langle\chi|\hat{x}|\chi\rangle^2 = \frac{\hbar}{2m\omega}\left((\lambda+\lambda^*)^2+1-(\lambda+\lambda^*)^2\right) \tag{6.92}$$

$$= \frac{\hbar}{2m\omega},$$

which is exactly the same as for the ground state $|\psi_0\rangle$!

While we won't do it here, you can verify that you find the same thing for the variance of momentum, σ_p^2. Thus, we find that eigenstates of the lowering operator $\hat{a}$ are minimal uncertainty states. Further, because the Heisenberg uncertainty principle is exclusively a quantum phenomenon, we say that states that saturate the bound are "maximally quantum."

6.4.1 Construction as an Eigenstate of the Annihilation Operator

As a final thing for this chapter, let's explicitly construct this state $|\chi\rangle$. To do this, we can simply write it as some linear combination of energy eigenstates $|\psi_n\rangle$, assuming completeness. Thus, we can write

$$|\chi\rangle = \sum_{n=0}^{\infty} \beta_n |\psi_n\rangle = \sum_{n=0}^{\infty} \beta_n \frac{(\hat{a}^\dagger)^n}{\sqrt{n!}} |\psi_0\rangle , \tag{6.93}$$

where the $\{\beta_n\}$ coefficients are some complex numbers and in the second equality we used the explicit form of $|\psi_n\rangle$ as n powers of $\hat{a}^\dagger$ acting on the ground state $|\psi_0\rangle$. As earlier, we note that the sum over powers of $\hat{a}^\dagger$ just represents the Taylor expansion of some function of $\hat{a}^\dagger$, $f(\hat{a}^\dagger)$. That is, we can write

$$|\chi\rangle = \sum_{n=0}^{\infty} \beta_n \frac{(\hat{a}^\dagger)^n}{\sqrt{n!}} |\psi_0\rangle = f(\hat{a}^\dagger)|\psi_0\rangle . \tag{6.94}$$

Our goal is then to determine what this function $f(\hat{a}^\dagger)$ is.

As an eigenstate of $\hat{a}$, $|\chi\rangle$ satisfies

$$\hat{a}|\chi\rangle = \hat{a}f(\hat{a}^\dagger)|\psi_0\rangle = \lambda|\chi\rangle = \lambda f(\hat{a}^\dagger)|\psi_0\rangle . \tag{6.95}$$

To continue, let's work in the $\hat{a}^\dagger$ basis. In this basis, as we have established earlier, the lowering operator $\hat{a}$ is a derivative of $\hat{a}^\dagger$:

$$\hat{a} = \frac{d}{d\hat{a}^\dagger} . \tag{6.96}$$

Therefore, noting that the ground state $|\psi_0\rangle$ is independent of the raising operator $\hat{a}^\dagger$, this eigenvalue equation becomes a differential equation for $f(\hat{a}^\dagger)$:

$$af(\hat{a}^\dagger)|\psi_0\rangle = \lambda f(\hat{a}^\dagger)|\psi_0\rangle \quad \Rightarrow \quad \frac{d}{d\hat{a}^\dagger} f(\hat{a}^\dagger) = \lambda f(\hat{a}^\dagger) . \tag{6.97}$$

This differential equation is easy to solve for $f(\hat{a}^\dagger)$:

$$f(\hat{a}^\dagger) = \beta_0 e^{\lambda \hat{a}^\dagger} , \tag{6.98}$$

for some normalization constant β_0. Therefore, this state is

$$|\chi\rangle = \beta_0 e^{\lambda \hat{a}^\dagger} |\psi_0\rangle . \tag{6.99}$$

The number β_0 can be found by enforcing normalization of $|\chi\rangle$. We want

$$\langle \chi|\chi\rangle = 1 = \langle \psi_0| \beta_0^* e^{\lambda^* \hat{a}} \beta_0 e^{\lambda \hat{a}^\dagger} |\psi_0\rangle = |\beta_0|^2 \langle \psi_0| e^{\lambda^* \hat{a}} e^{\lambda \hat{a}^\dagger} |\psi_0\rangle . \tag{6.100}$$

Now, to continue, we can work in the $\hat{a}^\dagger$ basis in which we know that the annihilation operator is just a derivative with respect to $\hat{a}^\dagger$. That is:

$$e^{\lambda^* \hat{a}} e^{\lambda \hat{a}^\dagger} = e^{\lambda^* \frac{d}{d\hat{a}^\dagger}} e^{\lambda \hat{a}^\dagger} = e^{\lambda(\hat{a}^\dagger + \lambda^*)} = e^{|\lambda|^2} e^{\lambda \hat{a}^\dagger} , \tag{6.101}$$

because the exponentiated derivative with respect to $\hat{a}^\dagger$ is just the translation operator. Thus, the normalization constant satisfies

$$1 = |\beta_0|^2 \langle \psi_0 | e^{|\lambda|^2} e^{\lambda \hat{a}^\dagger} | \psi_0 \rangle = |\beta_0|^2 e^{|\lambda|^2}, \tag{6.102}$$

because $\langle \psi_0 | \hat{a}^\dagger | \psi_0 \rangle = 0$. Thus, the correctly normalized state is

$$|\chi\rangle = e^{-\frac{|\lambda|^2}{2}} e^{\lambda \hat{a}^\dagger} | \psi_0 \rangle. \tag{6.103}$$

Apparently, these maximally quantum states can be created by acting this exponentiated $\hat{a}^\dagger$ operator. Such states are called **coherent states**, and in position space represent a Gaussian ($\psi_0(x)$) displaced from $x = 0$ and $p = 0$ by an amount determined by the eigenvalue of $\hat{a}$, λ. Coherent states appear all over in physics: in optics,[1] condensed matter or low-temperature physics,[2] and particle physics,[3] and they have interesting properties that make them ideal for the model quantum state.[4]

Example 6.3   With this abstract form of a coherent state in Eq. (6.103), defined through the action of the raising operator on the ground state of the harmonic oscillator, let's see if we can express it in a more interpretable form in position space.

Solution

The first thing we will do is to re-write the raising operator in position space. From Eq. (6.77), the position space representation of the raising operator

$$\hat{a}^\dagger = -\sqrt{\frac{\hbar}{2m\omega}} \frac{d}{dx} + \sqrt{\frac{m\omega}{2\hbar}} x = -\sqrt{\frac{\hbar}{2m\omega}} e^{\frac{m\omega}{2\hbar} x^2} \frac{d}{dx} e^{-\frac{m\omega}{2\hbar} x^2}, \tag{6.104}$$

where the derivative is understood to act on everything to the right. The exponentiated raising operator that appears in the expression for the coherent state can be expanded in its Taylor series in position space. This nicely simplifies to

$$e^{\lambda \hat{a}^\dagger} = \sum_{n=0}^\infty \frac{\lambda^n}{n!} (\hat{a}^\dagger)^n = \sum_{n=0}^\infty \frac{\lambda^n}{n!} \left(-\sqrt{\frac{\hbar}{2m\omega}} e^{\frac{m\omega}{2\hbar} x^2} \frac{d}{dx} e^{-\frac{m\omega}{2\hbar} x^2} \right)^n \tag{6.105}$$

$$= e^{\frac{m\omega}{2\hbar} x^2} \sum_{n=0}^\infty \frac{1}{n!} \left(-\lambda \sqrt{\frac{\hbar}{2m\omega}} \right)^n \frac{d^n}{dx^n} e^{-\frac{m\omega}{2\hbar} x^2}.$$

---

[1] E. C. G. Sudarshan, "Equivalence of semiclassical and quantum mechanical descriptions of statistical light beams," *Phys. Rev. Lett.* **10**, 277–279 (1963); R. J. Glauber, "Coherent and incoherent states of the radiation field," *Phys. Rev.* **131**, 2766–2788 (1963).

[2] J. Bardeen, L. N. Cooper, and J. R. Schrieffer, "Theory of superconductivity," *Phys. Rev.* **108**, 1175–1204 (1957).

[3] V. Chung, "Infrared divergence in quantum electrodynamics," *Phys. Rev.* **140**, B1110–B1122 (1965); P. P. Kulish and L. D. Faddeev, "Asymptotic conditions and infrared divergences in quantum electrodynamics," *Theor. Math. Phys.* **4**, 745 (1970).

[4] A number of historical physics papers that discuss the uses of coherent states are collected in J. R. Klauder and B-S. Skagerstam, *Coherent States: Applications in Physics and Mathematical Physics*, World Scientific (1985).

In writing this expression, we have used the associativity of the derivative as a linear operator. For example, consider the term at quadratic order in derivatives:

$$(\hat{a}^\dagger)^2 = \left(-\sqrt{\frac{\hbar}{2m\omega}}\, e^{\frac{m\omega}{2\hbar}x^2}\frac{d}{dx}e^{-\frac{m\omega}{2\hbar}x^2}\right)\left(-\sqrt{\frac{\hbar}{2m\omega}}\, e^{\frac{m\omega}{2\hbar}x^2}\frac{d}{dx}e^{-\frac{m\omega}{2\hbar}x^2}\right) \tag{6.106}$$
$$= \frac{\hbar}{2m\omega}e^{\frac{m\omega}{2\hbar}x^2}\frac{d}{dx}\frac{d}{dx}e^{-\frac{m\omega}{2\hbar}x^2},$$

where again, we always assume that the derivatives act on everything to the right.

Then, in this form, the infinite sum in Eq. (6.105) can be evaluated to find

$$e^{\lambda\hat{a}^\dagger} = e^{\frac{m\omega}{2\hbar}x^2}e^{-\lambda\sqrt{\frac{\hbar}{2m\omega}}\frac{d}{dx}}e^{-\frac{m\omega}{2\hbar}x^2}. \tag{6.107}$$

Recall that the ground-state wavefunction in position space is, from Eq. (6.75):

$$\psi_0(x) = Ne^{-\frac{m\omega}{2\hbar}x^2}, \tag{6.108}$$

for some normalization constant N. Therefore, putting the pieces together, the coherent state $|\chi\rangle$ defined by a complex eigenvalue λ of the lowering operator $\hat{a}$ can be expressed in position space as

$$\chi(x) = \langle x|e^{-\frac{|\lambda|^2}{2}}e^{\lambda\hat{a}^\dagger}|\psi_0\rangle = Ne^{-\frac{|\lambda|^2}{2}}e^{\frac{m\omega}{2\hbar}x^2}e^{-\lambda\sqrt{\frac{\hbar}{2m\omega}}\frac{d}{dx}}e^{-\frac{m\omega}{\hbar}x^2} \tag{6.109}$$
$$= Ne^{-\frac{|\lambda|^2}{2}}e^{\frac{m\omega}{2\hbar}x^2}e^{-\frac{m\omega}{\hbar}\left(x-\lambda\sqrt{\frac{\hbar}{2m\omega}}\right)^2}.$$

To get the second line, we note that the exponentiated derivative operator is just the translation operator where the translation Δx is

$$\Delta x = -\lambda\sqrt{\frac{\hbar}{2m\omega}}. \tag{6.110}$$

Completing the square in the exponent, we can express the coherent state in the most compact form on position space:

$$\chi(x) = Ne^{\frac{\lambda^2-|\lambda|^2}{2}}e^{-\frac{m\omega}{2\hbar}\left(x-\lambda\sqrt{\frac{2\hbar}{m\omega}}\right)^2}. \tag{6.111}$$

For real-valued λ, a coherent state is just a spatial translation of the ground state of the harmonic oscillator away from the origin $x = 0$, as we observed above.

6.4.2 Time Dependence of Coherent States

For a general eigenvalue λ of $\hat{a}$, a coherent state is not an energy eigenstate of the harmonic oscillator Hamiltonian, and so it will exhibit non-trivial time dependence. The time dependence of a coherent state can be found by acting the unitary time translation operator on the initial coherent state. So, for the state defined in Eq. (6.103), its general form at a later time t is

$$|\chi(t)\rangle = e^{-i\frac{t\hat{H}}{\hbar}}e^{-\frac{|\lambda|^2}{2}}e^{\lambda\hat{a}^\dagger}|\psi_0\rangle = e^{-i\omega t\left(\hat{a}^\dagger\hat{a}+\frac{1}{2}\right)}e^{-\frac{|\lambda|^2}{2}}e^{\lambda\hat{a}^\dagger}|\psi_0\rangle, \tag{6.112}$$

where we have inserted the explicit expression for the harmonic oscillator Hamiltonian on the right. From one perspective, we could just be done now, but it is much more enlightening to simplify the expression of the state as much as possible. Let's first pull all of the non-operator exponent factors to the left:

$$|\chi(t)\rangle = e^{-\frac{|\lambda|^2 + i\omega t}{2}} e^{-i\omega t \hat{a}^\dagger \hat{a}} e^{\lambda \hat{a}^\dagger} |\psi_0\rangle. \tag{6.113}$$

Now, if the raising and lowering operators $\hat{a}^\dagger$ and $\hat{a}$ were just *numbers*, then we could combine the exponents on the right through simple addition. However, they are not numbers and they do not commute, so we have to be careful about how to combine them.

To simplify further, we will use a trick we have employed before. We will consider working in the $\hat{a}^\dagger$ basis for which the lowering operator is just a derivative:

$$\hat{a} = \frac{d}{d\hat{a}^\dagger}. \tag{6.114}$$

Additionally, we note that we can write, by definition of the exponential and logarithm functions:

$$\hat{a}^\dagger = e^{\log \hat{a}^\dagger}. \tag{6.115}$$

Again, transcendental functions of an operator shouldn't scare you; it's just shorthand for the appropriate Taylor series. Then, the exponents that we want to simplify can be expressed as

$$e^{-i\omega t \hat{a}^\dagger \hat{a}} e^{\lambda \hat{a}^\dagger} = \exp\left[-i\omega t\, \hat{a}^\dagger \frac{d}{d\hat{a}^\dagger}\right] \exp\left[\lambda e^{\log \hat{a}^\dagger}\right] \tag{6.116}$$

$$= \exp\left[-i\omega t\, \frac{d}{d\log \hat{a}^\dagger}\right] \exp\left[\lambda e^{\log \hat{a}^\dagger}\right]$$

$$= \exp\left[\lambda e^{\log \hat{a}^\dagger - i\omega t}\right].$$

In the last line, we have used the fact that the exponentiated derivative $d/d\log \hat{a}^\dagger$ is just the translation operator in $\log \hat{a}^\dagger$. Therefore, the exponentials simplify to

$$e^{-i\omega t \hat{a}^\dagger \hat{a}} e^{\lambda \hat{a}^\dagger} = \exp\left[\lambda e^{-i\omega t} \hat{a}^\dagger\right]. \tag{6.117}$$

It then follows that the coherent state evaluated at a general time t is

$$|\chi(t)\rangle = e^{-\frac{|\lambda|^2 + i\omega t}{2}} \exp\left[\lambda e^{-i\omega t} \hat{a}^\dagger\right] |\psi_0\rangle. \tag{6.118}$$

Apparently all that time evolution does to a coherent state is to change the complex phase of the lowering operator eigenvalue λ: $\lambda \to \lambda e^{-i\omega t}$. You'll study more consequences of this simple result and its connection to classical mechanics in Exercise 6.4.

Exercises

6.1 Coherent states are potentially an interesting basis to consider in which to express states on the Hilbert space. Are they a good basis, satisfying qualities that we desire of a basis on the Hilbert space?

   (a) Consider two coherent states specified by eigenvalues of the lowering operator $\hat{a}$, λ and η. Call these coherent states $|\psi_\lambda\rangle$ and $|\psi_\eta\rangle$, respectively. What is their inner product, $\langle\psi_\lambda|\psi_\eta\rangle$? Do there exist values of λ and η for which the coherent states are orthogonal?

   (b) Are coherent states complete? Completeness would require that the sum over the outer product of all coherent states is the identity operator:

$$\mathbb{I} = \int_{-\infty}^{\infty}\int_{-\infty}^{\infty} d\mathrm{Re}(\lambda)\, d\mathrm{Im}(\lambda)\, |\psi_\lambda\rangle\langle\psi_\lambda|. \qquad (6.119)$$

   Summing over all coherent states means that we sum over all possible states with distinct real and imaginary parts of the eigenvalue λ of the annihilation operator. Is this equality true, assuming the completeness of the energy eigenstates of the harmonic oscillator?

   Hint: Express the coherent state $|\psi_\lambda\rangle$ in terms of the energy eigenstates of the harmonic oscillator.

6.2 In this chapter, we had only expressed eigenstates of the harmonic oscillator Hamiltonian through repeated action of the raising operator, $\hat{a}^\dagger$. This gives us a concrete algorithm for expressing the energy eigenstates in position space, which we will do here.

   (a) In Eq. (6.75), we had found the ground state of the harmonic oscillator in position space, up to a normalization constant N:

$$\psi_0(x) = Ne^{-\frac{m\omega}{2\hbar}x^2}. \qquad (6.120)$$

   What is N?

   (b) In Example 6.2, we constructed the first excited state of the harmonic oscillator in position space. Extending that procedure, determine the wave-functions for the second and third excited states of the harmonic oscillator, $\psi_2(x)$ and $\psi_3(x)$.

6.3 In this chapter, we noted that the Hilbert space of the harmonic oscillator corresponds to all those states that can be accessed from the ground state through action by an analytic function of $\hat{a}^\dagger$ about $\hat{a}^\dagger = 0$. We'll study a bit more why this is the case in this problem.

   (a) In the basis formed from eigenstates of the harmonic oscillator Hamiltonian, write down the raising operator $\hat{a}^\dagger$ as a matrix. Do this in two ways: first, as an explicit row–column matrix and second, as a sum over outer products of eigenstates of the Hamiltonian. For the explicit row–column matrix, just write down the first three rows and three columns from the upper-leftmost element.

(b) With this representation of the raising operator as a matrix in the basis of energy eigenstates of the harmonic oscillator, express the position and momentum operators $\hat{x}$ and $\hat{p}$ as matrices in this same basis.

(c) What is the determinant of the raising operator $\hat{a}^\dagger$ on the space of energy eigenstates?

(d) At a point where a function is non-analytic, we can express the function through an extension of the Taylor series to include negative powers of the argument. Such an expansion is called a **Laurent series**, and, if we assume that $g(\hat{a}^\dagger)$ is non-analytic at $\hat{a}^\dagger = 0$, its Laurent expansion can be expressed as

$$g(\hat{a}^\dagger) = \sum_{n=-\infty}^{\infty} c_n (\hat{a}^\dagger)^n = \cdots + \frac{c_{-2}}{(\hat{a}^\dagger)^2} + \frac{c_{-1}}{(\hat{a}^\dagger)^2} + c_0 + c_1 \hat{a}^\dagger + c_2 (\hat{a}^\dagger)^2 + \cdots,$$

(6.121)

where the c_n are some complex-valued numbers. For the raising operator on the space spanned by the energy eigenstates of the harmonic oscillator, can this general Laurent expansion exist? Are there constraints we must impose on the coefficients c_n to ensure that $g(\hat{a}^\dagger)$ exists?

6.4 We studied coherent states, which we had identified as eigenstates of the lowering operator, $\hat{a}$. A coherent state $|\psi\rangle$ is defined by the eigenvalue equation

$$\hat{a}|\psi\rangle = \lambda|\psi\rangle,$$

(6.122)

where λ is a complex number. In terms of the harmonic oscillator's ground state $|\psi_0\rangle$, this coherent state can be expressed as

$$|\psi\rangle = e^{-\frac{|\lambda|^2}{2}} e^{\lambda \hat{a}^\dagger} |\psi_0\rangle.$$

(6.123)

In this problem, we will study the properties of these coherent states.

(a) Calculate the variances of the position and momentum operators $\hat{x}$ and $\hat{p}$ on the general time-evolved coherent state $|\psi(t)\rangle$. What is the Heisenberg uncertainty principle now?

Hint: Be sure to use the result of Eq. (6.118).

(b) The time evolution of a coherent state is eerily familiar to that of a classical particle. Classically, if you start at rest at a position Δx displaced from the center of the harmonic oscillator potential, what is your position $x(t)$ and momentum $p(t)$ as a function of time? How do these classical results compare to the time evolution of the expectation values of the position $\hat{x}$ and momentum $\hat{p}$ operators on the time-evolved state $|\psi(t)\rangle$?

(c) Why don't we consider eigenstates of the raising operator, $\hat{a}^\dagger$? What is wrong with a state $|\chi\rangle$ that satisfies

$$\hat{a}^\dagger|\chi\rangle = \eta|\chi\rangle,$$

(6.124)

for some complex number η?

6.5 **Supersymmetry** is a proposed extension of the symmetries of space-time beyond that of just familiar Lorentz transformations and translations. A supersymmetry transformation interchanges fermions and bosons, which are particles of differ-ent spins and correspondingly different statistical properties. While there is no evidence for supersymmetry in Nature, it is nevertheless an interesting proposal that has produced new insights into familiar phenomena.

Supersymmetric quantum mechanics brings some of the ideas of the particle physics supersymmetry to non-relativistic quantum mechanics.[5] In this problem, we will study features of this formulation of quantum mechanics. The first step is to identify two operators A and $A^\dagger$ such that

$$\hat{A}^\dagger \hat{A} = \hat{H} - E_0, \tag{6.125}$$

where $\hat{H}$ is the familiar Hamiltonian and E_0 is its ground-state energy (i.e., just a number). $\hat{A}$ and $\hat{A}^\dagger$ are Hermitian conjugates of one another and in general can be expressed as

$$\hat{A} = \frac{i}{\sqrt{2m}}\hat{p} + W(\hat{x}), \qquad \hat{A}^\dagger = -\frac{i}{\sqrt{2m}}\hat{p} + W(\hat{x}), \tag{6.126}$$

where $\hat{p}$ and $\hat{x}$ are the familiar momentum and position operators. Here, $W(\hat{x})$ is called the superpotential and is a real function of $\hat{x}$.

(a)   With these relationships, determine the potential $V(\hat{x})$ in terms of the superpotential $W(\hat{x})$.

(b)   With this definition of $\hat{A}$ and $\hat{A}^\dagger$, what is their commutator, $[\hat{A}, \hat{A}^\dagger]$?

(c)   We can instead consider the *anti-commutator* $\{\hat{A}, \hat{A}^\dagger\}$ defined as

$$\{\hat{A}, \hat{A}^\dagger\} = \hat{A}\hat{A}^\dagger + \hat{A}^\dagger \hat{A}. \tag{6.127}$$

What is this?

The anti-commutator is present in the algebra that defines supersymmetry and its nice properties with these definitions is why this is called "super-symmetric quantum mechanics." The algebra of supersymmetry consists of both commutation and anti-commutation relations and is referred to as a "graded Lie algebra." We'll discuss Lie algebras in great detail starting in a few chapters.

(d)   Now, connecting back to the harmonic oscillator, what is its superpotential? When we constructed the raising and lowering operators $\hat{a}^\dagger$ and $\hat{a}$, we just used their commutator $[\hat{a}, \hat{a}^\dagger]$. Why could we get away with that, and didn't need to use their anti-commutator?

(e)   *Challenging!* What is the superpotential for the infinite square well?

6.6   In Example 6.1, we introduced the Hermitian number and phase operators $\hat{N}$ and $\hat{\Theta}$ constructed from the raising and lowering operators, $\hat{a}^\dagger$ and $\hat{a}$. In this exercise, we will study more properties of those operators and construct their commutation relation. Recall that the relationship with the raising and lowering operators is

$$\hat{a}^\dagger = \sqrt{\hat{N}}e^{-i\hat{\Theta}}, \qquad \hat{a} = e^{i\hat{\Theta}}\sqrt{\hat{N}}. \tag{6.128}$$

---

[5] An extensive review of supersymmetric quantum mechanics can be found in F. Cooper, A. Khare, and U. Sukhatme, "Supersymmetry and quantum mechanics," *Phys. Rep.* **251**, 267–385 (1995).

(a) Express an eigenstate $|n\rangle$ of the number operator $\hat{N}$ in terms of the continuous eigenstates $|\theta\rangle$ of the phase operator $\hat{\Theta}$ on $\theta \in [0, 2\pi)$.

Hint: Write the state $|n\rangle$ as a continuous sum over phase eigenstates like

$$|n\rangle = \int_0^{2\pi} d\theta \, c_\theta |\theta\rangle \,, \tag{6.129}$$

for some complex coefficients c_θ.

(b) What is the commutation relation of the number and phase operators? *Nota Bene*: Remember that $n = 0$ subtlety from Example 6.1? What is the corresponding uncertainty principle for the number and phase operators?

6.7 We had shown that any state $|\psi\rangle$ on the Hilbert space of the harmonic oscillator is described by some analytic function $f(\hat{a}^\dagger)$ of the raising operator acting on the ground state $|\psi_0\rangle$:

$$|\psi\rangle = f(\hat{a}^\dagger)|\psi_0\rangle \,. \tag{6.130}$$

Because both $|\psi_0\rangle$ and $|\psi\rangle$ are on the Hilbert space and normalized, this might suggest that the resulting operator $f(\hat{a}^\dagger)$ is unitary. However, we have only required $f(\hat{a}^\dagger)$ to act on one special state in the Hilbert space, $|\psi_0\rangle$, and not a general state. So is it unitary?

(a) Consider a general state $|\psi\rangle$ on the Hilbert space of the harmonic oscillator. Now, act the operator $f(\hat{a}^\dagger)$ as defined in Eq. (6.56) on the state $|\psi\rangle$; call the result $|\xi\rangle$:

$$|\xi\rangle = f(\hat{a}^\dagger)|\psi\rangle \,. \tag{6.131}$$

Is $|\xi\rangle$ a physical state and on the Hilbert space, as well? That is, is $f(\hat{a}^\dagger)$ unitary?

(b) If $f(\hat{a}^\dagger)$ is not unitary, can you modify it so it is? Note that this modification cannot affect its action on the ground state $|\psi_0\rangle$. If $f(\hat{a}^\dagger)$ is unitary, then it can be expressed as the exponential of a Hermitian operator $\hat{T}$:

$$f(\hat{a}^\dagger) = e^{i\hat{T}} \,. \tag{6.132}$$

What is $\hat{T}$?

(c) For a coherent state, we had identified this analytic function as (up to nomalization)

$$f(\hat{a}^\dagger) = e^{\lambda \hat{a}^\dagger} \,. \tag{6.133}$$

Is this unitary? If not, can you modify it to make it unitary such that its action on the ground-state wavefunction $|\psi_0\rangle$ is unchanged?

6.8 We happen to find that the ground state of the harmonic oscillator (and the coherent states in general) is a minimum uncertainty state, but can we do the converse? That is, can we directly determine what state saturates the Heisenberg uncertainty principle directly? In the following, we will assume that the state $|\psi\rangle$ on the Hilbert space is the state that saturates the Heisenberg uncertainty such that

$$\sigma_x^2 \sigma_p^2 = \frac{\hbar^2}{4} \,. \tag{6.134}$$

(a) For simplicity, we will first assume that the mean values of both position and momentum on this state are 0: $\langle\psi|\hat{x}|\psi\rangle = \langle\psi|\hat{p}|\psi\rangle = 0$. Thus this state satisfies

$$\langle\psi|\hat{x}^2|\psi\rangle\langle\psi|\hat{p}^2|\psi\rangle = \frac{\hbar^2}{4}. \qquad (6.135)$$

Because this state minimizes the uncertainty relation, any other state will necessarily have a larger uncertainty. Let's express another state as a deviation from $|\psi\rangle$ by some other ket $|\phi\rangle$:

$$|\psi_\epsilon\rangle = |\psi\rangle + \epsilon\,|\phi\rangle, \qquad (6.136)$$

and $\epsilon \ll 1$. First, if $|\psi\rangle$ is normalized, what must the inner product of $|\psi\rangle$ and $|\phi\rangle$ be if the state $|\psi_\epsilon\rangle$ is still on the Hilbert space? Just work to linear order in ϵ.

(b) If $|\psi\rangle$ is the state that minimizes the uncertainty, then the derivative of uncertainty with respect to the state vanishes for $|\psi\rangle$. We can define this derivative on the product of variances as

$$\frac{d}{d|\psi\rangle}\langle\psi|\hat{x}^2|\psi\rangle\langle\psi|\hat{p}^2|\psi\rangle = \lim_{\epsilon\to 0}\frac{\langle\psi_\epsilon|\hat{x}^2|\psi_\epsilon\rangle\langle\psi_\epsilon|\hat{p}^2|\psi_\epsilon\rangle - \langle\psi|\hat{x}^2|\psi\rangle\langle\psi|\hat{p}^2|\psi\rangle}{\epsilon}.$$
$$(6.137)$$

Evaluate this derivative for the states defined above.

(c) If this derivative vanishes, determine the linear equation that ket $|\psi\rangle$ must satisfy, in terms of the variances σ_x^2 and σ_p^2.

Hint: Don't forget the result you derived in part (a).

(d) Using Eq. (6.135), show that $|\psi\rangle$ must be the ground state of the harmonic oscillator; that is, it is an eigenstate of the harmonic oscillator Hamiltonian with the same eigenvalue.

(e) Repeat parts (b)–(d) with non-zero values for the expectation values $\langle\psi|\hat{x}|\psi\rangle$ and $\langle\psi|\hat{p}|\psi\rangle$. What equation must $|\psi\rangle$ satisfy now? Have we seen it before?

7 Quantum Mechanical Example: The Free Particle

In this chapter we will discuss the **free particle**, for which the potential is simply 0 for all $x \in (-\infty, \infty)$. That is, the Hamiltonian is just the kinetic energy:

$$\hat{H} = \frac{\hat{p}^2}{2m},$$
(7.1)

and thus energy eigenstates are also momentum eigenstates. That is, for some momentum eigenstate $|p\rangle$ with eigenvalue p, the energy of such a state is

$$\hat{H}|p\rangle = \frac{\hat{p}^2}{2m}|p\rangle = \frac{p^2}{2m}|p\rangle = E_p|p\rangle.$$
(7.2)

So, in position space, we can express the time dependence of the momentum eigenstate as

$$\langle x|p\rangle = e^{-i\frac{E_p t}{\hbar}} e^{i\frac{px}{\hbar}} = e^{-\frac{i}{\hbar}\left(\frac{p^2}{2m}t - px\right)}.$$
(7.3)

To write this, all we have done is multiply the energy eigenstate ($e^{-i\frac{E_p t}{\hbar}}$) and momentum eigenstate ($e^{i\frac{px}{\hbar}}$), with energy E_p for some momentum p.

In just a few lines, we have therefore diagonalized the free-particle Hamiltonian and established its general time evolution, so it may not be clear what else could be done. However, we will see that these eigenstates have problems if one wants to provide them with a physical interpretation. Identification of these problems, however, forces us to be deliberate in the way that we define their physical properties. That said, a free particle is still a good approximation to numerous systems and is relevant when studying the general phenomena of scattering of a particle off a localized potential. In this chapter, we will show that the linearity of quantum mechanics enables any scattering process to be broken into its individual momentum eigenstate parts whose dynamics are completely determined by the S-matrix.

7.1 Two Disturbing Properties of Momentum Eigenstates

This momentum eigenstate has some weird and disturbing properties, two of which we will review here. Addressing and ultimately resolving them will be central to understanding the physical states on the Hilbert space of the free particle, so it's a good place to start.

7.1.1 Normalization of Free-Particle States

First, we have said that the potential is 0 for all $x \in (-\infty, \infty)$, so a particle in this momentum eigenstate is allowed to be anywhere on the real position axis. The momentum eigenstate oscillates everywhere, at every possible position, which is problematic for its normalization. Recall that a state is only in the Hilbert space (i.e., the space of physical states of a system) if it is normalizable. Let's see if this state is normalizable. To do this, we need to evaluate the inner product of the momentum eigenstate with itself, $\langle p|p \rangle$. Before we do this, let's recall the inner product of two momentum eigenstates $|p\rangle$ and $|p'\rangle$. Suppressing temporal dependence, we have

$$\langle p|p' \rangle = \int_{-\infty}^{\infty} dx \, \langle p|x \rangle \langle x|p' \rangle = \int_{-\infty}^{\infty} dx \, e^{-i\frac{(p-p')x}{\hbar}} = 2\pi\hbar \, \delta(p-p'), \qquad (7.4)$$

as we established in Sec. 2.6. As discussed there, two momentum eigenstates with different values of momentum are orthogonal, but the normalization is awkward. If we then set $p' = p$, the argument of the Dirac δ-function is 0, for which it takes infinite value:

$$\langle p|p \rangle = \infty \, !?! \qquad (7.5)$$

Note that the magnitude of the momentum eigenstate is unity for all positions x, so we just integrate 1 over all real numbers, What is the result? Infinite? If so, there is no way for this momentum eigenstate to be normalizable, and therefore such a state does not live in the Hilbert space. This is fascinating: it means that a momentum eigenstate of a free particle is not a possible physical state in which a particle can exist. However, this does not mean that the notion of momentum space and momentum eigenstates is irrelevant; instead, it means that states in the Hilbert space must be constructed with non-trivial linear combinations of multiple momentum eigenstates.

7.1.2 Velocity of the Free Particle as a Wave

There's another strange feature of the momentum eigenstate. Let's write the momentum eigenstate using Euler's formula, to express it as a sum of sinusoidal functions:

$$\langle x|p \rangle = e^{-\frac{i}{\hbar}(E_p t - px)} = \cos\frac{E_p t - px}{\hbar} - i\sin\frac{E_p t - px}{\hbar}. \qquad (7.6)$$

Now, from your study of waves in introductory physics, you know that the argument of a sinusoidal function describing the wave can also be expressed as

$$\text{wave} \sim \cos(\omega t - kx), \qquad (7.7)$$

where ω is the angular frequency and k is the wavenumber of the wave. With this analogy, note that the angular frequency ω of the momentum eigenstate is

$$\omega = \frac{E_p}{\hbar} = \frac{2\pi}{T} = 2\pi f, \qquad (7.8)$$

where T is the period and f is the frequency of the wave. Further, the wavenumber k would be

$$k = \frac{p}{\hbar} = \frac{2\pi}{\lambda}, \tag{7.9}$$

where λ is the wavelength. Note that these identifications demonstrate that the energy and momentum can be written as

$$E_p = \hbar\omega = hf, \qquad\qquad p = \frac{2\pi\hbar}{\lambda} = \frac{h}{\lambda}. \tag{7.10}$$

You might recognize these expressions from a course on modern physics; in particular, $\lambda = h/p$ is the familiar de Broglie wavelength of the particle with momentum p.

Given an angular frequency ω and wavenumber k, we can determine the velocity of the wave by taking their ratio:

$$v = \lambda f = \frac{\omega}{k} = \frac{E_p}{\hbar}\frac{\hbar}{p} = \frac{E_p}{p} = \frac{\frac{p^2}{2m}}{p} = \frac{p}{2m}, \tag{7.11}$$

where on the right we have plugged in the expression of energy E_p in terms of the momentum eigenvalue p. This resulting velocity is a bit strange: if this were just a particle of mass m, its momentum would be $p = mv$, or its velocity v would be

$$v = \frac{p}{m}. \tag{7.12}$$

Somehow the velocity of this single wave specified by momentum p is half of that. Why different by this factor of two? The velocity of a wave just found from the ratio of angular frequency to the wavenumber is called the **phase velocity**, as it is the velocity of a single wave with a unique phase (i.e., a unique argument of the sinusoidal function). As we have argued that a momentum eigenstate cannot be a state in the Hilbert space, this would imply that the phase velocity is not the physical velocity of a free particle. So, what is the physical velocity?

7.2 Properties of the Physical Free Particle

Let's see if we can fix this behavior of both non-normalizability and the factor of two in the phase velocity. If a momentum eigenstate has these problems, let's now consider a linear combination of momentum eigenstates in position space, a wavefunction denoted by $\psi(x,t)$. We can write this as

$$\psi(x,t) = \int_{-\infty}^{\infty} \frac{dp}{\sqrt{2\pi\hbar}} g(p) e^{-\frac{i}{\hbar}(E_p t - px)}, \tag{7.13}$$

for some function of momentum $g(p)$. In going forward, we'll just study properties of the wavefunction at $t = 0$, so we have

$$\psi(x, t = 0) \equiv \psi(x) = \int_{-\infty}^{\infty} \frac{dp}{\sqrt{2\pi\hbar}} g(p) e^{i\frac{px}{\hbar}}. \tag{7.14}$$

We'll add back the time dependence when we study its velocity later. If such a wavefunction is to be in the Hilbert space, it must be normalizable; in particular, it must be L^2-normalized. Let's see what constraints this imposes on the function $g(p)$. We require

$$
\begin{aligned}
1 &= \int_{-\infty}^{\infty} dx\, \psi^*(x)\psi(x) = \int_{-\infty}^{\infty} dx \int_{-\infty}^{\infty} \frac{dp}{\sqrt{2\pi\hbar}}\, g^*(p)\, e^{-i\frac{px}{\hbar}} \int_{-\infty}^{\infty} \frac{dk}{\sqrt{2\pi\hbar}}\, g(k)\, e^{i\frac{kx}{\hbar}} \\
&= \frac{1}{2\pi\hbar} \int_{-\infty}^{\infty} dx \int_{-\infty}^{\infty} dp \int_{-\infty}^{\infty} dk\, g^*(p)\, g(k)\, e^{-i\frac{(p-k)x}{\hbar}} \\
&= \int_{-\infty}^{\infty} dp \int_{-\infty}^{\infty} dk\, g^*(p)\, g(k)\, \delta(p-k) \\
&= \int_{-\infty}^{\infty} dp\, g^*(p)\, g(p)\,.
\end{aligned}
\tag{7.15}
$$

To do these integrals, we used the definition of the δ-function:

$$
\int_{-\infty}^{\infty} dx\, e^{-i\frac{(p-k)x}{\hbar}} = 2\pi\delta\left(\frac{p-k}{\hbar}\right) = 2\pi\hbar\, \delta(p-k)\,.
\tag{7.16}
$$

Compactly, then, we have shown that the normalization of the wavefunction in position space implies and requires the normalization of the wavefunction in momentum space:

$$
1 = \int_{-\infty}^{\infty} dx\, |\psi(x)|^2 = \int_{-\infty}^{\infty} dp\, |g(p)|^2\,.
\tag{7.17}
$$

The wavefunction in momentum space, $g(p)$, is the Fourier transform of the wavefunction in position space, $\psi(x)$. This result is called **Plancherel's theorem** and states that both the wavefunction and its Fourier transform must be L^2-normalizable if one of them is. In particular, the wavefunction $\psi(x)$ (or $g(p)$) must have **compact support** for this normalization to be possible. Compact support means that $\psi(x)$ only has non-zero values (or not excessively small values) over a finite range in x. That is, the wavefunction must look something like that shown in Fig. 7.1. It may also possibly be translated away from $x = 0$, or possibly double-humped, etc., but all of the interesting part of $\psi(x)$ is contained in a finite domain. Such an object is clearly not a wave, as it is a localized disturbance/bump in probability amplitude. We therefore refer to such a wavefunction as a **wave packet**, as it is a linear combination of many waves (i.e., a collection of waves in a packet).

Additionally, the normalizability or compact support of the wavefunction is essential for the momentum operator $\hat{p}$ to be Hermitian. Further, because the Hamiltonian is exclusively the kinetic energy operator for the free particle, its Hermitivity relies on the

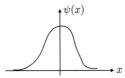

Fig. 7.1   Illustration of an example normalizable wave packet of the free particle.

momentum operator to be Hermitian. In Sec. 3.3.1, we showed that the momentum operator is Hermitian only if all states in the Hilbert space vanish sufficiently fast at the spatial boundaries. In the case of the free particle, the spatial boundaries are $x \to \pm\infty$, and so the wavefunction must vanish sufficiently rapidly toward infinity to ensure that momentum is Hermitian. Perhaps surprisingly and unintuitively, the momentum operator is not Hermitian on the full space spanned by the momentum eigenstates.

Okay, normalization of the free particle has been corrected; what about the velocity of the wave packet? A useful way to think about the wave packet is in analogy to a collection of point particles as shown in Fig. 7.2. With a collection of such particles, what would we call its "velocity"? Well, there is a preferred, single position that describes the global motion of the collection of particles: its center-of-mass x_{cm}. Recall that the center-of-mass is

$$x_{\mathrm{cm}} = \frac{\sum_i x_i m_i}{\sum_i m_i}, \tag{7.18}$$

where x_i is the position of the ith particle and m_i is its mass.

One way to think about the mass of a particle is as a relative probability: a particle with a large mass has more effect over the location of the center-of-mass than does a light particle. So, when thinking about the wavefunction $\psi(x,t)$, positions x with a large value of $|\psi(x,t)|^2$ affect the location of the mean or expected value of position more than places with low probability density. This suggests that, for a wavefunction, the expectation value of position $\langle \hat{x} \rangle$ is analogous to the center-of-mass of some classical system of masses.

We can therefore define the velocity v_{cm} of the wave packet to be the velocity of the expectation value of position. This velocity is just the time derivative:

$$v_{\mathrm{cm}} \equiv \frac{d\langle \hat{x} \rangle}{dt} = \frac{i}{\hbar} \langle [\hat{H}, \hat{x}] \rangle, \tag{7.19}$$

where, on the right, we have used the expression we derived for time dependence of expectation values from Sec. 4.2. Note also that the position operator $\hat{x}$ has no explicit time dependence, so no partial derivative with respect to time appears in this formula.

Let's now evaluate the commutation relation of the Hamiltonian $\hat{H}$ and position $\hat{x}$. We first note that the Hamiltonian, in general, is a function of the momentum operator:

$$\hat{H} \equiv \hat{H}(\hat{p}). \tag{7.20}$$

We'll leave the explicit functional form unevaluated until the end. Recall the canonical commutation relation, $[\hat{x}, \hat{p}] = i\hbar$, from which we showed, in general, that momentum $\hat{p}$ can be expressed as the derivative of position:

Fig. 7.2   A collection of particles from which we can define its group velocity.

$$\hat{p} = -i\hbar \frac{d}{d\hat{x}}. \tag{7.21}$$

The converse is also true: we can instead express the position operator as a derivative of momentum:

$$\hat{x} = i\hbar \frac{d}{d\hat{p}}. \tag{7.22}$$

With this implied form for the position operator, the general commutator of position with the Hamiltonian is

$$[\hat{x}, \hat{H}] = \left[i\hbar \frac{d}{d\hat{p}}, \hat{H}(\hat{p}) \right] = i\hbar \frac{d\hat{H}}{d\hat{p}}. \tag{7.23}$$

Using this result in the expression for the velocity of the center-of-probability of the wave packet, we find

$$v_{cm} = \frac{d\langle \hat{x} \rangle}{dt} = \frac{i}{\hbar} \langle [\hat{H}, \hat{x}] \rangle = \frac{i}{\hbar} \left(-i\hbar \left\langle \frac{d\hat{H}}{d\hat{p}} \right\rangle \right) = \left\langle \frac{d\hat{H}}{d\hat{p}} \right\rangle. \tag{7.24}$$

Now this expression is particularly interesting. Recall that the Hamiltonian is just the total energy, so, in terms of energy and momentum eigenvalues E_p and p, this velocity is

$$v_{cm} = \frac{dE_p}{dp} = \frac{d(\hbar\omega)}{d(\hbar k)} = \frac{d\omega}{dk}, \tag{7.25}$$

the derivative of the angular velocity ω with respect to the wavenumber k. This velocity is called the **group velocity** because it is the velocity of an honest normalizable, physical state that is necessarily a "group" of momentum eigenstates.

If we used the fact that the energy of a momentum eigenstate is

$$E_p = \frac{p^2}{2m}, \tag{7.26}$$

we find the group velocity of our wave packet to be

$$v_{group} = \frac{dE_p}{dp} = \frac{1}{2m} \frac{dp^2}{dp} = \frac{p}{m}, \tag{7.27}$$

which is indeed the velocity that we know and love. This functional dependence of the angular velocity on the wavenumber that determined the group velocity is called the **dispersion relation**. Apparently the expected relationship between the particle's group velocity and momentum, $v = p/m$, only works for a quadratic dispersion relation, where $\omega \propto k^2$.[1]

7.3 Scattering Phenomena

In this section, we introduce the theory of **scattering**, starting from an understanding of the free particle. The idea of scattering is the following. Let's assume that we can model

---

[1] This quadratic dispersion relation is a signature of non-relativistic, free particles. Dispersion relations of numerous functional forms exist for different waves in different media; for example, for a highly relativistic particle, its energy and momentum are related as $E = |\vec{p}|c$, a linear dispersion relation.

some system as a potential $V(x)$ that is **localized** around $x = 0$. That is, for $|x| \to \infty$, $V(x) \to 0$ sufficiently fast so that it is negligible. We won't discuss the mathematical properties of "sufficiently fast" in detail, but in the language developed for the wave packet earlier this chapter, the potential should have compact support around $x = 0$. We can draw a picture of this as illustrated in Fig. 7.3. There's some craziness in the potential that's going on around $x = 0$, but we can translate far enough away such that the potential at our location x is just $V(x) = 0$.

Now, let's imagine that we have prepared a wave packet at $t = 0$ far to the left (negative x) of where the potential is non-zero. This wave packet is given a right-moving velocity, so that it will eventually hit the potential, as shown in Fig. 7.4. Scattering theory asks the question: what happens when the wave packet hits the potential? Specifically, scattering theory asks the question what happens *long* after the wave packet hits the potential and continues traveling onward. In this scenario, if we wait a long enough time, $t \to \infty$, part of the wave will be **transmitted** (i.e., continue moving right toward $x = +\infty$) and part will be **reflected** (i.e., bounce off the potential and move left toward $x = -\infty$), as shown in Fig. 7.5. Scattering theory answers the questions of how much of the initial wave packet (i.e., how much probability amplitude) is reflected and how much is transmitted.

Note that scattering theory only asks questions when the wave packets are far away from the region where the potential is non-zero. Thus, a nice way to represent the wave packets is through linear combinations of momentum eigenstates, as we introduced earlier. We would write the initial, $t = 0$, wave packet as

 Fig. 7.3   Illustration of a localized potential whose energy vanishes sufficiently far away from the spatial origin.

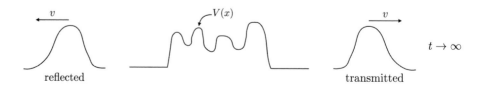

Fig. 7.4   An initial right-moving wave packet before it interacts with a localized non-zero potential.

Fig. 7.5   The reflected and transmitted wave packets long after scattering with a localized potential.

$$\psi(x, t = 0) = \int_{-\infty}^{\infty} \frac{dx}{\sqrt{2\pi\hbar}} g(p) e^{i\frac{px}{\hbar}}, \tag{7.28}$$

for some L^2-normalized complex-valued function $g(p)$. Now, we could ask how this initial wave packet reflects and transmits off a potential $V(x)$ for each and every $g(p)$, but there are a continuous infinity of them. So, we will use the linear combination to our advantage. We imagine sending in a momentum eigenstate from the left to hit the potential. While technically unphysical because momentum eigenstates are not in the Hilbert space, if we know how every momentum eigenstate scatters off the potential we can reconstruct how the wave packet would scatter, by linearity.

So, this is what we will do. We will scatter the momentum eigenstate

$$\psi_p(x, t) = e^{-\frac{i}{\hbar}\left(E_p t - px\right)}, \tag{7.29}$$

on the potential $V(x)$ and determine how much reflects and transmits. Note that I have written the argument of the exponential as $E_p t$ minus px, corresponding to moving right for $p > 0$. Now, the picture we have of this scattering is illustrated in Fig. 7.6. Because the reflected and transmitted waves are also only defined in the regions where $V(x) = 0$, we can write them as

$$\text{reflected: } \psi_R(x, t) = A_R e^{-\frac{i}{\hbar}\left(E_p t + px\right)}, \ x \to -\infty, \tag{7.30}$$

$$\text{transmitted: } \psi_T(x, t) = A_T e^{-\frac{i}{\hbar}\left(E_p t - px\right)}, \ x \to +\infty,$$

for amplitudes A_R, A_T. Note the different sign for momentum in reflection: it is traveling to the left. In the regions where $V(x) = 0$, momentum eigenstates are also energy eigenstates, and where $V(x) \neq 0$, energy eigenstates are still relevant, but momentum is less well-defined. Because our initial wave packet is prepared in the region where $V(x) = 0$, it's therefore more natural to relate everything to a fixed energy. For a fixed total energy E, the magnitude of momentum p is of course

$$|p| = \sqrt{2m\left(E - V(x)\right)}. \tag{7.31}$$

We have implicitly assumed that the potential $V(x)$ is time-independent, and so the energy of the scattered state is the same everywhere for all time. Thus, the time evolutions of the initial, reflected, and transmitted waves are identical and completely determined by the total energy, which is independent of the potential. Thus, as a further simplification, we will ignore the temporal dependence of the states, as that can be

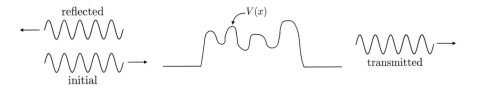

Fig. 7.6   The initial, reflected, and transmitted momentum eigenstates off a localized potential.

added on at the end. With this in mind, the initial (I), reflected, and transmitted states we consider are:

$$\psi_I(x) = e^{i\frac{px}{\hbar}}\,, \tag{7.32}$$

$$\psi_R(x) = A_R e^{-i\frac{px}{\hbar}}\,,$$

$$\psi_T(x) = A_T e^{i\frac{px}{\hbar}}\,.$$

Our goal will be to determine A_R and A_T given a potential $V(x)$.

7.3.1 General Approach to Scattering in One Spatial Dimension

Our general approach to determining the reflection and transmission amplitudes will be the following. We assume that the potential $V(x)$ is non-zero only within the region $0 < x < a$. This assumption allows us to write the exact result for the state at the positions $x = 0$ and $x = a$. As the initial state comes in from and the reflected wave goes out to $x \to -\infty$, their linear combination determines the state for all positions $x < 0$ for which the potential is 0:

$$\psi_{x<0}(x) = e^{i\frac{px}{\hbar}} + A_R e^{-i\frac{px}{\hbar}}\,. \tag{7.33}$$

Next, the transmitted wave continues beyond $x = a$ out toward $x \to \infty$:

$$\psi_{x>a}(x) = A_T e^{i\frac{px}{\hbar}}\,. \tag{7.34}$$

To relate these states at different positions, we can use the momentum operator to translate the $\psi_{x<0}(x)$ state from its boundary of applicability at $x = 0$ to the boundary of the $\psi_{x>a}(x)$ state.

Actually performing this translation is a bit subtle, however. As stated earlier, we consider global energy conservation, as nothing in the system of interest has explicit time dependence. However, in the region where the potential $V(x)$ is non-zero, momentum is in general not conserved. Another way to say this is that in the region where $V(x) \neq 0$, the momentum operator has explicit position dependence. This has consequences for how translations are implemented from $x = 0$ to $x = a$. We would like to construct the linear, unitary operator $\hat{U}_p(x, x+a)$ that translates $\psi_{x<0}(x = 0)$ by a distance a to produce $\psi_{x>a}(x = a)$:

$$\hat{U}_p(x, x+a)\psi_{x<0}(x)\big|_{x=0} = \psi_{x>a}(x = a)\,. \tag{7.35}$$

From our earlier analysis, we might expect that we can write the unitary operator as

$$\hat{U}_p(x, x+a) = e^{i\frac{a\hat{p}}{\hbar}} = e^{a\frac{d}{dx}}\,. \tag{7.36}$$

However, this form implicitly assumed that the momentum operator $\hat{p}$ was position-independent, so that we could say $\hat{p}(x_1) = \hat{p}(x_2)$, for two positions x_1 and x_2. This can only be true if momentum is conserved, but if momentum is not conserved as is the case with non-zero potential, then we must be more careful.

So, to construct this general unitary translation operator we will just take baby steps all the way from $x = 0$ to $x = a$, correcting the value of the momentum operator at each step. Recall that we had identified that translation by an infinitesimal step from x to $x + \Delta x$ is implemented by the operator

$$\hat{U}_p(x, x + \Delta x) = 1 + \frac{i}{\hbar} \Delta x \, \hat{p}(x). \tag{7.37}$$

In this operator, we have explicitly denoted that the momentum operator is evaluated at the position x. To take a step of size $2\Delta x$, we implement the operator twice, being careful about where momentum is evaluated:

$$\hat{U}_p(x + \Delta x, x + 2\Delta x)\hat{U}_p(x, x + \Delta x) = \left(1 + \frac{i}{\hbar} \Delta x \, \hat{p}(x + \Delta x)\right)\left(1 + \frac{i}{\hbar} \Delta x \, \hat{p}(x)\right). \tag{7.38}$$

Operators act to the right, and so the rightmost translation moves from x to $x + \Delta x$ and then one moves from $x + \Delta x$ to $x + 2\Delta x$. Note that the ordering here is important: because the momentum operator $\hat{p}$ depends on position, momenta evaluated at different positions in general do not commute.

We can continue this process arbitrarily to move between x and a position a finite distance away. For the process at hand, we want to move a total distance of a, so we take N steps and take $N \to \infty$ at the end. That is:

$$\hat{U}_p(x, x + a) \tag{7.39}$$
$$= \lim_{N \to \infty} \hat{U}_p\left(x + (N-1)\frac{a}{N}, x + N\frac{a}{N}\right) \cdots \hat{U}_p\left(x + \frac{a}{N}, x + 2\frac{a}{N}\right) \hat{U}_p\left(x, x + \frac{a}{N}\right)$$
$$= \lim_{N \to \infty} \prod_{n=1}^{N} \hat{U}_p\left(x + (N-n)\frac{a}{N}, x + (N-n+1)\frac{a}{N}\right)$$
$$= \lim_{N \to \infty} \prod_{n=1}^{N} \left(1 + \frac{i}{\hbar}\frac{a}{N}\hat{p}_n\right).$$

In the final line, we have used the shorthand $\hat{p}_n$ for the momentum evaluated at the position

$$\hat{p}_n \equiv \hat{p}\left(x + (N-n)\frac{a}{N}\right). \tag{7.40}$$

Now, we could just stop here, but it is useful to do the infinite product explicitly, at least to get a sense of what this general translation operator looks like. To do this, we will order the terms in the product by the number of momentum operators that occur. For example, there is a term that contains no momentum operators – this just comes from multiplying all of the "1"s in the parentheses together:

$$\hat{U}_p(x, x + a) = 1 + \cdots, \tag{7.41}$$

where the $\cdots$ hides terms with one, two, or more powers of the momentum operators.

Here, we'll work to explicitly construct the next two terms in the product. For the term with one power of momentum, we take one momentum operator in the product and multiply it by $N - 1$ "1"s. We must do this for every factor of momentum, so this produces the sum

$$\hat{U}_p(x, x+a) = 1 + \lim_{N\to\infty} \frac{i}{\hbar} \sum_{n=1}^{N} \frac{a}{N} \hat{p}_n + \cdots \tag{7.42}$$

$$= 1 + \lim_{N\to\infty} \frac{i}{\hbar} \sum_{n=1}^{N} \frac{a}{N} \hat{p}\left(x + (N-n)\frac{a}{N}\right) + \cdots,$$

$$= 1 + \frac{i}{\hbar} \int_x^{x+a} dx'\, \hat{p}(x') + \cdots.$$

To write the sum as an integral, we note that, in the limit that $N \to \infty$, $a/N \to dx$, and momentum $\hat{p}$ is evaluated at every position between x and $x+a$. Next, there is a term that contains two factors of the momentum operator. For this term, we have to be careful about the ordering: momentum operators to the left are evaluated at later spatial positions than momentum operators to the right:

$$\hat{U}_p(x, x+a) \tag{7.43}$$

$$= 1 + \frac{i}{\hbar} \int_x^{x+a} dx'\, \hat{p}(x') + \lim_{N\to\infty} \left(\frac{i}{\hbar}\right)^2 \sum_{1 \le m < n \le N} \frac{a}{N}\frac{a}{N} \hat{p}_m \hat{p}_n + \cdots$$

$$= 1 + \frac{i}{\hbar} \int_x^{x+a} dx'\, \hat{p}(x')$$

$$+ \lim_{N\to\infty} \left(\frac{i}{\hbar}\right)^2 \sum_{1 \le m < n \le N} \frac{a}{N}\frac{a}{N} \hat{p}\left(x + (N-m)\frac{a}{N}\right) \hat{p}\left(x + (N-n)\frac{a}{N}\right) + \cdots$$

$$= 1 + \frac{i}{\hbar} \int_x^{x+a} dx'\, \hat{p}(x') + \left(\frac{i}{\hbar}\right)^2 \int_x^{x+a} dx'\, \hat{p}(x') \int_x^{x'} dx''\, \hat{p}(x'') + \cdots.$$

Note the bounds of the nested integrals: this ensures that the momentum operator furthest right is evaluated at a smaller position value than the momentum operator on the left.

One can continue this construction for every term in the infinite product, generating an infinite sum of increasing nested, ordered integrals of the momentum operator evaluated at different positions on the domain $[x, x+a]$. This sum of nested integrals is similar in form to the **Dyson series** in which one can express the solution to the Schrödinger equation as an infinite series of nested integrals of the Hamiltonian evaluated at ordered times.[2] Again, we stated that we have to be careful with the order because if momentum is not conserved, then momentum operators at different positions in general do not commute. However, if momentum is conserved and momentum is independent of position, this translation operator simplifies significantly. Assuming that the momentum operator is independent of position, the translation operator reduces to

[2] F. J. Dyson, "The radiation theories of Tomonaga, Schwinger, and Feynman," *Phys. Rev.* **75**, 486–502 (1949).

$$\hat{U}_p(x, x+a) = 1 + \frac{i}{\hbar} \int_x^{x+a} dx' \, \hat{p}(x') + \left(\frac{i}{\hbar}\right)^2 \int_x^{x+a} dx' \, \hat{p}(x') \int_x^{x'} dx'' \, \hat{p}(x'') + \cdots$$

$$= 1 + \frac{i}{\hbar} \hat{p} \int_x^{x+a} dx' + \left(\frac{i}{\hbar} \hat{p}\right)^2 \int_x^{x+a} dx' \int_x^{x'} dx'' + \cdots$$

$$= 1 + \frac{i}{\hbar} a\hat{p} + \frac{1}{2}\left(\frac{i}{\hbar} a\hat{p}\right)^2 + \cdots$$

$$= e^{i\frac{a\hat{p}}{\hbar}}, \tag{7.44}$$

the exponential form with which we are familiar.

Connecting back to where we started, the most general form of the translation operator in Eq. (7.43) can be used to relate the initial and reflected waves to the left of the potential to the transmitted wave to the right of the potential:

$$\hat{U}_p(x, x+a)\psi_{x<0}(x)\big|_{x=0} = \psi_{x>a}(x=a). \tag{7.45}$$

This is a linear equation and can in principle be solved for the reflection and transmission amplitudes using standard techniques, if the action of the momentum operator $\hat{p}$ on the initial and reflected waves is known. Earlier, we had stated that we assume conservation of energy so that as the particle passes through the potential, its total energy remains E. Then, because total energy is just the sum of kinetic and potential energies, we can solve for the magnitude of momentum at position x:

$$|\hat{p}(x)| = \sqrt{2m(E - V(x))}. \tag{7.46}$$

This result can then be used to determine the action of the translation operator. We'll study this in a simple, explicit example next.

7.3.2 The Finite Square Potential

To illustrate the physics of scattering, we'll consider the simple, yet rich, potential off of which an initial wave is reflected and transmitted. We'll just assume that the potential is constant and non-zero on $x \in [0, a]$:

$$V(x) = \begin{cases} V_0, & 0 < x < a, \\ 0, & x < 0, \ x > a, \end{cases} \tag{7.47}$$

for some constant value V_0. This potential is shown in Fig. 7.7. We'll discuss a classical analogy for this potential later.

The first thing we'll do is to determine the action of the translation operator on the left-position wave. We'll consider translation by a general distance Δx and then specify to $\Delta x = a$ later. With the assumption that $\Delta x \leq a$, we claim that

$$\hat{U}_p(x, x+\Delta x)\psi_{x<0}(x)\big|_{x=0} = \hat{U}_p(x, x+\Delta x)\left[e^{i\frac{xp}{\hbar}} + A_R e^{-i\frac{xp}{\hbar}}\right]\Big|_{x=0} \tag{7.48}$$

$$= Be^{i\frac{\Delta x \sqrt{2m(E-V_0)}}{\hbar}} + Ce^{-i\frac{\Delta x \sqrt{2m(E-V_0)}}{\hbar}}$$

$$= Be^{i\frac{\Delta x \sqrt{p^2 - 2mV_0}}{\hbar}} + Ce^{-i\frac{\Delta x \sqrt{p^2 - 2mV_0}}{\hbar}},$$

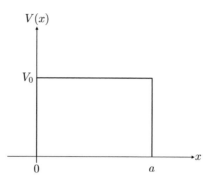

Fig. 7.7   An illustration of the finite square potential which is only non-zero in the range $x \in [0, a]$ where it takes the constant value V_0.

where B and C are some complex coefficients. This form for the action of the translation operator just assumes that it is linear: the result is a linear combination of the opposite-sign momentum eigenstates in the region where the potential is non-zero. To determine these coefficients, we will take Δx to be infinitesimal and expand to linear order in Δx. First, the action of the translation operator is

$$
\hat{U}_p(x, x + \Delta x) \left[e^{i\frac{xp}{\hbar}} + A_R e^{-i\frac{xp}{\hbar}} \right]\Big|_{x=0} \to \left(1 + \frac{i}{\hbar} \Delta x \, \hat{p} \right) \left[e^{i\frac{xp}{\hbar}} + A_R e^{-i\frac{xp}{\hbar}} \right]\Big|_{x=0}
$$

$$
= 1 + A_R + \frac{i}{\hbar} \Delta x \, (p - p A_R) \,. \tag{7.49}
$$

To evaluate the term at order Δx, we simply note that $\hat{p}$ picks out the eigenvalue of momentum of the object that it acts on. Next, we Taylor expand the linear combination of momentum eigenstates in the region where the potential is non-zero:

$$
B e^{i\frac{\Delta x \sqrt{p^2 - 2mV_0}}{\hbar}} + C e^{-i\frac{\Delta x \sqrt{p^2 - 2mV_0}}{\hbar}} \tag{7.50}
$$

$$
= B \left(1 + i\frac{\Delta x \sqrt{p^2 - 2mV_0}}{\hbar} + \cdots \right) + C \left(1 - i\frac{\Delta x \sqrt{p^2 - 2mV_0}}{\hbar} + \cdots \right)
$$

$$
= B + C + \frac{i}{\hbar} \Delta x \left(\sqrt{p^2 - 2mV_0} \, B - \sqrt{p^2 - 2mV_0} \, C \right) + \cdots \,.
$$

Now, we equate terms at the same order in Δx, which gives us two linear equations for B and C:

$$
1 + A_R = B + C \,, \tag{7.51}
$$

$$
p - p A_R = \sqrt{p^2 - 2mV_0} \, B - \sqrt{p^2 - 2mV_0} \, C \,.
$$

To connect with the transmitted wave, we need to translate the initial and reflected waves to the other boundary of the potential, to where $x = a$. In analogy with what we did around $x = 0$, we now consider translation by $a + \Delta x$, and Taylor expand to linear order in Δx. We have

$$\hat{U}_p(a, a+\Delta x)\psi_{x<0}(x)\big|_{x=0} = \hat{U}_p(a, a+\Delta x)\hat{U}_p(x, x+a)\psi_{x<0}(x)\big|_{x=0} \qquad (7.52)$$

$$= \hat{U}_p(\Delta x)\psi_{x>a}(a)$$

$$= B e^{i\frac{(a+\Delta x)\sqrt{p^2-2mV_0}}{\hbar}} + C e^{-i\frac{(a+\Delta x)\sqrt{p^2-2mV_0}}{\hbar}}.$$

First, we can Taylor expand the action of the translation operator on the wave $\psi_{x>a}(a)$:

$$\hat{U}_p(x, x+\Delta x)\psi_{x>a}(a) \rightarrow \left(1 + \frac{i}{\hbar}\Delta x \hat{p}\right) A_T\, e^{i\frac{ap}{\hbar}} \qquad (7.53)$$

$$= A_T\, e^{i\frac{ap}{\hbar}} + \frac{i}{\hbar}\Delta x\, p A_T\, e^{i\frac{ap}{\hbar}}.$$

Next, we Taylor expand the wave in the region where the potential is non-zero:

$$B e^{i\frac{(a+\Delta x)\sqrt{p^2-2mV_0}}{\hbar}} + C e^{-i\frac{(a+\Delta x)\sqrt{p^2-2mV_0}}{\hbar}} \qquad (7.54)$$

$$= B e^{i\frac{a\sqrt{p^2-2mV_0}}{\hbar}} + C e^{-i\frac{a\sqrt{p^2-2mV_0}}{\hbar}}$$

$$+ \frac{i}{\hbar}\Delta x\left(\sqrt{p^2-2mV_0}\, B e^{i\frac{a\sqrt{p^2-2mV_0}}{\hbar}} - \sqrt{p^2-2mV_0}\, C e^{-i\frac{a\sqrt{p^2-2mV_0}}{\hbar}}\right) + \cdots.$$

As before, we equate the terms in these two expansions at the same order in Δx, which produces:

$$A_T\, e^{i\frac{ap}{\hbar}} = B e^{i\frac{a\sqrt{p^2-2mV_0}}{\hbar}} + C e^{-i\frac{a\sqrt{p^2-2mV_0}}{\hbar}}, \qquad (7.55)$$

$$p A_T\, e^{i\frac{ap}{\hbar}} = \sqrt{p^2-2mV_0}\, B e^{i\frac{a\sqrt{p^2-2mV_0}}{\hbar}} - \sqrt{p^2-2mV_0}\, C e^{-i\frac{a\sqrt{p^2-2mV_0}}{\hbar}}.$$

With Eqs. (7.51) and (7.55), we then have four linear equations for four unknown coefficients: A_R, A_T, B, and C. I won't provide the details for how to solve for the reflection and transmission amplitudes, but when the dust settles, one finds

$$A_R = \frac{mV_0}{p^2 - mV_0 + ip\sqrt{p^2-2mV_0}\cot\left(\frac{a}{\hbar}\sqrt{p^2-2mV_0}\right)}, \qquad (7.56)$$

$$A_T = \frac{p\sqrt{p^2-2mV_0}\, e^{-i\frac{ap}{\hbar}}}{p\sqrt{p^2-2mV_0}\cos\left(\frac{a}{\hbar}\sqrt{p^2-2mV_0}\right) - i(p^2-mV_0)\sin\left(\frac{a}{\hbar}\sqrt{p^2-2mV_0}\right)}.$$

While these expressions are big and unwieldy, they importantly are non-zero, generically. At this stage, there are a few observations to make regarding these expressions. First, by conservation of probability, the sum of the absolute squares of the reflected and transmitted amplitudes must be unity:

$$|A_R|^2 + |A_T|^2 = 1. \qquad (7.57)$$

You can show this is true for the expressions above. We call these squared amplitudes the reflection R and transmission T coefficients:

$$R \equiv |A_R|^2, \qquad\qquad T \equiv |A_T|^2, \qquad (7.58)$$

and they represent the fraction of the initial wave that was reflected or transmitted, respectively.

Second, note that the reflection amplitude A_R is 0 if $V_0 = 0$: that is, if there is no intermediate potential, then nothing reflects off it. Further, for generic values of momentum and $V_0 \neq 0$, A_R is in general non-zero, and so something is always reflected off the potential. However, there are special values of momentum p for which nothing is reflected, corresponding to the cotangent factor in the denominator diverging. The cotangent diverges when its argument is an integer multiple of π (where the tangent vanishes), and so A_R is 0 for values of p_n for which

$$\frac{a}{\hbar}\sqrt{p_n^2 - 2mV_0} = n\pi, \tag{7.59}$$

where n is a natural number. We can rearrange this to solve for p_n, assuming $p > 0$:

$$p_n = \sqrt{\frac{n^2\pi^2\hbar^2}{a^2} + 2mV_0}. \tag{7.60}$$

The corresponding kinetic energy of the wave as it passes over the potential is then

$$K = E - V_0 = \frac{p_n^2}{2m} - V_0 = \frac{n^2\pi^2\hbar^2}{2ma^2}. \tag{7.61}$$

This is exactly the kinetic energy of a particle in the infinite square well that we studied a couple of chapters ago. Apparently, when the wave fits nicely into the width of the potential, the reflection amplitude vanishes and we say that the potential becomes transparent: the transmission coefficient is 1 for such values of momentum p_n. Indeed, for the values of p_n in Eq. (7.60), the transmission amplitude reduces to

$$A_T \to e^{-i\frac{ap_n}{\hbar}}, \tag{7.62}$$

which has unit magnitude.

Third, the transmission amplitude is only 0 if the momentum $p = 0$. Otherwise, A_T is non-zero and so some wave is always transmitted. In particular, A_T is non-zero even if the value of the potential is larger than the initial energy of the wave:

$$V_0 > E = \frac{p^2}{2m}. \tag{7.63}$$

This doesn't seem like it is possible classically, and is referred to as **quantum tunneling**. The wave tunnels through a potential, even if it would seem to have insufficient energy. Note, however, that as it tunnels, the amplitude exponentially decays. If $E < V_0$, then the wave in the region where the potential is non-zero is

$$\hat{U}_p(x, x+\Delta x)\,\psi_{x<0}(x)\big|_{x=0} \sim e^{\frac{i}{\hbar}\Delta x\sqrt{p^2 - 2mV_0}} = e^{-\Delta x\frac{\sqrt{2mV_0 - p^2}}{\hbar}}. \tag{7.64}$$

The further you have to tunnel, the more you are penalized, and correspondingly the smaller A_T will be. In particular, if you have to tunnel through a very long potential such that $a \to \infty$, then $A_T \to 0$.

These limits of the transmission and reflection amplitudes we have identified are illustrated in Fig. 7.8. At sufficiently low momenta, the squared transmission amplitude $|A_T|^2$ is very small, as the wave must tunnel through the potential to eventually travel to $x \to +\infty$. The momenta at which and below quantum tunneling occurs is labeled by p_0, corresponding to setting $n = 0$ in Eq. (7.60). The other labeled momenta, p_n for $n > 0$, correspond to those points at which the potential becomes transparent, and no reflection occurs. At sufficiently large momentum, the probability of reflection is highly suppressed because the height of the barrier is negligible compared to the momentum of the wave.

Example 7.1 A simple model for radioactive decay of an unstable nuclear isotope is as follows.[3] Consider the potential illustrated in Fig. 7.9, in which there is a hard, infinite barrier at the spatial origin, a finite barrier of width Δx starting at $x = a$ with height V_0, and the potential everywhere else 0. We can model this nucleus as a particle initially located in the "bound" region with energy E, where $0 < x < a$. As time evolves, however, more of the nucleus's wavefunction will leak out of this region, tunnel through the finite barrier, and decay into the region where $x > a + \Delta x$. Once a large fraction of the wavefunction has exited the bound region, say $1/2$ of the probability of the particle's position, then we can say that the nucleus decayed. In this example, we would like to determine, or at least estimate, the time it takes for the nucleus to decay.

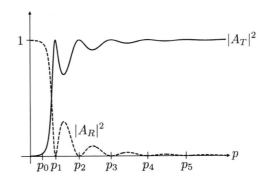

Fig. 7.8 Plots of the absolute squared transmission (solid) and reflection (dashed) amplitudes as a function of the momentum p of the wave scattered off the finite barrier potential. The identified momenta p_n correspond to those momenta from Eq. (7.60). For $p < p_0$, the wave tunnels through the potential barrier to be transmitted.

[3] This simple model was effectively introduced by George Gamow and by Ronald Gurney and Edward Condon in 1928, who noted that the exponential distribution of radioactive decay that had been discovered empirically two decades earlier was easily explained as a tunneling phenomenon within quantum mechanics. See G. Gamow, "Zur Quantentheorie des Atomkernes," *Z. Phys.* **51**, 204–212 (1928); R. W. Gurney and E. U. Condon, "Wave mechanics and radioactive disintegration," *Nature* **122**(3073), 439–439 (1928); R. W. Gurney and E. U. Condon, "Quantum mechanics and radioactive disintegration," *Phys. Rev.* **33**, 127–140 (1929).

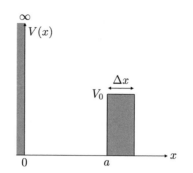

A simple potential as a model for radioactive decay. Initially, the unstable nucleus would be located in the region between $x = 0$ and $x = a$, but then its wavefunction would leak out past the finite barrier into the semi-infinite region beyond $x = a + \Delta x$.

Solution

First, assuming that the particle has a fixed energy E in the bound region, it also has a well-defined magnitude of momentum p:

$$p^2 = 2mE, \tag{7.65}$$

where m is the mass of the bound particle. Just like for the particle in the infinite square well, the particle in the nucleus will bounce back and forth in the bound region. Its bounce will be perfectly elastic at the $x = 0$ barrier, but at the $x = a$ barrier, a little bit of the wavefunction will leak out to $x \to \infty$ when it hits the finite barrier. Every time the particle hits this barrier, the fraction f of its wavefunction that leaks out is given by the absolute squared transmission amplitude, $|A_T|^2$. This can be directly calculated from Eq. (7.56) with the assumption that $E < V_0$. We find

$$|A_T|^2 = \frac{8E(V_0 - E)}{V_0^2 \left(\cosh\left(\frac{2\Delta x}{\hbar} \sqrt{2m(V_0 - E)} \right) - 1 \right) + 8E(V_0 - E)}. \tag{7.66}$$

This can be made more illuminating by making an approximation relevant for radioactive decay with a long half-life. We will assume that the energy of the particle in the well is significantly less than the height of the potential barrier, $E \ll V_0$, and so we can expand the transmission amplitude in this limit. Further, we will also assume that the barrier is high and wide, and so the argument of the hyperbolic cosine is large. Then, in these limits the transmission amplitude reduces to the exponential form we had identified above:

$$|A_T|^2 \to \frac{16E}{V_0} e^{-\frac{2\Delta x}{\hbar} \sqrt{2mV_0}}. \tag{7.67}$$

So, if every time the particle hits the finite barrier it loses this fraction of its wavefunction off to $x \to \infty$, then for a large fraction of the wavefunction to be lost and the nucleus decay, it needs to hit the barrier about $1/|A_T|^2$ times.

The time it takes for the nucleus to decay is therefore the product of the number of times the particle hits the finite barrier with the time between hitting the barrier. Again, in the bound region, the particle's squared momentum is $p^2 = 2mE$, and it has to travel across the well of width a twice from one hit of the finite barrier to the next. The time t between hits is therefore

$$t = \frac{2ma}{|p|} = a\sqrt{\frac{2m}{E}}. \tag{7.68}$$

Then, the characteristic time τ, or **lifetime**, of the nucleus to decay is

$$\tau = \frac{t}{|A_T|^2} \approx \frac{aV_0}{16E}\sqrt{\frac{2m}{E}}\, e^{\frac{2\Delta x}{\hbar}\sqrt{2mV_0}}. \tag{7.69}$$

This expression for the lifetime of an unstable nucleus describes some basic properties of radioactivity. A more complete description of α-decay in which a nucleus disintegrates by emission of an α particle or a helium nucleus was considered by Gamow. In this model, the square finite potential barrier is exchanged for the repulsive Coulomb potential between the α particle and the nucleus, and the resulting dependence of the lifetime on the energy of the α particle agrees remarkably well with data.

There are a few interesting limits of the expression Eq. (7.69) to note. First, if the height of the potential barrier $V_0 \to \infty$, then the characteristic time $\tau \to \infty$. That is, it takes an infinite amount of time for the nucleus to escape the bound region; or the nucleus is stable, and never decays. This makes sense because if $V_0 \to \infty$, then the nucleus just finds itself in the infinite square well, and can never leave. Similarly, if the width of the barrier $\Delta x \to \infty$, then the nucleus can never completely tunnel through the barrier, never reaching the free region. Hence, $\tau \to \infty$ in this limit, as well.

7.4 The S-matrix

In the previous section, we introduced the phenomenon of scattering: incident momentum eigenstates on a localized potential that then scatter, either transmitting or reflecting. One can argue, as we will in this section, that scattering is the most fundamental way we interact with the world in general. Let's focus on vision, for example. Light from the sun travels essentially unimpeded (i.e., free) until it hits something localized, like a blade of grass. Some amount of that light is transmitted or absorbed by the grass, and the rest is reflected into your eye, say. Your brain performs complicated analysis of the reflected light to interpret it as, indeed, a blade of grass. That is, we don't *see* the grass: we see and reinterpret the light reflected from the grass.

This idea is also explored in particle physics experiments, like the Large Hadron Collider at CERN, in Geneva, Switzerland. The goal of the Large Hadron Collider, or LHC, is to determine how particles interact at length scales of about 10^{-20} m. This is way too small to look at with light: visible light has a wavelength of hundreds of nanometers, about a trillion times larger than the distance the LHC probes. Instead, the LHC accelerates protons to very high energies and collides them head on.

Then, after collision, numerous particles are produced, according to the interactions that happened at the 10^{-20} m distance scale. Enormous particle detectors are built around these collisions or interaction points to measure all (or nearly all) of the particles produced. Then, the goal of the scientist is the inverse problem: given the initial proton momenta and the momenta of the final particles, what interactions happened in between? A picture of this is shown in Fig. 7.10. The blob in the middle represents the interactions responsible for turning the protons into whatever the final particles are.

While we don't immediately know what the blob is, quantum mechanics already tells us a lot about it. First, we can represent the initial state of the two protons as a Dirac ket, called an **in state**:

$$|\text{protons}\rangle \equiv |\text{in}\rangle. \tag{7.70}$$

As a quantum state in the Hilbert space, it must be normalized:

$$\langle \text{in}|\text{in}\rangle = 1. \tag{7.71}$$

Further, the collection of final-state particles can also be represented as a ket, called an **out state**:

$$|\text{final particles}\rangle \equiv |\text{out}\rangle. \tag{7.72}$$

Again, as a state in the Hilbert space, it must be normalized:

$$\langle \text{out}|\text{out}\rangle = 1. \tag{7.73}$$

These in and out states are both states in the same Hilbert space, as they happen in the same universe. Therefore, there is some unitary operator $\hat{S}$ that transforms the in state to an out state:

$$\hat{S}|\text{in}\rangle = |\text{out}\rangle. \tag{7.74}$$

Conservation of probability requires that $\hat{S}^\dagger \hat{S} = \mathbb{I}$, the identity operator. The unitary matrix $\hat{S}$ is called the **S-matrix**, short for scattering matrix.[4] Note that the elements of

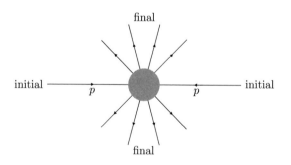

Fig. 7.10   A simplified representation of the scattering of two initial protons at the Large Hadron Collider into a large number of observed final-state particles.

[4] The S-matrix was first used in a restricted sense by John Wheeler in J. A. Wheeler, "On the mathematical description of light nuclei by the method of resonating group structure," *Phys. Rev.* **52**, 1107–1122 (1937),

the scattering matrix are observable: I know the in state (as I prepared it) and I measure the out state, so I can infer $\hat{S}$. A particle physicist's goal is to take the S-matrix and use it to determine the physics located at 10^{-20} m.

In this book, while we won't discuss the S-matrix relevant for the LHC, we will consider scattering off a localized, quantum mechanical potential as introduced in the previous section. Recall that we considered the scattering of an initial right-moving wave of momentum p incident on a localized potential, shown back in Fig. 7.6. We had calculated the amplitudes of the reflected and transmitted waves and from them constructed transmission T and reflection R coefficients. In principle, the dependence of T and R on the initial wave's momentum p can be used to determine what the potential is. This is somewhat analogous to taking a Fourier transform of the potential, as we know about its response to a wave of a fixed momentum, and from it, we want to determine its spatial profile.

Here, we will consider the more general case in which waves from both the left and right are incident on the potential, as shown in Fig. 7.11. With this identification, we can express the incoming and outgoing waves as the sum of momentum eigenstates:

$$\psi_{\text{in}}(x) = A_{\text{in},L}\, e^{i\frac{px}{\hbar}} + A_{\text{in},R}\, e^{-i\frac{px}{\hbar}}, \tag{7.75}$$

$$\psi_{\text{out}}(x) = A_{\text{out},R}\, e^{i\frac{px}{\hbar}} + A_{\text{out},L}\, e^{-i\frac{px}{\hbar}},$$

for some complex coefficients $A_{\text{in},L}$, $A_{\text{in},R}$, $A_{\text{out},R}$, and $A_{\text{out},L}$. The S-matrix of this configuration expresses the unitary transformation of the incoming amplitudes $A_{\text{in},L}$, $A_{\text{in},R}$ into the outgoing amplitudes $A_{\text{out},R}$, $A_{\text{out},L}$:

$$\begin{pmatrix} A_{\text{out},R} \\ A_{\text{out},L} \end{pmatrix} = \hat{S} \begin{pmatrix} A_{\text{in},L} \\ A_{\text{in},R} \end{pmatrix}. \tag{7.76}$$

In this representation, the S-matrix is a 2×2 matrix. We have already identified two of the entries of the S-matrix from our analysis of scattering in the previous section. If there is just a wave incident from the left, this is as if $A_{\text{in},R} = 0$ and the transformation of such an initial state into the outgoing waves is controlled by the first column of the S-matrix:

$$\hat{S} = \begin{pmatrix} A_T & S_{RR} \\ A_R & S_{RL} \end{pmatrix}. \tag{7.77}$$

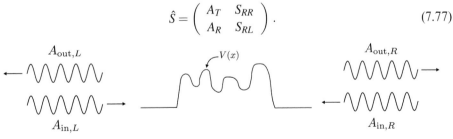

Fig. 7.11 To define the S-matrix for scattering of momentum eigenstates off a localized potential, we consider waves incident on the potential from both the left and the right.

but then generalized by Werner Heisenberg in W. Heisenberg, "The observable quantities in the theory of elementary particles," *Z. Phys.* **120**, 513–673 (1943).

Here, A_T and A_R are the transmission and reflection amplitudes that we introduced in the previous section and S_{RR} and S_{RL} are two as-of-yet undetermined entries of the S-matrix. First, note that the entry S_{RL} represents the transmission amplitude for an initial wave coming in from the right, rather than the left. A transmitted wave travels in the same direction as the initial wave, regardless of the specific direction of the initial wave. Thus, the configuration of an initial wave from the right and its transmitted wave going left is just exactly the spatial opposite of an initial wave coming from the left and its transmitted wave going right. This suggests that the transmission amplitude of a left-moving wave is identical to that of a right-moving wave, so that $S_{RL} = A_T$. Then, the S-matrix is

$$\hat{S} = \begin{pmatrix} A_T & S_{RR} \\ A_R & A_T \end{pmatrix}, \tag{7.78}$$

with only one undetermined entry, S_{RR}.

This final entry can be determined by demanding the unitarity of the S-matrix. The product of the S-matrix and its Hermitian conjugate is

$$\hat{S}^\dagger \hat{S} = \begin{pmatrix} A_T^* & A_R^* \\ S_{RR}^* & A_T^* \end{pmatrix} \begin{pmatrix} A_T & S_{RR} \\ A_R & A_T \end{pmatrix} \tag{7.79}$$

$$= \begin{pmatrix} |A_T|^2 + |A_R|^2 & A_T^* S_{RR} + A_R^* A_T \\ A_T S_{RR}^* + A_R A_T^* & |S_{RR}|^2 + |A_T|^2 \end{pmatrix}.$$

If this is to equal the identity matrix, then we must demand that the off-diagonal entries vanish. This occurs when

$$A_T^* S_{RR} + A_R^* A_T = 0, \tag{7.80}$$

or the final entry of the S-matrix is

$$S_{RR} = -\frac{A_T}{A_T^*} A_R^*. \tag{7.81}$$

It is interesting to note that this entry has magnitude equal to that of just the reflection coefficient, $|S_{RR}| = |A_R|$, but differs in phase. Putting it together, the S-matrix is thus

$$\hat{S} = \begin{pmatrix} A_T & -\frac{A_T}{A_T^*} A_R^* \\ A_R & A_T \end{pmatrix}. \tag{7.82}$$

This matrix is unitary by virtue of the fact that the sum of the reflection and transmission coefficients is unity: $|A_T|^2 + |A_R|^2 = 1$.

7.5 Three Properties of the S-matrix

To concretely demonstrate interesting features of the S-matrix, we'll use the reflection and transmission amplitudes that we derived earlier for the finite square potential.

Fig. 7.12 Illustration of the finite square potential, in the limit that its width $a \to 0$, but its area remains constant.

Recall that the potential was non-zero in the region where $x \in [0,a]$ with height given by V_0. The reflection and transmission amplitudes that we identified were

$$A_R = \frac{mV_0}{p^2 - mV_0 + ip\sqrt{p^2 - 2mV_0}\cot\left(\frac{a}{\hbar}\sqrt{p^2 - 2mV_0}\right)}, \tag{7.83}$$

$$A_T = \frac{p\sqrt{p^2 - 2mV_0}\,e^{-i\frac{ap}{\hbar}}}{p\sqrt{p^2 - 2mV_0}\cos\left(\frac{a}{\hbar}\sqrt{p^2 - 2mV_0}\right) - i(p^2 - mV_0)\sin\left(\frac{a}{\hbar}\sqrt{p^2 - 2mV_0}\right)}.$$

While one can work directly with these expressions for the amplitudes, they are unwieldy and obscure the physics we wish to study. We'll take a limit of the potential in which the amplitudes will simplify and very directly illustrate interesting features of the S-matrix. We will take the limit in which the width of the potential $a \to 0$ while at the same time increasing V_0 so as to keep the integral of the potential over positions constant. That is, we take

$$a \to 0 \quad \text{with} \quad aV_0 \equiv \alpha = \text{constant}, \tag{7.84}$$

which is illustrated in Fig. 7.12. In this limit, the reflection and transmission amplitudes simplify to

$$A_R \xrightarrow{a \to 0} \frac{-i\frac{m\alpha}{\hbar}}{p + i\frac{m\alpha}{\hbar}} = \frac{-i\frac{maV_0}{\hbar}}{p + i\frac{maV_0}{\hbar}}, \tag{7.85}$$

$$A_T \xrightarrow{a \to 0} \frac{p}{p + i\frac{m\alpha}{\hbar}} = \frac{p}{p + i\frac{maV_0}{\hbar}}.$$

From our general construction above, the S-matrix for scattering off this narrow potential is

$$\hat{S} = \begin{pmatrix} A_T & -\frac{A_T}{A_T^*}A_R^* \\ A_R & A_T \end{pmatrix} = \frac{p}{p + i\frac{maV_0}{\hbar}}\begin{pmatrix} 1 & -i\frac{maV_0}{\hbar p} \\ -i\frac{maV_0}{\hbar p} & 1 \end{pmatrix}. \tag{7.86}$$

Note that this S-matrix is both unitary and symmetric: in the $a \to 0$ limit, the potential is symmetric under position $x \to -x$.

Before studying its properties, we'll note two interesting limits of the S-matrix's elements. First, if the momentum $p \to \infty$, the kinetic energy of the wave is much greater than the energy of the potential, and so the particle flies right by, with no effect on its motion. That is, the S-matrix reduces to the identity matrix, in the limit that $p \to \infty$. At sufficiently high momenta, the incoming and outgoing states are identical. Next, consider the limit as $p \to 0$. In this limit, the S-matrix reduces to a matrix with zero entries on the diagonal, and -1 in the off-diagonal elements:

$$\hat{S} \xrightarrow{p \to 0} \begin{pmatrix} 0 & -1 \\ -1 & 0 \end{pmatrix}. \tag{7.87}$$

With zeros on the diagonal, this means that no wave is transmitted, there is only reflection. Further, the fact that the off-diagonal entries are -1 means that the amplitude of the reflected wave is negative to that of the incoming wave. That is, upon interaction with the potential, the wave flips over. When a rope attached to a wall is wiggled, the incoming and outgoing waves from the wall are out of phase by π: if a crest hits the wall, then a trough exits the wall. This $p \to 0$ limit of the scattered wave is exactly analogous to this rope.

Example 7.2　We constructed the S-matrix as encoding the reflection and transmission amplitudes for individual momentum eigenstates, but of course these are not physical states on the Hilbert space. In this example, we will use these results for the scattering of momentum eigenstates off this localized potential illustrated in Fig. 7.12 and determine the reflection and transmission amplitudes for scattering of a physical wave packet. The form of the wave packet we will consider is, in momentum space:

$$g(p) = \frac{e^{-i\frac{px_0}{\hbar}}}{(\pi \sigma_p^2)^{1/4}} e^{-\frac{(p - p_0)^2}{2\sigma_p^2}}. \tag{7.88}$$

The center of the wave packet is located at $x = x_0$ and as $x_0 \to -\infty$, the center of the wave packet is moved infinitely far to the left of the potential. Here, p_0 is an initial momentum with $p_0 > 0$, meaning that the packet moves right and σ_p is a measure of the width or spread of the wave packet about p_0. This wave packet is plotted in momentum space in Fig. 7.13. More about this example will be explored in Exercise 7.4.

Solution

The wave packet $g(p)$ encodes the initial amplitudes of the different momentum eigenstates that contribute to it. In our analysis of scattering off a potential, we considered the scattering of momentum eigenstates assuming that the amplitude of the initial wave was unity. So, all we need to do to determine the transmitted and reflected wave packets in momentum space is to multiply the appropriate reflection or transmission amplitude by the initial wave packet.

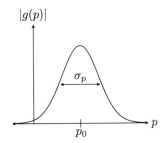

Fig. 7.13 An illustration of the momentum space wave packet of Eq. (7.88), with center located at $p = p_0$ and width or spread about p_0 of σ_0.

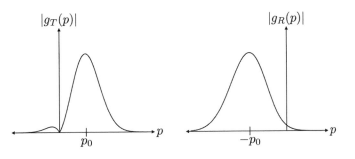

Fig. 7.14 An illustration of the transmission and reflection momentum space wave packets of Eqs. (7.89) (left) and (7.90) (right), with centers located at $p = \pm p_0$.

For the transmitted wave packet, $g_T(p)$, this is simple because the transmitted wave packet travels in the same direction as the initial wave packet. In terms of the individual momentum eigenstates, the initial and transmitted wave packets have the same sign of momentum, from Eq. (7.32). Then, the transmitted wave packet in momentum space is

$$g_T(p) = A_T\, g(p) = \frac{p}{p + i\frac{maV_0}{\hbar}} \frac{e^{-i\frac{px_0}{\hbar}}}{(\pi\sigma_p^2)^{1/4}} e^{-\frac{(p-p_0)^2}{2\sigma_p^2}}. \tag{7.89}$$

Now, for the reflected wave packet, the analysis is slightly different. The reflected wave travels in the opposite direction as the initial wave packet, so we must negate the momentum, again, as illustrated in Eq. (7.32). So, the reflected wave packet in momentum space $g_R(p)$ is

$$g_R(p) = A_R\, g(-p) = \frac{-i\frac{maV_0}{\hbar}}{p + i\frac{maV_0}{\hbar}} \frac{e^{i\frac{px_0}{\hbar}}}{(\pi\sigma_p^2)^{1/4}} e^{-\frac{(p+p_0)^2}{2\sigma_p^2}}. \tag{7.90}$$

These transmitted and reflected wave packets after scattering are plotted in Fig. 7.14.

Further, assuming that the initial wave packet is normalized, we can verify that the sum of the transmitted and reflected wave packets is also normalized. Their coherent sum squared is then

$$|g_T(p) + g_R(p)|^2 = |g_T(p)|^2 + |g_R(p)|^2 + g_T^*(p)g_R(p) + g_T(p)g_R^*(p) \tag{7.91}$$

$$= \frac{p^2}{p^2 + \left(\frac{maV_0}{\hbar}\right)^2} \frac{1}{(\pi^2 \sigma_p^2)^{1/2}} e^{-\frac{(p-p_0)^2}{\sigma_p^2}} + \frac{\left(\frac{maV_0}{\hbar}\right)^2}{p^2 + \left(\frac{maV_0}{\hbar}\right)^2} \frac{1}{(\pi^2 \sigma_p^2)^{1/2}} e^{-\frac{(p+p_0)^2}{\sigma_p^2}}$$

$$+ 2 \sin\left(\frac{2px_0}{\hbar}\right) \frac{p \frac{maV_0}{\hbar}}{p^2 + \left(\frac{maV_0}{\hbar}\right)^2} \frac{1}{(\pi^2 \sigma_p^2)^{1/2}} e^{-\frac{p^2+p_0^2}{\sigma_p^2}}.$$

To write this, I used relationships between imaginary exponential functions and trigonometric functions. In this form, the interference term on the third line seems problematic for normalization. However, for the initial wave packet to actually be completely in the region where the potential is 0, we must take the initial position $x_0 \to -\infty$. As the argument of a sinusoidal function grows, that function oscillates more and more rapidly, and its average, or integral, over any finite domain will approach 0. Therefore, as $x_0 \to -\infty$, the interference term does not contribute to the normalization of the wave packet.

Then, the normalization is encoded in the first two terms, and we note that

$$\int_{-\infty}^{\infty} dp \left(|g_T(p)|^2 + |g_R(p)|^2\right) = \int_{-\infty}^{\infty} dp \left(|g_T(p)|^2 + |g_R(-p)|^2\right) \tag{7.92}$$

$$= \int_{-\infty}^{\infty} dp \left(|A_T|^2 + |A_R|^2\right) |g(p)|^2 = \int_{-\infty}^{\infty} dp \, |g(p)|^2,$$

because the sum of the absolute squares of the transmission and reflection amplitudes is unity. Therefore, as long as the initial wave packet is normalized, so too are the sum of the reflected and transmitted wave packets after scattering.

7.5.1 The Optical Theorem

We had mentioned earlier that the S-matrix is just the identity matrix if the wave passes right through the potential. In such a case, no proper *scattering* took place. With this in mind, it is useful to re-write the S-matrix in a way that explicitly isolates the identity matrix component:

$$\hat{S} = \mathbb{I} + i\hat{\mathcal{M}}, \tag{7.93}$$

where $\hat{\mathcal{M}}$ can be called the **interaction matrix**. A wave is only reflected off the potential if $\hat{\mathcal{M}} \neq 0$. For the example at hand, we can determine $\hat{\mathcal{M}}$ straightforwardly:

$$i\hat{\mathcal{M}} = \hat{S} - \mathbb{I} = \frac{p}{p + i\frac{maV_0}{\hbar}} \begin{pmatrix} 1 & -i\frac{maV_0}{\hbar p} \\ -i\frac{maV_0}{\hbar p} & 1 \end{pmatrix} - \begin{pmatrix} 1 & 0 \\ 0 & 1 \end{pmatrix} \tag{7.94}$$

$$= \frac{-i\frac{maV_0}{\hbar}}{p + i\frac{maV_0}{\hbar}} \begin{pmatrix} 1 & 1 \\ 1 & 1 \end{pmatrix}.$$

This interaction matrix induces a maximal mixing between the reflected and transmitted waves: because all entries of the matrix $\hat{\mathcal{M}}$ are identical, reflected and transmitted waves have equal amplitude. We'll focus on particular properties of this matrix in a second, but for now, let's identify a very general constraint that the interaction matrix satisfies.

We can re-express the unitarity of the S-matrix in terms of the interaction matrix. We have

$$\hat{S}^\dagger \hat{S} = \mathbb{I} = \left(\mathbb{I} - i\hat{\mathcal{M}}^\dagger\right)\left(\mathbb{I} + i\hat{\mathcal{M}}\right) = \mathbb{I} + i\left(\hat{\mathcal{M}} - \hat{\mathcal{M}}^\dagger\right) + \hat{\mathcal{M}}^\dagger\hat{\mathcal{M}}. \qquad (7.95)$$

Now, if this is to hold, $\hat{\mathcal{M}}$ must satisfy

$$2\frac{\hat{\mathcal{M}} - \hat{\mathcal{M}}^\dagger}{2i} = 2\,\mathrm{Im}(\hat{\mathcal{M}}) = \hat{\mathcal{M}}^\dagger\hat{\mathcal{M}}. \qquad (7.96)$$

For the S-matrix to be unitary, the interaction matrix must satisfy this non-trivial relationship: its imaginary part is equal to its squared magnitude. This general relationship is called the **optical theorem** and its interpretation is the following. $\hat{\mathcal{M}}^\dagger\hat{\mathcal{M}}$ is just the total probability for a non-trivial scattering off the potential to occur; the transparent transmission component has been explicitly subtracted. The interpretation of $\mathrm{Im}(\hat{\mathcal{M}})$ is a bit more subtle. Note from the example in Eq. (7.94) that $\hat{\mathcal{M}}$ is exclusively a real-valued matrix if the potential is completely transparent, in the limit $p \to \infty$. Thus, $\mathrm{Im}(\hat{\mathcal{M}})$ represents the amount of transparency lost through scattering. By unitarity of the S-matrix, the loss of transparency must be compensated by an increase in reflection, as encoded in Eq. (7.96).

This demonstration of the optical theorem is illustrated in Fig. 7.15. A disc is placed in front of a perfectly absorbing screen and light is shined on the disc. The disc casts a shadow on the screen because its presence prohibits the transmission of light to that location on the screen. Instead, light is reflected off the disc, traveling back in the direction from which it originated. The area of the shadow is equal to the loss of transparency due to the presence of the disc, $2\,\mathrm{Im}(\hat{\mathcal{M}})$. Correspondingly, the area that the reflected light covers is $\hat{\mathcal{M}}^\dagger\hat{\mathcal{M}}$, and these must be equal because the same disc is responsible for both the shadow and the reflection.

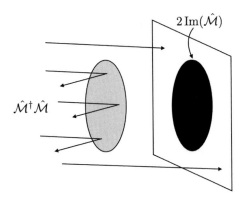

Fig. 7.15   Illustration of the optical theorem. Light rays (lines with arrows) are shined on a disc located in front of a perfectly absorbing screen. The area of the shadow on the screen is equal to the area of the reflected light.

7.5.2 Scattering Phase

With the identification of the interaction matrix, we can re-express the S-matrix in an interesting way. As a unitary matrix, the S-matrix can be expressed as an exponentiated Hermitian matrix $\hat{T}$:

$$\hat{S} = e^{i\hat{T}}. \tag{7.97}$$

To determine $\hat{T}$, we can take the logarithm of both sides and find

$$i\hat{T} = \log \hat{S} = \log\left(\mathbb{I} + i\hat{\mathcal{M}}\right), \tag{7.98}$$

where on the right we have expressed the S-matrix in terms of the interaction matrix. In the form on the right, we can explicitly calculate the logarithm through its Taylor expansion:

$$\log\left(\mathbb{I} + i\hat{\mathcal{M}}\right) = \sum_{n=1}^{\infty} (-1)^{n+1} \frac{(i\hat{\mathcal{M}})^n}{n} = i\hat{\mathcal{M}} + \frac{\hat{\mathcal{M}}^2}{2} - i\frac{\hat{\mathcal{M}}^3}{3} + \cdots. \tag{7.99}$$

For the example in Eq. (7.94), this sum can be evaluated simply. To evaluate its logarithm, we need to first evaluate the nth power of the interaction matrix:

$$(i\hat{\mathcal{M}})^n = \left(\frac{-i\frac{maV_0}{\hbar}}{p + i\frac{maV_0}{\hbar}}\right)^n \begin{pmatrix} 1 & 1 \\ 1 & 1 \end{pmatrix}^n. \tag{7.100}$$

To do this, we need to evaluate the nth power of the matrix with all entries equal to 1. Let's first see what its square is:

$$\begin{pmatrix} 1 & 1 \\ 1 & 1 \end{pmatrix}\begin{pmatrix} 1 & 1 \\ 1 & 1 \end{pmatrix} = 2\begin{pmatrix} 1 & 1 \\ 1 & 1 \end{pmatrix}. \tag{7.101}$$

Multiplying by the matrix again just multiplies by another factor of 2. This suggests that the nth power is just

$$\begin{pmatrix} 1 & 1 \\ 1 & 1 \end{pmatrix}^n = 2^{n-1}\begin{pmatrix} 1 & 1 \\ 1 & 1 \end{pmatrix}. \tag{7.102}$$

This can be proved inductively. We have already verified the $n = 1$ and $n = 2$ cases, so let's assume n. Then, let's multiply by the matrix again:

$$\begin{pmatrix} 1 & 1 \\ 1 & 1 \end{pmatrix}\begin{pmatrix} 1 & 1 \\ 1 & 1 \end{pmatrix}^n = 2^{n-1}\begin{pmatrix} 1 & 1 \\ 1 & 1 \end{pmatrix}\begin{pmatrix} 1 & 1 \\ 1 & 1 \end{pmatrix} = 2^n\begin{pmatrix} 1 & 1 \\ 1 & 1 \end{pmatrix}, \tag{7.103}$$

which completes the induction and proves the relationship.

This result allows us to very simply express the logarithm of the S-matrix as

$$\log\left(\mathbb{I} + i\hat{\mathcal{M}}\right) = \sum_{n=1}^{\infty} (-1)^{n+1} \frac{(i\hat{\mathcal{M}})^n}{n} \tag{7.104}$$

$$= \sum_{n=1}^{\infty} (-1)^{n+1} \frac{2^{n-1}}{n}\left(\frac{-i\frac{maV_0}{\hbar}}{p + i\frac{maV_0}{\hbar}}\right)^n \begin{pmatrix} 1 & 1 \\ 1 & 1 \end{pmatrix}$$

$$= \frac{1}{2} \log \left(1 + \frac{-2i\frac{maV_0}{\hbar}}{p + i\frac{maV_0}{\hbar}} \right) \begin{pmatrix} 1 & 1 \\ 1 & 1 \end{pmatrix}$$

$$= \frac{1}{2} \log \left(\frac{p - i\frac{maV_0}{\hbar}}{p + i\frac{maV_0}{\hbar}} \right) \begin{pmatrix} 1 & 1 \\ 1 & 1 \end{pmatrix}.$$

In this form, it might be a bit confusing as to what a logarithm of a complex number is; however, we can interpret and identify it by noting that a general complex number z can be expressed as the product of a magnitude r and a phase ϕ, or **argument**:

$$\log z = \log(re^{i\phi}) = \log r + i\phi. \tag{7.105}$$

Additionally, recall that the argument ϕ is defined as the angle above the positive real axis. Thus, it can be evaluated through a ratio of the imaginary (vertical) to the real (horizontal) components:

$$\tan \phi = \frac{\text{Im}(z)}{\text{Re}(z)}. \tag{7.106}$$

The logarithm of a complex number can then be defined as

$$\log z = \log(re^{i\phi}) = \log r + i\phi = \log r + i \tan^{-1} \frac{\text{Im}(z)}{\text{Re}(z)}. \tag{7.107}$$

Applying this result to the logarithm of interest, we first note that the absolute value is unity:

$$r = \left| \frac{p - i\frac{maV_0}{\hbar}}{p + i\frac{maV_0}{\hbar}} \right| = 1, \tag{7.108}$$

so $\log r = 0$. Then, the only contribution to the logarithm is through its argument, which requires its real and imaginary parts to evaluate. These are

$$\text{Re} \left(\frac{p - i\frac{maV_0}{\hbar}}{p + i\frac{maV_0}{\hbar}} \right) = \frac{1}{2} \left(\frac{p - i\frac{maV_0}{\hbar}}{p + i\frac{maV_0}{\hbar}} + \frac{p + i\frac{maV_0}{\hbar}}{p - i\frac{maV_0}{\hbar}} \right) = \frac{p^2 - \left(\frac{maV_0}{\hbar}\right)^2}{p^2 + \left(\frac{maV_0}{\hbar}\right)^2}, \tag{7.109}$$

$$\text{Im} \left(\frac{p - i\frac{maV_0}{\hbar}}{p + i\frac{maV_0}{\hbar}} \right) = \frac{1}{2i} \left(\frac{p - i\frac{maV_0}{\hbar}}{p + i\frac{maV_0}{\hbar}} - \frac{p + i\frac{maV_0}{\hbar}}{p - i\frac{maV_0}{\hbar}} \right) = \frac{2p\frac{maV_0}{\hbar}}{p^2 + \left(\frac{maV_0}{\hbar}\right)^2}.$$

It then follows that the logarithm we need is

$$\log \left(\frac{p - i\frac{maV_0}{\hbar}}{p + i\frac{maV_0}{\hbar}} \right) = i \tan^{-1} \frac{2p\frac{maV_0}{\hbar}}{p^2 - \left(\frac{maV_0}{\hbar}\right)^2}. \tag{7.110}$$

Putting it all together, the Hermitian matrix $\hat{T}$ that can be exponentiated to determine the S-matrix is

$$\hat{T} = \frac{1}{2} \tan^{-1} \left(\frac{2p\frac{maV_0}{\hbar}}{p^2 - \left(\frac{maV_0}{\hbar}\right)^2} \right) \begin{pmatrix} 1 & 1 \\ 1 & 1 \end{pmatrix}. \tag{7.111}$$

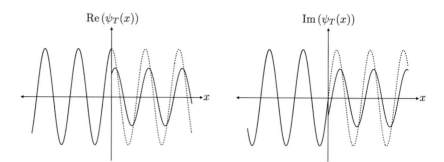

Fig. 7.16 Plots of the real and imaginary parts of an initial wave from the left incident on the high, narrow potential barrier, and the corresponding transmitted wave described by the S-matrix of Eq. (7.86) (solid). The dashed wave to the right of the spatial origin represents the initial wave if it continued right without scattering off the potential.

Note that this is indeed a Hermitian matrix: all elements are real and it is a symmetric matrix.

In this form, it is interesting to consider how the limiting behavior is manifest. First, as $p \to \infty$, we argued that the potential should be irrelevant and transparent, and so the S-matrix is just the identity. That is, the Hermitian matrix $\hat{T}$ should be 0. Indeed, as $p \to \infty$, we see that the argument of $\tan^{-1}$ approaches 0 from the positive values, for which $\hat{T}$ does indeed vanish. On the other hand, as $p \to 0$, the argument of $\tan^{-1}$ approaches 0 from negative values, demonstrating that the corresponding angle approaches 2π. The factor of $1/2$ in front of the inverse tangent means that the phase in this exponential form approaches π, as $p \to 0$, indicating that the reflected wave amplitude is opposite to that of the incident wave, as we identified earlier.

To illustrate the change of phase as the wave passes through this narrow potential, we have illustrated the initial wave from the left and the corresponding transmitted wave in Fig. 7.16. In this figure, we have plotted both the real and imaginary parts of the transmitted waves, according to the S-matrix of Eq. (7.86). The amplitude of the transmitted wave has decreased compared to the initial wave, as required by unitarity because there is some amount of the wave that is reflected (that is not plotted). The difference in the phase of the initial wave and the transmitted wave is also apparent. The potential barrier effectively slows the wave down where it is non-zero, resulting in a transmitted wave that is effectively "behind" the initial wave, were it to continue right with no impedance.

7.5.3 Poles of the S-matrix are Bound States

As a final property of the S-matrix, let's go back to its interpretation as encoding the relative amplitudes of the out states with respect to the in states. From the S-matrix elements, we can construct the waveform for all values of position x. For example, when $x < 0$, the wave can be written as a linear combination of the initial, right-moving wave, its reflection, and the transmission of the incoming wave from the right:

$$\psi_{x<0}(x) \propto e^{i\frac{px}{\hbar}} + (A_R + A_T)e^{-i\frac{px}{\hbar}} . \tag{7.112}$$

In the region on the right, for $x > a$, the wave can be written similarly, but as a linear combination of the left-moving wave incoming from the right, its reflection, and the transmission of the incoming wave from the left:

$$\psi_{x>a}(x) \propto e^{-i\frac{px}{\hbar}} + \left(-\frac{A_T}{A_T^*}A_R^* + A_T \right) e^{i\frac{px}{\hbar}} . \tag{7.113}$$

In these expressions, note the sign of the exponential factor that denotes the direction of propagation of the wave.

Now, let's plug in our expressions for these reflection and transmission amplitudes for our narrow well. We note that

$$A_R + A_T = -\frac{A_T}{A_T^*}A_R^* + A_T = \frac{p - i\frac{maV_0}{\hbar}}{p + i\frac{maV_0}{\hbar}} . \tag{7.114}$$

With this identification, the waves in the regions $x < 0$ and $x > 0$ are

$$\psi_{x<0}(x) \propto e^{i\frac{px}{\hbar}} + \frac{p - i\frac{maV_0}{\hbar}}{p + i\frac{maV_0}{\hbar}} e^{-i\frac{px}{\hbar}} , \tag{7.115}$$

$$\psi_{x>0}(x) \propto e^{-i\frac{px}{\hbar}} + \frac{p - i\frac{maV_0}{\hbar}}{p + i\frac{maV_0}{\hbar}} e^{i\frac{px}{\hbar}} .$$

These can be combined into a single expression for the wave, noting that the signs of the exponents can be expressed as absolute values of position, for all x:

$$\psi(x) \propto e^{-i\frac{p|x|}{\hbar}} + \frac{p - i\frac{maV_0}{\hbar}}{p + i\frac{maV_0}{\hbar}} e^{i\frac{p|x|}{\hbar}} . \tag{7.116}$$

Further, because we only care about proportionality to the wave, we can multiply both terms by the denominator factor:

$$\psi(x) \propto \left(p + i\frac{maV_0}{\hbar} \right) e^{-i\frac{p|x|}{\hbar}} + \left(p - i\frac{maV_0}{\hbar} \right) e^{i\frac{p|x|}{\hbar}} . \tag{7.117}$$

As we have stated many times before, momentum eigenstates are not normalizable on $x \in (-\infty, \infty)$ and so do not live in the Hilbert space. However, a momentum eigenstate is one for which the momentum eigenvalue p is a real number. If p is instead an imaginary number, $p = ik$ for real k, then our analysis of the normalization of states is different. With this imaginary replacement, the wave transforms to

$$\psi(x) \propto \left(k + \frac{maV_0}{\hbar} \right) e^{\frac{k|x|}{\hbar}} + \left(k - \frac{maV_0}{\hbar} \right) e^{-\frac{k|x|}{\hbar}} . \tag{7.118}$$

The exponential phase factors have transformed into either exponential growth or decay. If this is to be a normalizable wavefunction, then the exponential growth as $|x| \to \infty$ must be eliminated. For example, if $k > 0$, then we must set the coefficient of the $e^{k|x|/\hbar}$ term to 0:

$$k = -\frac{maV_0}{\hbar} . \tag{7.119}$$

This may not look like it can be positive, but recall that for a barrier potential $V_0 > 0$, and so $V_0 < 0$ for a well potential. Then, we should think of $V_0 = -|V_0|$ and indeed

$$k = \frac{ma|V_0|}{\hbar} > 0. \tag{7.120}$$

So, for this special value of imaginary momentum, the scattering momentum eigenstate actually transforms into a normalizable state that, thus, lives in the Hilbert space. Going back to where we started, this value of k corresponds to the value of momentum p for which the reflection and transmission amplitudes diverge. That is:

$$p = i\frac{ma|V_0|}{\hbar} \tag{7.121}$$

is the value of momentum for which the S-matrix develops a **pole** or divergence, which corresponds to a state in the Hilbert space. Such a state has compact support, or has no probability to exist at infinite position and thus it is a **bound state**. This bound-state wavefunction is then

$$\psi(x) \propto e^{-\frac{ma|V_0||x|}{\hbar^2}}. \tag{7.122}$$

In the limit we took of the potential, we restricted the region in which the potential was non-zero to be isolated at $x = 0$, but in such a way that the potential had a finite integral over positions. Thus, the potential became a Dirac δ-function, where

$$V(x) = aV_0\,\delta(x), \tag{7.123}$$

as shown in Fig. 7.17. If we consider a position away from $x = 0$, then the potential is 0 and the energy eigenvalue equation is

$$\frac{\hat{p}^2}{2m}\psi(x) = -\frac{\hbar^2}{2m}\frac{d^2}{dx^2}\psi(x) = E\psi(x). \tag{7.124}$$

Taking the second derivatives of the wavefunction we have identified through the pole of the S-matrix, we find the energy of the state to be

$$E = -\frac{\hbar^2}{2m}\frac{\frac{d^2}{dx^2}\psi(x)}{\psi(x)} = -\frac{\hbar^2}{2m}\frac{m^2a^2V_0^2}{\hbar^4} = -\frac{ma^2V_0^2}{2\hbar^2}. \tag{7.125}$$

(a) Illustration of the δ-function potential, with the solid black line representing the negative infinite potential value at $x = 0$. (b) The bound-state wavefunction of the δ-function potential.

Because we only found one pole in the S-matrix for scattering from this δ-function potential, it has only one bound state, with energy equal to that found above. This bound state's position space functional form is plotted in Fig. 7.17.

Exercises

7.1 Let's first consider a free particle, whose wavefunction can be expressed as

$$\psi(x,t) = \int_{-\infty}^{\infty} \frac{dp}{\sqrt{2\pi\hbar}} g(p) e^{-i\frac{E_p t - px}{\hbar}}, \qquad (7.126)$$

where $g(p)$ is a complex-valued, L^2-normalized function of momentum p and $E_p = p^2/2m$, the kinetic energy.

(a) Assume that this wavefunction is a coherent state at time $t = 0$:

$$\hat{a}|\psi\rangle = \lambda|\psi\rangle, \qquad (7.127)$$

where $\hat{a}$ is the lowering operator we introduced with the harmonic oscillator and λ is a complex number. What differential equation must $g(p)$ satisfy for this wavefunction to be a coherent state? Using the results of Sec. 6.2.2, can you explicitly solve this differential equation?

(b) What is the speed of the center-of-probability of this initial coherent state, at time $t = 0$? What is the acceleration of the center-of-probability for any time t? From these results, provide an interpretation of λ.

(c) We had demonstrated that the eigenstates of $\hat{a}^\dagger$ had problems in the harmonic oscillator. For the free particle, what are the states that are eigenstates of the raising operator, $\hat{a}^\dagger$? Are they allowed in this case?

7.2 For this problem, you will use properties of the reflection A_R and transmission A_T amplitudes that we derived in this chapter:

$$A_R = \frac{mV_0}{k^2 - mV_0 + ik\sqrt{k^2 - 2mV_0}\cot\left(\frac{a}{\hbar}\sqrt{k^2 - 2mV_0}\right)}, \qquad (7.128)$$

$$A_T = \frac{ke^{-i\frac{ka}{\hbar}}\sqrt{k^2 - 2mV_0}}{k\sqrt{k^2 - 2mV_0}\cos\left(\frac{a}{\hbar}\sqrt{k^2 - 2mV_0}\right) - i(k^2 - mV_0)\sin\left(\frac{a}{\hbar}\sqrt{k^2 - 2mV_0}\right)}.$$

We'll explore the properties of these amplitudes for complex-valued momentum $k \in \mathbb{C}$.

(a) Show that the poles of the reflection and transmission amplitudes are located at the same value of k.

(b) Under what conditions on a and V_0 will there be no poles in the reflection and transmission amplitudes? What does this mean for the well, physically, from our discussion of bound states?

(c) In this chapter, we had studied the limit of these amplitudes when the potential becomes narrow, but let's now consider the limit of fixed width a but very deep bottom of the potential: $V_0 < 0$ and $|V_0| \to \infty$. Where are the poles in these amplitudes now, in this limit? Have we seen this before?

7.3   From the construction of the S-matrix, we had identified the interaction matrix as that which encodes the non-trivial reflection and lack of transparency off a potential. In this problem, we'll provide another interpretation of the optical theorem and find a constraint on the eigenvalues of the interaction matrix.

(a)   Prove that any matrix $\mathbb{A}$ can be expressed as the sum of two Hermitian matrices $\mathbb{H}_1, \mathbb{H}_2$ as $\mathbb{A} = \mathbb{H}_1 + i\mathbb{H}_2$.

(b)   Using this result, we can write the interaction matrix as $\hat{\mathcal{M}} = \mathbb{X} + i\mathbb{Y}$, for two Hermitian matrices $\mathbb{X}, \mathbb{Y}$. Call the eigenvalues of these matrices x_n and y_n, respectively, for n that ranges over the dimension of the interaction matrix. Show that these eigenvalues lie on the circle in the (x_n, y_n) plane centered at $(0, 1)$ with radius 1. The representation of this circle is called an **Argand diagram**.

(c)   Explicitly determine the eigenvalues for the interaction matrix $\hat{\mathcal{M}}$ of the narrow potential barrier provided in Eq. (7.94). Do they live on the Argand circle? Can any point on the Argand circle be realized, for particular values of momentum?

7.4   In this chapter, we considered the scattering of momentum eigenstates off localized potentials. Of course, momentum eigenstates are not states in the Hilbert space, and so we have to consider L^2-normalizable linear combinations of momentum eigenstates as honest, physical states. In this exercise, we extend what was done in Example 7.2 where we determined the transmission and reflection amplitudes for an initial wave packet scattering off a potential that only has support near $x = 0$. The initial wave packet we consider in this problem has a profile in momentum space of

$$g(p) = \frac{e^{-i\frac{px_0}{\hbar}}}{(\pi\sigma_p^2)^{1/4}} e^{-\frac{(p-p_0)^2}{2\sigma_p^2}}, \qquad (7.129)$$

exactly as in that example.

(a)   Verify that this wave packet is L^2-normalized.

(b)   Determine the initial wavefunction in position space $\psi(x)$ through Fourier transforming the momentum space representation.

(c)   What is the group velocity, the velocity of the center-of-probability, of this wave packet in the region where there is 0 potential?

(d)   What are the transmitted and reflected wavefunctions in position space?

(e)   What do the transmitted and reflected wavefunctions become in the limit that the initial wave packet becomes very narrow about momentum p_0, $\sigma_p \to 0$? Does this make sense from the analysis of this chapter?

7.5   We introduced an operator $\hat{U}_p(x, x + \Delta x)$ in this chapter that represented generalized translations for which momentum was not assumed to be conserved. We used that operator to determine the reflection and transmission amplitudes for scattering off the step potential, but we only needed the first few orders of its Taylor expansion for our purposes there. In this problem, we will identify some more properties of this translation operator and explicitly evaluate it for a finite translation and a given potential.

(a) From the definition of $\hat{U}_p(x, x + \Delta x)$ as a translation operator and its connection to momentum, determine the differential equation in the displacement Δx that $\hat{U}_p(x, x + \Delta x)$ satisfies, for a fixed total energy E and a spatially varying potential $V(x)$.

(b) Let's consider the step potential of this chapter, with general width a and height V_0. Determine the translation operator $\hat{U}_p(x_0, x_1)$ in position space that translates the wavefunction across the entire potential, so that $x_0 < 0$ and $x_1 > a$.

    Hint: Remember how translation operators compose.

(c) Now, let's consider the narrow-potential limit we introduced in this chapter for which $a \to 0$ and $V_0 \to \infty$, but with $aV_0 = \alpha = $ constant. What is the translation operator $\hat{U}_p(x_0, x_1)$ in this case? Is the limit sensible?

7.6 We have stressed that the wavefunction of a quantum mechanical particle does not directly represent the trajectory of that particle through space. This is very unlike the way in which we typically formulate classical mechanics, in which Newton's laws, for example, directly quantify the response of a particle's trajectory to external forces. Additionally, classical mechanics exhibits a "smoothness" in which particle trajectories are continuous and differentiable functions in space and time. In this problem, we will study properties of quantum mechanical trajectories and attempt to determine ways in which they are similar or different to that of classical trajectories. This problem has been analyzed in the literature,[5] and here we will explore some aspects of that paper's conclusions.

The central quantity we will use to study the trajectory of a quantum mechanical particle is its dimension. We define it in the following way. Imagine dividing up the trajectory in steps of size Δx. Then, the sum of all such steps is the total length ℓ of the trajectory. In general, note that ℓ depends on the resolution length Δx. However, we can introduce a length that is independent of Δx, which we call the Hausdorff length L:

$$L = \lim_{\Delta x \to 0} \ell (\Delta x)^{D-1}, \tag{7.130}$$

where D is a quantity chosen so that L is independent of resolution Δx. The quantity D is called the **Hausdorff dimension** of the curve,[6] and if $D \neq 1$, the curve is a **fractal**: continuous but everywhere non-smooth.[7] Classical particle trajectories have Hausdorff dimension $D = 1$; that is, they are smooth curves. What is the dimension of a quantum mechanical trajectory?

---

[5] L. F. Abbott and M. B. Wise, "The dimension of a quantum mechanical path," *Am. J. Phys.* **49**, 37–39 (1981).

[6] F. Hausdorff, "Dimension und äußeres Maß," *Math. Ann.* **79**, 157–179 (1918).

[7] B. B. Mandelbrot, "How long is the coast of Britain? Statistical self-similarity and fractional dimension," *Science* **156**(3775), 636–638 (1967).

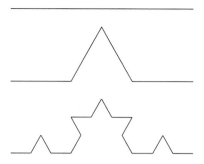

Fig. 7.18   Illustration of the first three steps of the construction of the Koch snowflake.

(a) We have just stated that smooth curves have Hausdorff dimension $D = 1$. Argue that this must be the case.

(b) A classical example of a continuous but everywhere non-smooth curve is the Koch snowflake.[8] It can be iteratively constructed and the first few steps of the construction are shown in Fig. 7.18. The procedure continues *ad infinitum*, with each straight segment replaced by four segments in a kinked shape, with each segment being one-third the length of the segment at the previous step. From the definition provided above, determine the Hausdorff dimension for the Koch snowflake.

(c) With this warm-up, let's determine the Hausdorff dimension of the trajectory of a quantum mechanical particle of mass m. We will assume that the average momentum of the particle is 0, $\langle \hat{p} \rangle = 0$. Using the Heisenberg uncertainty principle, determine a relationship between the step size Δx and a time step Δt. Assume that the Heisenberg uncertainty is saturated. Now, if the total elapsed time of the trajectory is $T = N\Delta t$, what is its length ℓ? Then, what is the Hausdorff dimension?

(d) Let's relax the assumption that the average momentum is 0. Let's assume that $\langle \hat{p} \rangle = p_0$ and, from the steps in part (c), what is the Hausdorff dimension of the trajectory of such a quantum mechanical particle now? Does the $p_0 \to \infty$ limit match with your expectation from the correspondence principle?
Hint: Given $\langle \hat{p} \rangle$, can you determine $\langle \hat{p}^2 \rangle$ in terms of the variance of momentum?

(e) Now redo parts (c) and (d) from the perspective of the one-dimensional free particle wavefunction, instead of the Heisenberg uncertainty principle. See the Abbott and Wise paper for hints.

7.7 We demonstrated that the S-matrix is unitary and that its poles correspond to L^2-normalizable eigenstates of the Hamiltonian, at least in the case of the narrow step potential. Assuming this is true in general for an arbitrary localized and

---

[8] H. von Koch, "Sur une courbe continue sans tangente, obtenue par une construction géométrique élémentaire," *Ark. Mat. Astron. Fys.* **1**, 681–702 (1904).

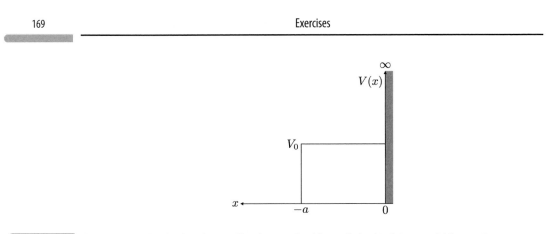

Fig. 7.19   Plot of a potential with a finite barrier of height V_0 and width a and a hard, infinite potential for $x > 0$.

compact potential, there are a few things that we immediately know about the structure of the S-matrix.

(a) Can an S-matrix have poles at real values of momentum p? If yes, provide an example; if no, explain why not.

(b) What is the large-momentum limit of the S-matrix? That is, as $|p| \to \infty$, what is the "worst" scaling with p that the S-matrix can exhibit?

7.8   Consider the potential illustrated in Fig. 7.19, where there is a finite potential barrier of height V_0 and width a, that terminates at $x = 0$ at which the potential becomes infinite for all $x > 0$. Consider a plane wave of momentum p incident on this potential from the left, toward $x \to -\infty$.

(a) Write down the S-matrix for scattering off this potential. What is the transmission amplitude?

(b) What is the phase accumulated from the reflection of an incident wave as it passes over the finite barrier and returns traveling to the left?

8

Rotations in Three Dimensions

With scattering and its formalism in the previous chapter, we pivot to studying quantum mechanics in more than one dimension. In this and the next chapter, we will work to generalize our formulation of quantum mechanics to be more realistic (i.e., actually account for the multiple spatial dimensions that we experience in our universe). In this chapter, we will introduce a profound consequence of living in multiple spatial dimensions and a study of angular momentum in quantum mechanics. We will start by setting the stage for describing rotations and later, construct the complete theory of angular momentum, at least as much as we need here.

8.1 Review of One-Dimensional Transformations

To start, let's quantify the transformations of quantum mechanics in one spatial dimension and see what that can get us. First, we restrict ourselves to linear operations as this is quantum mechanics, and further to transformations that keep us in the Hilbert space (i.e., unitary operations). In one spatial and one time dimension, we had identified two such transformations corresponding to translation through space or time. Translation in space is accomplished by the unitary operator

$$\hat{U}_p(x) = e^{i\frac{x\hat{p}}{\hbar}}, \tag{8.1}$$

which translates a distance x, according to the Hermitian momentum operator $\hat{p}$. Translation in time is accomplished with the unitary operator

$$\hat{U}_H(t) = e^{-i\frac{t\hat{H}}{\hbar}}, \tag{8.2}$$

where t is the time translation and $\hat{H}$ is the Hermitian Hamiltonian operator.

These transformations were the most relevant for our purposes in the last several chapters, but other operations on space and time can be defined. For example, one can define an operator that produces a linear combination of the space and time coordinates. Such a mixing of space and time is a **Lorentz transformation** and is the starting point for relativistic quantum mechanics, but beyond the scope of this book. Another operation that exclusively acts on position is that of **dilation**, or a scale transformation. The unitary dilation operator $\hat{U}_d(\lambda)$ acts on position x as

$$\hat{U}_d(\lambda)x = e^{\lambda}x, \tag{8.3}$$

and λ is some real number. This can also be expressed as an exponentiated Hermitian operator $\hat{D}$ where

$$\hat{D} = \hat{x}\hat{p} - \frac{i\hbar}{2}, \tag{8.4}$$

which you can verify is Hermitian. Other operations on the spatial coordinate can be defined as well, but they can all be expressed in terms of some combined action of the one-dimensional momentum and position operators.

8.2 Rotations in Two Dimensions in Quantum Mechanics

If we consider two spatial dimensions, things get much more interesting. First, in two spatial dimensions x and y, say, we can translate in either dimension, so we have two components of a Hermitian momentum vector $\hat{\vec{p}} = (\hat{p}_x, \hat{p}_y)$, where in position space

$$\hat{p}_x = -i\hbar \frac{\partial}{\partial x}, \qquad\qquad \hat{p}_y = -i\hbar \frac{\partial}{\partial y}. \tag{8.5}$$

Because partial derivatives commute, note that $\hat{p}_x$ and $\hat{p}_y$ commute with each other: $[\hat{p}_x, \hat{p}_y] = 0$.

However, unlike in one spatial dimension, these translations aren't everything. Just like we could mix space and time, we can in principle mix the x dimension and the y dimension in some linear combination. If, additionally, we want that linear combination to be unitary and keep positions real-valued, we are uniquely led to the possible transformation. If we call the position vector $\vec{r}$ where

$$\vec{r} = \begin{pmatrix} x \\ y \end{pmatrix}, \tag{8.6}$$

then the only possible real, unitary transformation with determinant 1 is

$$\hat{U}(\theta) = \begin{pmatrix} \cos\theta & -\sin\theta \\ \sin\theta & \cos\theta \end{pmatrix}, \tag{8.7}$$

for some angle θ. As we have studied, this is indeed unitary. As a unitary operator, of course it can be written as the exponentiation of some Hermitian operator we will call $\hat{L}$. What is this $\hat{L}$?

8.2.1 The Angular Momentum Operator

We want to write

$$\hat{U}(\theta) = e^{i\frac{\theta\hat{L}}{\hbar}} = \mathbb{I} + i\frac{\theta\hat{L}}{\hbar} + \cdots, \tag{8.8}$$

where we have Taylor expanded in angle θ on the right. To determine $\hat{L}$, we can just Taylor expand the action of $\hat{U}(\theta)$ on the vector $\vec{r}$:

$$\hat{U}(\theta)\vec{r} = \begin{pmatrix} x\cos\theta - y\sin\theta \\ x\sin\theta + y\cos\theta \end{pmatrix} = \begin{pmatrix} x - \theta y + \cdots \\ y + \theta x + \cdots \end{pmatrix} = \vec{r} + \theta \begin{pmatrix} -y \\ x \end{pmatrix} + \cdots \qquad (8.9)$$
$$= \left(\mathbb{I} + i\frac{\theta\hat{L}}{\hbar} \right)\vec{r} + \cdots.$$

Equating these expressions, we apparently must have

$$i\frac{\theta\hat{L}}{\hbar}\vec{r} = i\frac{\theta\hat{L}}{\hbar} \begin{pmatrix} x \\ y \end{pmatrix} = \theta \begin{pmatrix} -y \\ x \end{pmatrix}, \qquad (8.10)$$

or, by canceling θ and multiplying by $-i\hbar$:

$$\hat{L} \begin{pmatrix} x \\ y \end{pmatrix} = -i\hbar \begin{pmatrix} -y \\ x \end{pmatrix}. \qquad (8.11)$$

Now, this may look strange, but you can easily write down a 2×2 matrix for $\hat{L}$ that accomplishes this. However, I want to do something different. Let's focus on the first entry of this linear equation:

$$\hat{L}x = i\hbar y. \qquad (8.12)$$

What linear operations allow us to turn x into y? Differentiating with respect to x returns 1:

$$\frac{\partial}{\partial x}x = 1, \qquad (8.13)$$

so we just multiply by $i\hbar y$:

$$\hat{L}x = i\hbar y \frac{\partial}{\partial x}x = i\hbar y. \qquad (8.14)$$

Similarly, we can turn y into $-i\hbar x$:

$$\hat{L}y = -i\hbar x \frac{\partial}{\partial y}y = -i\hbar x. \qquad (8.15)$$

Because the partial derivatives annihilate the "wrong" coordinate

$$\frac{\partial}{\partial x}y = \frac{\partial}{\partial y}x = 0, \qquad (8.16)$$

we can combine these actions into a single operator:

$$\hat{L} = -i\hbar x \frac{\partial}{\partial y} + i\hbar y \frac{\partial}{\partial x} = \hat{x}\hat{p}_y - \hat{y}\hat{p}_x. \qquad (8.17)$$

In this re-writing, we have expressed the operator $\hat{L}$ in terms of the momentum and position operators. Now, even from freshman physics this should look very familiar: this operator $\hat{L}$ is just **angular momentum** expressed like an operator cross product. We say in quantum mechanics that angular momentum generates rotations in space. Note also that $\hat{L}$ is Hermitian because $\hat{x}$, $\hat{y}$, $\hat{p}_x$, and $\hat{p}_y$ are Hermitian, and the commutator of position and the "wrong" momentum is 0 (e.g., $[\hat{x}, \hat{p}_y] = 0$).

8.2.2 Rotations as a Group

We'll come back to angular momentum and get to three dimensions later in this chapter, but in this section, I want to focus on the unitary rotation matrix $\hat{U}(\theta)$:

$$\hat{U}(\theta) = \begin{pmatrix} \cos\theta & -\sin\theta \\ \sin\theta & \cos\theta \end{pmatrix}. \tag{8.18}$$

Acting on the vector $\vec{r}$, this of course rotates $\vec{r}$ by angle θ, as shown in Fig. 8.1. In the figure, we have

$$\vec{r}' = \begin{pmatrix} x\cos\theta - y\sin\theta \\ x\sin\theta + y\cos\theta \end{pmatrix}. \tag{8.19}$$

What if we further rotate by ϕ with the matrix $\hat{U}(\phi)$? We have

$$\hat{U}(\phi)\hat{U}(\theta)\vec{r} = \begin{pmatrix} \cos\phi & -\sin\phi \\ \sin\phi & \cos\phi \end{pmatrix} \begin{pmatrix} \cos\theta & -\sin\theta \\ \sin\theta & \cos\theta \end{pmatrix} \begin{pmatrix} x \\ y \end{pmatrix} \tag{8.20}$$

$$= \begin{pmatrix} \cos\phi\cos\theta - \sin\phi\sin\theta & -\cos\phi\sin\theta - \sin\phi\cos\theta \\ \sin\phi\cos\theta + \cos\phi\sin\theta & -\sin\phi\sin\theta + \cos\phi\cos\theta \end{pmatrix} \begin{pmatrix} x \\ y \end{pmatrix}$$

$$= \begin{pmatrix} \cos(\phi+\theta) & -\sin(\phi+\theta) \\ \sin(\phi+\theta) & \cos(\phi+\theta) \end{pmatrix} \begin{pmatrix} x \\ y \end{pmatrix}.$$

In the final line we used angle addition formulas, and actually, this is a way to remember them. Not surprisingly, rotating a two-dimensional vector an angle θ and then ϕ is equivalent to just rotation by the total angle $\theta + \phi$, as shown in Fig. 8.1.

Further, it didn't matter what order we performed the rotations, ϕ first or θ first, as

$$\hat{U}(\theta)\hat{U}(\phi) = \hat{U}(\phi)\hat{U}(\theta) = \hat{U}(\phi+\theta) = \hat{U}(\theta+\phi). \tag{8.21}$$

That is, we say that rotations of two-dimensional vectors are **commutative**. We have also identified the multiplication law of rotations, as expressed above. This rich structure of rotations actually forms a mathematical structure called a **group**. A group is a set of

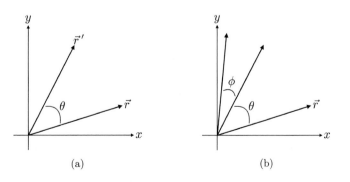

(a)   (b)

Fig. 8.1 (a) Rotation of a two-dimensional vector $\vec{r}$ by an angle θ. (b) Sequential rotation of the vector $\vec{r}$ by an angle θ first, and then an angle ϕ.

objects that has a multiplication operator "·" and that set and multiplication satisfy the following four properties:

(a) *Closure.* If a and b are in the group, then $a \cdot b$ is in the group.
(b) *Existence of an identity.* There exists an element of the group 1 such that $1 \cdot a = a \cdot 1 = a$ for another group element a.
(c) *Existence of an inverse.* For each element a in the group, there exists an element a^{-1} for which $a \cdot a^{-1} = a^{-1} \cdot a = 1$, the identity element.
(d) *Associativity.* For three group elements a, b, c, multiplication is associative: $a \cdot (b \cdot c) = (a \cdot b) \cdot c$.

I claim that the set of all two-dimensional rotations form a group. We call this two-dimensional rotation group **Abelian**, because all elements commute via matrix multiplication, named after Niels Abel, a nineteenth-century Norwegian mathematician. Commutativity is not a requirement to be a group, and we'll soon see an example where that is not the case.

To end this section, I want to emphasize something. If you know the multiplication law for a group, you know everything there is to know about the group. This feature of groups enables many different representations of how that multiplication law is manifest. We saw how rotations acted on the two-dimensional vector $\vec{r}$, but we can represent a point in the x–y plane in many ways. For example, we could say that x and y are the real and imaginary parts of a complex number z: $z = x + iy$. We can equivalently write this as

$$z = re^{i\alpha}, \tag{8.22}$$

for some length $r = \sqrt{x^2 + y^2}$ and phase α, as illustrated in Fig. 8.2.

How do we rotate this representation of the vector? Well, all we need to do is multiply by an appropriate exponential phase! To rotate by an angle θ, we just do

$$re^{i\alpha} \to re^{i\alpha}e^{i\theta} = re^{i(\alpha+\theta)}. \tag{8.23}$$

Rotating again by ϕ is just as trivial:

$$re^{i\alpha} \to re^{i\alpha}e^{i\theta} \to re^{i\alpha}e^{i\theta}e^{i\phi} = re^{i\alpha}e^{i(\theta+\phi)}. \tag{8.24}$$

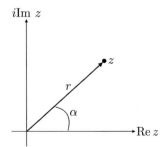

Fig. 8.2 The representation of a two-dimensional vector as a complex number z, with length r and angle α above the real axis.

Let's call

$$\hat{A}(\theta) = e^{i\theta}, \tag{8.25}$$

so that we have shown

$$\hat{A}(\theta)\hat{A}(\phi) = \hat{A}(\phi)\hat{A}(\theta) = \hat{A}(\theta + \phi) = \hat{A}(\phi + \theta). \tag{8.26}$$

But this is the identical multiplication law that we found for the 2×2 rotation matrix! Thus, as groups, we say that the action of multiplication by an exponential phase is equivalent to that of 2×2 rotation matrices. The former group is called **U(1)** (1×1 unitary matrices) and the latter group is called **SO(2)** (special orthogonal 2×2 matrices), so we find that

$$U(1) \simeq SO(2), \tag{8.27}$$

where "$\simeq$" means equivalent as groups. We have thus found two different **representations** of the same group.

8.3 Rotations in Three Dimensions Warm-Up: Two Surprising Properties

Two-dimensional rotations only get us so far because of course we live in three dimensions. On the surface, this distinction might seem to be a triviality, and just require a bit more bookkeeping to keep track of the multiple axes about which one can rotate in higher dimension. However, rotations in three dimensions exhibit some very counterintuitive properties that result in much richer physical phenomena than might be expected. We will discuss two of these properties here, and then work to construct the general theory of three-dimensional rotations in the next section.

8.3.1 Non-commutativity

One thing we noted when introducing groups was that the property of commutativity was not implied by the four group properties. We happened to see that two-dimensional rotations do commute, just like multiplication of familiar numbers. However, just like two generic matrices $\mathbb{A}$ and $\mathbb{B}$ do not commute: $\mathbb{AB} \neq \mathbb{BA}$, we shouldn't expect that the subsequent action of rotations in three dimensions commutes. So, to make this concrete, let's consider a three-dimensional object and just see how it rotates. A coffee cup is a convenient prop because the handle serves as a nice reference for the orientation of the cup. Now, we can rotate this thing around however we want and you are able to put it in whatever orientation you want by appropriate rotation.

What we will do now is consider what happens when we perform two subsequent rotations in different orders. One rotation, call it $\mathbb{R}_1$, rotates the cup $90°$ about the axis that passes through the center of the cup when thought of as a cylinder:

$$(8.28)$$

The other rotation we will perform, call it $\mathbb{R}_2$, is a 90° rotation of the cup about an axis that passes through the "walls" of the cup:

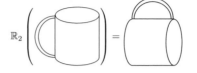

$$(8.29)$$

Okay, let's see what happens if we perform $\mathbb{R}_1$ first, and then $\mathbb{R}_2$:

$$\mathbb{R}_2\left(\mathbb{R}_1\left(\right)\right) = \mathbb{R}_2\left(\right) = \qquad (8.30)$$

with the bottom of the cup facing us. What about $\mathbb{R}_2$ first and then $\mathbb{R}_1$?

$$\mathbb{R}_1\left(\mathbb{R}_2\left(\right)\right) = \mathbb{R}_1\left(\right) = \qquad (8.31)$$

Are the implementations of these rotations in different orders the same? No! Apparently, three-dimensional rotations, unlike rotations in two dimensions, are not commutative or are **non-Abelian**:

$$\mathbb{R}_2\mathbb{R}_1 \neq \mathbb{R}_1\mathbb{R}_2. \qquad (8.32)$$

Again, we didn't require it to be Abelian and didn't necessarily expect it to be, but it is perhaps somewhat shocking to see it so explicitly.

8.3.2 $4\pi = 0$

The next property of three-dimensional rotations will be extremely surprising, I hope. To set the stage, let's go back to just two-dimensional rotations. As you well know, if you rotate by 2π, you get back to where you started, as exhibited by, say, you rotating in place when standing. But wait: you do get back where you start in the sense that a rotation by 2π returns you to your original orientation, but this is very different than equating a 2π rotation with no rotation at all. Indeed, if I rotate in place by 2π, my angle is then 2π; if I go another 2π, my angle is 4π, etc. By continuing to rotate in the same direction I can never get back to no rotation (i.e., 0 angle). A concrete way to track this is to wind a string around a cylinder or cardboard roll. If you wind the string by 2π, then indeed a point on the string gets back to where you started, but you can

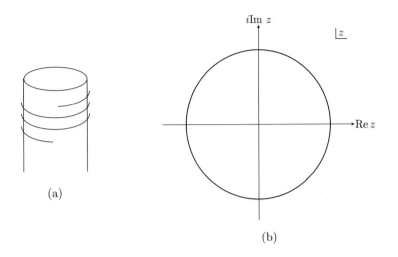

Fig. 8.3 (a) Illustration of a string wound around a cylinder three times. (b) The unit circle in the complex plane.

never unwind the string by wrapping in the same direction. In Fig. 8.3, a string has been wrapped three times around the cylinder. This is a concrete manifestation of the property that the group of two-dimensional rotations is not **simply connected**. There exist orbits in the group space (i.e., trajectories from repeated action of the group elements) that, while they return the object to the original orientation, are not equivalent to the identity, no rotation at all. This is not so surprising: we had identified the group of two-dimensional rotations as SO(2)$\simeq$U(1), and U(1) can be identified with the unit circle in the complex plane as illustrated in Fig. 8.3. A circle clearly has a hole in it, so another colloquial way to think about connectedness and simple vs. not simple is (simply): if the group or set of interest has a hole in it, then it is not simply connected. This is also why the cylinder works to track how rotations accumulate as you wind the string.

This was a long-winded way around. What about three-dimensional rotations? Well, you might think that essentially the same argument would hold for three dimensions and therefore three-dimensional rotations would not be simply connected either. However, in three dimensions one has "above-ness" and "below-ness" as an additional orientation between objects and this will make all the difference.

To test the connectedness of the set of all three-dimensional rotations, there's a profound demonstration that you can do with your body, rather than a cylinder and a string. We'll use your arm like the string to keep track of the orbit that you perform as you rotate. So, here's what you do. Put your arm out, palm up, like you are balancing a plate on it. In what follows, always keep your palm up: you don't want to drop the plate! Now, rotate your hand about a vertical axis that passes through your hand. Pass your hand below your arm, and rotate it through 2π. If you did it right, your palm should be in the original orientation, but your arm is all twisted! Apparently, just like two-dimensional rotations, in three dimensions $2\pi \neq 0$.

However, we aren't done. With your hand in this twisted orientation, continue to rotate your hand in the same direction as earlier, but now pass your hand above your arm. Once you've gone around another 2π, you should find that your hand is in the same, original, orientation, but your arm is untwisted! Note the importance of being able to rotate above and below; you can't do that in two dimensions. In total, you rotated your hand through an angle of 4π (2π twice), and everything gets back to where it started. Apparently, then, $4\pi = 0$ for three-dimensional rotations. This further demonstrates that the set of three-dimensional rotations is simply connected: we can always rotate by 4π and get back to where we started, as if nothing ever happened. By the way, this demonstration is called the "plate trick," "belt trick," and several other names.

8.4 Unitary Operators for Three-Dimensional Rotations

We'll come back to the consequences of the observation that $4\pi = 0$ for three-dimensional rotations. For now, we'll just file it away with the non-Abelian property, as well. We now turn our attention to explicit construction of the unitary operator that implements rotations in three dimensions. Let's call this operator $\hat{U}$, as it is unitary. Further, in three dimensions, you can rotate about three independent, orthogonal axes $\hat{x}$, $\hat{y}$, and $\hat{z}$. We'll call the angles we rotate about the appropriate axes θ_x, θ_y, and θ_z, and can form a three-dimensional vector from them:

$$\vec{\theta} = \begin{pmatrix} \theta_x \\ \theta_y \\ \theta_z \end{pmatrix}. \tag{8.33}$$

Then, $\hat{U}$ is a function of $\vec{\theta}$: $\hat{U} = \hat{U}(\vec{\theta})$.

As always, because $\hat{U}$ is unitary, it can be expressed as the exponential of a Hermitian operator $\hat{\vec{L}}$, where we have

$$\hat{U}(\vec{\theta}) = e^{i\frac{\vec{\theta} \cdot \hat{\vec{L}}}{\hbar}} = e^{\frac{i}{\hbar}(\theta_x \hat{L}_x + \theta_y \hat{L}_y + \theta_z \hat{L}_z)}, \tag{8.34}$$

for three Hermitian operators $\hat{L}_x$, $\hat{L}_y$, and $\hat{L}_z$. Actually, we already identified $\hat{L}_z$ as a rotation about the $\hat{z}$-axis and equivalent to rotation of the x–y plane:

$$\hat{L}_z = \hat{x}\hat{p}_y - \hat{y}\hat{p}_x. \tag{8.35}$$

"x," "y," and "z" are just labels, so we can also write down $\hat{L}_x$ and $\hat{L}_y$ by permuting x, y, z:

$$\hat{L}_x = \hat{y}\hat{p}_z - \hat{z}\hat{p}_y, \qquad\qquad \hat{L}_y = \hat{z}\hat{p}_x - \hat{x}\hat{p}_z. \tag{8.36}$$

Note that the x, y, z order is important in these expressions: including the subscript of $\hat{L}$, the first term has $x - y - z$ in that cyclic order. Of course, this is nothing more than our familiar angular momentum, but operator-ized for quantum mechanics. If we defined $\hat{\vec{r}}$ and $\hat{\vec{p}}$ as the position and momentum vector operators

$$\hat{\vec{r}} = \begin{pmatrix} \hat{x} \\ \hat{y} \\ \hat{z} \end{pmatrix}, \qquad\qquad \hat{\vec{p}} = \begin{pmatrix} \hat{p}_x \\ \hat{p}_y \\ \hat{p}_z \end{pmatrix}, \qquad (8.37)$$

then the three-dimensional angular momentum operator $\hat{\vec{L}}$ is

$$\hat{\vec{L}} = \hat{\vec{r}} \times \hat{\vec{p}}, \qquad (8.38)$$

the familiar cross product. So, our unitary rotation matrix $\hat{U}(\vec{\theta})$ can be written as

$$\hat{U}(\vec{\theta}) = e^{i\frac{\vec{\theta}\cdot\hat{\vec{L}}}{\hbar}} = e^{i\frac{\vec{\theta}\cdot(\hat{\vec{r}}\times\hat{\vec{p}})}{\hbar}}. \qquad (8.39)$$

We just need to specify the three numbers of $\vec{\theta}$, and we can perform any rotation we want.

Does the set of all $\hat{U}(\vec{\theta})$s form a group? Let's just check each of the four properties. First, is the identity in this set? The identity operator corresponds to no rotation, $\vec{\theta} = 0$, so this is indeed a "rotation." One property down.

What about the existence of an inverse? For $\hat{U}(\vec{\theta})$, what is its inverse? To get back to the identity, we just rotate in the opposite direction, by $-\vec{\theta}$. This is still a rotation, so if $\hat{U}(\vec{\theta})$ is in the group, then so too is $\hat{U}(-\vec{\theta})$. Further, from unitarity and its exponential form, note that

$$\hat{U}(-\vec{\theta}) = e^{i\frac{(-\vec{\theta})\cdot\hat{\vec{L}}}{\hbar}} = e^{-i\frac{\vec{\theta}\cdot\hat{\vec{L}}}{\hbar}} = \hat{U}(\theta)^{\dagger}, \qquad (8.40)$$

because $\vec{\theta}$ is a real vector and $\hat{\vec{L}}$ is a Hermitian operator. This must have been true because unitarity required that

$$\hat{U}(\vec{\theta})\hat{U}(\vec{\theta})^{\dagger} = \mathbb{I}. \qquad (8.41)$$

Next, what about associativity? We're considering linear operators, so we can always imagine that they are matrices. Matrix multiplication is associative, so we're good there.

The final requirement of a group is closure: for two angle vectors $\vec{\theta}, \vec{\phi}$, the product of their corresponding unitary rotation operators must be in the group:

$$\hat{U}(\vec{\theta})\hat{U}(\vec{\phi}) = \hat{U}(\vec{\gamma}), \qquad (8.42)$$

for a third angle vector $\vec{\gamma}$. Does this impose constraints on the angular momentum operators $\hat{L}_x$, $\hat{L}_y$, and $\hat{L}_z$? What is the multiplication rule for three-dimensional matrices? Can we express it in generality like we did with two-dimensional rotations?

8.4.1 The Lie Algebra of Rotations

Let's see what we find out from explicitly multiplying the unitary rotation operators. We have the requirement that

$$\hat{U}(\vec{\theta})\hat{U}(\vec{\phi}) = \hat{U}(\vec{\gamma}) \quad \Rightarrow \quad e^{i\frac{\vec{\theta}\cdot\hat{\vec{L}}}{\hbar}} e^{i\frac{\vec{\phi}\cdot\hat{\vec{L}}}{\hbar}} = e^{i\frac{\vec{\gamma}\cdot\hat{\vec{L}}}{\hbar}}. \qquad (8.43)$$

Now, if the factors in the exponent were just numbers, then we could simply add the exponents on the left to determine $\vec{\gamma}$, and would find that $\vec{\gamma} = \vec{\theta} + \vec{\phi}$. However, we know that this cannot be true because we had demonstrated that three-dimensional rotations are non-Abelian and do not commute. For operators that do not commute, we cannot simply add exponents; we must be more careful.

Lucky for us, this multiplication law has been worked out for non-commutative operators and it's called the **Baker–Campbell–Hausdorff formula**, or just the BCH formula for brevity.[1] For the case at hand, it reads

$$e^{i\frac{\vec{\theta}\cdot\hat{\vec{L}}}{\hbar}} e^{i\frac{\vec{\phi}\cdot\hat{\vec{L}}}{\hbar}} = \exp\left(i\frac{(\vec{\theta}+\vec{\phi})\cdot\hat{\vec{L}}}{\hbar} + \frac{1}{2\hbar^2}[\vec{\theta}\cdot\hat{\vec{L}}, \vec{\phi}\cdot\hat{\vec{L}}] + \cdots\right) = e^{i\frac{\vec{\gamma}\cdot\hat{\vec{L}}}{\hbar}}. \tag{8.44}$$

The ellipses denote nested commutators, like $[\vec{\theta}\cdot\hat{\vec{L}}, [\vec{\theta}\cdot\hat{\vec{L}}, \vec{\phi}\cdot\hat{\vec{L}}]]$, involving three, four, or more of the angular momentum operators. This is an absolute mess! I want to emphasize again that if this product is to be a rotation matrix, then it must be able to be written as on the right, with the exponent as a simple, real, linear combination of the $\hat{L}_x$, $\hat{L}_y$, and $\hat{L}_z$ angular momentum operators.

For this to be true, we must require a very non-trivial non-linear relationship of the angular momentum operators. If we enforce that the commutator of two angular momenta (a *non-linear* combination) reduces to a linear sum of angular momenta, then the resulting exponent is linear and everything works out. Thus, we require

$$[\vec{\theta}\cdot\hat{\vec{L}}, \vec{\phi}\cdot\hat{\vec{L}}] = i\vec{\alpha}\cdot\hat{\vec{L}}, \tag{8.45}$$

where $\vec{\alpha}$ is some other vector of angles. Actually, this can be expressed much more compactly and in generality exclusively in terms of the angular momentum basis that we have constructed: $\hat{L}_x$, $\hat{L}_y$, and $\hat{L}_z$. For $\hat{L}_i, \hat{L}_j \in \{\hat{L}_x, \hat{L}_y, \hat{L}_z\}$, we must have that

$$[\hat{L}_i, \hat{L}_j] = i\left(\alpha_x\hat{L}_x + \alpha_y\hat{L}_y + \alpha_z\hat{L}_z\right), \tag{8.46}$$

for some real numbers $\alpha_x, \alpha_y, \alpha_z$. In general, this Hermitian operator basis and this non-linear identity are called the **Lie algebra**, after Sophus Lie, a Norwegian mathematician. One point to emphasize is that the Lie algebra completely specifies the multiplication law of the (Lie) group. If you demonstrate that two sets of Hermitian operators have the same Lie algebra (i.e., the same constants $\vec{\alpha}$ called **structure constants**), then they are the same group. We will come back to this point later in much more detail.

So, with this in mind, is it actually true that our set of angular momentum operators form a Lie algebra, and hence the unitary rotation matrices are closed? Given our definitions of $\hat{L}_x$, $\hat{L}_y$, and $\hat{L}_z$ in terms of position and momentum operators, we can just explicitly calculate some commutators. We'll just calculate the commutator

---

[1] It is named after Henry Frederick Baker, John Edward Campbell, and Felix Hausdorff (H. F. Baker, "Further applications of matrix notation to integration problems," *Proc. Lond. Math. Soc.* **1**(1), 347–360 (1901); J. E. Campbell, "On a law of combination of operators bearing on the theory of continuous transformation groups," *Proc. Lond. Math. Soc.* **1**(1), 381–390 (1896), **1**(1), 14–32 (1897); F. Hausdorff, "Die symbolische Exponentialformel in der Gruppentheorie," *Leipz. Ber.* **58**, 19–48 (1906)), but a first exact formula was constructed by Eugene Dynkin (E. B. Dynkin, "Calculation of the coefficients in the Campbell–Hausdorff formula," *Dokl. Akad. Nauk. SSSR (NS)*, **57**, 323–326 (1947)).

$[\hat{L}_x, \hat{L}_y]$, and then show results for the other commutators. This commutator, with our definitions of these operators, is

$$[\hat{L}_x, \hat{L}_y] = [\hat{y}\hat{p}_z - \hat{z}\hat{p}_z, \hat{z}\hat{p}_x - \hat{x}\hat{p}_z] \tag{8.47}$$
$$= (\hat{y}\hat{p}_z - \hat{z}\hat{p}_z)(\hat{z}\hat{p}_x - \hat{x}\hat{p}_z) - (\hat{z}\hat{p}_x - \hat{x}\hat{p}_z)(\hat{y}\hat{p}_z - \hat{z}\hat{p}_z)$$
$$= \hat{x}\hat{p}_y(\hat{z}\hat{p}_z - \hat{p}_z\hat{z}) - \hat{y}\hat{p}_x(\hat{z}\hat{p}_z - \hat{p}_z\hat{z}).$$

In simplifying this expression, I used the fact that momenta commute $[\hat{p}_i, \hat{p}_j] = 0$, positions commute $[\hat{i}, \hat{j}] = 0$, and positions and momentum of different dimensions commute $[\hat{i}, \hat{p}_j] = 0$, for $i \neq j$. With this in mind, I was able to group the remaining terms in the way presented above.

Now, noting that $[\hat{z}, \hat{p}_z] = i\hbar$, the commutator of $\hat{L}_x$ and $\hat{L}_y$ reduces to

$$[\hat{L}_x, \hat{L}_y] = i\hbar\hat{x}\hat{p}_y - i\hbar\hat{y}\hat{p}_x = i\hbar\hat{L}_z. \tag{8.48}$$

Amazingly, this quadratic operator combination reduces to just a linear combination of the basis angular momentum operators. With the same techniques, you can compute the complete commutation relation of the Lie algebra:

$$[\hat{L}_x, \hat{L}_y] = i\hbar\hat{L}_z, \tag{8.49}$$
$$[\hat{L}_z, \hat{L}_x] = i\hbar\hat{L}_y,$$
$$[\hat{L}_y, \hat{L}_z] = i\hbar\hat{L}_x.$$

This can compactly be written as

$$[\hat{L}_i, \hat{L}_j] = i\hbar \sum_{k=1}^{3} \epsilon_{ijk}\hat{L}_k, \tag{8.50}$$

where ϵ_{ijk} is called the totally antisymmetric symbol or the **Levi-Civita symbol**, for which

$$\epsilon_{123} = \epsilon_{312} = \epsilon_{231} = -\epsilon_{213} = -\epsilon_{321} = -\epsilon_{132} = 1, \tag{8.51}$$

and $\epsilon_{ijk} = 0$ if any of the i, j, or k indices are repeated. These commutation relations are called the Lie algebra for $\mathfrak{su}(\mathbf{2})$.

Again, with emphasis, the Lie algebra defines the multiplication rule of the group of unitary operators through the BCH formula. Via exponentiation, two groups that have the same Lie algebra have the same multiplication law and are therefore the same group; properly, they are **isomorphic** as groups. Further, the angular momentum operators we have considered thus far act on three-dimensional position vectors. That is, to rotate the vector $\vec{r}^{\mathsf{T}} = (x\ y\ z)$, we act with a 3×3 matrix $\hat{U}(\vec{\theta})$:

$$\vec{r}' = \hat{U}(\vec{\theta})\vec{r} = \exp\left[\frac{i}{\hbar}\left(\theta_x\hat{L}_x + \theta_y\hat{L}_y + \theta_z\hat{L}_z\right)\right]\begin{pmatrix} x \\ y \\ z \end{pmatrix}. \tag{8.52}$$

We thus call this manifestation of the rotation group in three dimensions a three-dimensional representation, because it rotates a three-dimensional vector.

8.4.2 The Two-Dimensional Representation

Now, you might think that this is the only possible representation of rotations in three dimensions, because what else would you rotate? However, if we find another representation with the same Lie algebra, it's the same group! With this in mind, let's consider two-dimensional matrices and see if we can construct $\mathfrak{su}(2)$. First, these matrices must be Hermitian as we want them to exponentiate to a unitary operator on the Hilbert space. At the end of Chap. 3, we had identified the 2×2 matrices that form a basis for all 2×2 Hermitian matrices: the Pauli sigma matrices. Recall that they are

$$\sigma_1 = \begin{pmatrix} 0 & 1 \\ 1 & 0 \end{pmatrix}, \quad \sigma_2 = \begin{pmatrix} 0 & -i \\ i & 0 \end{pmatrix}, \quad \sigma_3 = \begin{pmatrix} 1 & 0 \\ 0 & -1 \end{pmatrix}. \tag{8.53}$$

What are the commutation relations of these matrices? Let's just evaluate $[\sigma_1, \sigma_2]$ explicitly here. We have

$$[\sigma_1, \sigma_2] = \sigma_1 \sigma_2 - \sigma_2 \sigma_1 = \begin{pmatrix} 0 & 1 \\ 1 & 0 \end{pmatrix}\begin{pmatrix} 0 & -i \\ i & 0 \end{pmatrix} - \begin{pmatrix} 0 & -i \\ i & 0 \end{pmatrix}\begin{pmatrix} 0 & 1 \\ 1 & 0 \end{pmatrix} \tag{8.54}$$

$$= 2i \begin{pmatrix} 1 & 0 \\ 0 & -1 \end{pmatrix} = 2i\sigma_3 .$$

Fascinating! These σ-matrices indeed form a Lie algebra. It's not quite the same commutation relation as the $\mathfrak{su}(2)$ algebra we identified earlier, but that is easily remedied. Let's call the **spin** operators

$$\hat{S}_x = \frac{\hbar}{2}\sigma_1 , \quad \hat{S}_y = \frac{\hbar}{2}\sigma_2 , \quad \hat{S}_z = \frac{\hbar}{2}\sigma_3 . \tag{8.55}$$

Then, we have that $[\hat{S}_x, \hat{S}_y] = i\hbar \hat{S}_z$, and similarly for the other commutators:

$$[\hat{S}_z, \hat{S}_x] = i\hbar \hat{S}_y , \qquad\qquad [\hat{S}_y, \hat{S}_z] = i\hbar \hat{S}_x . \tag{8.56}$$

Compactly, this is $[\hat{S}_i, \hat{S}_j] = i\hbar \sum_{k=1}^{3} \epsilon_{ijk} \hat{S}_k$, exactly the $\mathfrak{su}(2)$ algebra.

We correspondingly call this representation of the rotation group the two-dimensional representation because these 2×2 matrices would act to rotate two-component vectors, properly called **spinors**. Very strange that we can rotate a two-component object in full three dimensions, but that is what the math tells us.

Let's attempt to make more sense of these "spin" operators. We'll use them to perform a rotation about the z-axis, and see what we find. The unitary matrix that implements a rotation by ϕ about the z-axis is

$$\hat{U}(\phi) = e^{i\frac{\phi \hat{S}_z}{\hbar}} = e^{i\frac{\phi\sigma_3}{2}} = \mathbb{I} + i\frac{\phi\sigma_3}{2} - \frac{1}{2}\left(\frac{\phi}{2}\right)^2 \sigma_3^2 + \cdots \tag{8.57}$$

$$= \mathbb{I}\cos\frac{\phi}{2} + i\sigma_3 \sin\frac{\phi}{2} = \begin{pmatrix} e^{i\frac{\phi}{2}} & 0 \\ 0 & e^{-i\frac{\phi}{2}} \end{pmatrix}.$$

This matrix is fascinating and manifests the other property of three-dimensional rotations we had identified previously. Let's perform a rotation by $\phi = 2\pi$ which should, naïvely, just rotate the object back to where we started. However, the unitary matrix that implements this rotation is

$$\hat{U}(2\pi) = \begin{pmatrix} e^{i\frac{2\pi}{2}} & 0 \\ 0 & e^{-i\frac{2\pi}{2}} \end{pmatrix} = \begin{pmatrix} -1 & 0 \\ 0 & -1 \end{pmatrix} = -\mathbb{I}, \qquad (8.58)$$

so this rotation by 2π negates the object! As we argued earlier, indeed a rotation by 2π in three dimensions is not equivalent to doing nothing. By contrast, if we rotate by 4π, we do get back to where we started:

$$\hat{U}(4\pi) = \begin{pmatrix} e^{i\frac{4\pi}{2}} & 0 \\ 0 & e^{-i\frac{4\pi}{2}} \end{pmatrix} = \begin{pmatrix} 1 & 0 \\ 0 & 1 \end{pmatrix} = \mathbb{I}, \qquad (8.59)$$

as we established with the plate trick.

This two-dimensional representation of the rotation group is also called the spin-1/2 representation because a rotation by 2π rotates a spin-1/2 object by half of that, or π. The three-dimensional representation we started with is called the spin-1 representation because a rotation by 2π rotates a spin-1 object by 1 times that. There are representations of the rotation group indexed by spins of every non-negative integer and half-integer, representing the number of rotations that the object of spin-s completes for a 2π rotation (e.g., spin-2 rotates twice for a 2π rotation). Note that there's no such thing as "spin-1/3," because all objects must return to their original orientation after rotation by 4π.

8.5 The Angular Momentum Hilbert Space

In this section, we will make the relationship between the value of the spin and the Hilbert space spanned by eigenstates of the angular momentum operators more precise. To do this, we will use the formalism we employed when studying the quantum harmonic oscillator long ago.

The first thing to note about $\hat{L}_x$, $\hat{L}_y$, and $\hat{L}_z$ is that they don't commute, so they cannot be simultaneously diagonalized. Have no fear; we will just consider $\hat{L}_z$ as diagonal, and $\hat{L}_x$ and $\hat{L}_y$ whatever they need to be to satisfy the commutation relations. That is, we will study the eigenspace of $\hat{L}_z$. Note that everything we will derive about $\hat{L}_z$ will hold for $\hat{L}_x$ and $\hat{L}_y$ (if in a different basis) because there is nothing special about the z-axis; we could have labeled any axis as the z-axis. Because there are three angular momentum operators and they all do not commute, we can only choose one of them to study the eigensystem. $\hat{L}_z$ alone is called the "maximally commuting subalgebra" or the **Cartan subalgebra** of the Lie algebra, after Élie Cartan, a French mathematician. Note that $\hat{L}_z$ does indeed commute with itself, $[\hat{L}_z, \hat{L}_z] = 0$, so eigenstates of $\hat{L}_z$ are eigenstates of, well, $\hat{L}_z$.

Now, with the remaining $\hat{L}_x$ and $\hat{L}_y$ we would like to construct another subalgebra. We would like the elements of this subalgebra to be eigenstates of the commutator, when commuted with elements of the Cartan subalgebra, so that their action is as simple as possible. We will later see why this is so nice. That is, for two operators $\hat{L}_+$ and $\hat{L}_-$, we would like them to satisfy

$$[\hat{L}_+, \hat{L}_z] = \alpha \hat{L}_+, \qquad\qquad [\hat{L}_-, \hat{L}_z] = \beta \hat{L}_-, \qquad\qquad (8.60)$$

for some complex numbers α and β. Note that $\hat{L}_+$ and $\hat{L}_-$ cannot be $\hat{L}_x$ and $\hat{L}_y$, if for no other reason than their commutators are not "eigen-commutators," as, for example

$$[\hat{L}_x, \hat{L}_z] = -i\hbar \hat{L}_y. \qquad\qquad (8.61)$$

So we'll need to work a bit harder to determine $\hat{L}_+$ and $\hat{L}_-$.

The Lie algebra is a vector space, so we can construct $\hat{L}_+$ and $\hat{L}_-$ as a general linear combination of $\hat{L}_x$ and $\hat{L}_y$. That is, we will consider

$$\hat{L}_+ = a_x \hat{L}_x + a_y \hat{L}_y, \qquad\qquad \hat{L}_- = b_x \hat{L}_x + b_y \hat{L}_y, \qquad\qquad (8.62)$$

for some constants a_x, a_y, b_x, b_y. The normalization of the eigensystem is undefined, so we can with no loss of generality just set $a_x = b_x = 1$. This fixing of the normalization of $\hat{L}_+$ and $\hat{L}_-$ does not affect the "eigenvalues" α, β in the commutators above.

So, let's see what these coefficients might be. We have the commutator

$$[\hat{L}_+, \hat{L}_z] = [\hat{L}_x + a\hat{L}_y, \hat{L}_z] = [\hat{L}_x, \hat{L}_z] + a[\hat{L}_y, \hat{L}_z] \qquad\qquad (8.63)$$
$$= -i\hbar \hat{L}_y + i\hbar a \hat{L}_x = \alpha \hat{L}_+ = \alpha \hat{L}_x + \alpha a \hat{L}_y.$$

So, we simply match coefficients of $\hat{L}_x$ and $\hat{L}_y$ to determine a and α, appropriately. We find the two equations for the coefficients of $\hat{L}_x$ and $\hat{L}_y$ separately:

$$\text{coefficient of } \hat{L}_x: \quad \alpha = i\hbar a, \qquad\qquad (8.64)$$
$$\text{coefficient of } \hat{L}_y: \quad \alpha a = -i\hbar. \qquad\qquad (8.65)$$

Combining these equations, we find that $a^2 = -1$, so $a = \pm i$. If we then define

$$\hat{L}_+ = \hat{L}_x + i\hat{L}_y, \qquad\qquad \hat{L}_- = \hat{L}_x - i\hat{L}_y, \qquad\qquad (8.66)$$

they have the following commutators:

$$[\hat{L}_+, \hat{L}_z] = -\hbar \hat{L}_+, \qquad\qquad [\hat{L}_-, \hat{L}_z] = \hbar \hat{L}_-. \qquad\qquad (8.67)$$

Note also that their commutator together is

$$[\hat{L}_+, \hat{L}_-] = 2\hbar \hat{L}_z, \qquad\qquad (8.68)$$

which I will leave to you to prove.

Now, let's see the power of this formalism. Consider the state $|\psi\rangle$, which is an eigenstate of $\hat{L}_z$:

$$\hat{L}_z |\psi\rangle = c\hbar |\psi\rangle, \qquad\qquad (8.69)$$

where c is some real number, and we put $\hbar$ there to account for the correct units ($\hbar$ has units of angular momentum). Consider the state on which we act $\hat{L}_+$ on $|\psi\rangle$: $\hat{L}_+ |\psi\rangle$. Is this an eigenstate of $\hat{L}_z$? Let's just calculate it. We have

$$\hat{L}_z \hat{L}_+ |\psi\rangle = \left(\hat{L}_+ \hat{L}_z + [\hat{L}_z, \hat{L}_+]\right)|\psi\rangle = \hat{L}_+ c\hbar |\psi\rangle + \hbar \hat{L}_+ |\psi\rangle \qquad\qquad (8.70)$$
$$= (c+1)\hbar \hat{L}_+ |\psi\rangle.$$

So yes, $\hat{L}_+|\psi\rangle$ is an eigenstate of $\hat{L}_z$, with eigenvalue larger than that of just $|\psi\rangle$ by one $\hbar$. What about $\hat{L}_-$ on $|\psi\rangle$? We have

$$\hat{L}_z\hat{L}_-|\psi\rangle = \left(\hat{L}_-\hat{L}_z + [\hat{L}_z,\hat{L}_-]\right)|\psi\rangle = \hat{L}_-c\hbar|\psi\rangle - \hbar\hat{L}_-|\psi\rangle \qquad (8.71)$$
$$= (c-1)\hbar\hat{L}_-|\psi\rangle\,.$$

So $\hat{L}_-$ lowers the eigenvalue of $|\psi\rangle$ by one unit of $\hbar$. I hope the "+" and "−" subscripts on these operators make some sense now. $\hat{L}_+$ and $\hat{L}_-$ are exactly analogous to the raising and lowering operators we had constructed for the quantum harmonic oscillator Hamiltonian. We can construct the eigenstates of $\hat{L}_z$, that is, its representation, in an exactly analogous way.

Note that acting $\hat{L}_+$ on $|\psi\rangle$ n times changes the eigenvalue of $\hat{L}_z$ by $+n\hbar$:

$$\hat{L}_z(\hat{L}_+)^n|\psi\rangle = \left((\hat{L}_+)^n\hat{L}_z + [\hat{L}_z,(\hat{L}_+)^n]\right)|\psi\rangle = c\hbar(\hat{L}_+)^n|\psi\rangle + n\hbar(\hat{L}_+)^n|\psi\rangle \qquad (8.72)$$
$$= (c+n)\hbar(\hat{L}_+)^n|\psi\rangle\,.$$

This follows from the commutation relation

$$[(\hat{L}_+)^n,\hat{L}_z] = -n\hbar(\hat{L}_z)^n\,, \qquad (8.73)$$

which can be proved by induction, but I won't do that here. So I can keep raising and raising the $\hat{L}_z$ eigenvalue in steps of $\hbar$ by acting with $\hat{L}_+$ on $|\psi\rangle$. Correspondingly, I can keep lowering and lowering the eigenvalue in steps of $\hbar$ by acting with $\hat{L}_-$ on $|\psi\rangle$ as

$$\hat{L}_z(\hat{L}_-)^n|\psi\rangle = (c-n)\hbar(\hat{L}_-)^n|\psi\rangle\,, \qquad (8.74)$$

which I leave you to show.

If we restrict our attention to *finite*-dimensional representations of the rotation group, like the two-dimensional spin-1/2 or three-dimensional spin-1 we had seen before, we must enforce two properties on the set of eigenvectors of $\hat{L}_z$. First, there must be a minimal eigenvalue of $\hat{L}_z$ on some state $|\psi_{\min}\rangle$ for which

$$\hat{L}_-|\psi_{\min}\rangle = 0\,. \qquad (8.75)$$

There must also be a maximal eigenvalue of $\hat{L}_z$ on some state $|\psi_{\max}\rangle$ such that

$$\hat{L}_+|\psi_{\max}\rangle = 0\,. \qquad (8.76)$$

On these two states, let's call the minimum and maximum eigenvalues $c_{\min}$ and $c_{\max}$:

$$\hat{L}_z|\psi_{\min}\rangle = c_{\min}\hbar|\psi_{\min}\rangle\,, \qquad (8.77)$$
$$\hat{L}_z|\psi_{\max}\rangle = c_{\max}\hbar|\psi_{\max}\rangle\,. \qquad (8.78)$$

From our analysis of the raising and lowering operators, the difference between $c_{\min}$ and $c_{\max}$ must be some integer n: $c_{\min} + n = c_{\max}$. That is, n applications of $\hat{L}_+$ on $|\psi_{\min}\rangle$ returns a state proportional to $|\psi_{\max}\rangle$:

$$(\hat{L}_+)^n|\psi_{\min}\rangle \propto |\psi_{\max}\rangle\,. \qquad (8.79)$$

These maximum and minimum eigenstates are normalized and, if they are distinct, orthogonal to one another:

$$\langle \psi_{min} | \psi_{min} \rangle = \langle \psi_{max} | \psi_{max} \rangle = 1, \qquad\qquad \langle \psi_{min} | \psi_{max} \rangle = 0. \qquad (8.80)$$

We will use these observations in the next section to explicitly determine the possible numerical values of c_{min} and c_{max}, but it is useful here to pause and see what this abstract formalism means in some simple examples.

Example 8.1 We have already studied the two-dimensional representation, or spin-1/2. What does this representation look like in this language? "Two-dimensional" means that there are just two basis states, $|\psi_{min}\rangle$ and $|\psi_{max}\rangle$, and $c_{min} + 1 = c_{max}$. We had identified $\hat{L}_z$ with a Pauli sigma matrix, namely

$$\hat{L}_z = \frac{\hbar}{2}\sigma_3 = \frac{\hbar}{2}\begin{pmatrix} 1 & 0 \\ 0 & -1 \end{pmatrix}. \qquad (8.81)$$

In this matrix basis representation, its eigenvectors are

$$|\psi_{min}\rangle = \begin{pmatrix} 0 \\ 1 \end{pmatrix}, \quad \text{with eigenvalue} \quad -\frac{\hbar}{2}, \qquad (8.82)$$

$$|\psi_{max}\rangle = \begin{pmatrix} 1 \\ 0 \end{pmatrix}, \quad \text{with eigenvalue} \quad \frac{\hbar}{2}.$$

So, indeed, these eigenvalues differ by one $\hbar$ unit of angular momentum and further $-c_{min} = c_{max} = 1/2$. In fact, the maximum and minimum eigenvalues of $\hat{L}_z$ encode the "total" angular momentum of the state; here spin-"1/2." In the next section, we'll see how to prove this in generality.

Further, the raising and lowering operators for this spin-1/2 representation are just another linear combination of the Pauli matrices. We had identified $\hat{L}_x$ and $\hat{L}_y$ as

$$\hat{L}_x = \frac{\hbar}{2}\sigma_1 = \frac{\hbar}{2}\begin{pmatrix} 0 & 1 \\ 1 & 0 \end{pmatrix}, \qquad \hat{L}_y = \frac{\hbar}{2}\sigma_2 = \frac{\hbar}{2}\begin{pmatrix} 0 & -i \\ i & 0 \end{pmatrix}, \qquad (8.83)$$

so the raising and lowering operators are

$$\hat{L}_+ = \hat{L}_x + i\hat{L}_y = \frac{\hbar}{2}\left[\begin{pmatrix} 0 & 1 \\ 1 & 0 \end{pmatrix} + i\begin{pmatrix} 0 & -i \\ i & 0 \end{pmatrix}\right] = \hbar\begin{pmatrix} 0 & 1 \\ 0 & 0 \end{pmatrix}, \qquad (8.84)$$

$$\hat{L}_- = \hat{L}_x - i\hat{L}_y = \frac{\hbar}{2}\left[\begin{pmatrix} 0 & 1 \\ 1 & 0 \end{pmatrix} - i\begin{pmatrix} 0 & -i \\ i & 0 \end{pmatrix}\right] = \hbar\begin{pmatrix} 0 & 0 \\ 1 & 0 \end{pmatrix}.$$

When acting on the eigenvectors, we find

$$\hat{L}_+ |\psi_{max}\rangle = \hbar\begin{pmatrix} 0 & 1 \\ 0 & 0 \end{pmatrix}\begin{pmatrix} 1 \\ 0 \end{pmatrix} = 0, \qquad (8.85)$$

$$\hat{L}_- |\psi_{max}\rangle = \hbar\begin{pmatrix} 0 & 0 \\ 1 & 0 \end{pmatrix}\begin{pmatrix} 1 \\ 0 \end{pmatrix} = \hbar\begin{pmatrix} 0 \\ 1 \end{pmatrix} = \hbar|\psi_{min}\rangle.$$

Additionally, when acting on $|\psi_{min}\rangle$, we find, as expected:

$$\hat{L}_-|\psi_{min}\rangle = 0, \qquad\qquad \hat{L}_+|\psi_{min}\rangle = \hbar|\psi_{max}\rangle. \qquad (8.86)$$

Example 8.2   Previously, we had also studied the spin-1 or three-dimensional representation of angular momentum, but here we'll consider an even simpler representation than that of spin-1/2: the one-dimensional representation. What does this mean? Well, there is only one state $|\psi\rangle$ in the Hilbert space and it is annihilated by both $\hat{L}_+$ and $\hat{L}_-$:

$$\hat{L}_+|\psi\rangle = \hat{L}_-|\psi\rangle = 0, \qquad (8.87)$$

and thus $c_{min} = c_{max}$. Rather trivially, we also have

$$\hat{L}_+\hat{L}_-|\psi\rangle = 0 = \left(\hat{L}_-\hat{L}_+ + [\hat{L}_+, \hat{L}_-]\right)|\psi\rangle = 2\hbar\hat{L}_z|\psi\rangle, \qquad (8.88)$$

using the commutation relation of $\hat{L}_+$ and $\hat{L}_-$. Apparently, then, $\hat{L}_z$ also annihilates $|\psi\rangle$ and so $c_{min} = c_{max} = 0$. Thus, we call this representation the spin-0 representation, or even the **trivial representation** of the rotation group. Now, it would seem like the commutation relation of angular momentum operators couldn't be satisfied, but it is entirely consistent that $\hat{L}_x = \hat{L}_y = \hat{L}_z = 0$ and therefore it is indeed true, though trivially so, that

$$[\hat{L}_i, \hat{L}_j] = i\hbar \sum_{k=1}^{3} \epsilon_{ijk}\hat{L}_k. \qquad (8.89)$$

Hence the trivial representation.

8.6 Specifying Representations: The Casimir of Rotations

In this section, we will identify properties of the representations of the rotation group in general, and establish an operator which quantifies a representation, independent of the particular state we are considering.

Recall that when we first discussed linear operators and matrices, we had identified two quantities or sets that are basis independent: the eigenvalues of that operator and its dimension (when thought of as a matrix). We've discussed eigenvalues extensively and for rotations, we can quantify states by their eigenvalue under the action of $\hat{L}_z$, and can move between states of different eigenvalues with $\hat{L}_+$ and $\hat{L}_-$.

We've spent less time discussing the dimension of an operator as in many or even most cases we studied in this book, operators have been infinite dimensional, so this is less helpful. For example, the Hamiltonian of the infinite square well is infinite dimensional: the energy eigenvalues are unbounded from above. For the rotation group, the dimension of the representation is relevant as it is finite. (By the way, the rotation group has finite-dimensional representations while time translation or energy does not, in general, because the manifold of rotations is compact while the manifold of temporal translations is non-compact.) So, our goal in this section is to construct such a "dimension-counting" operator on the Lie algebra of angular momentum.

Let's enumerate the properties of this dimension-counting operator. First, let's give it a name: C_R for representation R of the rotation group. It is also called the quadratic **Casimir** more generally. Why "quadratic" will be shown soon and why Casimir, well, it had to be named after someone and that someone was Dutch physicist Hendrik Casimir.[2] Next, if this Casimir is just a measure of the dimension of a representation then it must return the same value on any state in that representation. If a matrix always returns the same eigenvalue for any vector it acts on, then that matrix is necessarily proportional to the identity matrix. So, for an n-dimensional representation, the Casimir is

$$C_R \equiv C_R \mathbb{I}_{n \times n}, \tag{8.90}$$

where $\mathbb{I}_{n \times n}$ is the $n \times n$ identity matrix. Because it is proportional to $\mathbb{I}$, C_R is just a number, so we typically ignore the identity matrix factor.

Another way to state this property is the following. Different states in a given representation are related to one another through the action of rotation (i.e., by action with $\hat{L}_x$, $\hat{L}_y$, or $\hat{L}_z$). If the Casimir is to be the same for every state in a representation, then it must be unaffected by rotation; equivalently, it must commute with every angular momentum operator. This property suggests a way forward to constructing C_R. First, C_R cannot be strictly a linear combination of $\hat{L}_x$, $\hat{L}_y$, and $\hat{L}_z$, simply because these operators don't commute with one another. If we think more broadly and consider a vector $\vec{v}$, we can ask what properties of it are rotation-invariant. The reason why this analogy works is that the Lie algebra spanned by $\hat{L}_x$, $\hat{L}_y$, and $\hat{L}_z$ is a vector space, just one additionally equipped with the Lie bracket $[,]$ (i.e., the commutator). A vector $\vec{v}$ has a magnitude and direction and clearly the direction changes under rotation. However, the magnitude does not, and, for a three-dimensional vector $\vec{v}$ this magnitude is

$$|\vec{v}|^2 = v_x^2 + v_y^2 + v_z^2. \tag{8.91}$$

This would suggest that the quantity

$$\hat{L}^2 \equiv C_R = \hat{L}_x^2 + \hat{L}_y^2 + \hat{L}_z^2 \tag{8.92}$$

is rotationally invariant and could be the Casimir that we want. So let's see if it indeed commutes with every element of the Lie algebra.

We can express the Casimir as a sum over angular momentum operators as

$$\sum_{i=1}^{3} \hat{L}_i^2 = \hat{L}_x^2 + \hat{L}_y^2 + \hat{L}_z^2, \tag{8.93}$$

---

[2] H. B. G. Casimir, "Rotation of a rigid body in quantum mechanics," PhD thesis, Universiteit Leiden (1931).

and its commutator with some $\hat{L}_j$ is

$$
\begin{aligned}
\left[\hat{L}_j, \sum_{i=1}^{3} \hat{L}_i^2\right] &= \sum_{i=1}^{3} [\hat{L}_j, \hat{L}_i^2] = \sum_{i=1}^{3} \left(\hat{L}_j \hat{L}_i^2 - \hat{L}_i^2 \hat{L}_j\right) \\
&= \sum_{i,k=1}^{3} \left(\hat{L}_i \hat{L}_j \hat{L}_i - i\hbar \epsilon_{ijk} \hat{L}_k \hat{L}_i - \hat{L}_i^2 \hat{L}_j\right) \\
&= \sum_{i,k=1}^{3} \left(\hat{L}_j \hat{L}_i^2 - i\hbar \epsilon_{ijk} \hat{L}_i \hat{L}_k - i\hbar \epsilon_{ijk} \hat{L}_k \hat{L}_i - \hat{L}_i^2 \hat{L}_j\right) \\
&= -i\hbar \sum_{i,k=1}^{3} \epsilon_{ijk} \left(\hat{L}_i \hat{L}_k + \hat{L}_k \hat{L}_i\right).
\end{aligned}
$$

(8.94)

Through this expression, I used the commutation relations of the angular momentum operators. Now, in the final form this is "manifestly" 0. Recall that ϵ_{ijk} is totally antisymmetric: $\epsilon_{ijk} = -\epsilon_{kji}$. However, the remaining sum over angular momentum operators is symmetric: $\hat{L}_i \hat{L}_k + \hat{L}_k \hat{L}_i$. So, when I sum over i and k, for every positive term there will be a negative term that exactly cancels it. Therefore, $\hat{L}^2$ commutes with all angular momentum operators:

$$
[\hat{L}^2, \hat{L}_j] = \left[\sum_{i=1}^{3} \hat{L}_i^2, \hat{L}_j\right] = 0.
$$

(8.95)

Again, the only operator that commutes with everything is the identity, or proportional to it. So, indeed $\hat{L}^2$ takes the same value on any state in a given representation. Thus, it is the quadratic Casimir.

What is its value? To determine C_R, let's re-express $\hat{L}^2$ with the raising and lowering operators $\hat{L}_+$ and $\hat{L}_-$. Recall that

$$
\hat{L}_+ = \hat{L}_x + i\hat{L}_y, \qquad\qquad \hat{L}_- = \hat{L}_x - i\hat{L}_y,
$$

(8.96)

and so

$$
\hat{L}_x^2 + \hat{L}_y^2 = \frac{\hat{L}_+ \hat{L}_- + \hat{L}_- \hat{L}_+}{2}.
$$

(8.97)

Thus, the Casimir is

$$
\hat{L}^2 = \frac{\hat{L}_+ \hat{L}_- + \hat{L}_- \hat{L}_+}{2} + \hat{L}_z^2.
$$

(8.98)

Let's calculate the Casimir on the special states $|\psi_{\min}\rangle$ and $|\psi_{\max}\rangle$ that we defined as the "boundaries" of the representation. Starting with $|\psi_{\min}\rangle$, recall that

$$
\hat{L}_- |\psi_{\min}\rangle = 0,
$$

(8.99)

so it's convenient to put all $\hat{L}_-$ operators on the right in $\hat{L}^2$. That is:

$$
\hat{L}^2 = \frac{\hat{L}_+ \hat{L}_- + \hat{L}_- \hat{L}_+}{2} + \hat{L}_z^2 = \hat{L}_+ \hat{L}_- + \frac{1}{2}[\hat{L}_-, \hat{L}_+] + \hat{L}_z^2 = \hat{L}_+ \hat{L}_- - \hbar \hat{L}_z + \hat{L}_z^2,
$$

(8.100)

where I used the commutator of $\hat{L}_-$ and $\hat{L}_+$. Now, acting on $|\psi_{\min}\rangle$, we find

$$\hat{L}^2|\psi_{\min}\rangle = \left(\hat{L}_+\hat{L}_- - \hbar\hat{L}_z + \hat{L}_z^2\right)|\psi_{\min}\rangle = \left(c_{\min}^2\hbar^2 - c_{\min}\hbar^2\right)|\psi_{\min}\rangle \tag{8.101}$$
$$= -\hbar^2 c_{\min}(1 - c_{\min})|\psi_{\min}\rangle,$$

where I have used that the eigenvalue of $|\psi_{\min}\rangle$ under $\hat{L}_z$ is $c_{\min}\hbar$.

Next, let's evaluate $\hat{L}^2$ on $|\psi_{\max}\rangle$. Now, it is convenient to express L^2 with all $\hat{L}_+$ operators to the right as $\hat{L}_+|\psi_{\max}\rangle = 0$. That is:

$$\hat{L}^2 = \frac{\hat{L}_+\hat{L}_- + \hat{L}_-\hat{L}_+}{2} + \hat{L}_z^2 = \hat{L}_-\hat{L}_+ + \frac{1}{2}[\hat{L}_+,\hat{L}_-] + \hat{L}_z^2 = \hat{L}_-\hat{L}_+ + \hbar\hat{L}_z + \hat{L}_z^2. \tag{8.102}$$

Acting on $|\psi_{\max}\rangle$, we find

$$\hat{L}^2|\psi_{\max}\rangle = \left(\hat{L}_-\hat{L}_+ + \hbar\hat{L}_z + \hat{L}_z^2\right)|\psi_{\max}\rangle = \left(c_{\max}^2\hbar^2 + c_{\max}\hbar^2\right)|\psi_{\max}\rangle \tag{8.103}$$
$$= \hbar^2 c_{\max}(1 + c_{\max})|\psi_{\max}\rangle.$$

This must be the same eigenvalue as that when acting on $|\psi_{\min}\rangle$, so demanding that $c_{\max} > c_{\min}$ forces that

$$c_{\min} = -c_{\max}. \tag{8.104}$$

We had already observed this for the one- and two-dimensional representations of the rotation group; now we've proved it in general.

We also stated that the dimension of the representation is $n+1$, where $n = c_{\max} - c_{\min} = 2c_{\max}$, using the result above. Thus, the dimension of a representation R is

$$\dim_R = n + 1 = 2c_{\max} + 1. \tag{8.105}$$

In previous sections, we had established from general arguments that representations of the rotation group are quantified by a non-negative integer or half-integer, depending on how many rotations the representation went through for a total 2π rotation. We called these "spin-ℓ" for $\ell = 0, 1/2, 1, 3/2, \ldots$, and note that the number of states of a given spin (i.e., its dimension) is $\dim_\ell = 2\ell + 1$. For example, $\ell = 1/2$ is the $\dim_{1/2} = 2 \cdot \frac{1}{2} + 1 = 2$-dimensional representation. We've also just derived that $\dim_\ell = n + 1 = 2c_{\max} + 1$, therefore

$$c_{\max} = \ell = -c_{\min}. \tag{8.106}$$

That is, states in a representation of the rotation group are indexed by their eigenvalues of $\hat{L}_z$, and the eigenvalues range over

$$-\ell, -\ell + 1, \ldots, \ell - 1, \ell, \tag{8.107}$$

a total of $2\ell + 1$ values.

Finally, we can evaluate the Casimir in terms of ℓ as

$$C_\ell = L^2 = \hbar^2\ell(\ell + 1), \tag{8.108}$$

or, with $\dim_\ell = 2\ell + 1$:

$$C_\ell = \hbar^2 \frac{\dim_\ell^2 - 1}{4}. \tag{8.109}$$

So indeed the Casimir encodes information about the dimension of the representation of interest.

One final comment to make about the Casimir: it is not in the Lie algebra of rotations. The Lie algebra is a vector space of angular momentum operators and the Casimir is a quadratic function of the operators $\hat{L}_x$, $\hat{L}_y$, and $\hat{L}_z$. However, the Casimir is a Hermitian operator as it is proportional to the identity matrix $\mathbb{I}$ and that constant of proportionality is a real number. Thus the Casimir is measurable. That is, we can determine the total angular momentum of some system through experiment. Indeed, this makes sense as it is basis-independent data of our system.

Example 8.3 An ideal model of a two-state quantum system is of an electrically charged, spin-1/2 particle immersed in a uniform magnetic field. The spin-1/2 particle therefore has a magnetic moment and that correspondingly interacts with the external magnetic field. Let's consider a magnetic field $\vec{B} = B_0 \hat{z}$ and let's put a spin-1/2 electron in it. Then, the Hamiltonian of such an electron in the magnetic field is

$$\hat{H} = \frac{e\hbar}{2m_e} B_0 \begin{pmatrix} 1 & 0 \\ 0 & -1 \end{pmatrix}, \tag{8.110}$$

where $e = 1.6 \times 10^{-19}$ C is the fundamental electric charge of the electron, and m_e is the mass of the electron. This physical system is illustrated in Fig. 8.4. The first factor in the Hamiltonian is called the **Bohr magneton** μ_B and is simply the magnetic moment of the electron:

$$\mu_B \equiv \frac{e\hbar}{2m_e}. \tag{8.111}$$

With this set-up, we would like to calculate the expectation value of the particle's spin operators $\hat{S}_x$, $\hat{S}_y$, and $\hat{S}_z$. In particular, we will determine the time evolution of the expectation values here, and you will study this further in Exercise 8.5.

Solution

Recall from Eq. (4.38) that the expression for the time evolution of an operator that itself does not depend explicitly on time is

$$\frac{d}{dt}\langle \hat{S}_z \rangle = \frac{i}{\hbar} \langle [\hat{H}, \hat{S}_z] \rangle. \tag{8.112}$$

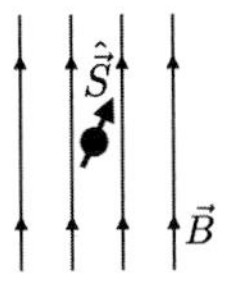

Fig. 8.4 Illustration of an electrically charged, spin-1/2 particle (black dot) immersed in a uniform magnetic field $\vec{B}$. The direction of the spin operator $\hat{\vec{S}}$, and hence the magnetic moment of the particle is illustrated by the thick arrow.

The operator $\hat{S}_z$ is

$$\hat{S}_z = \frac{\hbar}{2}\sigma_3 = \frac{\hbar}{2}\begin{pmatrix} 1 & 0 \\ 0 & -1 \end{pmatrix}, \tag{8.113}$$

which, like $\hat{H}$, is diagonal. Therefore, $[\hat{H}, \hat{S}_z] = 0$ and the expectation value of $\hat{S}_z$ of the electrically charge, spinning particle is constant in time.

By contrast, the expectation value of $\hat{S}_x$ has non-trivial time dependence. Now, the operator $\hat{S}_x$ is

$$\hat{S}_x = \frac{\hbar}{2}\sigma_1 = \frac{\hbar}{2}\begin{pmatrix} 0 & 1 \\ 1 & 0 \end{pmatrix}. \tag{8.114}$$

Its commutator with the Hamiltonian is

$$[\hat{H}, \hat{S}_x] = \left[\frac{e\hbar}{2m_e}B_0\sigma_3, \frac{\hbar}{2}\sigma_1\right] = \frac{eB_0}{m_e}[\hat{S}_z, \hat{S}_x] = i\frac{e\hbar B_0}{m_e}\hat{S}_y. \tag{8.115}$$

Therefore, the time derivative of the expectation value of $\hat{S}_x$ is

$$\frac{d}{dt}\langle\hat{S}_x\rangle = \frac{i}{\hbar}\langle[\hat{H}, \hat{S}_x]\rangle = -\frac{eB_0}{m_e}\langle\hat{S}_y\rangle. \tag{8.116}$$

With

$$\hat{S}_y = \frac{\hbar}{2}\sigma_2 = \frac{\hbar}{2}\begin{pmatrix} 0 & -i \\ i & 0 \end{pmatrix} \tag{8.117}$$

we can similarly show that the commutator with the Hamiltonian is

$$[\hat{H}, \hat{S}_y] = \frac{eB_0}{m_e}[\hat{S}_z, \hat{S}_y] = -i\frac{e\hbar B_0}{m_e}\hat{S}_x. \tag{8.118}$$

Then, the time derivative of the expectation value of $\hat{S}_y$ is

$$\frac{d}{dt}\langle\hat{S}_y\rangle = \frac{i}{\hbar}\langle[\hat{H}, \hat{S}_y]\rangle = \frac{eB_0}{m_e}\langle\hat{S}_x\rangle. \tag{8.119}$$

Therefore, under time evolution, the expectation values of the operators $\hat{S}_x$ and $\hat{S}_y$ transform into one another. These time-evolution equations can be decoupled by taking a second derivative. Note that, for example

$$\frac{d^2}{dt^2}\langle\hat{S}_x\rangle = -\frac{eB_0}{m_e}\frac{d}{dt}\langle\hat{S}_y\rangle = -\left(\frac{eB_0}{m_e}\right)^2\langle\hat{S}_x\rangle. \tag{8.120}$$

Similarly, for $\hat{S}_y$ we find the same equation:

$$\frac{d^2}{dt^2}\langle\hat{S}_y\rangle = \frac{eB_0}{m_e}\frac{d}{dt}\langle\hat{S}_x\rangle = -\left(\frac{eB_0}{m_e}\right)^2\langle\hat{S}_y\rangle. \tag{8.121}$$

These second-order differential equations describe sinusoidal oscillation of the expectation values with angular frequency $\omega = \frac{eB_0}{m_e}$. Further, by Eqs. (8.116) and (8.119), the phase of these expectation values differs by $\pi/2$, and so this describes **precession** of the spin vector operator $\vec{\hat{S}}$ about the $\hat{z}$-axis. This is exactly the phenomenon expected classically, satisfying Eherenfest's theorem.

8.7 Quantum Numbers and Conservation Laws

We can summarize the results established in the previous section as follows. A representation of the rotation group is specified by an integer or half-integer ℓ called the spin or angular momentum of the representation. For a given ℓ, there are $2\ell+1$ states in the representation; hence $\dim_\ell = 2\ell+1$ is the dimension of the representation. Further, the individual states in a representation are specified by their eigenvalue under the z-component of angular momentum, $\hat{L}_z$. This eigenvalue is labeled by an integer or half-integer m that lies in the range from $-\ell$ to ℓ: $-\ell \leq m \leq \ell$. Note that there are indeed $2\ell+1$ such values for m. Thus a particular eigenstate of the three-dimensional rotation group can be labeled with ℓ and m as $|\ell,m\rangle$ such that

$$\hat{L}^2|\ell,m\rangle = C_\ell|\ell,m\rangle = \hbar^2\ell(\ell+1)|\ell,m\rangle, \qquad (8.122)$$
$$\hat{L}_z|\ell,m\rangle = \hbar m|\ell,m\rangle.$$

C_ℓ is the Casimir operator that measures ℓ, independent of the particular state indexed by m.

By the way, these representations of the rotation group indexed by ℓ are called **irreducible representations** or irreps, as an irrep is completely closed under the action of $\hat{L}_x$, $\hat{L}_y$, or $\hat{L}_z$. It is "irreducible" because it is the smallest collection of states of a given ℓ for which it is closed.

In this section, we are going to seriously evaluate the statement that I just made above, that states are labeled by ℓ and m, eigenvalues of $\hat{L}^2$ and $\hat{L}_z$. We have been labeling states since we first studied an example quantum system, the infinite square well, several chapters ago. At that time, we basically just stated things without considering the interpretation, and it is high time to correct that.

8.7.1 Time Evolution-Preserving Conserved Charges

So what is our goal here? Well, it's all about naming. Shakespeare was correct to ask what's in a name, and the particular word "rose" isn't special, but some name for it is important. If you say "rose," then everyone around you knows what you are talking about. Further, your name is important: the act of addressing someone personally is a very intimate action and demonstrates an almost sacred knowledge about that person. Now, you may be thinking that we're getting a bit too romantic for quantum mechanics, but we have to address this issue of naming here, as well. A more mathematically well-defined way to think about this question is as the minimal data necessary to know completely and precisely the state of a quantum system.

We can attack this problem by thinking about what properties we want in a quantum "name." First, for the case of addressing you by name, it must be something that we can all agree on. For example, your name doesn't change if you are seated versus running to get lunch. Someone who sees you running would shout the same name as your instructor of this class would to get your attention. If this were not the case, then we must have an infinite, or at least enormous, collection of names for every possible activity we might be doing. In my case: Teaching-Andrew, Sleeping-Andrew, Relaxing-Andrew, etc.

In the quantum analogy, we want our state to be labeled in a way that is independent of basis on the Hilbert space. That is, we can all agree on what the state is, irrespective of what "coordinates" we happen to use to describe the space of states. We've also stated that the only basis-independent data are eigenvalues of operators that act on the Hilbert space. Thus, the basis-independent way to talk about a state is as the eigenstate of some operator, or collection of operators. Thinking back to you as an individual, your name is the eigenvalue of the "name" operator, and *you* are the eigenstate!

Okay, so let's say we have some state $|\psi\rangle$ that is an eigenstate of a collection of operators $\{\hat{A}_i\}_{i=1}^n$ such that

$$\hat{A}_i|\psi\rangle = a_i|\psi\rangle, \qquad (8.123)$$

and a_i is the eigenvalue. In our analogy, we would call the collection of eigenvalues $\{a_i\}_{i=1}^n$ of a state its "name." If you say that collection of eigenvalues then everyone can agree on the precise state $|\psi\rangle$ you are talking about. For example, if I tell you that $|\psi\rangle$ has eigenvalues ℓ and m under the Casimir and $\hat{L}_z$ rotation operators, then you know that the state is $|\psi\rangle = |\ell, m\rangle$. As another example, if I tell you that state $|\psi\rangle$ has eigenvalue λ under the action of the annihilation operator $\hat{a}$, then you know that $|\psi\rangle$ satisfies

$$\hat{a}|\psi\rangle = \lambda|\psi\rangle, \qquad (8.124)$$

and can be expressed as

$$|\psi\rangle = e^{-\frac{|\lambda|^2}{2}} e^{\lambda\hat{a}^\dagger}|\psi_0\rangle, \qquad (8.125)$$

where $|\psi_0\rangle$ is the ground state of the quantum harmonic oscillator.

We call the collection of eigenvalues of operators on some state $|\psi\rangle$ the **quantum numbers** of that state. A sufficient collection of quantum numbers uniquely specifies the state. However, this may still be less than ideal for defining the state for all time t. A general state $|\psi\rangle$ changes in time through action with the Hamiltonian as

$$|\psi(t)\rangle = e^{-i\frac{t\hat{H}}{\hbar}}|\psi\rangle, \qquad (8.126)$$

and $|\psi(t)\rangle$ might not have the same eigenvalue of some operator $\hat{A}$ that it did at time $t = 0$.

Let's make this concrete. Let's assume that $|\psi\rangle$ is an eigenstate of operator $\hat{A}$ with eigenvalue a at time $t = 0$:

$$\hat{A}|\psi\rangle = a|\psi\rangle. \qquad (8.127)$$

Further, we assume that operator $\hat{A}$ doesn't depend on time explicitly. The eigenvalue equation at a later time t is then

$$e^{-i\frac{t\hat{H}}{\hbar}}\hat{A}|\psi\rangle = a e^{-i\frac{t\hat{H}}{\hbar}}|\psi\rangle = \left(e^{-i\frac{t\hat{H}}{\hbar}}\hat{A}e^{i\frac{t\hat{H}}{\hbar}}\right)e^{-i\frac{t\hat{H}}{\hbar}}|\psi\rangle, \qquad (8.128)$$

where we just inserted "1" between $\hat{A}$ and $|\psi\rangle$ in the third equality. So, at a general time t, the state

$$|\psi(t)\rangle = e^{-i\frac{t\hat{H}}{\hbar}}|\psi\rangle \qquad (8.129)$$

is an eigenstate of the time-evolved operator

$$e^{-i\frac{t\hat{H}}{\hbar}}\hat{A}e^{i\frac{t\hat{H}}{\hbar}}. \qquad (8.130)$$

This is in general not just $\hat{A}$, so it makes our job of labeling $|\psi(t)\rangle$ at a general time more challenging. The operator under which $|\psi(t)\rangle$ is an eigenstate is changing in time. In general we can't all agree on what this operator is because it depends on when we start our clock.

However, $\hat{A}$ is a good operator by which to name the state $|\psi(t)\rangle$ for all time if time evolution does not affect $\hat{A}$:

$$e^{-i\frac{t\hat{H}}{\hbar}}\hat{A}e^{i\frac{t\hat{H}}{\hbar}} = \hat{A}. \qquad (8.131)$$

This is true if and only if $\hat{A}$ commutes with the Hamiltonian $\hat{H}$:

$$[\hat{H},\hat{A}] = 0 \quad \Rightarrow \quad e^{-i\frac{t\hat{H}}{\hbar}}\hat{A}e^{i\frac{t\hat{H}}{\hbar}} = \hat{A}. \qquad (8.132)$$

If this is true, then we call the eigenvalues of $\hat{A}$ "good quantum numbers," in the same sense that a dog is a "good boy" because it stays when told (I guess!). To recapitulate, those states that we all agree on and are unambiguously specified for all time are identified by a set of quantum numbers $\{a_i\}_{i=1}^{n}$ which are eigenvalues of the state $|\psi\rangle$ under the action of operators $\{\hat{A}_i\}_{i=1}^{n}$ that commute with the Hamiltonian, $[\hat{H},\hat{A}_i] = 0$, for all $i = 1, 2, \ldots, n$.

We've seen this commutator with the Hamiltonian before, in the time derivative of an expectation value:

$$\frac{d\langle\hat{A}\rangle}{dt} = \frac{i}{\hbar}\langle[\hat{H},\hat{A}]\rangle. \qquad (8.133)$$

Apparently, if $[\hat{H},\hat{A}] = 0$ then the expectation value of $\hat{A}$ is constant or **conserved** in time. Actually, if $[\hat{H},\hat{A}] = 0$, we can simultaneously diagonalize the Hamiltonian and $\hat{A}$ and so every eigenvalue of $\hat{A}$ is also constant in time. This is the statement of conservation laws in quantum mechanics. For a Hermitian operator $\hat{A}$, the measurable quantity it corresponds to is conserved if and only if $[\hat{H},\hat{A}] = 0$. Sometimes $\hat{A}$ is then called a **charge** because it is conserved, just like familiar electric charge.

8.7.2 Symmetries-Preserving Physical Laws

In classical mechanics, you were introduced to conservation laws from a very general perspective, namely, through **Noether's theorem**.[3] Noether's theorem states that a transformation that leaves the action S unchanged (i.e., a **symmetry**) has a corresponding conservation law. This vanishing commutator $[\hat{H},\hat{A}] = 0$ is the Noether's theorem of quantum mechanics. Noether's theorem really states that the laws of physics are

---

[3] E. Noether, "Invariant variation problems," *Gott. Nachr.* **1918**, 235–257 (1918).

unaffected by some transformation if and only if there is a corresponding conservation law. So far, we have only interpreted $[\hat{H}, \hat{A}] = 0$ through the lens of a conservation law; that is, $\hat{A}$ corresponds to a conserved, measurable quantity. However, we could also think of this from another perspective: of $\hat{H}$ commuting with $\hat{A}$.

If $\hat{A}$ is a Hermitian operator, then we can construct a unitary operator through exponentiation, as we have seen many times before:

$$\hat{U}_A = e^{i\hat{A}}. \tag{8.134}$$

The action of $\hat{U}_A$ thus keeps a state in the Hilbert space $\mathcal{H}$, but changes it to a new state on the Hilbert space:

$$|\psi'\rangle = \hat{U}_A|\psi\rangle, \tag{8.135}$$

for some $|\psi\rangle \in \mathcal{H}$. Thus, $\hat{U}_A$ is a transformation that maps the Hilbert space to itself: $\hat{U}_A : \mathcal{H} \to \mathcal{H}$.

Now, what is the "law" of quantum mechanics? As of now, the Schrödinger equation is as close as we get to a law:

$$i\hbar\frac{d}{dt}|\psi\rangle = \hat{H}|\psi\rangle. \tag{8.136}$$

What transformations can we perform on the Schrödinger equation that leave it unchanged? Well, the transformation must be unitary to maintain normalization. So, I can write it as $\hat{U}_A = e^{i\hat{A}}$, for some Hermitian $\hat{A}$. We can act with $\hat{U}_A$ on both sides of the Schrödinger equation:

$$\hat{U}_A\left(i\hbar\frac{d}{dt}|\psi\rangle\right) = \hat{U}_A\left(\hat{H}|\psi\rangle\right) = i\hbar\frac{d}{dt}\left(\hat{U}_A|\psi\rangle\right) = \left(\hat{U}_A\hat{H}\hat{U}_A^\dagger\right)\hat{U}_A|\psi\rangle. \tag{8.137}$$

On the right, we have used the assumption that $\hat{A}$ is time-independent and put a factor of the identity operator between $\hat{H}$ and $|\psi\rangle$, where

$$\mathbb{I} = \hat{U}_A\hat{U}_A^\dagger, \tag{8.138}$$

as $\hat{U}_A$ is unitary.

The state $\hat{U}_A|\psi\rangle = |\psi'\rangle$ is just some other state on the Hilbert space, so the Schrödinger equation is unchanged as a law of physics under the transformation induced by $\hat{A}$ if and only if

$$\hat{U}_A\hat{H}\hat{U}_A^\dagger = e^{i\hat{A}}\hat{H}e^{-i\hat{A}} = \hat{H}. \tag{8.139}$$

This is only true if $\hat{A}$ and $\hat{H}$ commute: $[\hat{A}, \hat{H}] = 0$. Thus, as in classical mechanics, there exists a one-to-one map between conservation laws (those observables that are unchanged by time evolution) and symmetry transformations (maps of the Hilbert space to itself) that maintain the laws of physics.

Exercises

8.1   In this chapter, we studied the Lie algebra of rotations in great detail, but didn't construct the Lie group through exponentiation, except in some limited examples. In this problem, we'll study the structure of the honest rotation group, focusing on its two-dimensional representation.

(a)   A general element of the two-dimensional representation of the rotation group $\mathbb{R}$ can be expressed through exponentiation of the spin-1/2 matrices as

$$\mathbb{U} = e^{\frac{i}{\hbar}(\theta_x \hat{S}_x + \theta_y \hat{S}_y + \theta_z \hat{S}_z)}, \qquad (8.140)$$

where $\theta_x, \theta_y, \theta_z$ are real-valued angles of rotation about their respective axis. By construction, such a matrix is unitary, as the spin operators are Hermitian. Show that the determinant of this matrix is 1.

Hint: Remember, the determinant of a matrix is basis-independent.

(b)   With part (a), this shows that the group of rotations is isomorphic to the group of 2×2 unitary matrices with unit determinant, which is called **SU(2)**. We can construct a general complex-valued 2×2 unitary matrix in the following way. Let's first construct two orthonomal complex vectors, $\vec{v}_1$ and $\vec{v}_2$:

$$\vec{v}_1 = \begin{pmatrix} e^{i\xi_1} \cos\theta \\ e^{i\xi_2} \sin\theta \end{pmatrix}, \qquad \vec{v}_2 = \begin{pmatrix} -e^{i\xi_3} \sin\theta \\ e^{i\xi_4} \cos\theta \end{pmatrix}, \qquad (8.141)$$

for some real angle θ and real phases $\xi_1, \xi_2, \xi_3, \xi_4$. Show that these vectors are unit normalized and determine the constraint on the phases such that they are orthogonal: $\vec{v}_1^\dagger \vec{v}_2 = 0$. With this constraint on $\xi_1, \xi_2, \xi_3, \xi_4$ imposed, show that the matrix formed from these vectors as columns is unitary. That is, show that

$$\mathbb{U} = \begin{pmatrix} \vec{v}_1 & \vec{v}_2 \end{pmatrix} \qquad (8.142)$$

is a unitary matrix.

(c)   Next, impose the unit determinant constraint on the matrix $\mathbb{U}$. What further restriction does this impose on the phases $\xi_1, \xi_2, \xi_3, \xi_4$?

(d)   A general complex-valued 2×2 matrix can be represented as

$$\mathbb{U} = \begin{pmatrix} a_{11} + ib_{11} & a_{12} + ib_{12} \\ a_{21} + ib_{21} & a_{22} + ib_{22} \end{pmatrix}, \qquad (8.143)$$

where a_{ij} and b_{ij} are real numbers. Determine all values of the a_{ij}s and b_{ij}s for the 2×2 unitary matrix with unit determinant that you constructed in part (c).

(e)   In terms of a_{11}, b_{11}, a_{21}, and b_{21}, express the normalization of vector $\vec{v}_1$, $\vec{v}_1^\dagger \vec{v}_1 = 1$. What space does this describe? The parametrization of this space with the vectors as in Eq. (8.141) is called the **Hopf fibration**[4] and demonstrates that the group SU(2) has a very beautiful geometric structure.

8.2   The **Jacobi identity** is a requirement of the commutation relations of a Lie algebra that ensures the corresponding Lie group is associative. For elements $\hat{A}, \hat{B}, \hat{C}$ of a Lie algebra, the Jacobi identity is

$$[\hat{A}, [\hat{B}, \hat{C}]] + [\hat{C}, [\hat{A}, \hat{B}]] + [\hat{B}, [\hat{C}, \hat{A}]] = 0. \tag{8.144}$$

In this expression, $[\hat{A}, \hat{B}]$ is called the Lie bracket of the Lie algebra. We have only studied Lie brackets that correspond to the familiar commutator, but it is possible to consider other definitions that satisfy the Jacobi identity.

(a)   Show that the Jacobi identity is satisfied for the Lie algebra of three-dimensional rotations, with Lie bracket

$$[\hat{L}_i, \hat{L}_j] = i\hbar \sum_{k=1}^{3} \epsilon_{ijk} \hat{L}_k. \tag{8.145}$$

(b)   Show that the Jacobi identity is satisfied for any Lie algebra for which the Lie bracket is then just the commutator, that is, for

$$[\hat{A}, \hat{B}] = \hat{A}\hat{B} - \hat{B}\hat{A}. \tag{8.146}$$

8.3   The **Killing form** of a Lie algebra provides the definition of normalization of operators in a particular representation of the Lie algebra. For representation R of $\mathfrak{su}(2)$, the Killing form is

$$\text{tr}\left[\hat{L}_i^{(R)} \hat{L}_j^{(R)} \right] = k_R \, \delta_{ij}, \tag{8.147}$$

where tr denotes the trace (i.e., the sum of diagonal elements) and $\hat{L}_i^{(R)}$ is an element in the representation R of the Lie algebra. The quantity k_R depends on the representation. For a representation R of dimension D and Casimir C_R, where

$$C_R \mathbb{I}_D = \left(\hat{L}_x^{(R)} \right)^2 + \left(\hat{L}_y^{(R)} \right)^2 + \left(\hat{L}_z^{(R)} \right)^2, \tag{8.148}$$

and $\mathbb{I}_D$ is the $D \times D$ identity matrix, express k_R in terms of D and C_R. What is k_R for a representation of spin ℓ?

8.4   We've studied coherent states in the context of the harmonic oscillator and the free particle, and our formulation of angular momentum suggests that there is a definition of coherent state in that case, too. We constructed the angular momentum raising and lowering operators, $\hat{L}_+$ and $\hat{L}_-$, and we could in principle define a coherent state to be an eigenstate of $\hat{L}_-$, for example.

Consider an eigenstate of the angular momentum lowering operator for spin ℓ:

$$\hat{L}_- |\psi\rangle = \lambda |\psi\rangle. \tag{8.149}$$

[4] H. Hopf, "Über die Abbildungen der dreidimensionalen Sphäre auf die Kugelfläche," *Math. Ann.* **104**, 637–665 (1931).

What is this state? What are the possible values of eigenvalue λ? Does such a state exist in the Hilbert space?

8.5   In Example 8.3, we introduced a simple quantum system involving an electrically charged, spin-1/2 particle immersed in a uniform magnetic field. In this exercise, we will continue the study that was started there, and calculate more expectation values and establish uncertainty principles for the particle's spin.

(a)   What is the uncertainty principle for energy and the x-component of the spin of the electron? That is, for variances σ_E^2 and $\sigma_{S_x}^2$ for the energy and x-component of the spin, respectively, what is their minimum product:

$$\sigma_E^2 \sigma_{S_x}^2 \geq ? \tag{8.150}$$

(b)   When does the lower bound in the uncertainty relation vanish? Show that this makes sense from when we know the variances vanish.

8.6   A very powerful technique for expressing amplitudes of scattering processes in quantum field theory, the harmonious marriage of special relativity and quantum mechanics, is through **spinor helicity**, in which every quantity is related to the eigenstates of the spin-1/2 $\hat{S}_z$ operator. This is exceptionally convenient, because eigenstates of spin-1/2 are just two-component spinors! What could be simpler.

In this problem, we will just study one identity that is often exploited in this business, called the Schouten identity. For four spin-1/2 states $|\psi\rangle, |\rho\rangle, |\chi\rangle, |\eta\rangle$, it states that

$$\langle\psi|\rho\rangle\langle\chi|\eta\rangle = \langle\psi|\eta\rangle\langle\chi|\rho\rangle + ((\rho|)^* i\sigma_2|\eta\rangle\langle\psi|i\sigma_2(|\chi\rangle))^*, \tag{8.151}$$

where σ_2 is the second Pauli spin matrix. Prove this equality. Note that the complex conjugation acts only on a single bra or ket in the final term.

8.7   The $\mathfrak{su}(2)$ algebra consists of three operators in the basis of its Lie algebra which are not mutually commutative. We could also consider a Lie algebra for which the basis is a single Hermitian operator, $\hat{A}$, but that necessarily commutes with itself, $[\hat{A}, \hat{A}] = 0$, so describes an Abelian Lie group. We can also imagine Lie algebras with more than three basis elements, but what about a Lie algebra with two elements, call them $\hat{A}$ and $\hat{B}$?

(a)   In general, we can write the commutation relation of the Lie algebra spanned by $\hat{A}$ and $\hat{B}$ as

$$[\hat{A}, \hat{B}] = i\alpha\hat{A} + i\beta\hat{B}, \tag{8.152}$$

where α, β are the real-valued structure constants of the algebra. Further, if $\hat{A}$ and $\hat{B}$ are orthogonal, this is expressed through the trace of the product of $\hat{A}$ and $\hat{B}$:

$$\text{tr}[\hat{A}\hat{B}] = 0. \tag{8.153}$$

Show that this orthogonality constraint requires that $\alpha = \beta = 0$, or that $\hat{A}$ and $\hat{B}$ necessarily commute. Thus, the smallest dimension of a non-Abelian Lie algebra is 3.

(b) What if $\hat{A}$ and $\hat{B}$ are not initially orthogonal, $\mathrm{tr}[\hat{A}\hat{B}] \neq 0$? Show that you can always define a new operator $\hat{C}$ in the Lie algebra which is orthogonal to $\hat{A}$; that is, an element of the Lie algebra spanned by $\hat{A}$ and $\hat{B}$ for which

$$\mathrm{tr}[\hat{A}\hat{C}] = 0. \tag{8.154}$$

8.8 We had illustrated that rotations in three dimensions are non-commutative or non-Abelian through the example of rotating a coffee cup about two orthogonal axes in two different orders. While a very visceral representation of rotations, we would like to provide concrete mathematics to demonstrate the same property. To do this, we will need to explicitly construct the three-dimensional representation of the rotation group.

(a) First, let's just write down the three rotation matrices for arbitrary rotations about the x-, y-, and z-axes individually. Using the rotation matrix in two dimensions, Eq. (8.7), as a guide, write down the three, three-dimensional matrices that correspond to rotation by an angle θ about the x-, y-, and z-axes individually. Call these matrices $\mathbb{U}_x(\theta)$, $\mathbb{U}_y(\theta)$, and $\mathbb{U}_z(\theta)$, respectively. *Hint*: A rotation about the x-axis, for example, mixes components of a vector in which plane?

(b) Now, for each of these rotation matrices, write them as the exponential of a Hermitian operator, as we did in Sec. 8.2. For example, write

$$\mathbb{U}_x(\theta) = e^{i\frac{\theta \hat{L}_x}{\hbar}}, \tag{8.155}$$

for Hermitian $\hat{L}_x$. That is, determine the elements of the three-dimensional representation of the Lie algebra of rotations.

(c) We explicitly compared what happens to a three-dimensional object when it is rotated about two orthogonal axes by $\pi/2$ in different orders. First, draw a picture of how an initial three-dimensional vector $\vec{v} = (0\ 0\ 1)^\mathsf{T} = \hat{z}$ is rotated in two ways: first $\pi/2$ about the x-axis, then $\pi/2$ about the y-axis, and then vice-versa. Call these resulting vectors $\vec{v}_{yx}$ and $\vec{v}_{xy}$, respectively. Note that the initial vector $\vec{v}$ is just the unit vector along the z-axis. Your drawing should show three vectors: the initial vector $\vec{v}$, and the two final vectors corresponding to action of the rotation in two different orders. What is the difference of the resulting vectors, $\vec{v}_{yx} - \vec{v}_{xy}$?

(d) Now, using the result of part (a), calculate the commutator of the two rotation matrices $[\mathbb{U}_y(\theta), \mathbb{U}_x(\theta)]$ with $\theta = \pi/2$. Act the commutator on the vector $\vec{v}$ from part (c):

$$[\mathbb{U}_y(\theta), \mathbb{U}_x(\theta)]\vec{v}. \tag{8.156}$$

Does this equal the difference vector that you wrote down in part (c), $\vec{v}_{yx} - \vec{v}_{xy}$?

9

The Hydrogen Atom

In the previous chapter, we studied the consequences of rotations and angular momentum in three spatial dimensions, building up to the topic of this chapter: the quantum mechanics of the hydrogen atom. Hydrogen is, of course, the lightest element of the periodic table and consists of a proton and an electron bound through electromagnetism. Our goal for studying this problem is to determine the bound-state energy eigenstates, just like we did with the infinite square well and harmonic oscillator. These energy eigenstates will then tell us how the proton and electron are positioned with respect to one another in space, as well as the energy levels and how energy is transferred when the hydrogen atom transforms from one energy level to another. As always, our goal is to diagonalize the Hamiltonian $\hat{H}$; that is, determine the states $|\psi\rangle$ and energies E such that

$$\hat{H}|\psi\rangle = E|\psi\rangle. \tag{9.1}$$

9.1 The Hamiltonian of the Hydrogen Atom

Our very first step in doing this is to determine what the Hamiltonian is in the first place! To make progress, we will make a number of simplifying assumptions. The first and biggest assumption is in regards to how we will treat electromagnetism within the Hamiltonian. Generally, we would imagine the proton–electron bound state as some configuration not unlike the orbit of the Earth around the Sun, illustrated in Fig. 9.1. The hydrogen atom is bound by electromagnetism (where I've drawn some field lines) and the proton and electron are continuously moving around with velocities $\vec{v}_p$ and $\vec{v}_e$, respectively. So, the distribution of electric and magnetic fields and their effect on how the proton and electron are bound continuously changes. Thus, we might expect that to determine how they are bound we would need the full weight of Maxwell's equations, notions like retarded potentials, etc. However, we also know that the information carried by electromagnetic fields (e.g., the force that they exert on a charge particle) travels at the speed of light c. So, if a proton and electron have relative velocity that is much less than c, $v_{ep} \ll c$, then, from the perspective of electromagnetism, we can approximate them as at rest.

Further, in addition to this small-speed approximation, we will assume that the electromagnetic field is actually completely classical, and has no corresponding Schrödinger equation. That is, we will assume that the electromagnetic field is "large"

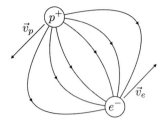

Fig. 9.1 An illustration of the hydrogen atom, composed of an electron e^- and proton p^+ bound in mutual orbit through electromagnetism. Some electric field lines are drawn.

compared to $\hbar$, and has no non-trivial uncertainty relations. With these assumptions, the potential that appears in the Hamiltonian is then just the potential energy we know and love for two charged point particles from introductory physics:

$$V(\vec{r}) = -\frac{e^2}{4\pi\epsilon_0}\frac{1}{|\vec{r}|}, \qquad (9.2)$$

where $\vec{r}$ is the position vector of the electron relative to the proton. Here, e is the fundamental electric charge, which is $e \simeq 1.6 \times 10^{-19}$ C in SI units. The proton carries electric charge $+e$ and the electron carries electric charge $-e$, hence the "$-$" sign in the potential: the proton and electron are attracted to one another.

The other part of the Hamiltonian is the kinetic energy $\hat{K}$, which we can simply express as the sum of kinetic energies of the proton and electron:

$$\hat{K} = \frac{\hat{p}_p^2}{2m_p} + \frac{\hat{p}_e^2}{2m_e}. \qquad (9.3)$$

Now, this is all well and good, but as written, this kinetic energy is sensitive to the momentum of both the proton and the electron and so relates a total of six quantities (two three-dimensional vectors). We can make progress in simplification by rephrasing the problem slightly. First, the total momentum of the hydrogen atom (proton+electron) is irrelevant and does not affect the Hamiltonian for determining the energy eigenstates of the bound system. Only the relative momentum of the proton and electron is important, so we can work in the frame in which

$$\hat{p}_H = \hat{p}_p + \hat{p}_e = 0, \qquad (9.4)$$

with $\hat{p}_p = -\hat{p}_e \equiv -\hat{p}$, so that the kinetic energy reduces to

$$\hat{K} = \frac{\hat{p}_p^2}{2m_p} + \frac{\hat{p}_e^2}{2m_e} = \frac{m_e + m_p}{2m_e m_p}\hat{p}^2. \qquad (9.5)$$

We call the combination of the electron and proton masses the **reduced mass** μ:

$$\mu = \frac{m_e m_p}{m_e + m_p}, \qquad (9.6)$$

a kind of "center-of-mass" mass for the system. Now, we can just work with μ in everything we do later, but it is useful to approximate this reduced mass given the known masses of the proton and electron. It turns out that the proton has a mass that is about 2000 times that of the electron:

$$m_p \approx 2000 m_e , \tag{9.7}$$

so we can approximate the reduced mass as

$$\mu = \frac{m_e m_p}{m_e + m_p} = \frac{m_e}{1 + \frac{m_e}{m_p}} \approx \frac{m_e}{1 + \frac{1}{2000}} \approx m_e . \tag{9.8}$$

Because of this huge difference between the electron and proton masses, we will just take the reduced mass to be the mass of the electron. With all of these assumptions, the Hamiltonian can be written as

$$\hat{H} = \frac{\hat{p}^2}{2m_e} - \frac{e^2}{4\pi\epsilon_0} \frac{1}{\hat{r}} , \tag{9.9}$$

where $\hat{r}$ is the "radius operator" that measures the distance between the proton and the electron. That is:

$$\hat{r} = \sqrt{\hat{x}^2 + \hat{y}^2 + \hat{z}^2} , \tag{9.10}$$

where $\hat{x}, \hat{y}, \hat{z}$ are the coordinate position operators for the electron with respect to the proton.

To go further, we would like to write the square of the momentum operator $\hat{p}^2$ in a nice way. As we are working in three dimensions, we need to think about a coordinate system in which to express $\hat{p}$. Let's go back to introductory physics and the description of circular motion. Recall that for a particle of mass m_e moving with respect to some identified axis, its kinetic energy can be expressed as

$$K = \frac{1}{2} m_e \dot{r}^2 + \frac{1}{2} I \omega^2 , \tag{9.11}$$

where r is the distance of the particle to the axis, I is its moment of inertia about that axis, and ω is its angular velocity about the axis. If we identify the radial momentum p_r as

$$p_r = m_e \dot{r} , \tag{9.12}$$

and the angular momentum $L = I\omega$, then we can express the kinetic energy as

$$K = \frac{p_r^2}{2m_e} + \frac{1}{2} I \left(\frac{L}{I} \right)^2 = \frac{p_r^2}{2m_e} + \frac{L^2}{2I} = \frac{p_r^2}{2m_e} + \frac{L^2}{2m_e r^2} . \tag{9.13}$$

In the final equality, we replaced $I = m_e r^2$, the moment of inertia of a particle of mass m_e a distance r from the axis.

Now, with this construction, in quantum mechanics we just put little hats above every function on phase space to promote them to operators on Hilbert space. That is, the kinetic energy operator is

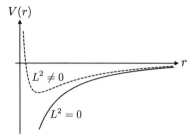

Fig. 9.2 Illustration of the radial potential experienced by the electron in hydrogen. The solid curve corresponds to zero angular momentum L and the dashed curve is the non-zero angular momentum potential.

$$\hat{K} = \frac{\hat{p}_r^2}{2m_e} + \frac{\hat{L}^2}{2m_e \hat{r}^2} \, , \qquad (9.14)$$

and so the Hamiltonian for hydrogen is

$$\hat{H} = \frac{\hat{p}_r^2}{2m_e} + \frac{\hat{L}^2}{2m_e \hat{r}^2} - \frac{e^2}{4\pi\epsilon_0} \frac{1}{\hat{r}} \, . \qquad (9.15)$$

In this form of the Hamiltonian $\hat{H}$, we have separated the radial kinetic energy from the rotational kinetic energy of the electron about the proton, as described by the angular momentum $\hat{L}$. In a classical two-body system that experiences electric or gravitational forces, we are familiar with the result that angular momentum is conserved. We will justify this quantum mechanically shortly. If classical angular momentum is conserved, then L^2 is just a constant number, and so the potential of the hydrogen atom can be identified as the sum of the electric potential and angular momentum terms. Classically, then, we would identify the potential energy as a function of distance r between the electron and proton $V(r)$ as

$$V(r) = \frac{L^2}{2m_e r^2} - \frac{e^2}{4\pi\epsilon_0} \frac{1}{r} \, . \qquad (9.16)$$

In Fig. 9.2 we illustrate the shape of this potential for zero angular momentum $L^2 = 0$ and for non-zero angular momentum $L^2 \neq 0$. For non-zero angular momentum, the potential has a global minimum and classically this would correspond to a circular orbit of the electron. We will see that even if $\hat{L}^2 = 0$, quantum mechanically stable orbits exist.

9.2 The Ground State of Hydrogen

With the Hamiltonian of the hydrogen atom established, an eigenstate $|\psi\rangle$ with eigenenergy E satisfies

$$\hat{H}|\psi\rangle = \left(\frac{\hat{p}_r^2}{2m_e} + \frac{\hat{L}^2}{2m_e \hat{r}^2} - \frac{e^2}{4\pi\epsilon_0} \frac{1}{\hat{r}} \right) |\psi\rangle = E|\psi\rangle \, . \qquad (9.17)$$

To solve this eigenvalue equation, we'll start with some observations and then establish the ground state, the lowest-energy state, explicitly. Because the Hamiltonian only depends on the radial distance $\hat{r}$, the radial momentum $\hat{p}_r$, and the squared angular momentum operator $\hat{L}^2$, every component of angular momentum commutes with the Hamiltonian:

$$[\hat{H},\hat{L}_i] = 0, \tag{9.18}$$

for $i = x, y, z$. Thus, energy eigenstates are equivalently angular momentum eigenstates, which can be labeled by two indices ℓ and m, the eigenvalues of the $\hat{L}^2$ and $\hat{L}_z$ operators, respectively. Angular momentum quantifies rotations on the two-dimensional sphere, which is parametrized by two angles, a polar angle θ and an azimuthal angle ϕ. That is, of the three spatial dimensions in which the state $|\psi\rangle$ can live, the dynamics in two of those dimensions are specified by the particular angular momentum state of $|\psi\rangle$. The dynamics in the third, radial dimension are distinct from angular momentum. However, as we are interested in the discrete set of bound states of the hydrogen atom, the dynamics in the radial dimension can necessarily be indexed by an integer, n. Thus, an eigenstate of the hydrogen Hamiltonian can be uniquely specified by three numbers, n, ℓ, m:

$$|\psi\rangle \equiv |n, \ell, m\rangle, \tag{9.19}$$

the radial number, the total angular momentum, and the z-component of angular momentum, respectively. We will choose n so that $n = 1$ is the ground state, and higher-energy states have $n > 1$. Again, this is simply for convenience because we know that normalized bound states of the hydrogen atom on Hilbert space must have a discrete spectrum.

Further, the value of the eigenenergies is independent of the z-component of angular momentum m, for fixed ℓ. A rotation implemented by $\hat{L}_z$, for example, does not change the Hamiltonian, as shown in Eq. (9.18). Thus, the action of $\hat{L}_z$ cannot affect the eigenvalues of the Hamiltonian. On the other hand, the total angular momentum $\hat{L}^2$ does affect the value of the energy, as it explicitly appears in the Hamiltonian. Radial dependence is also explicit in the Hamiltonian. Then, the values of the eigenenergies can only depend on n and ℓ:

$$E \equiv E_{n,\ell}. \tag{9.20}$$

Focusing on the ground state, it must have $\ell = 0$, because that is the smallest value of the total angular momentum, which appears as a positive contribution to the Hamiltonian. Also, by our definition above, the ground state has $n = 1$. Therefore, the equation for the ground state is

$$\left(\frac{\hat{p}_r^2}{2m_e} - \frac{e^2}{4\pi\epsilon_0} \frac{1}{\hat{r}} \right) |1,0,0\rangle = E_{1,0}|1,0,0\rangle, \tag{9.21}$$

in the notation established above.

To solve for the ground-state energy, we'll work in position space. The ground-state wavefunction can be expressed as

$$\psi_{1,0,0}(r) \equiv \langle \vec{r} | 1,0,0 \rangle, \tag{9.22}$$

with $|\vec{r}\rangle$ a position eigenstate in three dimensions. Because the state has no angular momentum, the wavefunction exclusively has dependence on the radial coordinate r. We would like to express the eigenvalue problem for the ground-state wavefunction as a differential equation exclusively in r which requires an appropriate expression for the squared radial momentum $\hat{p}_r^2$ in position space. Our procedure for determining this will be to just explicitly take derivatives in Cartesian coordinates x, y, z of a function $f(r)$, where $r = \sqrt{x^2 + y^2 + z^2}$, and then we can identify precisely what the operator $\hat{p}_r^2$ is in radial position space. So, we have

$$\hat{p}_r^2 f(r) = (\hat{p}_x^2 + \hat{p}_y^2 + \hat{p}_z^2) f(r) = -\hbar^2 \left(\frac{\partial^2}{\partial x^2} + \frac{\partial^2}{\partial y^2} + \frac{\partial^2}{\partial z^2} \right) f(r). \tag{9.23}$$

Let's now take the derivatives with respect to x. For the first derivative, we have

$$\frac{\partial}{\partial x} f(r) = \frac{\partial r}{\partial x} \frac{df}{dr} = \frac{x}{r} \frac{df}{dr}. \tag{9.24}$$

The second derivative is then

$$\frac{\partial^2}{\partial x^2} f(r) = \frac{\partial}{\partial x} \frac{x}{r} \frac{df}{dr} = \frac{y^2 + z^2}{r^3} \frac{df}{dr} + \frac{x^2}{r^2} \frac{d^2 f}{dr^2}. \tag{9.25}$$

The y and z derivatives are a simple relabeling:

$$\frac{\partial^2}{\partial y^2} f(r) = \frac{x^2 + z^2}{r^3} \frac{df}{dr} + \frac{y^2}{r^2} \frac{d^2 f}{dr^2}, \qquad \frac{\partial^2}{\partial z^2} f(r) = \frac{x^2 + y^2}{r^3} \frac{df}{dr} + \frac{z^2}{r^2} \frac{d^2 f}{dr^2}. \tag{9.26}$$

Therefore, the action of the squared radial momentum on a function of radius r is simply

$$\hat{p}_r^2 f(r) = -\hbar^2 \left(\frac{y^2 + z^2}{r^3} \frac{df}{dr} + \frac{x^2}{r^2} \frac{d^2 f}{dr^2} + \frac{x^2 + z^2}{r^3} \frac{df}{dr} + \frac{y^2}{r^2} \frac{d^2 f}{dr^2} + \frac{x^2 + y^2}{r^3} \frac{df}{dr} + \frac{z^2}{r^2} \frac{d^2 f}{dr^2} \right)$$

$$= -\hbar^2 \left(\frac{2}{r} \frac{df}{dr} + \frac{d^2 f}{dr^2} \right). \tag{9.27}$$

The ground-state eigenvalue equation is then the differential equation, for $r > 0$:

$$\left(-\frac{\hbar^2}{m_e} \frac{1}{r} \frac{d}{dr} - \frac{\hbar^2}{2m_e} \frac{d^2}{dr^2} - \frac{e^2}{4\pi\epsilon_0} \frac{1}{r} \right) \psi_{1,0,0}(r) = E_{1,0} \psi_{1,0,0}(r). \tag{9.28}$$

For normalizability, the wavefunction must vanish as $r \to \infty$, $\psi_{1,0,0}(r \to \infty) = 0$, and because this is the ground state, there can be no other nodes of the wavefunction for $0 < r < \infty$. With these pieces of information, we will find the ground state by making an ansatz for the wavefunction's functional form that satisfies these requirements. Again, while it may not be so elegant, if your guess works, then we've solved the problem at hand. A functional form for the wavefunction that satisfies these requirements is

$$\psi_{1,0,0}(r) \propto r^b e^{-r/a_0}, \tag{9.29}$$

for $a_0, b > 0$. Note that we just write $\propto$ and not $=$ because the wavefunction must be normalized. We can fix normalization later.

With this ansatz, let's just plug it into the eigenvalue equation and see what we find. First, the derivatives of the ansatz are

$$\left(\frac{2}{r}\frac{d}{dr} + \frac{d^2}{dr^2}\right) r^b e^{-r/a_0} = \left(\frac{b(b+1)}{r^2} - \frac{2(b+1)}{a_0 r} + \frac{1}{a_0^2}\right) r^b e^{-r/a_0}. \tag{9.30}$$

Plugging this into the eigenvalue differential equation and simplifying, we find

$$-\frac{\hbar^2}{2m_e}\left(\frac{b(b+1)}{r^2} - \frac{2(b+1)}{a_0 r} + \frac{1}{a_0^2}\right) - \frac{e^2}{4\pi\epsilon_0}\frac{1}{r} = E_{1,0}. \tag{9.31}$$

Now, we can just match terms at each power of r. As there is only one term proportional to $1/r^2$, it must vanish by itself to ensure that the wavefunction doesn't diverge at $r = 0$. Thus, we require that $b = 0$. With this established, the terms at $1/r$ order require

$$\frac{\hbar^2}{m_e a_0} = \frac{e^2}{4\pi\epsilon_0}, \tag{9.32}$$

or

$$a_0 = \frac{4\pi\epsilon_0 \hbar^2}{m_e e^2}. \tag{9.33}$$

To ensure that the units in the exponent of the wavefunction are correct, a_0 must have units of length. It is called the **Bohr radius**, and is the characteristic radial size of the orbit of the electron about the proton in the ground state of hydrogen.

Finally, the energy $E_{1,0}$ is set by the r^0 terms in the eigenvalue equation:

$$E_{1,0} \equiv E_0 = -\frac{\hbar^2}{2m_e a_0^2} = -\frac{m_e e^4}{2(4\pi\epsilon_0)^2 \hbar^2}. \tag{9.34}$$

If you evaluate the Bohr radius a_0 with the established values for the fundamental constants $\hbar$, ϵ_0, m_e, and e, you find

$$a_0 = 5.3 \times 10^{-11} \text{ m}. \tag{9.35}$$

The ground-state energy is

$$E_0 = -13.6 \text{ eV (electronvolts)}. \tag{9.36}$$

One **electronvolt** is the energy that an electron gains through an electric potential of one volt, $1 \text{ eV} = 1.6 \times 10^{-19}$ J. For some comparison, the wavelength of visible light is of the order of hundreds of nanometers, thousands of times larger than the Bohr radius. Also, the mean kinetic energy of an air molecule at room temperature is about $1/40$ eV.

The Bohr radius and the ground-state energy can be re-expressed in terms of the **fine-structure constant** α, a dimensionless, pure number that quantifies the strength of electromagnetism. The constant α is defined to be

$$\alpha = \frac{e^2}{4\pi\epsilon_0 \hbar c}, \tag{9.37}$$

and it has a value of about $1/137$. Dimensionless numbers are universal quantities in the sense that they are the same value in any unit system. First observed by Albert Michelson and Edward Morley in 1887,[1] it was introduced by Arnold Sommerfeld in 1916 in his study of the structure of hydrogen due to physics at a finer resolution than just the attractive Coulomb potential.[2] With the fine-structure constant, the Bohr radius is

$$a_0 = \frac{1}{\alpha} \frac{\hbar}{m_e c} = \frac{\lambda_C}{2\pi\alpha} , \qquad (9.38)$$

which is a factor of $1/(2\pi\alpha)$ larger than the **Compton wavelength** λ_C of the electron.[3] The ground-state energy of hydrogen is

$$E_0 = -\frac{\alpha^2}{2} m_e c^2 , \qquad (9.39)$$

which (up to a minus sign) is a factor of $\frac{\alpha^2}{2}$ smaller than the rest mass energy of the electron.

Let's now correct the normalization of this wavefunction, $\psi_{1,0,0}(r)$. This wavefunction describes the position of the electron in three-dimensional space, so it is normalized by integrating over all possible x, y, z positions:

$$1 = \langle 1,0,0 | 1,0,0 \rangle = \int_{-\infty}^{\infty} dx \int_{-\infty}^{\infty} dy \int_{-\infty}^{\infty} dz \, \psi_{1,0,0}^*(\vec{r}) \psi_{1,0,0}(\vec{r}) . \qquad (9.40)$$

Because the wavefunction depends exclusively on the radial distance from the proton, it is most convenient to express this normalization integral in spherical coordinates, where

$$x = r \sin\theta \cos\phi , \qquad (9.41)$$
$$y = r \sin\theta \sin\phi ,$$
$$z = r \cos\theta .$$

In these coordinates, the normalization integral can be expressed as

$$1 = \int_0^{\infty} r^2 \, dr \int_{-1}^{1} d(\cos\theta) \int_0^{2\pi} d\phi \, \psi_{1,0,0}^*(r) \psi_{1,0,0}(r). \qquad (9.42)$$

We can immediately do the integrals over the polar and azimuthal angles θ and ϕ, because the wavefunction is independent of them:

$$1 = 4\pi \int_0^{\infty} r^2 \, dr \, \psi_{1,0,0}^*(r) \psi_{1,0,0}(r) . \qquad (9.43)$$

---

[1] A. A. Michelson and E. W. Morley, "On a method of making the wavelength of sodium light the actual and practical standard of length," *Am. J. Sci.* **34**(204), 427–430 (1887).

[2] A. Sommerfeld, "Zur Quantentheorie der Spektrallinien," *Ann. Phys.* **356**(17), 1–94 (1916).

[3] A. H. Compton, "A quantum theory of the scattering of X-rays by light elements," *Phys. Rev.* **21**, 483–502 (1923).

Now, the ground-state wavefunction we identified is

$$\psi_{1,0,0}(r) = N e^{-r/a_0}, \tag{9.44}$$

where N is the normalization constant, and so the remaining integral to determine N is

$$1 = 4\pi N^2 \int_0^\infty dr\, r^2\, e^{-2r/a_0}. \tag{9.45}$$

To do this integral, one can just use integration by parts twice to eliminate the r^2 factor. Doing this, we find that

$$1 = 4\pi N^2 \frac{a_0^3}{4}, \tag{9.46}$$

or that the normalization constant is

$$N = \frac{1}{\sqrt{\pi a_0^3}}. \tag{9.47}$$

Thus, the normalized, ground-state wavefunction of the hydrogen atom is

$$\psi_{1,0,0}(r) = \frac{e^{-r/a_0}}{\sqrt{\pi a_0^3}}, \tag{9.48}$$

where a_0 is the Bohr radius.

This wavefunction is plotted in Fig. 9.3. Specifically, we plot the radial probability density $p(r)$ of the electron about the proton in the ground state of the hydrogen atom, where

$$p(r) = r^2 |\psi_{1,0,0}(r)|^2. \tag{9.49}$$

Probability densities always involve the absolute square of the wavefunction, and the extra factor of r^2 comes from the integration measure in spherical coordinates as noted for example in the normalization defined in Eq. (9.43). The radius at which the electron is most likely to be found, the global maximum of this probability density is when

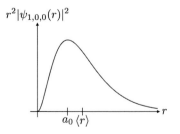

Fig. 9.3   A plot of the probability density of the electron's radius r from the proton in the ground state of the hydrogen atom. The most likely radius is the Bohr radius, a_0, which is at a slightly smaller radius than the expectation value $\langle r \rangle = 3a_0/2$.

$r = a_0$, the Bohr radius. However, the Bohr radius is not the expectation value of the electron's radius, $\langle r \rangle$. That is instead

$$\langle r \rangle = 4\pi \int_0^\infty dr\, r^3 \, |\psi_{1,0,0}(r)|^2 = \frac{4}{a_0^3} \int_0^\infty dr\, r^3 \, e^{-2r/a_0} = \frac{3}{2}a_0 , \tag{9.50}$$

or slightly larger than the Bohr radius.

9.3 Generating All Bound States: The Laplace–Runge–Lenz Vector

One way to proceed to establish all of the energy eigenstates of the hydrogen Hamiltonian, with any value of angular momentum, is to express Eq. (9.17) in position space as a differential equation, and then just solve it in full generality. This is very concrete and mechanical, but doesn't provide clear physical insight into the structure of bound states and how they are related. We will take a very different approach to diagonalizing the Hamiltonian, similar to how we solved the harmonic oscillator. To motivate it, we will go back to the *classical* Hamiltonian for the hydrogen atom, and then use canonical quantization to bring it back to quantum mechanics. This approach to solving the hydrogen atom was first done by Wolfgang Pauli,[4] essentially simultaneously with Erwin Schrödinger's analysis, which approached the problem from the point of view of the Schrödinger equation as a differential equation.[5]

9.3.1 The Classical $1/r$ Potential Problem

The classical Hamiltonian for the hydrogen atom is

$$H = \frac{|\vec{p}|^2}{2m_e} - \frac{e^2}{4\pi\epsilon_0}\frac{1}{r} = \frac{p_r^2}{2m_e} + \frac{L^2}{2m_e r^2} - \frac{e^2}{4\pi\epsilon_0}\frac{1}{r} , \tag{9.51}$$

which is of course the identical form of the quantum mechanical Hamiltonian, but is not an operator. In classical mechanics, time evolution of functions on phase space is determined by the Poisson bracket of that function with the Hamiltonian. For instance, the angular momentum vector $\vec{L}$ is conserved in the hydrogen atom, and so its time derivative is 0:

$$\frac{d\vec{L}}{dt} = \{\vec{L}, H\} = \sum_{i=1}^{3}\left(\frac{\partial H}{\partial r_i}\frac{\partial \vec{L}}{\partial p_i} - \frac{\partial H}{\partial p_i}\frac{\partial \vec{L}}{\partial r_i}\right) = 0 . \tag{9.52}$$

Here, r_i and p_i are the ith components of the position and momentum vectors of the electron, $\vec{r}$ and $\vec{p}$, respectively. It is straightforward to verify this relationship given hydrogen's Hamiltonian and the expression for the electron's angular momentum,

---

[4] W. Pauli, "Über das Wasserstoffspektrum vom Standpunkt der neuen Quantenmechanik," *Z. Phys.* **36**(5), 336–363 (1926).
[5] E. Schrödinger, "Quantisierung als Eigenwertproblem," *Ann. Phys.* **384**(4), 361–376 (1926).

$\vec{L} = \vec{r} \times \vec{p}$, which you will do in the exercises. As you well know from a course on introductory physics, knowing that angular momentum is conserved highly constrains the possible motion of the electron; indeed, its trajectory must lie in a plane.

More generally, the hydrogen atom's Hamiltonian is of the form that it exclusively depends on the distance between the proton and the electron in the atom, and not on an absolute orientation. In particular, the electric potential energy is a **central potential**, just like that for gravitation, though proportional to electric charge and not mass. Every system with a central potential conserves angular momentum, but this special inverse radial form of the potential, corresponding to an inverse square force, has an additional conserved quantity. Centuries ago, when scientists were studying the properties of gravitation, it was realized that a vector now called the **Laplace–Runge–Lenz vector** is also conserved.[6] The Laplace–Runge–Lenz vector is typically denoted as $\vec{A}$ and is defined to be

$$\vec{A} \equiv \frac{\vec{p} \times \vec{L}}{m_e} - \frac{e^2}{4\pi\epsilon_0}\hat{r} = \frac{\vec{p} \times (\vec{r} \times \vec{p})}{m_e} - \frac{e^2}{4\pi\epsilon_0}\hat{r}, \qquad (9.53)$$

where $\hat{r}$ is the unit radial vector. That $\vec{A}$ is conserved means that its Poisson bracket with the Hamiltonian vanishes:

$$\frac{d\vec{A}}{dt} = \{\vec{A}, H\} = \sum_{i=1}^{3}\left(\frac{\partial H}{\partial r_i}\frac{\partial \vec{A}}{\partial p_i} - \frac{\partial H}{\partial p_i}\frac{\partial \vec{A}}{\partial r_i}\right) = 0. \qquad (9.54)$$

Having both $\vec{L}$ and $\vec{A}$ conserved will be extremely powerful for diagonalizing the Hamiltonian of the hydrogen atom.

We first note that the angular momentum and the Laplace–Runge–Lenz vectors are orthogonal:

$$\vec{L} \cdot \vec{A} = \vec{A} \cdot \vec{L} = 0. \qquad (9.55)$$

Conservation of angular momentum means that the motion of the electron about the proton is confined to a plane. The direction of the angular momentum vector $\vec{L}$ is then necessarily perpendicular to this plane, because both $\vec{r}$ and $\vec{p}$ lie in the plane. By construction, the Laplace–Runge–Lenz vector $\vec{A}$ lies in the plane of the electron's motion,

---

[6] This is an example of Stigler's law of eponymy: that no scientific discovery is named after its original discoverers. While named after Pierre-Simon de Laplace, Carl Runge, and Wilhelm Lenz, this was first shown to be conserved by Jakob Hermann nearly a century before work by Laplace and nearly two centuries before either Runge or Lenz. More detail about the history of the Laplace–Runge–Lenz vector can be found in H. Goldstein, "Prehistory of the 'Runge–Lenz' vector," *Am. J. Phys.* **43**, 737 (1975); H. Goldstein, "More on the prehistory of the Laplace or Runge–Lenz vector," *Am. J. Phys.* **44**, 1123 (1976). Stigler's law of eponymy was introduced in S. M. Stigler, "Stigler's law of eponymy," *Trans. N. Y. Acad. Sci.* **39**, 147–157 (1980), in which Stephen Stigler tongue-in-cheek credits sociologist Robert Merton with its original discovery.

therefore it is also orthogonal to $\vec{L}$. This is actually a way to construct the Laplace–Runge–Lenz vector if you didn't know about its existence beforehand. Vectors $\vec{r}$ and $\vec{p}$ are perpendicular to $\vec{L}$, and the Laplace–Runge–Lenz vector is just a linear combination of $\vec{r}$ and $\vec{p}$. The coefficients of this linear combination can be determined by demanding that the Poisson bracket of $\vec{A}$ and the Hamiltonian vanishes. Indeed, note that we have

$$\vec{A} = \frac{\vec{p} \times (\vec{r} \times \vec{p})}{m_e} - \frac{e^2}{4\pi\epsilon_0}\hat{r} = \frac{|\vec{p}|^2}{m_e}\vec{r} - \frac{\vec{r}\cdot\vec{p}}{m_e}\vec{p} - \frac{e^2}{4\pi\epsilon_0}\hat{r} \qquad (9.56)$$
$$= \left(\frac{|\vec{p}|^2}{m_e} - \frac{e^2}{4\pi\epsilon_0}\frac{1}{r}\right)\vec{r} - \frac{\vec{r}\cdot\vec{p}}{m_e}\vec{p} = H\vec{r} + \frac{|\vec{p}|^2}{2m_e}\vec{r} - \frac{\vec{r}\cdot\vec{p}}{m_e}\vec{p}.$$

In the last equality, we have inserted the hydrogen atom's Hamiltonian.

As a conserved vector in the plane of the orbit of the electron, we can illustrate the Laplace–Runge–Lenz vector clearly to demonstrate this property. In Fig. 9.4, we show the orbit of the electron about the proton and at two points on the orbit, draw several relevant vectors. As required, the Laplace–Runge–Lenz vector $\vec{A}$ is identical at these two distinct points on the orbit, even though the radius, momentum, and cross product of the momentum and angular momentum of the electron vary over the orbit.

9.3.2 The Poisson Algebra of Angular Momentum and Laplace–Runge–Lenz

Additionally, there are interesting Poisson brackets between angular momentum and the Laplace–Runge–Lenz vector. The expression for the Laplace–Runge–Lenz vector with the Hamiltonian enables straightforward evaluation of $\{\vec{L},\vec{A}\}$. In components, note that the ith component of the angular momentum and Laplace–Runge–Lenz vectors are

$$L_i = \sum_{j,k=1}^{3} \epsilon_{ijk} r_j p_k, \qquad\qquad A_i = H r_i + \frac{|\vec{p}|^2}{2m_e} r_i - \frac{\vec{r}\cdot\vec{p}}{m_e} p_i. \qquad (9.57)$$

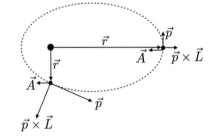

Fig. 9.4 Illustration of the orbit (dashed) of an electron (small dot) about the proton (large dot). At two points on the orbit, multiple vectors are drawn: the position with respect to the proton $\vec{r}$, the electron's momentum $\vec{p}$, the cross product of the momentum and angular momentum of the electron $\vec{p} \times \vec{L}$, and the Laplace–Runge–Lenz vector $\vec{A}$. For clarity of illustration, the vector $\vec{A}$ has been slightly displaced vertically at the point on the right.

The derivatives that we need to evaluate to calculate the Poisson bracket are

$$\frac{\partial L_i}{\partial r_j} = \sum_{k=1}^{3} \epsilon_{ijk} p_k, \qquad\qquad \frac{\partial L_i}{\partial p_j} = -\sum_{k=1}^{3} \epsilon_{ijk} r_k, \qquad (9.58)$$

and

$$\frac{\partial A_i}{\partial r_j} = \frac{\partial H}{\partial r_j} r_i + H\delta_{ij} + \frac{|\vec{p}|^2}{2m_e}\delta_{ij} - \frac{p_i p_j}{m_e}, \qquad (9.59)$$

$$\frac{\partial A_i}{\partial p_j} = \frac{\partial H}{\partial p_j} r_i + \frac{r_i p_j}{m_e} - \frac{r_j p_i}{m_e} - \frac{\vec{r}\cdot\vec{p}}{m_e}\delta_{ij}.$$

Now, we can just use these derivatives to evaluate the Poisson bracket of the vector components L_i and A_l:

$$\{L_i, A_l\} = \sum_{j=1}^{3}\left(\frac{\partial L_i}{\partial r_j}\frac{\partial A_l}{\partial p_j} - \frac{\partial L_i}{\partial p_j}\frac{\partial A_l}{\partial r_j}\right) \qquad (9.60)$$

$$= \sum_{j,k=1}^{3}\left[\epsilon_{ijk}p_k\left(\frac{\partial H}{\partial p_j}r_l + \frac{r_l p_j}{m_e} - \frac{r_j p_l}{m_e} - \frac{\vec{r}\cdot\vec{p}}{m_e}\delta_{lj}\right)\right.$$

$$\left. + \epsilon_{ijk}r_k\left(\frac{\partial H}{\partial r_j}r_l + H\delta_{lj} + \frac{|\vec{p}|^2}{2m_e}\delta_{lj} - \frac{p_l p_j}{m_e}\right)\right]$$

$$= \sum_{k=1}^{3}\epsilon_{ilk}\left(Hr_k + \frac{|\vec{p}|^2}{2m_e}r_k - \frac{\vec{r}\cdot\vec{p}}{m_e}p_k\right) + \{L_i, H\}r_l$$

$$+ \sum_{j,k=1}^{3}\left[\epsilon_{ijk}r_l p_j p_k - \frac{p_l}{m_e}\epsilon_{ijk}(r_j p_k + r_k p_j)\right].$$

In the final equality, we have just collected related terms from the second equality. The first term in parentheses on the final line is just the kth component of the Laplace–Runge–Lenz vector. All other terms actually vanish! For example, the Poisson bracket of angular momentum and the Hamiltonian vanishes by conservation of angular momentum. So, we find the very simple relationship that the Poisson bracket of angular momentum and the Laplace–Runge–Lenz vector is

$$\{L_i, A_j\} = \sum_{k=1}^{3}\epsilon_{ijk}A_k. \qquad (9.61)$$

With a similar approach, one can also calculate the Poisson bracket between different components of $\vec{A}$. This calculation is very long and tedious, so we won't do it here in the text. Nevertheless, doing this one finds that

$$\{A_i, A_j\} = -\sum_{k=1}^{3}\epsilon_{ijk}L_k\frac{2}{m_e}H. \qquad (9.62)$$

So, to summarize, we have identified the following Poisson brackets that involve the Laplace–Runge–Lenz vector $\vec{A}_i$:

$$\{\vec{A}, H\} = 0, \qquad \{L_i, A_j\} = \sum_{k=1}^{3} \epsilon_{ijk} A_k, \qquad \{A_i, A_j\} = -\sum_{k=1}^{3} \epsilon_{ijk} L_k \frac{2}{m_e} H. \qquad (9.63)$$

Additionally, by construction, we have that $\vec{A} \cdot \vec{L} = 0$.

9.4 Quantizing the Hydrogen Atom

With this classical mechanics description of the hydrogen atom and the Laplace–Runge–Lenz vector, we can quantize it to correspondingly describe the *quantum* hydrogen atom and thus determine its energy levels and all that.[7] Long ago, we had described the correspondence principle and canonical quantization which gave a precise, mathematical connection between Poisson brackets of functions on phase space in classical mechanics and commutation relations of Hermitian operators in quantum mechanics. Well, we have identified the Poisson brackets of the classical hydrogen atom, so we can just canonically quantize it to construct the quantum mechanical description of the hydrogen atom. The first thing we need to do is to promote all of the functions on phase space to operators that act on the Hilbert space. Both the Hamiltonian and angular momentum operators, $\hat{H}$ and $\hat{\vec{L}}$, we are well-familiar with, but the Laplace–Runge–Lenz vector is novel. We want this vector as an operator to be Hermitian, which requires being a bit careful with the order of operators that construct it. We define the Laplace–Runge–Lenz operator $\hat{\vec{A}}$ to be

$$\hat{\vec{A}} \equiv \frac{\hat{\vec{p}} \times \hat{\vec{L}} - \hat{\vec{L}} \times \hat{\vec{p}}}{2m_e} - \frac{e^2}{4\pi\epsilon_0} \frac{\hat{\vec{r}}}{\hat{r}}. \qquad (9.64)$$

This operator is Hermitian because of the careful ordering of the operators in the cross products: for example, $\hat{\vec{p}} \times \hat{\vec{L}}$ means that momentum always appears to the left of angular momentum in expanding this in components. Additionally, the cross product is antisymmetric, so in the classical limit the difference of the two cross products reduces to $\vec{p} \times \vec{L} - \vec{L} \times \vec{p} = 2\vec{p} \times \vec{L}$. Also, now $\hat{r}$ is the operator for radial distance of the electron from the proton.

The form of canonical quantization that we had defined long ago was that classical Poisson brackets are quantum commutators, up to a factor of $i\hbar$:

$$\{,\}_{\text{classical}} \rightarrow \frac{[,]_{\text{quantum}}}{i\hbar}. \qquad (9.65)$$

With this correspondence, we immediately know the commutators of these quantum operators from their classical Poisson brackets:

$$[\hat{\vec{L}}, \hat{H}] = 0, \qquad\qquad\qquad [\hat{\vec{A}}, \hat{H}] = 0, \qquad (9.66)$$

---

[7] This section closely follows M. Bander and C. Itzykson, "Group theory and the hydrogen atom," *Rev. Mod. Phys.* **38**, 330–345 (1966).

$$[\hat{L}_i, \hat{A}_j] = i\hbar \sum_{k=1}^{3} \epsilon_{ijk}\hat{A}_k, \qquad\qquad [\hat{A}_i, \hat{A}_j] = -i\hbar \sum_{k=1}^{3} \epsilon_{ijk}\hat{L}_k \frac{2}{m_e}\hat{H}.$$

Additionally, from our study of angular momentum, we know that the commutation relation of components of angular momentum is

$$[\hat{L}_i, \hat{L}_j] = i\hbar \sum_{k=1}^{3} \epsilon_{ijk}\hat{L}_k. \tag{9.67}$$

9.4.1 Squaring the Laplace–Runge–Lenz Operator

We need one more piece of information before we apply this framework to understanding the energy eigenstates of the Hamiltonian of hydrogen. In components, the Laplace–Runge–Lenz operator can be written as

$$\hat{A}_k = \sum_{i,j=1}^{3} \epsilon_{ijk} \frac{\hat{p}_i \hat{L}_j - \hat{L}_i \hat{p}_j}{2m_e} - \frac{e^2}{4\pi\epsilon_0}\frac{\hat{r}_k}{\hat{r}}. \tag{9.68}$$

Its square, being careful with the order of non-commuting operators, is

$$\hat{\vec{A}} \cdot \hat{\vec{A}} = \frac{e^4}{(4\pi\epsilon_0)^2} + \frac{1}{4m_e^2} \sum_{i,j,k,l,m=1}^{3} \epsilon_{ijk}\epsilon_{lmk}(\hat{p}_i\hat{L}_j - \hat{L}_i\hat{p}_j)(\hat{p}_l\hat{L}_m - \hat{L}_l\hat{p}_m) \tag{9.69}$$

$$- \frac{e^2}{4\pi\epsilon_0}\frac{1}{2m_e} \sum_{i,j,k=1}^{3} \epsilon_{ijk}\left((\hat{p}_i\hat{L}_j - \hat{L}_i\hat{p}_j)\frac{\hat{r}_k}{\hat{r}} + \frac{\hat{r}_k}{\hat{r}}(\hat{p}_i\hat{L}_j - \hat{L}_i\hat{p}_j)\right).$$

A fair warning to readers: the next few pages contain extensive algebra in manipulating and simplifying the expression for the square of the Laplace–Runge–Lenz operator. If you believe the manipulations are correct, you can skip to the very compact and simple result, in Eq. (9.86).

This can be simplified in several ways. For the first term involving momentum and angular momentum, it can be simplified by using a nice property of the Levi-Civita symbols. We have that

$$\sum_{k=1}^{3} \epsilon_{ijk}\epsilon_{lmk} = \delta_{il}\delta_{jm} - \delta_{im}\delta_{jl}. \tag{9.70}$$

This relationship can be proved by considering all possible values for $i, j, l, m \in \{1, 2, 3\}$ and showing that both sides are equal. For example, consider $i = l = 1$, $j = m = 2$. Then, we find

$$\sum_{k=1}^{3} \epsilon_{12k}\epsilon_{12k} = \epsilon_{123}\epsilon_{123} = 1 = \delta_{11}\delta_{22} - \delta_{12}\delta_{12}, \tag{9.71}$$

because $\delta_{11} = \delta_{22} = 1$ and $\delta_{12} = 0$.

With this simplification, the second term in the square of the Laplace–Runge–Lenz vector is

$$\sum_{i,j,k,l,m=1}^{3} \epsilon_{ijk}\epsilon_{lmk}(\hat{p}_i\hat{L}_j - \hat{L}_i\hat{p}_j)(\hat{p}_l\hat{L}_m - \hat{L}_l\hat{p}_m) \tag{9.72}$$

$$= \sum_{i,j,l,m=1}^{3} (\delta_{il}\delta_{jm} - \delta_{im}\delta_{jl})(\hat{p}_i\hat{L}_j\hat{p}_l\hat{L}_m - \hat{p}_i\hat{L}_j\hat{L}_l\hat{p}_m - \hat{L}_i\hat{p}_j\hat{p}_l\hat{L}_m + \hat{L}_i\hat{p}_j\hat{L}_l\hat{p}_m)$$

$$= \sum_{i,j=1}^{3} (\hat{p}_i\hat{L}_j\hat{p}_i\hat{L}_j - \hat{p}_i\hat{L}_j\hat{L}_i\hat{p}_j - \hat{L}_i\hat{p}_j\hat{p}_i\hat{L}_j + \hat{L}_i\hat{p}_j\hat{L}_i\hat{p}_j - \hat{p}_i\hat{L}_j\hat{p}_j\hat{L}_i$$

$$+ \hat{p}_i\hat{L}_j\hat{L}_j\hat{p}_i + \hat{L}_i\hat{p}_j\hat{p}_j\hat{L}_i - \hat{L}_i\hat{p}_j\hat{L}_j\hat{p}_i) \ .$$

This can be further simplified with the commutation relations of momentum and angular momentum. Explicitly, note that

$$[\hat{p}_i, \hat{L}_j] = [\hat{p}_i, \sum_{l,m=1}^{3} \epsilon_{lmj}\hat{r}_l\hat{p}_m] = \sum_{l,m=1}^{3} \epsilon_{lmj}\hat{p}_i\hat{r}_l\hat{p}_m - \sum_{l,m=1}^{3} \epsilon_{lmj}\hat{r}_l\hat{p}_m\hat{p}_i \tag{9.73}$$

$$= \sum_{l,m=1}^{3} \epsilon_{lmj}(\hat{r}_l\hat{p}_i - i\hbar\,\delta_{il})\hat{p}_m - \sum_{l,m=1}^{3} \epsilon_{lmj}\hat{r}_l\hat{p}_m\hat{p}_i$$

$$= i\hbar \sum_{k=1}^{3} \epsilon_{ijk}\hat{p}_k \ .$$

Then, with this commutator, let's move all momentum operators right and all angular momentum operators left:

$$\sum_{i,j,k,l,m=1}^{3} \epsilon_{ijk}\epsilon_{lmk}(\hat{p}_i\hat{L}_j - \hat{L}_i\hat{p}_j)(\hat{p}_l\hat{L}_m - \hat{L}_l\hat{p}_m) \tag{9.74}$$

$$= \sum_{i,j=1}^{3} (\hat{p}_i\hat{L}_j\hat{p}_i\hat{L}_j - \hat{p}_i\hat{L}_j\hat{L}_i\hat{p}_j - \hat{L}_i\hat{p}_j\hat{p}_i\hat{L}_j + \hat{L}_i\hat{p}_j\hat{L}_i\hat{p}_j - \hat{p}_i\hat{L}_j\hat{p}_j\hat{L}_i$$

$$+ \hat{p}_i\hat{L}_j\hat{L}_j\hat{p}_i + \hat{L}_i\hat{p}_j\hat{p}_j\hat{L}_i - \hat{L}_i\hat{p}_j\hat{L}_j\hat{p}_i)$$

$$= \sum_{i,j=1}^{3} \left[(\hat{L}_j\hat{p}_i + i\hbar\sum_{k=1}^{3}\epsilon_{ijk}\hat{p}_k)\hat{p}_i\hat{L}_j - (\hat{L}_j\hat{p}_i + i\hbar\sum_{k=1}^{3}\epsilon_{ijk}\hat{p}_k)\hat{L}_i\hat{p}_j \right.$$

$$- \hat{L}_i\hat{p}_j(\hat{L}_j\hat{p}_i + i\hbar\sum_{k=1}^{3}\epsilon_{ijk}\hat{p}_k) + \hat{L}_i(\hat{L}_i\hat{p}_j + i\hbar\sum_{k=1}^{3}\epsilon_{jik}\hat{p}_k)\hat{p}_j$$

$$- (\hat{L}_j\hat{p}_i + i\hbar\sum_{k=1}^{3}\epsilon_{ijk}\hat{p}_k)\hat{p}_j\hat{L}_i + (\hat{L}_j\hat{p}_i + i\hbar\sum_{k=1}^{3}\epsilon_{ijk}\hat{p}_k)\hat{L}_j\hat{p}_i$$

$$\left. - \hat{L}_i\hat{L}_j\hat{p}_j\hat{p}_i \right] + \hat{L}^2\hat{p}^2$$

$$= 3\hat{L}^2\hat{p}^2 + \sum_{i,j=1}^{3} \hat{L}_j\hat{p}_i\hat{L}_j\hat{p}_i - 2i\hbar \sum_{i,j,k=1}^{3} \epsilon_{ijk}\hat{p}_k\hat{L}_i\hat{p}_j$$

$$= 4\hat{L}^2\hat{p}^2 + 2\hbar^2 \sum_{i,j,k,l=1}^{3} \epsilon_{ijk}\epsilon_{kil}\hat{p}_l\hat{p}_j \ .$$

To perform this simplification, a number of identities were used, including $\hat{\vec{p}} \cdot \hat{\vec{L}} = \hat{\vec{L}} \cdot \hat{\vec{p}} = 0$ (because of the central potential nature of hydrogen), $[\hat{p}^2, \hat{L}_i] = 0$ (which follows from Eq. (9.73)), and $\epsilon_{ijk}\hat{p}_i\hat{p}_j = 0$ (because different momentum components commute). The final term involving the momentum operators can be simplified using the identity that

$$\sum_{i,j=1}^{3} \epsilon_{ijk}\epsilon_{ijl} = 2\delta_{kl}, \qquad (9.75)$$

which follows from Eq. (9.70). Thus, after all of this, we find the remarkably simple result that

$$\sum_{i,j,k,l,m=1}^{3} \epsilon_{ijk}\epsilon_{lmk}(\hat{p}_i\hat{L}_j - \hat{L}_i\hat{p}_j)(\hat{p}_l\hat{L}_m - \hat{L}_l\hat{p}_m) = 4\hat{p}^2(\hat{L}^2 + \hbar^2). \qquad (9.76)$$

At this point, we can be a bit less careful with the order of operators because the squared momentum and angular momentum operators commute, $[\hat{p}^2, \hat{L}^2] = 0$.

With this result, the square of the Laplace–Runge–Lenz operator is

$$\hat{\vec{A}} \cdot \hat{\vec{A}} \qquad (9.77)$$

$$= \frac{2}{m_e}\frac{\hat{p}^2}{2m_e}(\hat{L}^2 + \hbar^2) + \frac{e^4}{(4\pi\epsilon_0)^2}$$

$$- \frac{e^2}{4\pi\epsilon_0}\frac{1}{2m_e}\sum_{i,j,k=1}^{3} \epsilon_{ijk}\left((\hat{p}_i\hat{L}_j - \hat{L}_i\hat{p}_j)\frac{\hat{r}_k}{\hat{r}} + \frac{\hat{r}_k}{\hat{r}}(\hat{p}_i\hat{L}_j - \hat{L}_i\hat{p}_j)\right)$$

$$= \frac{e^4}{(4\pi\epsilon_0)^2} + \frac{2}{m_e}\hat{H}(\hat{L}^2 + \hbar^2) + \frac{2}{m_e}\frac{e^2}{4\pi\epsilon_0}\frac{1}{\hat{r}}(\hat{L}^2 + \hbar^2)$$

$$- \frac{e^2}{4\pi\epsilon_0}\frac{1}{2m_e}\sum_{i,j,k=1}^{3} \epsilon_{ijk}\left((\hat{p}_i\hat{L}_j - \hat{L}_i\hat{p}_j)\frac{\hat{r}_k}{\hat{r}} + \frac{\hat{r}_k}{\hat{r}}(\hat{p}_i\hat{L}_j - \hat{L}_i\hat{p}_j)\right),$$

where in the second equality, we inserted the expression for the hydrogen atom Hamiltonian $\hat{H}$. To simplify the final term in parentheses, we first note the following. The angular momentum operator $\hat{L}_k$ can be expressed in terms of the momentum and position operators as

$$\hat{L}_k = \sum_{i,j=1}^{3} \epsilon_{ijk}\hat{r}_i\hat{p}_j. \qquad (9.78)$$

Thus, the final term can be massaged into a form with no explicit position or momentum vector components with this relationship. For instance, the second term in parentheses can be massaged into the form

$$\sum_{i,j,k=1}^{3} \epsilon_{ijk}\frac{\hat{r}_k}{\hat{r}}(\hat{p}_i\hat{L}_j - \hat{L}_i\hat{p}_j) = \sum_{i,j,k=1}^{3} \epsilon_{ijk}\frac{\hat{r}_k}{\hat{r}}\left(\hat{p}_i\hat{L}_j - \hat{p}_j\hat{L}_i - i\hbar\sum_{l=1}^{3}\epsilon_{ijl}\hat{p}_l\right) \qquad (9.79)$$

$$= \frac{2}{\hat{r}}\hat{L}^2 - 2i\hbar\sum_{i=1}^{3}\frac{\hat{r}_i}{\hat{r}}\hat{p}_i.$$

The term involving the momentum operator $\hat{p}_k$ was simplified using the identity that

$$\sum_{i,j=1}^{3} \epsilon_{ijk}\epsilon_{ijl} = 2\delta_{kl}, \tag{9.80}$$

established above.

To simplify the first term in parentheses, note that the commutator of angular momentum and position is

$$[\hat{r}_i, \hat{L}_j] = [\hat{r}_i, \sum_{l,m=1}^{3} \epsilon_{lmj}\hat{r}_l\hat{p}_m] = \sum_{l,m=1}^{3} \left(\epsilon_{lmj}\hat{r}_i\hat{r}_l\hat{p}_m - \epsilon_{lmj}\hat{r}_l\hat{p}_m\hat{r}_i\right) \tag{9.81}$$

$$= \sum_{l,m=1}^{3} \left(\epsilon_{lmj}\hat{r}_i\hat{r}_l\hat{p}_m - \epsilon_{lmj}\hat{r}_l(\hat{r}_i\hat{p}_m - i\hbar\delta_{im})\right) = i\hbar\sum_{k=1}^{3} \epsilon_{ijk}\hat{r}_k.$$

Thus, we have the relationship that

$$\sum_{i,j,k=1}^{3} \epsilon_{ijk}(\hat{p}_i\hat{L}_j - \hat{L}_i\hat{p}_j)\hat{r}_k \tag{9.82}$$

$$= \sum_{i,j,k=1}^{3} \epsilon_{ijk}\left(\hat{p}_i\hat{r}_k\hat{L}_j - i\hbar\sum_{l=1}^{3}\epsilon_{kjl}\hat{p}_i\hat{r}_l - \hat{p}_j\hat{L}_i\hat{r}_k + i\hbar\sum_{l=1}^{3}\epsilon_{jil}\hat{p}_l\hat{r}_k\right)$$

$$= \sum_{i,j,k=1}^{3} \epsilon_{ijk}\left(\hat{r}_k\hat{p}_i\hat{L}_j - \hat{p}_j\hat{r}_k\hat{L}_i + i\hbar\sum_{l=1}^{3}\epsilon_{kil}\hat{p}_j\hat{r}_l\right)$$

$$= 2\hat{L}^2 + 2i\hbar\sum_{i=1}^{3}\hat{p}_i\hat{r}_i.$$

With this result, the last term in the square of the Laplace–Runge–Lenz vector is

$$\frac{e^2}{4\pi\epsilon_0}\frac{1}{2m_e}\sum_{i,j,k=1}^{3} \epsilon_{ijk}\left((\hat{p}_i\hat{L}_j - \hat{L}_i\hat{p}_j)\frac{\hat{r}_k}{\hat{r}} + \frac{\hat{r}_k}{\hat{r}}(\hat{p}_i\hat{L}_j - \hat{L}_i\hat{p}_j)\right) \tag{9.83}$$

$$= \frac{e^2}{4\pi\epsilon_0}\frac{1}{2m_e}\left(4\hat{L}^2\frac{1}{\hat{r}} - 2i\hbar\sum_{i=1}^{3}\left[\frac{\hat{r}_i}{\hat{r}}, \hat{p}_i\right]\right).$$

We're very close; we just have to evaluate the commutator of the unit radial vector with momentum. To do this, we will use a trick we've employed several times in the past. In position space, the momentum operator is just a spatial derivative, and so the commutator is just the divergence of the unit radial vector:

$$\sum_{i=1}^{3}\left[\frac{\hat{r}_i}{\hat{r}}, \hat{p}_i\right] = \sum_{i=1}^{3}\left[\frac{\hat{r}_i}{\hat{r}}, -i\hbar\frac{\partial}{\partial\hat{r}_i}\right] = i\hbar\sum_{i=1}^{3}\frac{\partial}{\partial\hat{r}_i}\frac{\hat{r}_i}{\hat{r}} = 2i\hbar\frac{1}{\hat{r}}, \tag{9.84}$$

which follows from noting that $\hat{r} = \sqrt{\hat{x}^2 + \hat{y}^2 + \hat{z}^2}$. With this result, we then find

$$\frac{e^2}{4\pi\epsilon_0}\frac{1}{2m_e}\sum_{i,j,k=1}^{3} \epsilon_{ijk}\left((\hat{p}_i\hat{L}_j - \hat{L}_i\hat{p}_j)\frac{\hat{r}_k}{\hat{r}} + \frac{\hat{r}_k}{\hat{r}}(\hat{p}_i\hat{L}_j - \hat{L}_i\hat{p}_j)\right) = \frac{2}{m_e}\frac{e^2}{4\pi\epsilon_0}\frac{1}{\hat{r}}(\hat{L}^2 + \hbar^2).$$

$$\tag{9.85}$$

Putting it all together, perhaps amazingly, the square of the Laplace–Runge–Lenz operator is exceptionally simple:

$$\hat{A}^2 \equiv \hat{\vec{A}} \cdot \hat{\vec{A}} = \frac{e^4}{(4\pi\epsilon_0)^2} + \frac{2}{m_e}\hat{H}(\hat{L}^2 + \hbar^2). \tag{9.86}$$

9.4.2 A New $\mathfrak{su}(2)$

At this stage, we have established a number of relevant results for the quantum hydrogen atom and its Laplace–Runge–Lenz operator. Let's put them together to establish the spectrum of the Hamiltonian. Angular momentum and the Laplace–Runge–Lenz operator satisfy the commutation relations

$$[\hat{L}_i, \hat{A}_j] = i\hbar \sum_{k=1}^{3} \epsilon_{ijk}\hat{A}_k, \tag{9.87}$$

$$[\hat{A}_i, \hat{A}_j] = -i\hbar \frac{2}{m_e}\hat{H} \sum_{k=1}^{3} \epsilon_{ijk}\hat{L}_k,$$

$$[\hat{L}_i, \hat{L}_j] = i\hbar \sum_{k=1}^{3} \epsilon_{ijk}\hat{L}_k.$$

Let's now consider an energy eigenstate $|\psi\rangle$ in the Hilbert space of the hydrogen atom, that is

$$\hat{H}|\psi\rangle = E|\psi\rangle, \tag{9.88}$$

for some energy E. Further, let's restrict $|\psi\rangle$ to be a bound state: because the potential of hydrogen is negative and vanishes as $r \to \infty$, this bound state has negative energy, $E < 0$. When acting on this particular state, we can replace all appearances of the Hamiltonian by E, and so the commutation relations when acting on this state are

$$[\hat{L}_i, \hat{A}_j]|\psi\rangle = i\hbar \sum_{k=1}^{3} \epsilon_{ijk}\hat{A}_k|\psi\rangle, \tag{9.89}$$

$$[\hat{A}_i, \hat{A}_j]|\psi\rangle = -i\hbar \frac{2}{m_e}E \sum_{k=1}^{3} \epsilon_{ijk}\hat{L}_k|\psi\rangle,$$

$$[\hat{L}_i, \hat{L}_j]|\psi\rangle = i\hbar \sum_{k=1}^{3} \epsilon_{ijk}\hat{L}_k|\psi\rangle.$$

Remember that the Hamiltonian commutes with the angular momentum operator $\hat{L}_i$. For compactness in what follows, we will drop the explicit ket $|\psi\rangle$, but remember that the manipulations only hold for an energy eigenstate.

Now, let's construct the two new operators $\hat{T}_i$ and $\hat{S}_i$ formed from a linear combination of angular momentum and the Laplace–Runge–Lenz operator:

$$\hat{T}_i \equiv \frac{1}{2}\left(\hat{L}_i + \sqrt{-\frac{m_e}{2E}}\hat{A}_i\right), \qquad \hat{S}_i \equiv \frac{1}{2}\left(\hat{L}_i - \sqrt{-\frac{m_e}{2E}}\hat{A}_i\right). \tag{9.90}$$

These operators are especially nice to consider because they actually commute with one another! Their commutation relations are

$$[\hat{T}_i, \hat{S}_j] = 0, \tag{9.91}$$

$$[\hat{T}_i, \hat{T}_j] = i\hbar \sum_{k=1}^{3} \epsilon_{ijk} \hat{T}_k,$$

$$[\hat{S}_i, \hat{S}_j] = i\hbar \sum_{k=1}^{3} \epsilon_{ijk} \hat{S}_k.$$

These follow simply from the commutation relations in Eq. (9.89). Additionally, because angular momentum and Laplace–Runge–Lenz are orthogonal, the squares of $\hat{T}$ and $\hat{S}$ are equal:

$$\hat{T}^2 = \hat{S}^2 = \frac{1}{4}\hat{L}^2 - \frac{m_e}{8E}\hat{A}^2 = -\frac{m_e}{8E}\frac{e^4}{(4\pi\epsilon_0)^2} - \frac{\hbar^2}{4}. \tag{9.92}$$

The final equality simply follows from inserting Eq. (9.86), with E substituted for $\hat{H}$.

These final expressions now yield a fascinating result and interpretation. The commutation relations of Eq. (9.91) describe two $\mathfrak{su}(2)$ Lie algebras that commute with each other. We had studied this Lie algebra in great detail in understanding angular momentum in Chap. 8, and we can simply apply what we learned there to the problem at hand. First, we had established that the quadratic Casimir of the $\mathfrak{su}(2)$ algebra has values indexed by an integer or half-integer t:

$$\hat{T}^2|\psi\rangle = t(t+1)\hbar^2|\psi\rangle, \tag{9.93}$$

where $t = 0, 1/2, 1, 3/2, 2, \dots$. Additionally, for a given value of t for $\hat{T}^2$, the Cartan subalgebra of Eq. (9.91) consists of the (say) $\hat{T}_z$ and $\hat{S}_z$ operators whose eigenvalues range in integer steps from $-t$ to t:

$$\hat{T}_z|\psi\rangle = m_T\hbar|\psi\rangle, \qquad\qquad \hat{S}_z|\psi\rangle = m_S\hbar|\psi\rangle, \tag{9.94}$$

where $m_T, m_S = -t, -t+1, \dots, t-1, t$ and m_T and m_S are independent of one another. Again, all of this follows directly from the discussion of angular momentum in the previous chapter because their Lie algebras are isomorphic; functionally, this means that they only differ in name and not substance.

So, using this result and the value of the square $\hat{T}^2$ from Eq. (9.92), we can now solve for the value of the energy E on an eigenstate of the Hamiltonian of hydrogen. That is, we have

$$\hat{T}^2 = t(t+1)\hbar^2 = -\frac{m_e}{8E}\frac{e^4}{(4\pi\epsilon_0)^2} - \frac{\hbar^2}{4}, \tag{9.95}$$

which is now just an algebraic equation for E. We find

$$E = -\frac{m_e e^4}{2(4\pi\epsilon_0)^2\hbar^2}\frac{1}{(2t+1)^2}. \tag{9.96}$$

This can be written in a slightly nicer form. Note that the value $2t + 1$ ranges over the natural numbers, for the range of t that is allowed. Thus, we can compactly notate $2t + 1 \equiv \tilde{n} = 1, 2, 3, \ldots$, and so the energy eigenvalues $E_{\tilde{n}}$ can be written as

$$E_{\tilde{n}} = -\frac{m_e e^4}{2(4\pi\epsilon_0)^2 \hbar^2} \frac{1}{\tilde{n}^2}.$$
(9.97)

For a given value of t, the eigenvalues of $\hat{T}_z$ and $\hat{S}_z$ can each range over $2t + 1 = \tilde{n}$ values. For the same value of t, states with distinct $\hat{T}_z$ and $\hat{S}_z$ eigenvalues are different states as they have a different set of quantum numbers, but the same energy eigenvalue. Because the eigenvalues of $\hat{T}_z$ and $\hat{S}_z$ are independent, then for a given value of $\tilde{n}$, there are a total of $\tilde{n}^2$ orthogonal states in the Hilbert space with the same energy eigenvalue. We say that the **degeneracy** of the $\tilde{n}$th-energy eigenstate is $\tilde{n}^2$.

Importantly, note that the ground-state energy, for $\tilde{n} = 1$, is identical to what we found by explicitly solving the Schrödinger equation in position space, Eq. (9.34). However, this algebraic formulation provides a general description of hydrogen in one fell swoop, and describes why the energy eigenvalues of the Hamiltonian are discrete and quantized like they are. The first six energy levels are displayed in Fig. 9.5 as dashed lines, compared to the functional form of the Coulomb potential of the electron. Higher energy levels get squeezed closer and closer to zero energy, and if the electron has a positive energy, then it is not bound into hydrogen with the proton.

9.4.3 A Hidden SO(4) in Hydrogen

The Lie algebra of Eq. (9.91) is richer than just two commuting $\mathfrak{su}(2)$ algebras. Such an algebra can be expressed as the direct product $\mathfrak{su}(2) \otimes \mathfrak{su}(2)$ and this Lie algebra is isomorphic to the Lie algebra $\mathfrak{so}(4)$:

$$\mathfrak{su}(2) \otimes \mathfrak{su}(2) \simeq \mathfrak{so}(4).$$
(9.98)

Through exponentiation, exactly as we described with rotations, the Lie algebra $\mathfrak{so}(4)$ generates the Lie group SO(4), the group of 4×4 orthogonal matrices with unit determinant. In our first foray into rotations, we had identified a rotation matrix as

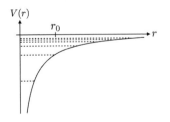

Illustration of the spacing of the first few energy eigenvalues in the hydrogen atom potential, denoted by the dashed horizontal lines. The potential with zero angular momentum of the electron $\hat{L}^2 = 0$ is also drawn for reference.

an element of SO(3); that is, those matrices that perform rotations on real, three-dimensional vectors. In a similar spirit, SO(4) is the group of matrices that perform rotations on real, four-dimensional vectors. That is, the group of symmetries of the hydrogen atom, for which both angular momentum and the Laplace–Runge–Lenz vectors are conserved, is isomorphic to SO(4).[8]

This is quite abstract, and it's not possible for us to imagine what four-dimensional rotations look like. However, the hydrogen atom is an example of a quantum system that exhibits **hidden symmetries**, or a set of conservation laws that is larger than what one imposes on the system. That is, we had constructed the Hamiltonian of hydrogen to conserve angular momentum, because it is only a function of the magnitudes of momentum and position. However, due to the specific $1/r$ form of the hydrogen atom potential, the Laplace–Runge–Lenz vector is also conserved, but its conservation was not something that we had designed from the beginning. Any Hamiltonian that is exclusively a function of the magnitude of momentum and position conserves angular momentum, but *only* if the potential is $1/r$ does that Hamiltonian also describe a system for which Laplace–Runge–Lenz vector is conserved. As we have seen, the existence of such hidden symmetries enables extremely powerful analysis that would not be possible without them.

While there are several examples of systems with hidden symmetries in physics, they are typically quite special and consist of limited examples, like the hydrogen atom. Hydrogen is further an example of an **integrable system**, or a system for which its symmetries or conserved quantities completely and uniquely determine its time evolution. The degrees of freedom of the classical hydrogen atom consist of the position of the electron, or a total of three real quantities at every time t. For hydrogen, we had identified that the Hamiltonian, angular momentum, and Laplace–Runge–Lenz vector are conserved. This would naïvely seem to be seven conserved quantities (two three-dimensional vectors and energy). However, we had identified two constraints on the Laplace–Runge–Lenz vector and its relationship to the Hamiltonian and angular momentum, namely that it is perpendicular to angular momentum and its magnitude is fixed:

$$\vec{A}\cdot\vec{L}=0, \qquad\qquad \vec{A}\cdot\vec{A}=\frac{e^4}{(4\pi\epsilon_0)^2}+\frac{2}{m_e}H\vec{L}\cdot\vec{L}, \qquad (9.99)$$

where we took the classical $\hbar \to 0$ limit of the square we had found in Eq. (9.86). Thus, along with the Hamiltonian and angular momentum, only one component of the Laplace–Runge–Lenz vector is a new conserved quantity, for a total of five, or one fewer conserved quantity than twice the number of degrees of freedom. Such a system with one fewer conserved quantity than twice the degrees of freedom is referred to as "maximally integrable."

[8] Vladimir Fock first showed that the hydrogen atom Hamiltonian took an exceptionally simple form in momentum space, projected onto a three-dimensional sphere (the set of all points in four dimensions equidistant from the center). Shortly thereafter, Valentine Bargmann demonstrated that Fock's analysis was equivalent to Pauli's analysis of hydrogen a decade earlier using the Laplace–Runge–Lenz vector. See V. Fock, "Zur Theorie des Wasserstoffatoms," *Z. Phys.* **98**, 145–154 (1935); V. Bargmann, "Zur Theorie des Wasserstoffatoms," *Z. Phys.* **99**, 576–582 (1936).

However, the vast majority of problems in physics are non-integrable, and depend sensitively on the initial conditions of a system. For example, the helium atom consisting of two electrons orbiting a nucleus is an example of a "three-body problem" that is non-integrable. Even treating the nucleus of helium as infinitely massive, the degrees of freedom of this problem consist of the three-dimensional position vectors for both electrons, or a total of six real quantities. Total energy and angular momentum are conserved, for a total of four conserved quantities, which is less than the number of degrees of freedom. Even if helium has a Laplace–Runge–Lenz–like vector, it would only add one more independent conserved quantity, for the same reasons as we observed in hydrogen. So, at best, the helium atom has six degrees of freedom, but only five conserved quantities, and thus is non-integrable. It is further well-known that three-body problems like helium or in gravitational mechanics can exhibit chaotic behavior classically and future trajectories are extremely sensitive to initial data.

9.5 Energy Eigenvalues with Spherical Data

The diagonalization of the Hamiltonian of the quantum hydrogen atom with the Laplace–Runge–Lenz operator is clearly very powerful, but there's a bit of a disconnect with what we had claimed for labeling energy eigenstates in the beginning of this chapter. We claimed that we could express an energy eigenstate as $|n, \ell, m\rangle$, where

$$\hat{H}|n, \ell, m\rangle = E_{n,\ell}|n, \ell, m\rangle, \tag{9.100}$$

and n is a radial quantum number, ℓ is the magnitude of angular momentum, and m is the z-component of angular momentum. With the Laplace–Runge–Lenz operator and its relationship with angular momentum, we had also demonstrated that energy eigenstates can be labeled by a natural number $2t + 1 = \tilde{n}$, the number of states in a representation of the $\hat{T}$ or $\hat{S}$ operators that form $\mathfrak{su}(2)$ algebras, and their values of z-components, m_T and m_S. That is:

$$\hat{H}|\tilde{n}, m_T, m_S\rangle = E_{\tilde{n}}|\tilde{n}, m_T, m_S\rangle, \tag{9.101}$$

where

$$E_{\tilde{n}} = -\frac{m_e e^4}{2(4\pi\epsilon_0)^2 \hbar^2} \frac{1}{\tilde{n}^2}. \tag{9.102}$$

We would like to determine the precise relationship between these two representations of energy eigenstates of the hydrogen atom Hamiltonian and ultimately understand the dependence of the energy eigenvalue $E_{n,\ell}$ on n and ℓ. Another way to say this is that we want to determine the functional dependence of $\tilde{n}$ on n and ℓ:

$$\tilde{n} \equiv \tilde{n}(n, \ell). \tag{9.103}$$

We had used the fact that all components of the angular momentum operator commuted with the Hamiltonian to establish that we could label states by their values of

ℓ and m, and now we will use the fact that the Laplace–Runge–Lenz operator commutes with the Hamiltonian to learn more about the radial quantum number n. To do this, we will determine how an energy eigenstate $|n,\ell,m\rangle$ is affected by the action of the Laplace–Runge–Lenz operator. Recall that the commutation relation of the Laplace–Runge–Lenz operator and angular momentum is

$$[\hat{L}_i, \hat{A}_j] = i\hbar \sum_{k=1}^{3} \epsilon_{ijk} \hat{A}_k. \qquad (9.104)$$

This has a similar form to the commutation relations of different components of angular momentum with itself, so motivates construction of raising and lowering Laplace–Runge–Lenz operators

$$\hat{A}_+ = \hat{A}_x + i\hat{A}_y, \qquad\qquad \hat{A}_- = \hat{A}_x - i\hat{A}_y. \qquad (9.105)$$

They raise and lower the eigenvalue of $\hat{L}_z$ by one unit of $\hbar$:

$$[\hat{L}_z, \hat{A}_+] = \hbar\hat{A}_+, \qquad\qquad [\hat{L}_z, \hat{A}_-] = -\hbar\hat{A}_-. \qquad (9.106)$$

Additionally, they do not commute with the angular momentum Casimir:

$$[\hat{L}^2, \hat{A}_+] = 2\hbar^2\hat{A}_+ + 2\hbar\hat{A}_+\hat{L}_z - 2\hbar\hat{A}_z\hat{L}_+, \qquad (9.107)$$
$$[\hat{L}^2, \hat{A}_-] = 2\hbar^2\hat{A}_- - 2\hbar\hat{A}_-\hat{L}_z + 2\hbar\hat{A}_z\hat{L}_-.$$

While I just quote the result here, it follows from application of Eq. (9.104).

We can use these results to determine how dependence on the radial and angular momentum quantum numbers appear in the energy eigenvalues. First, let's consider a state with general n, but $\ell = 0$: $|n,0,0\rangle$. For this case, we identify $n = \tilde{n}(n,0)$ so that its energy is

$$E_{n,0} = -\frac{m_e e^4}{2(4\pi\epsilon_0)^2\hbar^2} \frac{1}{n^2}. \qquad (9.108)$$

Now, let's see what the Laplace–Runge–Lenz raising operator does to this state. That is, we will consider the new state $\hat{A}_+|n,0,0\rangle$ and determine its properties. First and foremost, its energy is unchanged because the Laplace–Runge–Lenz operator commutes with the Hamiltonian:

$$\hat{H}\hat{A}_+|n,0,0\rangle = \hat{A}_+\hat{H}|n,0,0\rangle = E_{n,0}\hat{A}_+|n,0,0\rangle. \qquad (9.109)$$

Let's determine what its angular momentum quantum numbers are. Acting with $\hat{L}^2$, we have

$$\hat{L}^2\hat{A}_+|n,0,0\rangle = \left(\hat{A}_+\hat{L}^2 + [\hat{L}^2, \hat{A}_+]\right)|n,0,0\rangle \qquad (9.110)$$
$$= \left(2\hbar^2\hat{A}_+ + 2\hbar\hat{A}_+\hat{L}_z - 2\hbar\hat{A}_z\hat{L}_+\right)|n,0,0\rangle$$
$$= 2\hbar^2\hat{A}_+|n,0,0\rangle.$$

Now, recall that the eigenvalue of $\hat{L}^2$ is $\hbar^2\ell(\ell+1)$, and so the state $\hat{A}_+|n,0,0\rangle$ has value $\ell = 1$. Additionally, its z-component of angular momentum is $m = 1$:

$$\hat{L}_z \hat{A}_+ |n, 0, 0\rangle = \left(\hat{A}_+ \hat{L}_z + [\hat{L}_z, \hat{A}_+] \right) |n, 0, 0\rangle \qquad (9.111)$$
$$= \hbar \hat{A}_+ |n, 0, 0\rangle .$$

Then, the state $\hat{A}_+ |n, 0, 0\rangle$ is proportional to another energy eigenstate with one larger unit of angular momentum:

$$\hat{A}_+ |n, 0, 0\rangle \propto |n', 1, 1\rangle . \qquad (9.112)$$

Here, n' is another radial quantum number, in general different than the original n.

Let's continue to act with $\hat{A}_+$ on an energy eigenstate to see what state results. We'll focus on the action of $\hat{A}$ on a state with total and z-component of angular momentum ℓ, for simplicity. This new state's total angular momentum is

$$\hat{L}^2 \hat{A}_+ |n, \ell, \ell\rangle = \left(\hat{A}_+ \hat{L}^2 + [\hat{L}^2, \hat{A}_+] \right) |n, \ell, \ell\rangle \qquad (9.113)$$
$$= \hbar^2 \ell(\ell+1) \hat{A}_+ |n, \ell, \ell\rangle + \left(2\hbar^2 \hat{A}_+ + 2\hbar \hat{A}_+ \hat{L}_z - 2\hbar \hat{A}_z \hat{L}_+ \right) |n, \ell, \ell\rangle$$
$$= \hbar^2 (\ell+1)(\ell+2) \hat{A}_+ |n, \ell, \ell\rangle .$$

Further, its z-component of angular momentum is

$$\hat{L}_z \hat{A}_+ |n, \ell, \ell\rangle = \left(\hat{A}_+ \hat{L}_z + [\hat{L}_z, \hat{A}_+] \right) |n, \ell, \ell\rangle \qquad (9.114)$$
$$= \hbar(\ell+1) \hat{A}_+ |n, \ell, \ell\rangle .$$

Thus, $\hat{A}_+$ has increased the total and z-component of angular momentum by one unit:

$$\hat{A}_+ |n, \ell, \ell\rangle \propto |n', \ell+1, \ell+1\rangle , \qquad (9.115)$$

but the energy of the state is unaffected. So, we can keep raising and raising the total angular momentum by action of the Laplace–Runge–Lenz operator on the zero angular momentum state:

$$(\hat{A}_+)^\ell |n, 0, 0\rangle \propto |n', \ell, \ell\rangle , \qquad (9.116)$$

but you might see that there's a limit to this. The Laplace–Runge–Lenz operator does not affect the energy of the state and for fixed energy, there must be a maximal value of angular momentum $\ell_{\max}$, as the angular momentum adds a positive value to the energy. Acting with $\hat{A}_+$ more than $\ell_{\max}$ times on $|n, 0, 0\rangle$ annihilates the state:

$$(\hat{A}_+)^{\ell > \ell_{\max}} |n, 0, 0\rangle = 0 . \qquad (9.117)$$

These observations, along with our assumption that the radial quantum number $n \geq 1$, suggest that the integer $\tilde{n}$ that determines the energy level of the state is simply the sum of the radial and angular momentum quantum numbers:

$$\tilde{n}(n, \ell) = n + \ell . \qquad (9.118)$$

Then, the action of $\hat{A}_+$ increases ℓ, while decreasing n by one unit each to keep $\tilde{n}$ the same:

$$\hat{A}_+ |n, \ell, \ell\rangle \propto |n-1, \ell+1, \ell+1\rangle . \qquad (9.119)$$

If $\hat{A}_+$ acts on the zero angular momentum state $\tilde{n}$ times then it annihilates the state:

$$(\hat{A}_+)^{\tilde{n}} |\tilde{n}, 0, 0\rangle = 0 , \qquad (9.120)$$

as then the radial quantum number would have been lowered to 0. Further, we can determine the degeneracy of the energy eigenstates. As long as the sum $\tilde{n} = n + \ell$ is fixed, then all such states have the same energy. Note that the range of angular momentum values is $\ell = 0, 1, 2, \ldots, \tilde{n} - 1$, as n is at least 1. For a fixed value of ℓ, there are $2\ell + 1$ eigenstates of the $\hat{L}_z$ operator that all have the same energy. So, the total degeneracy of an energy eigenstate with fixed $\tilde{n} = n + \ell$ is the sum of all $\hat{L}_z$ eigenstates for each allowed value of ℓ:

$$\sum_{\ell=0}^{\tilde{n}-1} (2\ell + 1) = 2 \sum_{\ell=1}^{\tilde{n}-1} \ell + \tilde{n} = 2\frac{\tilde{n}(\tilde{n}-1)}{2} + \tilde{n} = \tilde{n}^2, \tag{9.121}$$

where we used that the sum of positive integers is

$$\sum_{i=1}^{n} i = \frac{n(n+1)}{2}. \tag{9.122}$$

This $\tilde{n}^2$ degeneracy is exactly the same that we had found from our analysis of the $\mathfrak{su}(2) \otimes \mathfrak{su}(2)$ algebra.

So, in summary, we have found that the energy eigenstates can be labeled by a radial number, total angular momentum, and z-component of angular momentum such that

$$\hat{H}|n, \ell, m\rangle = E_{n,\ell}|n, \ell, m\rangle, \tag{9.123}$$

where

$$E_{n,\ell} = -\frac{m_e e^4}{2(4\pi\epsilon_0)^2 \hbar^2} \frac{1}{(n+\ell)^2}, \tag{9.124}$$

for $n \in \mathbb{N}$, $\ell \in \{0, 1, 2, \ldots, \}$.

Example 9.1 We have evaluated the ground-state wavefunction of hydrogen in Eq. (9.48), and in this example, we will construct the first excited state $n = 2$, $\ell, m = 0$ wavefunction. This can then be used, along with the components of the Laplace–Runge–Lenz vector, to construct all wavefunctions with the same energy, which you will explore in Exercise 9.5.

Solution

With the knowledge of the result for the ground-state wavefunction, we will take the following form as an ansatz for the first excited-state wavefunction of hydrogen, $\psi_{2,0,0}(r)$:

$$\psi_{2,0,0}(\vec{r}) = N\left(\alpha - \frac{r}{a_0}\right) e^{-\beta r/a_0}, \tag{9.125}$$

where a_0 is the Bohr radius, N is the normalization factor, and α, β are some undetermined constants. This form is motivated by the fact that we expect that the first excited state has one node or point of zero value on the domain $0 < r < \infty$. For normalizability, we must require that $\beta > 0$. Because we are considering an $\ell = 0$ state, the wavefunction will only have dependence on the radius r from the proton. We will solve for the undetermined constants α, β in the same way that we did for the ground state: by demanding that it it an eigenstate of the hydrogen atom Hamiltonian.

The eigenvalue equation we need to solve is

$$\left(-\frac{\hbar^2}{m_e} \frac{1}{r} \frac{d}{dr} - \frac{\hbar^2}{2m_e} \frac{d^2}{dr^2} - \frac{e^2}{4\pi\epsilon_0} \frac{1}{r} \right) \psi_{2,0,0}(r) = E_{2,0} \psi_{2,0,0}(r) , \tag{9.126}$$

where now the energy is

$$E_{2,0} = -\frac{1}{4} \frac{m_e e^4}{2(4\pi\epsilon_0)^2 \hbar^2} = -\frac{1}{4} \frac{\hbar^2}{2m_e a_0^2} . \tag{9.127}$$

We can also re-write the Coulomb potential term with the Bohr radius as

$$\frac{e^2}{4\pi\epsilon_0} \frac{1}{r} = \frac{\hbar^2}{m_e a_0} \frac{1}{r} . \tag{9.128}$$

With these identifications, the eigenvalue equation in position space nicely reduces to

$$\left(-\frac{1}{r} \frac{d}{dr} - \frac{1}{2} \frac{d^2}{dr^2} - \frac{1}{a_0} \frac{1}{r} \right) \psi_{2,0,0}(r) = -\frac{1}{4} \frac{1}{2a_0^2} \psi_{2,0,0}(r) . \tag{9.129}$$

So, we just need to evaluate some derivatives.

The derivatives we need are

$$\left(-\frac{1}{r} \frac{d}{dr} - \frac{1}{2} \frac{d^2}{dr^2} - \frac{1}{a_0} \frac{1}{r} \right) \psi_{2,0,0}(r) \tag{9.130}$$

$$= \left(-\frac{1}{r} \frac{d}{dr} - \frac{1}{2} \frac{d^2}{dr^2} - \frac{1}{a_0} \frac{1}{r} \right) N \left(\alpha - \frac{r}{a_0} \right) e^{-\beta r/a_0}$$

$$= \left(\beta^2 r + a_0(2 - 4\beta - \alpha\beta^2) + \frac{2a_0^2}{r}(1 - \alpha + \alpha\beta) \right) \frac{N}{2a_0^3} e^{-\beta r/a_0} .$$

For this to equal

$$-\frac{1}{4} \frac{1}{2a_0^2} \psi_{2,0,0}(r) = -\frac{1}{4} \frac{1}{2a_0^2} N \left(\alpha - \frac{r}{a_0} \right) e^{-\beta r/a_0} , \tag{9.131}$$

we must enforce that the term proportional to $1/r$ vanishes by itself. That is:

$$1 - \alpha + \alpha\beta = 0 , \tag{9.132}$$

or

$$\beta = \frac{\alpha - 1}{\alpha} . \tag{9.133}$$

Setting the terms at order r^0 equal, we find

$$(2 - 4\beta - \alpha\beta^2) = \frac{3}{\alpha} - \alpha = -\frac{\alpha}{4} , \tag{9.134}$$

where in the second equality we inserted the expression for β from Eq. (9.133). Demanding that the value of α produces a positive value for β, we find the solution that $\alpha = 2$, and therefore $\beta = 1/2$. Remember that β must be positive for the wavefunction to be normalizable. With these results, the terms at order r^1 are automatically consistent.

Therefore, we find that the first excited radial-state wavefunction of hydrogen is

$$\psi_{2,0,0}(r) = N \left(1 - \frac{r}{2a_0} \right) e^{-r/2a_0} . \tag{9.135}$$

We can now fix normalization. We require that

$$1 = \langle 2,0,0|2,0,0\rangle = 4\pi N^2 \int_0^\infty dr\, r^2\, \psi_{2,0,0}(r)^* \psi_{2,0,0}(r) \tag{9.136}$$

$$= 4\pi N^2 \int_0^\infty dr\, r^2 \left(1 - \frac{r}{2a_0} \right)^2 e^{-r/a_0} = 8\pi a_0^3 N^2 .$$

Thus

$$N = \frac{1}{\sqrt{8\pi a_0^3}} , \tag{9.137}$$

and the normalized wavefunction is

$$\psi_{2,0,0}(r) = \frac{1}{\sqrt{8\pi a_0^3}} \left(1 - \frac{r}{2a_0} \right) e^{-r/2a_0} . \tag{9.138}$$

We can also verify that this is orthogonal to the ground-state wavefunction, $\psi_{1,0,0}(r)$, where

$$\psi_{1,0,0}(r) = \frac{e^{-r/a_0}}{\sqrt{\pi a_0^3}} . \tag{9.139}$$

Orthogonality requires that

$$0 = \langle 2,0,0|1,0,0\rangle = 4\pi \int_0^\infty dr\, r^2\, \psi_{2,0,0}(r)^* \psi_{1,0,0}(r) \tag{9.140}$$

$$= \frac{4\pi}{8\pi a_0^3} \int_0^\infty dr\, r^2 \left(1 - \frac{r}{2a_0} \right) e^{-3r/2a_0} ,$$

which is indeed true.

In Fig. 9.6, we have plotted the corresponding radial probability distribution for this first excited state. As expected, there is one node in the wavefunction in the region $0 < r < \infty$, which can be seen to occur from Eq. (9.138) at a radius of $2a_0$. The expectation value of the radius on this state $\langle r \rangle$ is significantly larger than the Bohr radius:

$$\langle r \rangle = 4\pi \int_0^\infty dr\, r^3 |\psi_{2,0,0}(r)|^2 = 6a_0 , \tag{9.141}$$

which is also illustrated on the figure.

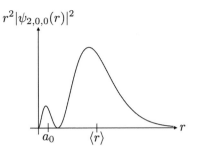

$$r^2|\psi_{2,0,0}(r)|^2$$

Fig. 9.6   A plot of the probability density of the electron's radius r from the proton in the first excited radial state of the hydrogen atom. The Bohr radius, a_0, is at a significantly smaller radius than the expectation value $\langle r \rangle = 6a_0$.

9.6 Hydrogen in the Universe

Our analysis of the energy eigenstates of hydrogen has innumerable consequences and in this section we focus on two applications to understanding the cosmos. Hydrogen is the most abundant element in the universe and as the first entry in the periodic table was also the first element to form in the early universe. Hydrogen therefore enables a powerful lever to interpret stars, galaxies, and other astrophysical phenomena, and a window to a time as close to the Big Bang as we could possibly observe.

9.6.1 Recombination and the Cosmic Microwave Background

In the time after the Big Bang, the universe was relatively small, high temperature, and densely filled with electrically charged particles, like electrons and protons. A dense system of charged particles is a **plasma** and because photons, particles of light, interact with electrically charged particles, the early universe was also opaque. Photons would bounce between the charged particles and after a very short time the direction of their trajectory would be essentially uncorrelated with their point of origin. As the universe expanded, it also cooled, meaning that the average kinetic energy of the electrons and protons decreased. As the universe continued to cool, their kinetic energy became comparable to the electrostatic potential energy of their mutual attraction and eventually the protons and electrons would bind, forming hydrogen. Because the ground-state energy of bound hydrogen is about -13.6 eV, this means that the kinetic energy of the protons and electrons when they bind is around 10 eV, or so.

This time in the history of the cosmos when free electrons and protons became bound into hydrogen is called **recombination**. Instead of energies, epochs in the universe are often expressed through their temperature. We had stated that room temperature means that the average kinetic energy of air molecules is about $1/40$ eV, and room temperature is about 300 K. Thus, recombination, corresponding to kinetic energies of about 10 eV, is 400 times that of room temperature, or about $100,000$ K. (Because of the statistical properties of objects in thermal equilibrium, recombination happened at a temperature a bit lower than this estimate.) At that point, the electrically charged electron–proton plasma disappeared into neutral hydrogen, so the photons that had been trapped bouncing around suddenly became free and were able to travel enormous distances with no impedance. This dramatic point of transparency is called the **surface of last scattering** and occurred about 300,000 years after the Big Bang.

The wavelength λ of light that has energy of about 10 eV, just low enough to not ionize hydrogen, breaking it into an individual electron and proton, can be determined by the relationship between energy E and frequency f of light:

$$E = 2\pi\hbar f = \frac{2\pi\hbar c}{\lambda}, \tag{9.142}$$

where c is the speed of light. Light of energy 10 eV therefore has a wavelength of

$$\lambda_{10\,\text{eV}} = 1.2 \times 10^{-7}\,\text{m}, \tag{9.143}$$

or about 120 nm. The universe, however, didn't stop expanding at recombination, and has continued to expand in the intervening 13 or so billion years since. As the universe expands, the wavelength of the light from the surface of last scattering expands with it, correspondingly decreasing the observed energy of this light and increasing its wavelength. Because *everywhere* in the universe was the surface of last scattering, this remnant radiation permeates the universe, and should be observable.

In the mid-twentieth century, with improved understanding of early universe cosmology, the first predictions of the energy of these last scattering photons that would be observed today were made. Astrophysical observations had identified a white-noise background that could not be accounted for from known phenomena, but had an effective temperature of about 2.7 K.[9] This was soon interpreted as precisely this remnant radiation from recombination, but stretched out due to the expansion of the universe. 2.7 K corresponds to an energy of about 2.5×10^{-4} eV or a wavelength of about a millimeter. Such a wavelength of electromagnetic radiation is in the microwave band and because it has a cosmological source and is a background to any astrophysical observation, it is therefore called the **cosmic microwave background**, or CMB. The connection of the CMB today to the maximum kinetic energy of photons such that bound hydrogen can exist in the early universe demonstrates that the radius of the universe has increased by about a factor of 1000 since recombination.

9.6.2 The 21-cm Line

The ubiquity of hydrogen throughout the universe means that it has numerous observable consequences that encode information such as relative hydrogen abundance or relative velocity to Earth of some astrophysical system. The fundamental phenomenon that connects energy levels in hydrogen to the frequency or wavelength of observed

---

[9] A. McKellar, "Molecular lines from the lowest states of diatomic molecules composed of atoms probably present in interstellar space," *Publ. Domin. Astrophys. Obs. Vict.*, 7, 251 (1941); A. A. Penzias and R. W. Wilson, "A measurement of excess antenna temperature at 4080-Mc/s," *Astrophys. J.* **142**, 419–421 (1965).

light is the following. A hydrogen atom can move between its energy levels by emitting or absorbing a photon of the correct frequency corresponding to the difference of the relevant energy levels. Energy is of course conserved and so if hydrogen is in an energy eigenstate it cannot spontaneously transform to a different, non-degenerate, eigenstate without accounting for the difference in energy somehow. Light carries energy and so a transformation from, say, the first excited state to the ground state of hydrogen can be accomplished by emission of a photon that has energy

$$\Delta E = 13.6 \left(\frac{1}{1^2} - \frac{1}{2^2} \right) \text{ eV} = 10.2 \text{ eV}. \tag{9.144}$$

Photons also carry angular momentum, they are spin-1 particles, and so the two states must also differ by one $\hbar$ unit of angular momentum.

In our description of the hydrogen atom, we really only concerned ourselves with the fact that the proton and electron are electrically charged, and are bound together by electromagnetism. Additionally, however, both the proton and electron carry angular momentum as spin-1/2 particles. A classical object that is both electrically charged and has angular momentum (i.e., spinning about its axis) would also have a magnetic dipole moment. With our canonical quantization prescription, this would suggest that both the proton and electron have magnetic dipole moments and as such, their magnetic fields can interact. Depending on the relative orientation of their dipoles, this new interaction can increase or decrease the energy of hydrogen. Because of the spin-1/2 nature of both the electron and the proton, there are two possible configurations quantum mechanically: either their spins are aligned or they are anti-aligned and the difference in angular momentum between those configurations is one $\hbar$. This suggests that flipping the spin of the electron in hydrogen from up to down emits a photon of a characteristic frequency or wavelength, and this can be observed in the cosmos to learn about stars, galaxies, and the dust in between.

In this section, we will analyze this spin-flip and explicitly calculate the energy of the emitted photon. In astronomy, it is referred to as the "21-cm line" because the wavelength of this light is 21 cm. The importance and ubiquity of this 21-cm line was exploited for potential communication with life outside of our solar system. The first probes sent on an extra-solar trajectory were the Pioneers and Voyagers, launched in the 1970s. On the Pioneer probes there is a plaque with symbolic representations of Earth, humanity, numbers, our knowledge of astrophysics, etc., and on the Voyagers, a "golden record" was sent with similar representations inscribed on it, as well as digitally recorded images and sounds. The back of the Voyager Golden Record is shown in Fig. 9.7. Some of the etchings represent the frequency to play the record, while the etching in the lower left is representative of the location of Earth with respect to known pulsars in the Milky Way galaxy. Distance scales in all of these etchings are made with respect to the 21-cm line in hydrogen, which is depicted in the bottom right of the record. Hydrogen is represented as a point proton orbited by an electron, and from the left to the right on that part of the record the spin of the electron has flipped. So, if *we* were the alien civilization that discovered a Voyager probe, we would need to know the wavelength of light from this spin flip in hydrogen to learn about humanity!

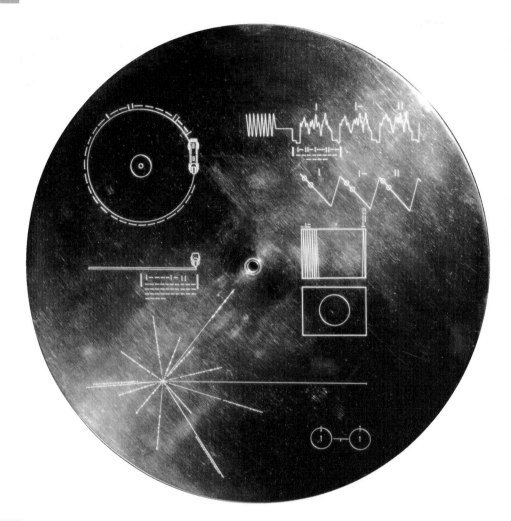

Fig. 9.7   Voyager Golden Record. The representation of the electron spin flip in hydrogen producing the 21-cm line is located at the bottom right of the record. Courtesy NASA/JPL-Caltech.

So, let's determine this energy. To do this, as we have done before, we'll start classical and then canonically quantize the system to determine the quantum mechanical result. First, the energy U of a magnetic dipole $\vec{m}$ immersed in a magnetic field $\vec{B}$ is

$$U = -\vec{m} \cdot \vec{B}. \tag{9.145}$$

In the case of the hydrogen atom, the relevant magnetic field is produced by the proton's magnetic dipole. The magnetic field of a magnetic dipole a distance $\vec{r}$ away is

$$\vec{B}(\vec{r}) = \frac{\mu_0}{4\pi} \left(\frac{3\vec{r}(\vec{r} \cdot \vec{m})}{|\vec{r}|^5} - \frac{\vec{m}}{|\vec{r}|^3} + \frac{8\pi}{3}\vec{m}\,\delta^{(3)}(\vec{r}) \right), \tag{9.146}$$

where μ_0 is the permeability of free space. We are assuming that the magnetic dipole is a point with no extent and the final term, proportional to the three-dimensional Dirac

δ-function, ensures that the magnetic field is divergenceless, $\vec{\nabla} \cdot \vec{B} = 0$. It then follows that the energy contained in the interaction of the magnetic fields of two magnetic dipoles $\vec{m}_e$ and $\vec{m}_p$ is

$$U = -\frac{\mu_0}{4\pi} \left(\frac{3(\vec{r} \cdot \vec{m}_e)(\vec{r} \cdot \vec{m}_p)}{|\vec{r}|^5} - \frac{\vec{m}_e \cdot \vec{m}_p}{|\vec{r}|^3} + \frac{8\pi}{3} \vec{m}_e \cdot \vec{m}_p \, \delta^{(3)}(\vec{r}) \right). \qquad (9.147)$$

To go further, we need to specify the state that the electron's orbit about the proton occupies. Unexcited hydrogen lives in its ground state, so that's the state in which we will consider the electron flipping its spin. An important feature of the orbital ground state of hydrogen is that it has zero angular momentum. As a consequence, the wavefunction of the ground state is spherically symmetric: the electron's orbit has no preferred orientation about the proton. This correspondingly means that the electron's magnetic moment vector is isotropic: it also has no preferred orientation about the proton. Thus, in the ground state of hydrogen or any zero angular momentum state, the expectation values of all components of the magnetic moment vector are equal:

$$\langle m_x \rangle_{\ell=0} = \langle m_y \rangle_{\ell=0} = \langle m_z \rangle_{\ell=0} \equiv \bar{m}. \qquad (9.148)$$

In the expression for the dipoles' potential energy, the two terms that lack the δ-function cancel on a state with zero angular momentum. Because there is no preferred orientation, we can always align our axes so that $\vec{r} = |\vec{r}|\hat{z}$, say. Then, with this orientation, the value of those terms on an $\ell = 0$ state are

$$\left\langle \frac{3(\vec{r} \cdot \vec{m}_e)(\vec{r} \cdot \vec{m}_p)}{|\vec{r}|^5} - \frac{\vec{m}_e \cdot \vec{m}_p}{|\vec{r}|^3} \right\rangle_{\ell=0} \qquad (9.149)$$

$$= \left\langle \frac{3 m_{e,z} m_{p,z}}{|\vec{r}|^3} - \frac{m_{e,x} m_{p,x} + m_{e,y} m_{p,y} + m_{e,z} m_{p,z}}{|\vec{r}|^3} \right\rangle_{\ell=0}$$

$$= \left\langle \frac{3 \bar{m}_e \bar{m}_p}{|\vec{r}|^3} - \frac{3 \bar{m}_e \bar{m}_p}{|\vec{r}|^3} \right\rangle_{\ell=0}$$

$$= 0.$$

With this observation, the interaction energy of the proton and electron dipoles simplifies to

$$U = -\frac{8\pi}{3} \frac{\mu_0}{4\pi} \vec{m}_e \cdot \vec{m}_p \, \delta^{(3)}(\vec{r}). \qquad (9.150)$$

We are now in a position to quantize this magnetic dipole interaction energy. To do this, we need to determine the quantum magnetic dipole moments for the electron and proton. For the electron, for example, this magnetic moment can be expressed in proportion to its spin operator $\hat{\vec{S}}_e$:

$$\hat{\vec{m}}_e = \gamma_e \hat{\vec{S}}_e = -g_e \frac{e}{2m_e} \hat{\vec{S}}_e. \qquad (9.151)$$

The proportionality constant γ_e is called the electron's **gyromagnetic ratio**. On the right, we have further written it out in terms of the electric charge, the mass of the electron, and the electron's **g-factor**, g_e. The g-factor is a dimensionless number that accounts for the deviation between the classical gyromagnetic ratio, $-e/(2m_e)$, and its quantum value. For the electron, g_e is very close to 2 as a spin-1/2 point particle, with deviations from 2 described in the quantum theory of electromagnetism, quantum electrodynamics. The magnetic moment of the proton can be expressed similarly:

$$\hat{\vec{m}}_p = g_p \frac{e}{2m_p} \hat{\vec{S}}_p, \tag{9.152}$$

where $\hat{\vec{S}}_p$ is the spin operator for the proton. The proton is a composite spin-1/2 particle, as it is itself the bound state of more fundamental particles called **quarks**, and so its spin comes from a combination of orbital angular momentum and the spins of its constituents. Correspondingly, its g-factor is quite large, with $g_p \simeq 5.6$. Then, the quantum Hamiltonian of this dipole–dipole interaction is

$$\hat{H} = -\frac{8\pi}{3} \frac{\mu_0}{4\pi} \hat{\vec{m}}_e \cdot \hat{\vec{m}}_p \, \delta^{(3)}\left(\hat{\vec{r}}\right) = \frac{8\pi}{3} \frac{g_e}{2} \frac{g_p}{2} \frac{\mu_0}{4\pi} \frac{e^2}{m_e m_p} \hat{\vec{S}}_e \cdot \hat{\vec{S}}_p \, \delta^{(3)}\left(\hat{\vec{r}}\right). \tag{9.153}$$

To determine its eigenvalues on the ground state of hydrogen, we need to determine the action of the three-dimensional δ-function on the ground state, and what the dot product of the proton and electron spins is. We'll start with the δ-function. In position space, the three-dimensional δ-function just restricts each coordinate to be 0, while integrating to unity:

$$1 = \int_{-\infty}^{\infty} dx \int_{-\infty}^{\infty} dy \int_{-\infty}^{\infty} dz \, \delta^{(3)}(\vec{r}) \tag{9.154}$$

$$= \int_{-\infty}^{\infty} dx \int_{-\infty}^{\infty} dy \int_{-\infty}^{\infty} dz \, \delta(x)\delta(y)\delta(z).$$

Hydrogen's ground-state wavefunction $\psi_{1,0,0}(r)$ is purely a function of radial distance from the proton, so we would like to express the δ-function in spherical coordinates for convenience. Changing coordinates in the integrand is simple and we must get the same value for the integral, that is

$$1 = \int_0^{\infty} r^2 \, dr \int_{-1}^{1} d(\cos\theta) \int_0^{2\pi} d\phi \, \delta^{(3)}(\vec{r}) \tag{9.155}$$

$$= \int_0^{\infty} r^2 \, dr \int_{-1}^{1} d(\cos\theta) \int_0^{2\pi} d\phi \, \frac{1}{4\pi r^2} \delta(r).$$

In the first equality, we have just expressed the integration measure in spherical coordinates. On the right, we note that if $x = y = z = 0$, then $r = 0$ too. Additionally, we must divide by $4\pi r^2$ to account for the total solid angle on a sphere and eliminate the r^2 factor of the measure. Putting this together, the expectation value of the three-dimensional δ-function on the ground state of hydrogen is

$$\langle 1,0,0|\delta^{(3)}\left(\hat{\vec{r}}\right)|1,0,0\rangle = \int_0^{\infty} r^2 \, dr \int_{-1}^{1} d(\cos\theta) \int_0^{2\pi} d\phi \, \frac{e^{-r/a_0}}{\sqrt{\pi a_0^3}} \frac{1}{4\pi r^2} \delta(r) \frac{e^{-r/a_0}}{\sqrt{\pi a_0^3}}$$

$$= \frac{1}{\pi a_0^3}. \tag{9.156}$$

Now, let's evaluate the dot product of the electron and proton spins, $\hat{\vec{S}}_e \cdot \hat{\vec{S}}_p$. To do this, we can complete the square as

$$\hat{\vec{S}}_e \cdot \hat{\vec{S}}_p = \frac{\left(\hat{\vec{S}}_e + \hat{\vec{S}}_p\right)^2 - \hat{\vec{S}}_e^2 - \hat{\vec{S}}_p^2}{2}. \tag{9.157}$$

Note that the spin operators of the electron and proton commute, because they correspond to measurements of distinct particles. Then, as spin-1/2 particles the square of their spin operators is

$$\hat{\vec{S}}_e^2 = \hat{\vec{S}}_p^2 = \frac{3}{4}\hbar^2. \tag{9.158}$$

Next, the square of the sum of their spins depends on the relative orientation of their spins. If their spins are aligned, that corresponds to a total of spin 1 ($1/2 + 1/2 = 1$), while if their spins are anti-aligned then they have a total spin 0 ($1/2 - 1/2 = 0$). Thus, the Casimir of the sum of their spins is

$$\left(\hat{\vec{S}}_e + \hat{\vec{S}}_p\right)^2 = \begin{cases} 2\hbar^2, & \text{spins aligned}, \\ 0, & \text{spins anti-aligned}. \end{cases} \tag{9.159}$$

Then, it follows that the dot product of the electron and proton spins is

$$\hat{\vec{S}}_e \cdot \hat{\vec{S}}_p = \begin{cases} \frac{\hbar^2}{4}, & \text{spins aligned}, \\ -\frac{3}{4}\hbar^2, & \text{spins anti-aligned}. \end{cases} \tag{9.160}$$

Now, putting it all together, the difference in the energy of the ground state of hydrogen when the proton and electron spins are aligned to when they are anti-aligned is

$$\langle 1,0,0|\hat{H}_{\uparrow\uparrow} - \hat{H}_{\uparrow\downarrow}|1,0,0\rangle \equiv \Delta E = \frac{8\pi}{3}\frac{g_e}{2}\frac{g_p}{2}\frac{\mu_0}{4\pi}\frac{e^2\hbar^2}{m_e m_p}\frac{1}{\pi a_0^3}. \tag{9.161}$$

If this energy corresponds to a photon emitted from hydrogen to accomplish this transition, it has a wavelength of

$$\lambda = \frac{2\pi\hbar c}{\Delta E} = \pi\frac{3}{4}\frac{2}{g_e}\frac{2}{g_p}\frac{4\pi}{\mu_0}\frac{m_e m_p}{e^2\hbar}a_0^3 c. \tag{9.162}$$

Plugging in the known values for all of these constants, we find that $\lambda \simeq 21$ cm, exactly the wavelength we observe from distant sources of hydrogen.

Example 9.2 The **Zeeman effect** is the phenomenon of splitting energy levels with different angular momentum due to the presence of a weak external magnetic field, named after Pieter Zeeman who first observed it.[10] The weak magnetic field breaks

---

[10] P. Zeeman, "Over de invloed eener magnetisatie op den aard van het door een stof uitgezonden licht; Over de invloed eener magnetisatie op den aard van het door een stof uitgezonden licht; On the influence of magnetism on the nature of the light emitted by a substance," *Verslagen Meded. Afd. Natuurk. Kon. Akad. Wetensch.* **5**, 181 (1896); P. Zeeman, "XXXII. On the influence of magnetism on the nature of the light emitted by a substance," *Lond. Edinb. Dublin Philos. Mag. J. Sci.* **43**(262), 226–239 (1897); P. Zeeman, "The effect of magnetisation on the nature of light emitted by a substance," *Nature* **55**, 347 (1897).

the degeneracy of the energy eigenstates because states of different angular momentum have different magnetic dipole moments. In this example, we'll see how this works and we will assume throughout that the external magnetic field is uniform and pointed in the z direction:

$$\vec{B} = B_0\hat{z}, \tag{9.163}$$

where B_0 is the magnitude of the magnetic field. You will further study this effect in Exercise 9.7.

Let's apply this magnetic field to the ground state of hydrogen. What is the difference between the energies when the electron has spin-up vs. spin-down in the ground state? For what value of the magnetic field B_0 does this separation equal that between the ground state and the first excited state of hydrogen?

Solution

As established above, the energy U of a magnetic dipole $\vec{m}$ immersed in a magnetic field $\vec{B}$ classically is

$$U = -\vec{m}\cdot\vec{B}. \tag{9.164}$$

This external magnetic field then projects out the z-component of the electron's magnetic moment due to its spin, according to Eq. (9.151). Then, the energy $\hat{H}$ due to the interaction of the electron's spin and the external magnetic field is

$$\hat{H} = g_e\frac{eB_0}{2m_e}\hat{S}_z = g_e\frac{e\hbar B_0}{4m_e}\begin{pmatrix} 1 & 0 \\ 0 & -1 \end{pmatrix}. \tag{9.165}$$

where $\hat{S}_z$ is the z-component of the electron's spin. In this form, the eigenvalues of the energy are obvious, and the energy difference ΔE between when the electron's spin is aligned or anti-aligned with the external magnetic field is

$$\Delta E = E_\uparrow - E_\downarrow = g_e\frac{e\hbar B_0}{2m_e}. \tag{9.166}$$

Plugging in known numbers, the prefactor of this energy difference is approximately

$$g_e\frac{e\hbar}{2m_e} \approx 1.2\times 10^{-4} \text{ eV/T}. \tag{9.167}$$

Recall that the energy difference between the ground state and the first excited state of hydrogen is

$$E_2 - E_1 = \frac{3m_ee^2}{8(4\pi\epsilon_0)^2\hbar^2} \approx 10.2 \text{ eV}. \tag{9.168}$$

So, for the Zeeman effect to be of comparable size to the energy difference between the lowest states in hydrogen, the external magnetic field must be enormous, thousands of Tesla. This therefore violates our assumption that the external magnetic field is "weak."

Exercises

9.1   Our analysis of the hydrogen atom simply extends to any element which has been ionized to have a single electron orbiting the nucleus. In this problem, we will consider such an atom, whose nucleus consists of Z protons and a single electron.

    (a)   What are the energy eigenstates of this ionized atom? What is its "Bohr radius" and how does it compare to that of hydrogen?

    (b)   What is the minimal value of Z such that the speed of the orbiting electron in the ground state is comparable to the speed of light, say, $v \sim c/2$? What element does this correspond to?

    (c)   The first corrections to the description of the hydrogen atom from the speed of the electron can be accounted for by the expression for the kinetic energy of a relativistic particle. Classically, this is

$$K = \sqrt{m_e^2 c^4 + |\vec{p}|^2 c^2} - m_e c^2 , \qquad (9.169)$$

where $\vec{p}$ is the momentum of the particle. By Taylor expanding this expression in the non-relativistic limit where $|\vec{p}| \ll m_e c$, show that the first non-zero term is independent of the speed of light c, and is the non-relativistic kinetic energy with which we are familiar. What is the next term in this Taylor expansion? Do relativistic corrections tend to increase or decrease the kinetic energy, as the speed of the electron becomes comparable to the speed of light?

9.2   In introducing the Laplace–Runge–Lenz vector and the canonically quantized hydrogen atom, there were a number of Poisson brackets for which we just stated the result, without explicit calculation. It's time to address that here.

    (a)   Evaluate the Poisson bracket of the angular momentum vector $\vec{L}$ and the hydrogen atom's Hamiltonian H and show explicitly that angular momentum is conserved (i.e., $\{H, \vec{L}\} = 0$).

    (b)   Evaluate the Poisson bracket of the Laplace–Runge–Lenz vector $\vec{A}$ and the hydrogen atom's Hamiltonian H and show explicitly that it is conserved (i.e., $\{H, \vec{A}\} = 0$).

    (c)   Evaluate the Poisson bracket of two components of the Laplace–Runge–Lenz vector and show that

$$\{A_i, A_j\} = -\sum_{k=1}^{3} \epsilon_{ijk} L_k \frac{2}{m_e} H . \qquad (9.170)$$

9.3   The **virial theorem** is a statement about the relationship between the kinetic and potential energies for a system bound in a potential that is a power of relative distance between particles. For such a system, both total energy and angular momentum are conserved, and the potential energy U is proportional to the kinetic energy K, with $U + K = E$, the total energy. In this problem, we will study the virial theorem and its consequences classically and quantum mechanically.

(a) Identify the proportionality relationship between the kinetic and potential energies and prove the virial theorem for a classical system of a pair of masses bound by a potential energy $U(r) = \frac{n}{|n|}kr^n$, where r is the distance between the masses, n is a non-zero real number, and k is a constant that has units of energy/distance^n.

Hint: Remember that angular momentum is conserved and think about Newton's second law for the system.

(b) Now, let's study the virial theorem in a quantum mechanical system. Consider a two-body system for which the potential is

$$V(\hat{r}) = \frac{n}{|n|}k\hat{r}^n, \qquad (9.171)$$

where $\hat{r}$ is the operator of the distance between the two masses. On an energy eigenstate $|\psi\rangle$, argue that the operator

$$\hat{O} = \hat{\vec{r}} \cdot \hat{\vec{p}} = \hat{x}\hat{p}_x + \hat{y}\hat{p}_y + \hat{z}\hat{p}_z \qquad (9.172)$$

has a time-independent expectation value. Using this observation, determine the relationship between the expectation values of the kinetic energy operator $\langle \hat{K} \rangle$ and the potential operator $\langle V(\hat{r}) \rangle$.

(c) Why is the operator $\hat{O} = \hat{\vec{r}} \cdot \hat{\vec{p}}$ used to prove the virial theorem quantum mechanically? What does this operator do?

Hint: Consider its action and eigenstates/values in position space.

(d) Explicitly show that the virial theorem holds for the ground state of the hydrogen atom by taking expectation values of the kinetic and potential energy operators.

(e) Explicitly show that the virial theorem holds for every energy eigenstate of the one-dimensional quantum harmonic oscillator. Does the virial theorem provide a new interpretation of the Heisenberg uncertainty principle for the harmonic oscillator?

9.4 The form of the potential of the hydrogen atom is of course special, because it originates from Coulomb's law of electric force through canonical quantization. However, we can imagine a general form of a two-body, central potential like

$$V(\hat{r}) = -\frac{k}{\hat{r}^n}, \qquad (9.173)$$

for some $n > 0$, and constant k that has units of energy$\times$distance^n. Here, $\hat{r}$ is the familiar distance operator in three dimensions introduced in this chapter. For what values of n does the corresponding Hamiltonian of such a potential have normalizable eigenstates? It's enough to just consider the case for which angular momentum is 0.

Hint: Can you identify the $r \to \infty$ and $r \to 0$ behavior of a potential eigenstate?

9.5 We explicitly constructed the ground-state and first excited-state wavefunctions for the hydrogen atom in this chapter, but we can go further. The Laplace–Runge–Lenz operator commutes with the Hamiltonian, but its action changes the angular momentum of a state so can be used to relate known eigenstate wavefunctions to presently unknown wavefunctions. We know what the energy eigenvalue of the nth state is, and that will also help us construct the energy eigenstate wavefunctions.

(a) Consider the nth-energy eigenstate with total angular momentum $\ell = 0$. What is a reasonable ansatz for the form of the eigenstate wavefunction $\psi_{n,0,0}(r)$, that is purely a function of the electron's distance from the proton?

Hint: What behavior must you have as $r \to 0$? How do you get the factor of n^2 in the energy eigenvalue? How many nodes should there be in the wavefunction of the nth-energy eigenstate?

(b) Using the wavefunction of the first excited state of hydrogen, $\psi_{2,0,0}(\vec{r})$, calculated in Example 9.1, we can then act with components of the Laplace–Runge–Lenz operator to change the value of the angular momentum of the state, while maintaining the energy eigenvalue. Here, we'll construct all states that have the same energy as the $n = 2$, $\ell = 0$ eigenstate. This simplifies our task because the angular momentum factors in the Laplace–Runge–Lenz operator annihilate this state. With this observation, determine the Laplace–Runge–Lenz operator with zero angular momentum; you should find

$$\hat{A}_k = -i\frac{\hbar}{m_e}\hat{p}_k - \frac{e^2}{4\pi\epsilon_0}\frac{\hat{r}_k}{\hat{r}}. \tag{9.174}$$

(c) Using this result, construct the energy eigenstate wavefunctions $\psi_{1,1,1}(\vec{r})$, $\psi_{1,1,0}(\vec{r})$, and $\psi_{1,1,-1}(\vec{r})$ from the action of $\hat{A}_+$, $\hat{A}_z$, and $\hat{A}_-$ on the wavefunction $\psi_{2,0,0}(r)$ of Eq. (9.138), respectively.

9.6 The visible light spectrum ranges from wavelengths of 380 nm (violet) to 750 nm (red). If a hydrogen atom were excited, what color light would it emit and therefore appear to our eye? Don't forget to consider all possible transitions between energy eigenstates.

9.7 In Example 9.2, we introduced the Zeeman effect as a splitting of the energy levels of states in hydrogen due to the presence of a weak, external magnetic field. In that example, we just studied the Zeeman effect applied to the ground state of hydrogen and considering the different effects of an electron with spin-up vs. spin-down in the magnetic field. As a charged particle orbiting the proton, the electron produces a magnetic moment and this will also introduce a splitting of states whose z components of orbital angular momentum, $\hat{L}_z$, differ. We will study this effect in this exercise. Just as in Example 9.2, assume that the external magnetic field points along the z direction, as defined in Eq. (9.163).

(a) Consider first the electron as a classical, electrically charged particle orbiting the proton with angular momentum $\vec{L}$. Show that the classical gyromagnetic ratio of this system is indeed

$$\gamma_{\text{class}} = -\frac{e}{2m_e}. \tag{9.175}$$

Hint: Recall the Biot–Savart law applied to this orbiting electron.

(b)   With this result, now consider the Zeeman effect on a general energy eigen-state of hydrogen, $|n,\ell,m\rangle$. What is the change in the energy of this state due to the external magnetic field? Are the energies still independent of the eigenvalue of the z-component of angular momentum, m?

9.8   To determine the eigenvalues of the Hamiltonian of hydrogen we constructed the operators $\hat{T}$ and $\hat{S}$ from angular momentum and the Laplace–Runge–Lenz oper-ator. The actual form of these operators in a specific basis was not so important, just that they satisfied mutually commuting $\mathfrak{su}(2)$ algebras. In this problem, we'll learn a bit more about these operators and how they act on energy eigenstates of hydrogen.

(a)   First, what is the action of the operators $\hat{T}_i$ and $\hat{S}_i$ on the ground state of hydrogen?

(b)   As we did with other $\mathfrak{su}(2)$ algebras, we can construct the raising and lowering operators

$$\hat{T}_+ = \hat{T}_x + i\hat{T}_y, \qquad\qquad \hat{T}_- = \hat{T}_x - i\hat{T}_y, \qquad (9.176)$$
$$\hat{S}_+ = \hat{S}_x + i\hat{S}_y, \qquad\qquad \hat{S}_- = \hat{S}_x - i\hat{S}_y.$$

Consider the action of these operators on a general energy eigenstate of hydrogen, the $|n,\ell,m\rangle$. What state or states labeled by radial, total angu-lar momentum and z-component of angular momentum are produced from acting the raising and lowering operators?

(c)   What is the action of the operators $\hat{T}_z$ and $\hat{S}_z$ on an energy eigenstate $|n,\ell,m\rangle$? Can you identify the linear combination that produces the z-component of the angular momentum operator, $\hat{L}_z$?

(d)   Now, what do these $\hat{T}_i$ and $\hat{S}_i$ operators look like in position space?

10 Approximation Techniques

In this chapter, we are going to step back from the exact solutions of the Schrödinger equation (i.e., diagonalization of the Hamiltonian) and introduce methods for approximation. We have nearly exhausted those problems for which there exists a known, exact, analytic solution to the Schrödinger equation, and much of modern research on quantum mechanics and its generalizations focuses around approaches to approximating the eigenvalues of the Hamiltonian. We will just barely scratch the surface of techniques for approximation in this chapter, focusing on four different methods. The techniques we introduce here are general techniques for determining approximations to eigenvalues of linear operators or matrices, but applied to the problem of solving a quantum system, namely, diagonalization of its Hamiltonian.

10.1 Quantum Mechanics Perturbation Theory

In this section, we will study the following problem: given an arbitrary potential $V(\hat{x})$, what are the eigenvalues and eigenstates of the Hamiltonian $\hat{H}$, where

$$\hat{H} = \frac{\hat{p}^2}{2m} + V(\hat{x})\,? \tag{10.1}$$

We'll restrict our attention to problems in one spatial dimension here. Everything we introduce can be extended to higher dimensions, but things are interesting enough in one dimension, so we'll stick to that. We've identified all sorts of fun tricks for doing this in special examples, but the general problem has no closed-form solution. So, we need an efficient way to estimate the eigenstates/eigenvalues. Further it's not just enough to estimate it, we want to have a framework for estimation that is **systematically improvable**, which means that we have an algorithm for improving the accuracy of the estimate. It just involves calculating more and more terms or something that we can do and is only limited by our time and calculational ability. The first technique we will employ is the systematic method for estimating nearly everything in physics: the Taylor expansion. Recall that the Taylor expansion is used to estimate an unknown functional value, say $f(x_0)$, for some function $f(x)$.

With this in mind, the problem we want to solve is

$$\hat{H}|\psi\rangle = E|\psi\rangle = \left(\frac{\hat{p}^2}{2m} + V(\hat{x})\right)|\psi\rangle, \tag{10.2}$$

for some eigenvalue E and eigenstate $|\psi\rangle$. Now, $V(\hat{x})$ is arbitrary, and to use the Taylor expansion we need some toehold, some familiar place to start from which we can work to estimate the unknown E and $|\psi\rangle$. The following expansion was introduced by Erwin Schrödinger at the same time as his equation,[1] with historical foundations 50 years earlier in Lord Rayleigh's study of sound waves.[2]

The trick for this method is the following: if $V(\hat{x})$ is similar or close to a potential whose eigenvalues are known, then we can make progress. Let's assume that we can write

$$V(\hat{x}) = V_0(\hat{x}) + \epsilon\, V'(\hat{x}), \tag{10.3}$$

where $V_0(\hat{x})$ is a simple potential (infinite square well, harmonic oscillator, or the like), $V'(\hat{x})$ is the "difference potential," and ϵ is a value that controls the effect of the perturbation away from $V_0(\hat{x})$. So, the eigenvalue problem we want to solve is

$$\hat{H}|\psi\rangle = E|\psi\rangle = \left(\frac{\hat{p}^2}{2m} + V_0(\hat{x}) + \epsilon\, V'(\hat{x}) \right)|\psi\rangle. \tag{10.4}$$

The quantity ϵ is our Taylor expansion parameter, just like Δx in the Taylor expansion of $f(x_0 + \Delta x)$. So, we can in general express the unknown quantities $|\psi\rangle$ and E as Taylor expansions in ϵ as

$$E = \sum_{n=0}^{\infty} \epsilon^n E^{(n)}, \qquad\qquad |\psi\rangle = \sum_{n=0}^{\infty} \epsilon^n |\psi^{(n)}\rangle, \tag{10.5}$$

and now $E^{(n)}$ and $|\psi^{(n)}\rangle$ are just terms in the expansion. To make progress, we just insert these ansätze for E and $|\psi\rangle$ into our Schrödinger equation and expand order-by-order in ϵ. Each order in ϵ will correspond to a new equation we can solve for those corresponding terms in the expansion.

10.1.1 First-Order Correction to Energies

Concretely, plugging these expressions in, we find

$$\left(\frac{\hat{p}^2}{2m} + V_0(\hat{x}) + \epsilon\, V'(\hat{x}) \right) \sum_{n=0}^{\infty} \epsilon^n |\psi^{(n)}\rangle = \sum_{m=0}^{\infty} \epsilon^m E^{(m)} \sum_{n=0}^{\infty} \epsilon^n |\psi^{(n)}\rangle. \tag{10.6}$$

Now, we expand both sides of the equation in ϵ, and identify the equations that hold at each order in ϵ. For example, only keeping terms that are coefficients of ϵ^0, we must have

$$\left(\frac{\hat{p}^2}{2m} + V_0(\hat{x}) \right)|\psi^{(0)}\rangle = E^{(0)}|\psi^{(0)}\rangle. \tag{10.7}$$

---

[1] E. Schrödinger, "Quantisierung als Eigenwertproblem," *Ann. Phys.* **384**(4), 361–376(6), 489–527, **385**(13), 437–490, **386**(18), 109–139 (1926); "Der stetige Übergang von der Mikro- zur Makromechanik," *Naturwiss.* **14**, 664–666 (1926).
[2] J. W. S. B. Rayleigh, *The Theory of Sound*, vol. 1, Macmillan & Co. (1877).

By construction, we assumed that we could solve this equation for $E^{(0)}$ and $|\psi^{(0)}\rangle$. Further, for compactness, we will denote this "unperturbed" Hamiltonian as

$$\hat{H}_0 \equiv \frac{\hat{p}^2}{2m} + V_0(\hat{x}). \tag{10.8}$$

Next, the ϵ^1 terms in the expansion produce the equation

$$\epsilon \hat{H}_0|\psi^{(1)}\rangle + \epsilon V'(\hat{x})|\psi^{(0)}\rangle = \epsilon E^{(0)}|\psi^{(1)}\rangle + \epsilon E^{(1)}|\psi^{(0)}\rangle. \tag{10.9}$$

Let's first isolate the first correction to the energy, $E^{(1)}$. To do this, let's hit everything with a bra $\langle\psi^{(0)}|$:

$$\langle\psi^{(0)}|\hat{H}_0|\psi^{(1)}\rangle + \langle\psi^{(0)}|V'(\hat{x})|\psi^{(0)}\rangle = \langle\psi^{(0)}|E^{(0)}|\psi^{(1)}\rangle + \langle\psi^{(0)}|E^{(1)}|\psi^{(0)}\rangle. \tag{10.10}$$

I've canceled out the overall factor of ϵ for cleanliness. Now, note that

$$\langle\psi^{(0)}|\hat{H}_0|\psi^{(1)}\rangle = E^{(0)}\langle\psi^{(0)}|\psi^{(1)}\rangle, \tag{10.11}$$

as $\langle\psi^{(0)}|$ is an eigenstate of the unperturbed Hamiltonian $\hat{H}_0$. Then, this term cancels from both sides of the equation. Further, we assume that $|\psi^{(0)}\rangle$ is normalized, $\langle\psi^{(0)}|\psi^{(0)}\rangle = 1$, so we find

$$E^{(1)} = \langle\psi^{(0)}|V'(\hat{x})|\psi^{(0)}\rangle. \tag{10.12}$$

That is, the first correction to the energy is the expectation value of the perturbing potential $V'(\hat{x})$ in the unperturbed state $|\psi^{(0)}\rangle$. We had encountered this expression, at least implicitly, in the previous chapter in studying the 21-cm line of hydrogen, due to a small, new spin–spin interaction energy.

10.1.2 First-Order Correction to Wavefunctions

Determining the first correction to the eigenstate is a bit trickier. Let's rearrange the first-order equation to read

$$\left(\hat{H}_0 - E^{(0)}\right)|\psi^{(1)}\rangle = \left(E^{(1)} - V'(\hat{x})\right)|\psi^{(0)}\rangle. \tag{10.13}$$

Now, remember that we want to solve for the ket $|\psi^{(1)}\rangle$, which is some vector in a Hilbert space. Also, the right side of this equality is another, known vector on this Hilbert space. $\hat{H}_0 - E^{(0)}$ is some operator that acts on the states of the Hilbert space, so this equation is nothing more than the familiar equation for a system of linear equations:

$$\mathbb{A}\vec{x} = \vec{b}, \tag{10.14}$$

where $\mathbb{A}$ is a matrix, $\vec{b}$ is a known vector, and we want to find $\vec{x}$. Of course, we know how to solve for $\vec{x}$: we just invert $\mathbb{A}$

$$\vec{x} = \mathbb{A}^{-1}\vec{b}. \tag{10.15}$$

This is only possible if, in fact, $\mathbb{A}$ is invertible and non-singular, and has a non-zero determinant. By our analogy, $\mathbb{A} = \hat{H}_0 - E^{(0)}$ in our problem, so we want to invert this:

$$\mathbb{A}^{-1} = \frac{1}{\hat{H}_0 - E^{(0)}} \,. \tag{10.16}$$

Does this actually exist?

Recall that $E^{(0)}$ is an eigenvalue of $\hat{H}_0$ and $\mathbb{A} = \hat{H}_0 - E^{(0)}\mathbb{I}$ is just the operator that we would consider to determine the eigenvalue:

$$(\hat{H}_0 - E^{(0)}\mathbb{I})|\psi^{(0)}\rangle = 0 \,. \tag{10.17}$$

That is, the operator $\hat{H}_0 - E^{(0)}$ has an eigenvector with zero eigenvalue, thus its determinant vanishes:

$$\det\left(\hat{H}_0 - E^{(0)}\right) = 0 \,. \tag{10.18}$$

Hence, we would (naïvely) say that $\hat{H}_0 - E^{(0)}$ is not invertible, so it would seem that we can't solve for $|\psi^{(1)}\rangle$. Or can we?

Let's write our Eq. (10.13) in a more illuminating way, using the completeness of the eigenstates of $\hat{H}_0$. We will denote an eigenstate as

$$\hat{H}_0|\psi_n^{(0)}\rangle = E_n^{(0)}|\psi_n^{(0)}\rangle \,, \tag{10.19}$$

where n denotes the nth-energy eigenvalue. Now, assuming that the eigenstates of $\hat{H}_0$ form an orthonormal and complete basis, we can re-write our equation for $|\psi^{(1)}\rangle$ in another useful manner. We have

$$\left(\hat{H}_0 - E_n^{(0)}\right)|\psi_n^{(1)}\rangle = \left(E_n^{(1)} - V'(\hat{x})\right)|\psi_n^{(0)}\rangle \,, \tag{10.20}$$

where we have denoted the eigenstate of $\hat{H}_0$ by an integer n. Also:

$$E_n^{(1)} = \langle\psi_n^{(0)}|V'(\hat{x})|\psi_n^{(0)}\rangle \,. \tag{10.21}$$

Now, by completeness we can write the identity operator as

$$\mathbb{I} = \sum_{n=0}^{\infty} |\psi_n^{(0)}\rangle\langle\psi_n^{(0)}| \,, \tag{10.22}$$

and the Hamiltonian $\hat{H}_0$ as

$$\hat{H}_0 = \sum_{n=0}^{\infty} E_n^{(0)}|\psi_n^{(0)}\rangle\langle\psi_n^{(0)}| \,. \tag{10.23}$$

Then, our equation for $|\psi^{(1)}\rangle$ becomes

$$\left(\hat{H}_0 - E_m^{(0)}\right)|\psi_m^{(1)}\rangle = \sum_{n=0}^{\infty} \left(E_n^{(0)} - E_m^{(0)}\right)|\psi_n^{(0)}\rangle\langle\psi_n^{(0)}|\psi_m^{(1)}\rangle \tag{10.24}$$

$$= E_m^{(1)} - \sum_{n=0}^{\infty} |\psi_n^{(0)}\rangle\langle\psi_n^{(0)}|V'(\hat{x})|\psi_m^{(0)}\rangle \,.$$

Now, in this form we recognize the solution to our problem of inverting the operator $\hat{H}_0 - E_m^{(0)}$. The only issue is that $|\psi_m^{(0)}\rangle$ is an eigenstate of $\hat{H}_0 - E_m^{(0)}$ with zero eigenvalue. So, in our new equation above, we can focus on individual terms and identify their consequence. First, if $n = m$, then we have the problematic term, but note that the right side of the first line vanishes, and the second line implies

$$E_m^{(1)} |\psi_m^{(0)}\rangle = \langle \psi_m^{(0)} |V'(\hat{x})| \psi_m^{(0)}\rangle |\psi_m^{(0)}\rangle , \qquad (10.25)$$

which is exactly what we found earlier for the value of $E_m^{(1)}$. Note that we are allowed to look at individual terms in the expression above because the $|\psi_n^{(0)}\rangle$ states are orthogonal, so $|\psi_n^{(0)}\rangle$ and $|\psi_m^{(0)}\rangle$ for $n \neq m$ have no components in common.

Now, we have shown that the $n = m$ term cancels, so for $n \neq m$ we find

$$\langle \psi_n^{(0)} |\psi_m^{(1)}\rangle = \frac{\langle \psi_n^{(0)} |V'(\hat{x})| \psi_m^{(0)}\rangle}{E_m^{(0)} - E_n^{(0)}} , \qquad (10.26)$$

which is the coefficient of $|\psi_m^{(1)}\rangle$ in the expansion of energy eigenstates of $\hat{H}_0$. That is, our first correction to the state is

$$|\psi_m^{(1)}\rangle = \sum_{\substack{n=0 \\ n \neq m}}^{\infty} \frac{\langle \psi_n^{(0)} |V'(\hat{x})| \psi_m^{(0)}\rangle}{E_m^{(0)} - E_n^{(0)}} |\psi_n^{(0)}\rangle . \qquad (10.27)$$

As long as all energy eigenvalues of $\hat{H}_0$ are distinct and non-degenerate, all coefficients in the expression are non-zero and finite, in general.

So, to summarize, for our Schrödinger equation written as

$$\left(\hat{H}_0 + \epsilon V'(\hat{x}) \right) |\psi\rangle = E |\psi\rangle , \qquad (10.28)$$

we can write the energy eigenvalue and eigenstate as

$$E = E_m^{(0)} + \epsilon \langle \psi_m^{(0)} |V'(\hat{x})| \psi_m^{(0)}\rangle + \mathcal{O}(\epsilon^2) , \qquad (10.29)$$

$$|\psi\rangle = |\psi_m^{(0)}\rangle + \epsilon \sum_{\substack{n=0 \\ n \neq m}}^{\infty} \frac{\langle \psi_n^{(0)} |V'(\hat{x})| \psi_m^{(0)}\rangle}{E_m^{(0)} - E_n^{(0)}} |\psi_n^{(0)}\rangle + \mathcal{O}(\epsilon^2) .$$

The "big O notation" $\mathcal{O}(\epsilon^2)$ means that our expression is correct, up to terms in the Taylor expansion in ϵ at and beyond ϵ^2 that we have ignored. One can work to calculate these terms, but we won't do that here.

Example 10.1   Let's see how this formalism works in an explicit example. Let's consider a potential that is the infinite square well that has a little bump in the center, as shown in Fig. 10.1. That is, the potential is

$$V(x) = \begin{cases} 0, & 0 < x < \frac{a}{4} , \; \frac{3a}{4} < x < a , \\ V_0 , & \frac{a}{4} < x < \frac{3a}{4} , \\ \infty , & x < 0 , \; x > a . \end{cases} \qquad (10.30)$$

The Schrödinger equation in position space is then

$$\left(-\frac{\hbar^2}{2m} \frac{d^2}{dx^2} + V_0 \Theta(x - a/4) \Theta(3a/4 - x) \right) \psi(x) = E \psi(x) , \qquad (10.31)$$

where $\Theta(x)$ is called the **Heaviside theta function** and is defined to be

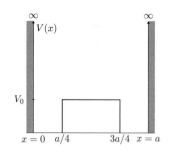

Fig. 10.1 The infinite square well potential of width a with a bump of height V_0 in the region $x \in [a/4, 3a/4]$.

$$\Theta(x) = \begin{cases} 0, & x < 0, \\ 1, & x > 0. \end{cases} \tag{10.32}$$

We can then identify the little bump as our perturbation potential:

$$V'(x) = V_0 \Theta(x - a/4)\Theta(3a/4 - x), \tag{10.33}$$

and let's see how the ground state of the well is affected by the bump.

Solution

Recall that the energies of the infinite square well are

$$E_n^{(0)} = \frac{(n+1)^2 \pi^2 \hbar^2}{2ma^2}, \tag{10.34}$$

for $n = 0, 1, 2, \ldots$, so the ground state is

$$E_0^{(0)} = \frac{\pi^2 \hbar^2}{2ma^2}. \tag{10.35}$$

Now, the first correction to the ground-state energy is

$$E_0^{(1)} = \langle \psi_0^{(0)} | V'(x) | \psi_0^{(0)} \rangle = \int_{a/4}^{3a/4} dx \sqrt{\frac{2}{a}} \sin\left(\frac{\pi x}{a}\right) V_0 \sqrt{\frac{2}{a}} \sin\left(\frac{\pi x}{a}\right) \tag{10.36}$$

$$= \frac{2 + \pi}{2\pi} V_0.$$

To evaluate this integral, I used that the ground-state wavefunction of the infinite square well is

$$\psi_0^{(0)}(x) = \sqrt{\frac{2}{a}} \sin\left(\frac{\pi x}{a}\right). \tag{10.37}$$

Then, to linear order in V_0, the small perturbation potential, the new ground-state energy of the infinite square well with a bump is

$$E_0 = \frac{\pi^2 \hbar^2}{2ma^2} + \frac{2 + \pi}{2\pi} V_0 + \mathcal{O}(V_0^2). \tag{10.38}$$

One can also calculate the first corrections to the ground-state wavefunction, but we'll leave that for an exercise.

10.2 The Variational Method

The perturbative method introduced in the previous section can be very powerful when you have a state in a known, diagonalized Hamiltonian to start from, and then expand about. However, what if no such state exists? Can we still make progress on a systematic method for estimating eigenvalues of the Hamiltonian? In this section and the next, we will introduce two methods which enable honest guessing and we'll prove results which will allow us to bound eigenvalues in a robust way. The first technique we will study is the **variational method**.

The idea is extremely simple. Let's consider a system described by some Hamiltonian $\hat{H}$. The question we would like to ask is what the ground-state energy of this Hamiltonian is. Solving this problem exactly might be challenging depending on what $\hat{H}$ is, but we can place an upper bound on the ground-state energy quite easily. Everything that follows in this section comes from a very simple property that we will prove. For any state $|\psi\rangle$ in a Hilbert space, the expectation value of the Hamiltonian on that state is bounded from below by the ground-state energy of $\hat{H}$, E_0. That is, with a sufficiently good estimate for the state $|\psi\rangle$, we can get close to the value of the ground-state energy, E_0. With a better estimate, we can get closer, and for many problems of interest getting the ground-state energy is sufficient. In fact, this technique is powerful enough that Nobel Prizes have been awarded for its implementation.[3]

So let's prove this. For a Hamiltonian $\hat{H}$, the expectation value of $\hat{H}$ on any state $|\psi\rangle$ in the Hilbert space is bounded from below by the ground-state energy E_0 of the Hamiltonian:

$$\langle \psi | \hat{H} | \psi \rangle \geq E_0 \,. \tag{10.39}$$

Proof Assuming orthonormality and completeness of the energy eigenstates of $\hat{H}$, we can expand $|\psi\rangle$ in the basis of energy eigenstates:

$$|\psi\rangle = \sum_{n=0}^{\infty} \beta_n |\psi_n\rangle \,, \tag{10.40}$$

where $|\psi_n\rangle$ is the nth energy eigenstate and β_n is a complex coefficient. Now, as $|\psi\rangle$ is in the Hilbert space, it is normalized $\langle \psi | \psi \rangle = 1$, or

$$\sum_{n=0}^{\infty} |\beta_n|^2 = 1 \,. \tag{10.41}$$

Computing the expectation value of $\hat{H}$ on $|\psi\rangle$, we have

$$\langle \psi | \hat{H} | \psi \rangle = \left(\sum_{m=0}^{\infty} \beta_m^* \langle \psi_m | \right) \hat{H} \left(\sum_{n=0}^{\infty} \beta_n |\psi_n\rangle \right) = \sum_{n,m=0}^{\infty} \beta_m^* \beta_n E_n \langle \psi_m | \psi_n \rangle \tag{10.42}$$

$$= \sum_{n=0}^{\infty} |\beta_n|^2 E_n \,,$$

[3] R. B. Laughlin, "Anomalous quantum Hall effect: An incompressible quantum fluid with fractionally charged excitations," *Phys. Rev. Lett.* **50**, 1395 (1983).

where we have used that

$$\hat{H}|\psi_n\rangle = E_n|\psi_n\rangle\,, \qquad\qquad \langle\psi_m|\psi_n\rangle = \delta_{mn}\,. \qquad (10.43)$$

Now, let's subtract E_0, the ground-state energy, from this expectation value:

$$\langle\psi|\hat{H}|\psi\rangle - E_0 = \sum_{n=0}^{\infty} |\beta_n|^2 E_n - E_0 = \sum_{n=0}^{\infty} |\beta_n|^2 (E_n - E_0) \qquad (10.44)$$

$$= \sum_{n=1}^{\infty} |\beta_n|^2 (E_n - E_0)\,.$$

In this sequence of equalities, we have used the normalization condition and removed the $n = 0$ term (which vanished anyway) in the final equality. As E_0 is the ground-state energy, we have that $E_n > E_0$ for $n \geq 1$, and so the final expression is manifestly ("obviously") non-negative. This proves that therefore

$$\langle\psi|\hat{H}|\psi\rangle - E_0 \geq 0\,, \qquad (10.45)$$

as we claimed.   □

So with a good enough guess, we can get close to the ground-state energy. But how do we go about guessing? Just throwing out random states and testing the expectation value $\langle\psi|\hat{H}|\psi\rangle$ is incredibly inefficient and there is no guarantee that any individual guess will be close to the ground-state energy at all. To use this result to its fullest potential, our plan will be to make a guess of a ground state $|\psi\rangle$ that contains parameters that we can minimize over. For quantum mechanical problems in one spatial dimension, we thus consider wavefunctions $\psi(x; \alpha_1, \alpha_2, \ldots, \alpha_n)$ where the α_i parameters specify the particular shape of the wavefunction, given a general functional form. Then, we can calculate the expectation value of the Hamiltonian as

$$\langle\psi|\hat{H}|\psi\rangle = \int dx\, \psi^*(x; \alpha_1, \alpha_2, \ldots, \alpha_n)\, \hat{H}\, \psi(x; \alpha_1, \alpha_2, \ldots, \alpha_n)\,. \qquad (10.46)$$

By our theorem above, for any choice of the parameters $\{\alpha_1, \alpha_2, \ldots, \alpha_n\}$ this is guaranteed to be at least the value of the ground-state energy. However, these parameters give us another handle to adjust the form of the wavefunction to minimize $\langle\psi|\hat{H}|\psi\rangle$ further. That is, with these parameters, we can minimize over them to get the least upper bound possible on the ground-state energy, given the chosen (i.e., guessed) functional form of the wavefunction $\psi(x; \alpha_1, \alpha_2, \ldots, \alpha_n)$.

This is pretty abstract and possibly confusing. Seeing it work in an explicit example can demonstrate that it's actually very simple.

Example 10.2 Let's test this out for the infinite square well, where we will set the width $a = 1$ of the well for simplicity. That is, the potential we consider is

$$V(x) = \begin{cases} 0, & 0 < x < 1\,, \\ \infty, & x < 0, x > 1\,. \end{cases} \qquad (10.47)$$

We know what the ground-state energy is in this case:

$$E_0 = \frac{\pi^2}{2}\frac{\hbar^2}{m} = 4.934802\ldots\frac{\hbar^2}{m},$$ (10.48)

where I have expanded the numerical value of $\pi^2/2$ to seven significant figures. Now, on to a guess for the ground-state wavefunction. We know that the wavefunction must vanish at both $x = 0, 1$, the boundaries of the well. Additionally, the well is symmetric about its central point $x = 1/2$, so we expect that the wavefunction would also be symmetric about $x = 1/2$. A guess for a wavefunction that vanishes at the endpoints and is symmetric about $x = 1/2$ is

$$\psi(x; \alpha) = Nx^{\alpha}(1 - x)^{\alpha},$$ (10.49)

where N is a normalization constant and α is a parameter we will vary to minimize the expectation value of the Hamiltonian on this state.

Solution

First, we will calculate the normalization N, for any α. We require that

$$1 = \int_0^1 dx\, \psi^*(x; \alpha)\psi(x; \alpha) = N^2 \int_0^1 dx\, x^{2\alpha}(1 - x)^{2\alpha}.$$ (10.50)

This integral can't be done with elementary functions for arbitrary α, but can be expressed in terms of the Euler Beta function $B(a, b)$. The Euler Beta function is defined to be

$$B(a, b) = \int_0^1 dx\, x^{a-1}(1 - x)^{b-1} = \frac{\Gamma(a)\Gamma(b)}{\Gamma(a + b)},$$ (10.51)

where $\Gamma(x)$ is the Euler **Gamma function** and is a continuous generalization of the factorial. Then, the integral we need, in terms of the Beta function, is

$$1 = N^2 \int_0^1 dx\, x^{2\alpha}(1 - x)^{2\alpha} = N^2 B(2\alpha + 1, 2\alpha + 1) = N^2 \frac{\Gamma(2\alpha + 1)^2}{\Gamma(4\alpha + 2)}.$$ (10.52)

Thus, it follows that the normalization factor N is

$$N = \frac{\sqrt{\Gamma(4\alpha + 2)}}{\Gamma(2\alpha + 1)}.$$ (10.53)

A plot of this guessed form of the ground-state wavefunction of the infinite square well is shown in Fig. 10.2 for two different values of α, compared to the true ground-state wavefunction of the infinite square well.

With this result, we can then calculate the expectation value of the Hamiltonian, $\langle\psi|\hat{H}|\psi\rangle$. Recall that, for the infinite square well, the Hamiltonian is

$$\hat{H} = \frac{\hat{p}^2}{2m} = -\frac{\hbar^2}{2m}\frac{d^2}{dx^2},$$ (10.54)

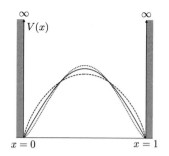

Fig. 10.2 Illustration of wavefunctions of the parametrized form of Eq. (10.49) with $\alpha = 0.75$ (dashed) and $\alpha = 1.5$ (dotted) compared to the true ground-state wavefunction of the infinite square well (solid).

in the position basis. Thus, the evaluation of the expectation value of $\hat{H}$ involves the second derivative of the wavefunction, where

$$\frac{d^2}{dx^2} N x^\alpha (1-x)^\alpha = \alpha N x^{\alpha-2}(1-x)^{\alpha-2}(\alpha-1+(2-4\alpha)x(1-x)) . \tag{10.55}$$

Then, the expectation value of the Hamiltonian is

$$\langle \psi | \hat{H} | \psi \rangle = -N^2 \frac{\hbar^2}{2m} \int_0^1 dx \left[x^\alpha (1-x)^\alpha \frac{d^2}{dx^2} x^\alpha (1-x)^\alpha \right] \tag{10.56}$$

$$= \frac{\alpha(1+4\alpha)}{2\alpha-1} \frac{\hbar^2}{m} ,$$

with $\alpha > 1/2$. Now, I don't expect you to just know how to evaluate this integral, as I used extensive identities of the Gamma function to simplify it this far. Also, note that this integral is actually only finite if $\alpha > 1/2$, which perhaps isn't so surprising. If $\alpha < 1/2$, then the second derivative of x^α has an exponent that is less than $-3/2$, and so the integrand has a non-integrable singularity at $x = 0$.

At any rate, this final result is necessarily bounded from below by the ground-state energy. To find the least upper bound given this form of the guess of the wavefunction, we then minimize this result over α. To do the minimization, we can just differentiate with respect to α and demand that it vanish:

$$\frac{d}{d\alpha} \frac{\alpha(1+4\alpha)}{2\alpha-1} = 0 = \frac{8\alpha^2 - 8\alpha - 1}{(1-2\alpha)^2} . \tag{10.57}$$

That is, $8\alpha^2 - 8\alpha - 1 = 0$ or

$$\alpha = \frac{8 \pm \sqrt{64+32}}{16} = \frac{1}{2} \pm \frac{\sqrt{6}}{4} . \tag{10.58}$$

As $\alpha > 1/2$ established earlier, we take the "+" root, so the exponent that minimizes the expectation value is

$$\alpha_{\min} = \frac{2+\sqrt{6}}{4} = 1.112372.... \tag{10.59}$$

Now, at this point, we can evaluate the expectation value of the Hamiltonian on this wavefunction:

$$\langle \psi | \hat{H} | \psi \rangle = \frac{\alpha_{min}(1 + 4\alpha_{min})}{2\alpha_{min} - 1} \frac{\hbar^2}{m} = \left(\frac{5}{2} + \sqrt{6} \right) \frac{\hbar^2}{m} = 4.94948\ldots \frac{\hbar^2}{m}. \tag{10.60}$$

Note that this is (very slightly!) larger than the true ground-state energy value from Eq. (10.48). The difference between this expectation value and the ground state is

$$\langle \psi | \hat{H} | \psi \rangle - E_0 = \left(\frac{5}{2} + \sqrt{6} - \frac{\pi^2}{2} \right) \frac{\hbar^2}{m} = 0.01468\ldots \frac{\hbar^2}{m}, \tag{10.61}$$

or less than 1% the value of the true ground state. So this method gets us to an excellent estimate relatively easily. By the way, everything we did here was analytic in evaluating integrals, but it can be even easier if you just numerically evaluate them.

10.3 The Power Method

In this section, we are going to introduce yet another approximation technique, called the **power method**. Its efficacy relies on the following observation regarding repeated action of any linear operator on a vector. For concreteness, consider a Hermitian matrix $\mathbb{A}$ and we will call its collection of eigenvectors $\{\vec{v}_i\}_i$ such that

$$\mathbb{A}\vec{v}_i = \lambda_i \vec{v}_i, \tag{10.62}$$

for real eigenvalue λ_i. Further, we will assume that the eigenvalues are ordered in magnitude such that

$$|\lambda_1| > |\lambda_2| > \cdots > |\lambda_i| > \cdots. \tag{10.63}$$

We assume for simplicity that all eigenvalues are unique, which is why we can write strict inequalities. Now, let's consider what happens when we act matrix $\mathbb{A}$ on an arbitrary vector $\vec{u}$ a number of times. Assuming that the eigenvectors of $\mathbb{A}$ form a complete basis, we can express $\vec{u}$ as a linear combination of $\{\vec{v}_i\}_i$:

$$\vec{u} = \sum_{i=1} \beta_i \vec{v}_i, \tag{10.64}$$

for some coefficients β_i. Now, acting $\mathbb{A}$ on $\vec{u}$ once pulls out a factor of eigenvalues for each term:

$$\mathbb{A}\vec{u} = \sum_{i=1} \beta_i \lambda_i \vec{v}_i. \tag{10.65}$$

Acting $\mathbb{A}$ on $\vec{u}$ n times pulls out n factors of eigenvalues in each term:

$$\mathbb{A}^n \vec{u} = \sum_{i=1} \beta_i \lambda_i^n \vec{v}_i. \tag{10.66}$$

Here's the trick: as we have assumed an ordering to the eigenvalues, for sufficiently large n, $\mathbb{A}^n \vec{u}$ gets closer and closer to parallel to $\vec{v}_1$, the eigenvector with largest eigenvalue. So, given any vector $\vec{u}$, we can get a better and better approximation to the eigenvector with largest eigenvalue.

Now, how does this help us with quantum mechanics, in diagonalizing the Hamiltonian? Again, we want to solve

$$\hat{H}|\psi\rangle = E|\psi\rangle, \tag{10.67}$$

for eigenstate $|\psi\rangle$ and eigenenergy E. Typically, we don't want to identify the largest eigenvalue of $\hat{H}$ or its corresponding state, and in most cases there isn't even a largest-energy eigenvalue. As with the variational method, we want to find the ground state, which, by definition, is the state with smallest eigenvalue of $\hat{H}$. Additionally, let's assume that $\hat{H}$ has only positive eigenvalues for the following discussion, though this can be relaxed. So why do I claim this power method can help us find the ground state?

If the Hamiltonian is invertible and has no eigenvalues of 0, then its inverse $\hat{H}^{-1}$ exists. If so, then we can multiply our traditional Schrödinger equation eigenvalue problem by $\hat{H}^{-1}$ on both sides:

$$\hat{H}|\psi\rangle = E|\psi\rangle \quad \Rightarrow \quad \hat{H}^{-1}|\psi\rangle = \frac{1}{E}|\psi\rangle. \tag{10.68}$$

That is, eigenstates of $\hat{H}$ are also eigenstates of $\hat{H}^{-1}$, but with eigenvalue that is inverted. So, while the ground state of $\hat{H}$ is the state with smallest eigenvalue, that same state has the largest eigenvalue of $\hat{H}^{-1}$. Using our earlier observation, if we take an arbitrary state on the Hilbert space $|\varphi\rangle$, acting $\hat{H}^{-1}$ sufficiently many times on $|\varphi\rangle$ will produce a state that becomes closer and closer to proportional to the ground state of $\hat{H}$. That is:

$$\left(\hat{H}^{-1}\right)^n |\varphi\rangle \propto |\psi_0\rangle, \tag{10.69}$$

the ground state of $\hat{H}$, for sufficiently large n. Note that the proportionality is not equality: every time $\hat{H}^{-1}$ acts, it pulls out factors of eigenvalues. So even if $|\varphi\rangle$ is initially normalized, $\left(\hat{H}^{-1}\right)^n |\varphi\rangle$ is not. However, fixing normalization is pretty trivial, so that's something we can correct later.

Example 10.3   While this seems pretty magical, the biggest challenge will be inverting the Hamiltonian in the first place. As we saw with the variational method, an example can be helpful in demonstrating the technique. Let's just consider our familiar infinite square well, again setting its width $a = 1$ for simplicity. That is, the potential is

$$V(x) = \begin{cases} 0, & 0 < x < 1, \\ \infty, & x < 0, \, x > 1. \end{cases} \tag{10.70}$$

In the well, the Hamiltonian is of course exclusively kinetic energy:

$$\hat{H} = \frac{\hat{p}^2}{2m} = -\frac{\hbar^2}{2m}\frac{d^2}{dx^2}, \tag{10.71}$$

in position space. To use this power method to find the ground state, we need to determine what the inverse of this Hamiltonian is; that is

$$\hat{H}^{-1} = \frac{2m}{\hat{p}^2}, \tag{10.72}$$

or, what is the inverse of the momentum operator $1/\hat{p}$?

Solution

The inverse momentum operator, $\hat{p}^{-1}$, is defined to return the identity operator when acting on $\hat{p}$. That is

$$(\hat{p}^{-1})\hat{p} = \mathbb{I}, \tag{10.73}$$

obviously. Let's work in position space and we need to introduce a test function $f(x)$ to keep us honest about the linearity of these operators. In position space, of course

$$\hat{p} = -i\hbar \frac{d}{dx}, \tag{10.74}$$

so we want to find $\hat{p}^{-1}$ such that

$$\hat{p}^{-1}(-i\hbar)\frac{d}{dx}f(x) = f(x). \tag{10.75}$$

The fundamental theorem of calculus tells us that the antiderivative (i.e., indefinite integral) is the inverse of the derivative. So we can write

$$\hat{p}^{-1} = \frac{i}{\hbar} \int^x dx', \tag{10.76}$$

so that

$$\left(\hat{p}^{-1}\right)\hat{p}f(x) = \frac{i}{\hbar}\int^x dx' \, (-i\hbar)\frac{d}{dx'}f(x') = f(x), \tag{10.77}$$

as required. As $f(x)$ was an arbitrary test function, this shows that indeed $(\hat{p}^{-1})\hat{p} = \mathbb{I}$.

Now, for the inverse of the infinite square well Hamiltonian, we need $(\hat{p}^{-1})^2$. Just like $\hat{p}^2$, we think of $(\hat{p}^{-1})^2$ as two consecutive actions of the operator $\hat{p}^{-1}$. Thus, if $\hat{p}^{-1}$ integrates once, $(\hat{p}^{-1})^2$ integrates twice, in composition. That is

$$(\hat{p}^{-1})^2 = \left(\frac{i}{\hbar}\right)^2 \int^x dx' \int^{x'} dx''. \tag{10.78}$$

To verify this, let's act on a test function $f(x)$ with $\hat{p}^2$ and $(\hat{p}^{-1})^2$ on the left:

$$\hat{p}^2(\hat{p}^{-1})^2 f(x) = (-i\hbar)^2 \left(\frac{i}{\hbar}\right)^2 \frac{d^2}{dx^2} \int^x dx' \int^{x'} dx'' \, f(x'') = \frac{d}{dx}\int^x dx'' \, f(x'')$$

$$= f(x), \tag{10.79}$$

through two applications of the fundamental theorem of calculus. Then, we have shown that

$$\hat{p}^2 (\hat{p}^{-1})^2 f(x) = f(x) = \mathbb{I} f(x), \tag{10.80}$$

and so we have indeed correctly identified $(\hat{p}^{-1})^2$. Finally, the inverse of the Hamiltonian of the infinite square well is

$$\hat{H}^{-1} = \frac{2m}{\hat{p}^2} = -\frac{2m}{\hbar^2} \int_0^x dx' \int^{x'} dx'', \tag{10.81}$$

in position space. Additionally, we have to remember that all states in the Hilbert space vanish at the endpoints of the well, at $x = 0$ and $x = 1$.

We now have our inverse Hamiltonian, so let's figure out an estimate for the ground-state wavefunction. Motivated by our consideration with the variational method of the previous section, let's make the ansatz that the ground-state wavefunction $\psi(x)$ is

$$\psi(x) \propto x(1-x). \tag{10.82}$$

As mentioned before, this vanishes at $x = 0, 1$ and is symmetric about the center of the well, $x = 1/2$. We only care about proportionality here because we can always divide out by the normalization factor $\langle \psi | \psi \rangle$. For this wavefunction guess, our estimate of the ground-state energy E_0 is

$$E_0 \leq \frac{\langle \psi | \hat{H} | \psi \rangle}{\langle \psi | \psi \rangle} = -\frac{\hbar^2}{2m} \frac{\int_0^1 dx\, x(1-x) \frac{d^2}{dx^2} x(1-x)}{\int_0^1 dx\, x^2(1-x)^2}. \tag{10.83}$$

These integrals are easy enough to evaluate. For the numerator, we find

$$\int_0^1 dx\, x(1-x) \frac{d^2}{dx^2} x(1-x) = -2 \int_0^1 dx\, x(1-x) = -\frac{1}{3}. \tag{10.84}$$

The denominator is

$$\int_0^1 dx\, x^2(1-x)^2 = \frac{1}{30}. \tag{10.85}$$

Thus, the estimate of the ground-state energy is

$$E_0 \leq \frac{\langle \psi | \hat{H} | \psi \rangle}{\langle \psi | \psi \rangle} = 5\frac{\hbar^2}{m}. \tag{10.86}$$

Recall that the exact value is

$$E_0 = \frac{\pi^2}{2} \frac{\hbar^2}{m} = 4.934802 \ldots \frac{\hbar^2}{m}, \tag{10.87}$$

so we're already pretty close.

The power method should produce a better estimate for the ground-state wavefunction and hence its energy. The power method states that $\hat{H}^{-1} | \psi \rangle$ is more parallel to the ground-state wavefunction than $| \psi \rangle$ itself. So let's calculate it!

We identified $\psi(x) \propto x(1-x)$, again not worrying about normalization. Then the action of $\hat{H}^{-1}$ on this is

$$\hat{H}^{-1}\psi(x) = \frac{2m}{\hat{p}^2}\psi(x) = -\frac{2m}{\hbar^2}\int_0^x dx' \int_0^{x'} dx'' \, x''(1-x'') \tag{10.88}$$

$$= -\frac{2m}{\hbar^2}\int_0^x dx' \left(\frac{(x')^2}{2} - \frac{(x')^3}{3} + c_1\right)$$

$$= -\frac{2m}{\hbar^2}\left(\frac{x^3}{6} - \frac{x^4}{12} + c_1 x + c_2\right),$$

where c_1 and c_2 are integration constants. We fix c_1 and c_2 by demanding that $\hat{H}^{-1}\psi(x)$ vanishes at the boundaries of the well. The boundary $x = 0$ forces $c_2 = 0$ and the boundary $x = 1$ forces $c_1 = -1/12$. Thus, we find that

$$\hat{H}^{-1}\psi(x) = \frac{2m}{\hbar^2}\left(\frac{x}{12} - \frac{x^3}{6} + \frac{x^4}{12}\right). \tag{10.89}$$

As we only care about proportionality, we can remove overall constant factors, so

$$\hat{H}^{-1}\psi(x) \propto x - 2x^3 + x^4 = x(1-x) + x^2(1-x)^2. \tag{10.90}$$

Now, with this improved estimate for the ground-state wavefunction, let's evaluate the updated estimate of the ground-state energy:

$$E_0 \leq \frac{\langle\psi|(\hat{H}^{-1})\hat{H}(\hat{H}^{-1})|\psi\rangle}{\langle\psi|(\hat{H}^{-1})(\hat{H}^{-1})|\psi\rangle} \tag{10.91}$$

$$= -\frac{\hbar^2}{2m}\frac{\int_0^1 dx\,(x - 2x^3 + x^4)\frac{d^2}{dx^2}(x - 2x^3 + x^4)}{\int_0^1 dx\,(x - 2x^3 + x^4)^2}.$$

We just evaluate the integrals in the numerator and denominator. For the numerator, we have

$$\int_0^1 dx\,(x - 2x^3 + x^4)\frac{d^2}{dx^2}(x - 2x^3 + x^4) = -\int_0^1 dx\,(x - 2x^3 + x^4)12x(1-x)$$

$$= -\frac{17}{35}. \tag{10.92}$$

The integral of the denominator is

$$\int_0^1 dx\,(x - 2x^3 + x^4)^2 = \frac{31}{630}, \tag{10.93}$$

and so the estimate for the ground-state energy is

$$E_0 \leq \left(-\frac{\hbar^2}{2m}\right)\left(-\frac{17}{35}\right)\left(\frac{630}{31}\right) = \frac{153}{31}\frac{\hbar^2}{m} = 4.93548\ldots\frac{\hbar^2}{m}. \tag{10.94}$$

This differs from the exact result by about one part in 10,000, which is an order of magnitude more accurate than what we predicted from the variational method!

10.4 The WKB Approximation

The translation operator $\hat{U}_p(x_0,x)$ that we introduced in Sec. 7.3.1 allows us to determine the spatial dependence of a wavefunction from knowledge of the potential $V(x)$ and its value at one point. Note that a wavefunction $\psi(x)$ can be expressed as

$$\psi(x) = \hat{U}_p(x_0,x)\,\psi(x_0) \tag{10.95}$$

$$= \left(1 + \frac{i}{\hbar}\int_{x_0}^{x} dx'\,\sqrt{2m(E-V(x'))} + \cdots\right)\psi(x_0),$$

where we have just explicitly written out the first two terms in the expression for $\hat{U}_p(x_0,x)$. As we emphasized in that chapter, due to spatial dependence of the potential, momentum is in general not conserved and so one needs to account for all terms in the expansion to actually render this an equality. However, assuming that the potential doesn't vary in space too rapidly, momentum is approximately conserved, and the expression for $\hat{U}_p(x_0,x)$ takes the form of an exponential:

$$\psi(x) \approx \exp\left[\frac{i}{\hbar}\int_{x_0}^{x} dx'\,\sqrt{2m(E-V(x'))}\right]\psi(x_0). \tag{10.96}$$

In this form, we are assuming that $x > x_0$, so momentum is positive, or to the right. In general, we would need a linear combination of right- and left-directed momentum, as we used in studying scattering in Chap. 7. This approximation of the wavefunction is closely related to the **WKB approximation**.[4] In this section, we'll study how this approximation of the wavefunction can be used to estimate bound-state eigenenergies.

In our studies of the wavefunctions for energy eigenstates, we noted that the ground-state wavefunction had no nodes (points of zero value) in the bulk of the well, the first excited state has one node, the second excited state has two nodes, etc. Explicitly, we had noted in Sec. 5.1.1 that the nth-energy eigenstate of the infinite square well corresponds to fitting $n/2$ wavelengths within the size of the well, according to the form of the energy eigenstate in position space:

$$\psi_n(x) = \sqrt{\frac{2}{a}}\sin\frac{n\pi x}{a}. \tag{10.97}$$

The wavefunction must vanish at the boundaries of the well for momentum to be Hermitian, and n half-wavelengths have $n+1$ nodes. So, the nth-energy eigenstate has $n-1$ nodes in the bulk of the well, away from the boundaries. Further, $\sin\frac{n\pi x}{a}$ is just a linear combination of two complex exponentials of the form presented in Eq. (10.96).

These observations suggest a quantization technique. "In the well" means that the particle's position is between the classical turning points; the points of maximum and

---

[4] Also called the JWKB approximation. H. Jeffreys, "On certain approximate solutions of linear differential equations of the second order," *Proc. Lond. Math. Soc.* **2**(1), 428–436 (1925); G. Wentzel, "Eine Verallgemeinerung der Quantenbedingungen für die Zwecke der Wellenmechanik," *Z. Phys.* **38**(6), 518–529 (1926); H. A. Kramers, "Wellenmechanik und halbzahlige Quantisierung," *Z. Phys.* **39**(10), 828–840 (1926); L. Brillouin, "La mécanique ondulatoire de Schrödinger; une méthode générale de resolution par approximations successives," *Compt. Rend. Hebd. Seances Acad. Sci.* **183**(1), 24–26 (1926).

minimum displacement allowed classically. For a ball placed in a potential at an initial point at rest, the classical turning points correspond to the points in the well where the ball assumes zero kinetic energy, as illustrated in Fig. 10.3. Having no nodes in this region means that the total phase acquired by the wavefunction between the classical turning points is π, exactly how $\sin\phi$ has no nodes for $0 < \phi < \pi$. Continuing, having one node in the well means that the wavefunction acquires a total phase of 2π between the classical turning points, just as $\sin\phi$ has one node for $0 < \phi < 2\pi$. One can continue this for higher-energy eigenstates and motivate the following quantization condition:

$$\frac{1}{\hbar} \int_{x_{\min}}^{x_{\max}} dx' \, \sqrt{2m\left(E - V(x')\right)} = n\pi \,, \tag{10.98}$$

where $x_{\min}, x_{\max}$ are the left and right classical turning points for a particle with total energy E in a potential $V(x)$, and $n = 1, 2, 3, \ldots$. This is called the **Bohr–Sommerfeld quantization** condition.[5]

While it was the way in which we motivated the Bohr–Sommerfeld quantization condition, we can verify explicitly that Eq. (10.98) reproduces the energy eigenvalues of the infinite square well. Within the well, the potential is 0, $V(x) = 0$, and the classical turning points are the boundaries of the well. For a well of width a, the classical turning points are then $x_{\min} = 0$, $x_{\max} = a$. So, the quantization condition for the infinite square well reads

$$\frac{1}{\hbar} \int_{0}^{a} dx' \, \sqrt{2mE_n} = \sqrt{\frac{2ma^2 E_n}{\hbar^2}} = n\pi \,, \tag{10.99}$$

where E_n is the nth-energy eigenstate. Thus, solving for the energy E_n we find

$$E_n = \frac{n^2 \pi^2 \hbar^2}{2ma^2} \,, \tag{10.100}$$

exactly the energy levels that we know and love.

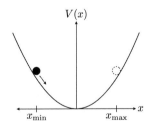

Fig. 10.3 Illustration of the classical turning points of a ball initially placed at rest at a position $x_{\min}$ on a potential $V(x)$. The ball rolls down the potential and stops briefly at a position $x_{\max}$ when it reaches the initial height, by conservation of energy. It then rolls back to the point $x_{\min}$, repeating this oscillation *ad infinitum*.

5 N. Bohr, "On the constitution of atoms and molecules," *Phil. Mag. Ser. 6* **26**, 1–24 (1913); N. Bohr, "On the constitution of atoms and molecules. 2. Systems containing only a single nucleus," *Phil. Mag. Ser. 6* **26**, 476 (1913); A. Sommerfeld, *Atombau und Spektrallinien*, Vieweg (1919).

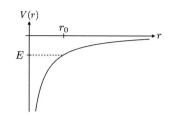

Fig. 10.4 Plot of the potential energy for an electron in hydrogen with zero angular momentum. A fixed energy $E < 0$ of the electron has a point of maximum displacement from the proton, denoted by r_0.

Example 10.4 The WKB approximation and the Bohr–Sommerfeld quantization condition can be used to calculate the energy levels of the hydrogen atom, as well, and was actually the system in which it was initially applied. In this example, we will determine the energy levels of hydrogen using Eq. (10.98).

Solution

The first thing we need to do is to establish the classical turning points. For concreteness, we assume that the electron has zero angular momentum, and its potential energy is plotted in Fig. 10.4. For a fixed energy $E < 0$, there is a maximum radius that the electron can assume, at which point it has zero kinetic energy, as illustrated. If we call this radius r_0, then the energy is

$$E = -\frac{e^2}{4\pi\epsilon_0}\frac{1}{r_0},\qquad(10.101)$$

just the electric potential energy at radius r_0. The electron is classically allowed to be at any radius less than r_0, so the Bohr–Sommerfeld quantization condition of the hydrogen atom becomes

$$\frac{1}{\hbar}\int_0^{r_0} dr\,\sqrt{2m_e(E-V(r))} = \sqrt{\frac{m_e e^2}{2\pi\epsilon_0\hbar^2}}\int_0^{r_0} dr\,\sqrt{\frac{1}{r}-\frac{1}{r_0}} = n\pi.\qquad(10.102)$$

To evaluate the integral that remains, we can make the change of variables

$$r = r_0\sin^2\phi,\qquad(10.103)$$

and we find

$$\int_0^{r_0} dr\,\sqrt{\frac{1}{r}-\frac{1}{r_0}} = \int_0^{\pi/2} d\phi\,2r_0\cos\phi\,\sin\phi\,\sqrt{\frac{1}{r_0\sin^2\phi}-\frac{1}{r_0}}\qquad(10.104)$$

$$= 2\sqrt{r_0}\int_0^{\pi/2} d\phi\,\cos^2\phi = \frac{\pi}{2}\sqrt{r_0}.$$

The Bohr–Sommerfeld condition is then

$$\sqrt{\frac{m_e e^2}{2\pi\epsilon_0\hbar^2}}\int_0^{r_0} dr\,\sqrt{\frac{1}{r}-\frac{1}{r_0}} = \pi\sqrt{\frac{m_e e^2 r_0}{8\pi\epsilon_0\hbar^2}} = n\pi.\qquad(10.105)$$

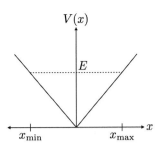

Fig. 10.5   Plot of the potential $V(x) = k|x|$, with a total energy E and the corresponding turning points illustrated.

The turning point radius is therefore

$$r_0 = \frac{8\pi\epsilon_0\hbar^2}{m_e e^2}n^2 = 2a_0 n^2, \tag{10.106}$$

where a_0 is the Bohr radius. The energy that corresponds to this turning point is

$$E = -\frac{e^2}{4\pi\epsilon_0}\frac{1}{r_0} = -\frac{m_e e^4}{2(4\pi\epsilon_0)^2\hbar^2}\frac{1}{n^2}, \tag{10.107}$$

exactly the energy eigenvalues we had calculated in Chap. 9.

Example 10.5   While the Bohr–Sommerfeld condition sometimes gets the energy eigenvalues exactly correct, it can also be used for systems where the exact solution is not known. In this example, we will estimate the energy eigenvalues of the potential

$$V(x) = k|x|, \tag{10.108}$$

where k is a constant with units of energy/distance. This potential is illustrated in Fig. 10.5.

Solution

The first thing we need to do to use the Bohr–Sommerfeld quantization condition is to determine the relationship between the total energy E and the classical turning points. Let's call the upper turning point $x_{\max} > 0$, then the total energy is

$$E = kx_{\max}. \tag{10.109}$$

Because the potential is a symmetric function of x, $V(x) = V(-x)$, the lower turning point $x_{\min}$ is just the opposite position, $x_{\min} = -x_{\max}$. This is also illustrated in Fig. 10.5. The quantization condition is then

$$n\pi = \frac{1}{\hbar}\int_{x_{\min}}^{x_{\max}} dx \sqrt{2m(E - V(x))} = \frac{2\sqrt{2mk}}{\hbar}\int_0^{x_{\max}} dx \sqrt{x_{\max} - x} \tag{10.110}$$

$$= \frac{4\sqrt{2mk}}{3\hbar}x_{\max}^{3/2}.$$

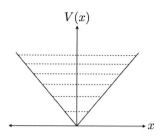

$V(x)$

x

Fig. 10.6   Illustration of the spacing of the first few energy eigenvalues in the linear potential, denoted by the dashed horizontal lines.

Now, inserting the expression for E in terms of $x_{\max}$ from Eq. (10.109), we have the quantization condition

$$n\pi = \frac{4\sqrt{2mk}}{3\hbar}\left(\frac{E}{k}\right)^{3/2}, \tag{10.111}$$

or

$$E = \frac{1}{2}\left(\frac{3\pi k\hbar n}{2\sqrt{m}}\right)^{2/3}. \tag{10.112}$$

So, the energy levels for this linear potential scale like $n^{2/3}$. The first six energy levels are illustrated as dashed lines in this linear potential in Fig. 10.6. This is interesting because the harmonic oscillator, or quadratic potential, has energy levels linear in n. You will study general power-law potentials more in Exercise 10.7, where you will see that the Bohr–Sommerfeld quantization will need to be corrected, and connect to a limit in which the infinite square well emerges.

Exercises

10.1   It's useful to see how our quantum perturbation theory works in a case that we can solve exactly. Let's consider a two-state system in which the Hamiltonian is

$$\hat{H}_0 = \begin{pmatrix} E_0 & 0 \\ 0 & E_1 \end{pmatrix}, \tag{10.113}$$

for energies $E_0 \leq E_1$. Now, let's consider adding a small perturbation to this Hamiltonian:

$$\hat{H}' = \begin{pmatrix} \epsilon & \epsilon \\ \epsilon & \epsilon \end{pmatrix}, \tag{10.114}$$

for some $\epsilon > 0$. The complete Hamiltonian of our system we are considering is $\hat{H} = \hat{H}_0 + \hat{H}'$.

(a) First, calculate the exact energy eigenstates and eigenvalues of the complete Hamiltonian. From your complete result, Taylor expand it to first order in ϵ.

(b) Now, use our formulation of quantum perturbation theory to calculate the first corrections to the energy eigenstates and eigenvalues. Do these corrections agree with the explicit expansion from part (a)?

(c) From the Taylor expansion in part (a), for what range of ϵ does the Taylor expansion converge?

10.2 Let's see how the variational method works in another application. Let's assume we didn't know the ground-state energy of the quantum harmonic oscillator and use the variational method to determine it. First, the Hamiltonian is

$$\hat{H} = \frac{\hat{p}^2}{2m} + \frac{m\omega^2}{2}\hat{x}^2 . \tag{10.115}$$

Note that this potential is symmetric about $x = 0$ and the position extends over the entire real line: $x \in (-\infty, \infty)$. This motivates the guess for the ground-state wavefunction

$$\psi(x; a) = N(a^2 - x^2) , \tag{10.116}$$

for a normalization constant N and a is the parameter that we will minimize over. Now, this wavefunction is only defined on $x \in [-a, a]$.

(a) First calculate the normalization factor N, as a function of a.

(b) Now, calculate the expectation value of the Hamiltonian on this state, $\langle \psi | \hat{H} | \psi \rangle$, as a function of the parameter a.

(c) Finally, minimize the expectation value over a to provide an upper bound on the ground-state energy, E_0. How does this estimate compare to the known, exact value?

10.3 While we introduced the variational method and the power method both as a way to approximate the ground state of some system, they both can be used to approximate excited states as well, with appropriate modification. For concreteness, in this problem we will attempt to use the variational method to identify the first excited state of the infinite square well, extending what we did in Example 10.2. As in that example, we will take the width of the well to be $a = 1$.

(a) A fundamental property of the first excited state ($n = 2$) with respect to the ground state ($n = 1$) of the infinite square well is that they are orthogonal. So, our ansatz for the first excited state should be orthogonal to our ansatz of the ground state, Eq. (10.49), over the well. With this in mind, let's make the ansatz that the first excited-state wavefunction is

$$\psi_2(x; \beta) = N x^\beta (1 - x)^\beta (1 - 2x) , \tag{10.117}$$

for some parameter β and normalization constant N. Show that this is indeed orthogonal to the ansatz we used for the ground state in Eq. (10.49).

(b) Determine the normalization constant N such that $\psi_2(x;\beta)$ is normalized over the well. You will likely need to use the definition of the Beta function from Eq. (10.51). Also, the Gamma function satisfies the identities

$$\Gamma(x+1) = x\Gamma(x), \qquad\qquad \Gamma(1) = 1. \qquad (10.118)$$

(c) Now, determine the expectation value of the infinite square well's Hamiltonian on this ansatz wavefunction, as a function of the parameter β. You should find

$$\langle \psi_2 | \hat{H} | \psi_2 \rangle = \frac{\hbar^2}{2m} \frac{6\beta(3+4\beta)}{2\beta - 1}. \qquad (10.119)$$

(d) Now, minimize over β to establish an estimate for the energy of the first excited state of the infinite square well. How does this compare to the exact result?

(e) Can you extend this procedure to the second excited state? What do you find in that case for an estimate of its energy?

10.4 Let's now study the power method for estimating the ground-state energy, applied to the quantum harmonic oscillator. For this problem, we will work with the Hamiltonian

$$\hat{H} = \hbar\omega(1 + \hat{a}^\dagger \hat{a}), \qquad (10.120)$$

where $\hat{a}$ and $\hat{a}^\dagger$ are the familiar lowering and raising operators, respectively. Note the slight difference between this Hamiltonian and what we would typically think of as the harmonic oscillator: all we've done here is shift the potential up by an amount $\hbar\omega/2$, which will make some of the mathematical manipulations simpler later. Importantly, this constant shift does not affect the eigenstates of the Hamiltonian, it just shifts the eigenvalues up by that same amount.

(a) Let's first calculate the inverse of this Hamiltonian, $\hat{H}^{-1}$. One answer is, of course, simply

$$\hat{H}^{-1} = \frac{1}{\hbar\omega} \frac{1}{1 + \hat{a}^\dagger \hat{a}}. \qquad (10.121)$$

However, this isn't so useful for determining how this operator acts on states. Instead, we can express the inverse Hamiltonian as a sum over products of $\hat{a}$ and $\hat{a}^\dagger$:

$$\hat{H}^{-1} = \frac{1}{\hbar\omega} \sum_{n=0}^{\infty} \beta_n (\hat{a}^\dagger)^n \hat{a}^n, \qquad (10.122)$$

for some coefficients β_n. Determine the coefficients β_n and thus the inverse Hamiltonian $\hat{H}^{-1}$ by demanding that $\hat{H}\hat{H}^{-1} = \mathbb{I}$, the identity operator. *Nota Bene*: This way of ordering the terms in $\hat{H}^{-1}$ is called **normal order**: all raising operators $\hat{a}^\dagger$ are to the left of all lowering operators $\hat{a}$ in each term in the sum.

(b) Let's use this inverse Hamiltonian to estimate the ground-state energy of the harmonic oscillator in which we start with a coherent state $|\chi\rangle$. Recall that $|\chi\rangle$ satisfies

$$\hat{a}|\chi\rangle = \lambda|\chi\rangle, \tag{10.123}$$

for some complex number λ. From this coherent state, what is the expectation value of the Hamiltonian, $\langle\chi|\hat{H}|\chi\rangle$?

(c) Now, let's use the inverse Hamiltonian to improve our estimate of the ground-state energy. What is your new estimate of the ground state after one application of $\hat{H}^{-1}$ on the coherent state? Recall that this estimate is

$$E_0 \leq \frac{\langle\chi|\hat{H}^{-1}\hat{H}\hat{H}^{-1}|\chi\rangle}{\langle\chi|\hat{H}^{-1}\hat{H}^{-1}|\chi\rangle} = \frac{\langle\chi|\hat{H}^{-1}|\chi\rangle}{\langle\chi|\hat{H}^{-1}\hat{H}^{-1}|\chi\rangle}. \tag{10.124}$$

You should ultimately find

$$E_0 \leq \frac{\langle\chi|\hat{H}^{-1}|\chi\rangle}{\langle\chi|\hat{H}^{-1}\hat{H}^{-1}|\chi\rangle} = \hbar\omega \frac{e^{|\lambda|^2} - 1}{\int_0^{|\lambda|^2} dx \frac{e^x - 1}{x}}. \tag{10.125}$$

The integral that remains can't be expressed in terms of elementary functions.

Hint: For the denominator factor, it might help to slide the identity $\mathbb{I}$ in between the two inverse Hamiltonians, where

$$\mathbb{I} = \sum_{n=0}^{\infty} |\psi_n\rangle\langle\psi_n|, \tag{10.126}$$

where $|\psi_n\rangle$ is the nth energy eigenstate.

(d) Now, just to get a sense of this approximation, verify that you get the exact result for the ground-state energy if $\lambda \to 0$. Also, evaluate the approximation of part (c) for $\lambda = 1$. The remaining integral evaluates to

$$\int_0^1 dx \frac{e^x - 1}{x} = 1.31790215\ldots. \tag{10.127}$$

How does this result compare to the initial estimate of part (b)?

10.5 The anharmonic oscillator is the quantum system that is a modification to the harmonic oscillator, including a term quartic in position:

$$\hat{H} = \frac{\hat{p}^2}{2m} + \frac{m\omega^2}{2}\hat{x}^2 + \lambda\hat{x}^4, \tag{10.128}$$

where $\lambda > 0$ is a parameter that controls the strength of the quartic term. A plot of the anharmonic oscillator potential is shown in Fig. 10.7. In this problem, we will use perturbation theory for estimating the effect of this quartic term on the energy eigenvalues of the oscillator.

(a) Determine the first-order in λ correction to the energy eigenvalue of the nth-energy eigenstate of the harmonic oscillator using quantum mechanics perturbation theory.

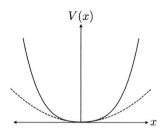

$$V(x)$$

$$x$$

Fig. 10.7 An illustration of the anharmonic oscillator potential (solid), compared to the harmonic oscillator potential (dashed).

Hint: It might help to express the quartic factor with the raising and lowering operators $\hat{a}^\dagger$ and $\hat{a}$.

(b) Now, let's briefly consider what happens if $\lambda < 0$. For this case, is there a first-energy eigenstate labeled by n for which its correction due to the quartic term becomes ill-defined or non-sensible?

(c) Back with $\lambda > 0$, determine the first correction in λ to the ground-state wavefunction of the harmonic oscillator.

(d) Calculate the variance σ_x^2 of the ground state of the anharmonic oscillator using the perturbative result of part (c). Does the quartic term in the potential tend to increase or decrease the width of the ground-state position space wavefunction?

10.6 We had introduced quantum mechanical perturbation theory as analogous to the Taylor expansion. As such, one might expect that as higher and higher orders are calculated, a better approximation of the true result is achieved. However, in general, quantum mechanical perturbation theory produces a non-convergent, asymptotic series in the expansion parameter. For the anharmonic oscillator Hamiltonian of Eq. (10.128), for example, an arbitrarily small, but negative, value for λ destabilizes the potential. Regardless of the size of λ, if it is negative then at large positions $|x| \to \infty$ the potential runs to negative infinity. This destabilization with $\lambda < 0$ is illustrated in Fig. 10.8. This suggests that the perturbation theory in λ has zero radius of convergence about $\lambda = 0$. This divergence of the perturbation theory of the anharmonic oscillator was first studied in detail by Carl Bender and Tai Tsun Wu.[6]

Recently, techniques identified by Bender and Wu for studying the anharmonic oscillator have been packaged into a plug-in for Mathematica and can be used to extend the perturbation theory for the anharmonic oscillator and other potentials to extremely high orders.[7] This program can be downloaded from `https://arxiv.org/abs/1608.08256` and the associated article contains several examples. Try this program out with the anharmonic oscillator and calculate very high orders ($\sim$100) in perturbation theory. What do you notice

[6] C. M. Bender and T. T. Wu, "Anharmonic oscillator," *Phys. Rev.* **184**, 1231–1260 (1969).

[7] T. Sulejmanpasic and M. Ünsal, "Aspects of perturbation theory in quantum mechanics: The BenderWu Mathematica ® package," *Comput. Phys. Commun.* **228**, 273–289 (2018) [arXiv:1608.08256 [hep-th]].

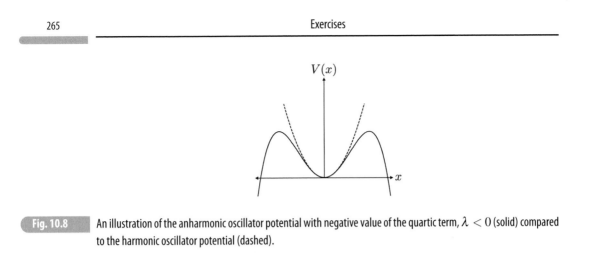

Fig. 10.8 An illustration of the anharmonic oscillator potential with negative value of the quartic term, $\lambda < 0$ (solid) compared to the harmonic oscillator potential (dashed).

about the approximations of, say, the ground-state energy and wavefunction as the order in perturbation theory increases? Think up your own potential and estimate its energy levels with the program.

10.7 The WKB approximation and the Bohr–Sommerfeld quantization condition worked perfectly for calculating the energy eigenvalues of the infinite square well and the hydrogen atom. Does it work for the other system we solved exactly, the quantum harmonic oscillator?

(a) Apply the Bohr–Sommerfeld quantization condition to the quantum harmonic oscillator. What does it predict for the energy eigenvalues? Is this correct?

(b) A difference between the infinite square well and the harmonic oscillator is that a particle can tunnel a short distance outside of the classically allowed region of the harmonic oscillator, while it cannot in the infinite square well. In the classically allowed region, the momentum is real-valued; that is, $E - V(x) > 0$. On the other hand, outside the harmonic oscillator well, the momentum becomes imaginary, because $E - V(x) < 0$ and we take its square root. Thus, when thinking about momentum as a general complex number, its phase or argument changes by $\pi/2$ in going from real to imaginary valued, exactly how $e^0 = 1$ is real while $e^{i\pi/2} = i$ is imaginary. This phase change at the boundary, beyond the phase of the wavefunction accumulated over the classically allowed region, suggests that a more general quantization condition is

$$\frac{1}{\hbar} \int_{x_{\min}}^{x_{\max}} dx' \sqrt{2m\left(E - V(x')\right)} = \left(n + \frac{1}{2}\right)\pi. \qquad (10.129)$$

This additional $\pi/2$ phase is called the **Maslov correction**.[8] Apply this quantization condition to the harmonic oscillator. Are the energy eigenvalues correctly predicted now?

---

[8] V. P. Maslov, *Theory of Perturbations and Asymptotic Methods* (in Russian), Izv. MGU Moscow (1965). [Translation into French, 1972.]

(c)   In principle, we should have included a Maslov correction to the calcula-
       tion of the hydrogen atom energy eigenvalues because the electron could
       have tunneled a small distance beyond the upper turning point at r_0, from
       Example 10.4. Why was no Maslov correction needed there?

(d)   Now, use the Bohr–Sommerfeld quantization condition with the Maslov
       correction to estimate the energy eigenvalues for a potential of the form
       $V(x) = k|x|^\alpha$, for $\alpha > 0$ and k some constant with units of energy/distance$^\alpha$.
       What do you find in the limit that $\alpha \to \infty$?

For the next problems, we'll use the power method to study the hydrogen atom
and extensions. Unlike the power method introduced in this chapter, there's no
need to invert the hydrogen atom's Hamiltonian to use the power method to
focus on the ground state. The ground-state energy of hydrogen has the largest-
in-magnitude value of any bound-energy eigenstate of hydrogen, so we can
simply apply increasing powers of the Hamiltonian itself on a state to return
a better estimate of the ground state.

10.8   We can quantitatively study this claim, that for the hydrogen atom, we do not
       need to invert the Hamiltonian to use the power method to estimate its ground-
       state energy. In this exercise, we'll just consider the power method applied to
       spherically symmetric energy eigenstates of hydrogen; that is, those states that
       can be expressed as $|n,0,0\rangle \equiv |n\rangle$ in the n,l,m quantum numbers, where $n \geq 1$.
       As established in Chap. 9, the state for which $n = 1$ is the ground state and the
       energy eigenvalues are

       $$\hat{H}|n\rangle = -\frac{m_e e^4}{2(4\pi\epsilon_0)^2 \hbar^2} \frac{1}{n^2}|n\rangle. \tag{10.130}$$

       (a)   Consider the generic, spherically symmetric state $|\chi\rangle$ of the hydrogen atom,
             which can be expressed as

             $$|\chi\rangle = \sum_{n=1}^{\infty} \beta_n |n\rangle, \tag{10.131}$$

             for some coefficients β_n. Estimate the ground-state energy of hydrogen after
             N applications of the power method applied to this state.

       (b)   For concreteness, let's take the coefficients in the expansion of the state $|\chi\rangle$
             to be

             $$\beta_n \propto \frac{1}{n}. \tag{10.132}$$

             What is the expectation value of the Hamiltonian on this state? Is it greater
             than the known, exact value of the ground-state energy?

       (c)   Determine the estimate of the ground-state energy of hydrogen after N
             applications of the power method on the state defined in part (b). Express
             the result in terms of the Riemann zeta function, which is defined to be

             $$\zeta(s) = \sum_{n=1}^{\infty} \frac{1}{n^s}, \tag{10.133}$$

             for $s > 1$.

(d)   Show that this estimate converges to the true ground-state energy as $N \to \infty$. Plot the value of the Nth application power method estimate as a function of N. For what N does the power method produce a result within 1% of the true ground-state energy?

10.9   In our introduction of the hydrogen atom, we assumed that the electron orbited the proton with a speed significantly smaller than the speed of light, c. However, we can include the effect of a finite speed of light into the description of hydrogen and see what its effect on the ground-state energy is.

(a)   The relativistic, but classical, kinetic energy of the electron can be expressed as

$$K = \sqrt{m_e^2 c^4 + |\vec{p}|^2 c^2} - m_e c^2, \qquad (10.134)$$

where m_e is the mass of the electron and $\vec{p}$ is the momentum of the electron. Taylor expand this expression for the relativistic kinetic energy to quartic order in the electron's momentum; that is, up to and including terms that go like $|\vec{p}|^4$. For this Taylor-expanded kinetic energy, promote it to a Hermitian operator $\hat{K}$ on Hilbert space, and express it in position space.

(b)   With this $\hat{K}$ operator, construct the relativistically corrected Hamiltonian for the ground state of the hydrogen atom. Modify the Hamiltonian for hydrogen from Sec. 9.2, and note that it can be expressed as

$$\hat{H} = \hat{H}_0 + \hat{H}_{\text{rel}}, \qquad (10.135)$$

where $\hat{H}_0$ is the unperturbed Hamiltonian and $\hat{H}_{\text{rel}}$ is the relativistic correction. From the ground state of the unperturbed Hamiltonian $|\psi\rangle$, use the power method to estimate the relativistically corrected ground-state energy, $E_{0,\text{rel}}$, where

$$E_{0,\text{rel}} \simeq \frac{\langle \psi | \hat{H}\hat{H}\hat{H} | \psi \rangle}{\langle \psi | \hat{H}\hat{H} | \psi \rangle}. \qquad (10.136)$$

Hint: Remember that $\hat{H}_0 |\psi\rangle = E_0 |\psi\rangle$, where E_0 is the ground-state energy of non-relativistic hydrogen, Eq. (9.34).

(c)   Now, use quantum mechanical perturbation theory to first order to estimate the ground-state energy of relativistic hydrogen. How does the perturbation theory result compare to that of the power method?

11

The Path Integral

In this chapter, we will introduce the **path integral**, which is a formulation of quantum mechanics equivalent to the Schrödinger equation, but has a profoundly distinct interpretation. Further, the path integral is very easily extended to incorporate special relativity, which is very challenging and inconvenient within the context of the Schrödinger equation. So, what is the idea of this path integral? Our goal will be to calculate the amplitude for a quantum mechanical particle that starts at position x_i at time $t = 0$ and ends at position x_f at some later time $t = T > 0$. In some sense, this question is analogous to what you ask in an introduction to kinematics in introductory physics; however, its analysis in quantum mechanics will prove to be a bit more complicated than that in the first week of your first course in physics.

11.1 Motivation and Interpretation of the Path Integral

Let's draw a picture of what we're working with, as shown in Fig. 11.1. In this picture, we have drawn time on the abscissa and space on the ordinate axes. I have placed the initial and final positions, x_i and x_f, appropriately. We want to find the probability amplitude for ending up at x_f at $t = T$ given that we start at x_i at $t = 0$. So, here's what we will do. We can just write down the answer first, then we will work to unpack it. Let's express the state of the particle at $t = 0$ as $|x_i, t = 0\rangle$, which is an eigenstate of the position operator:

$$\hat{x}|x_i, t = 0\rangle = x_i|x_i, t = 0\rangle, \tag{11.1}$$

Fig. 11.1 Space-time diagram of a particle's initial position x_i at $t = 0$ and final position x_f at $t = T$.

with eigenvalue of that initial position. Next we can write down the final state of the particle at $t = T$ as $|x_f, t = T\rangle$, which is also an eigenstate of the position operator:

$$\hat{x}|x_f, t = T\rangle = x_f |x_f, t = T\rangle, \qquad (11.2)$$

but evaluated at $t = T$.

Then, the probability amplitude for a particle to start at x_i at $t = 0$ and end up at x_f at $t = T$ is the inner product of these initial and final position eigenstates. However, we have to be somewhat careful in comparing them, because they are evaluated at different times. Indeed, if we thought of these position eigenstates as living in the same position basis of the Hilbert space, then their inner product would be a δ-function:

$$\langle x_f | x_i \rangle = \delta(x_i - x_f). \qquad (11.3)$$

While this result can be interpreted as orthogonality of position eigenstates, another interpretation is that a particle cannot travel at infinite speed. The states $|x_i\rangle$ and $|x_f\rangle$ can only be compared in this way if they are evaluated at the same time, and if no time has elapsed in going from x_i to x_f, then those positions had better be equal so that the particle's speed is always finite. If the positions are evaluated at different times, as is the case with the states $|x_i, t = 0\rangle$ and $|x_f, t = T\rangle$, then we have to time-evolve the initial point to the final point, corresponding to allowing the particle to actually move in space. We know how to move from $t = 0$ to $t = T$: we translate in time with the unitary time translation operator, the exponentiated Hamiltonian. Then, the transition amplitude from position x_i at $t = 0$ to x_f at $t = T$ is

$$\langle x_f, t = T \mid x_i, t = 0 \rangle \equiv \langle x_f | e^{-i\frac{T\hat{H}}{\hbar}} | x_i \rangle, \qquad (11.4)$$

where in the final line we suppress all time dependence of the states because it is explicitly included in the exponentiated Hamiltonian operator. So, this is one answer: we just act the Hamiltonian on our position eigenstates and evaluate the inner product.

However, let's see if we can express it in another, ultimately much more useful and powerful, way. The idea will be the following. Going back to our picture of the initial and final states, we will divide the total time T into N steps with small time intervals of

$$\Delta t = \frac{T}{N}, \qquad (11.5)$$

as shown in Fig. 11.2. We will denote the corresponding nth time step as $t_n = n\Delta t$, so that $t = 0$ is $n = 0$ and $t = T$ is the Nth step. Further, at each of these time steps, the particle can have any position $x_n \in (-\infty, \infty)$ at the nth time step. I have drawn the first four possible positions x_1, x_2, x_3, x_4 at times t_1, t_2, t_3, t_4 in the figure for illustration.

Now, I want to emphasize something: given this set-up, all we require of the particle is that $x_i = x_0$ at $t = 0$ and $x_f = x_N$ at $t = T$. That's it. We only measure its initial and final positions and perform no measurements in the range $0 < t < T$, exclusive of its endpoints. Because time evolution is a probabilistic process in quantum mechanics, anything that does not expressly have zero probability of occurring in that intermediate time will happen, according to its probability. Also, because we make no measurements

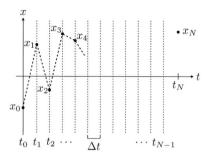

  Depiction of one possible path of positions from initial position x_0 to final position x_N on a temporal grid.

in the intermediate times, we have to sum together the probability amplitudes for any-thing in the intermediate times to occur. This is familiar from expanding a state in a complete sum of energy eigenstates, given an initial state at $t = 0$. We write the state at time t as

$$|\psi(t)\rangle = \sum_{n=0}^{\infty} \beta_n e^{-i\frac{E_n t}{\hbar}} |\psi_n\rangle, \tag{11.6}$$

where $|\psi_n\rangle$ are energy eigenstates and the β_n are properly the probability amplitudes for the wavefunction to be in any energy eigenstate at $t = 0$. The exponential phase factors of energy just control how the probability amplitudes evolve in time.

Now, with these observations, anything with non-zero probability can occur and that we make no measurements in the intermediate time means that to calculate the transition amplitude $\langle x_f, t = T | x_i, t = 0 \rangle$, we have to sum over all possible (i.e., non-zero probability) trajectories that the particle can take between these points weighted by their probability amplitudes. For example, let's draw just three possible trajectories in Fig. 11.3. Path I ($>$) is just a straight line and what we would expect for a classical particle experiencing no external forces. That would just be the answer classically, but there are other paths that have non-zero probability. Path II ($\gg$) travels from x_i to x_f in a curved trajectory in this (t, x) plane. In space its trajectory would actually consist of several loops; can you tell how many? Finally, example path III ($\ggg$) is really wild, and the particle goes all the way out to the Andromeda galaxy before coming back to point x_f. Again, unless something expressly forbids a trajectory from happening, it has non-zero probability and must be included in the sum.

This construction of "sum over paths" or "sum over histories" will produce the path integral, another way to express the dynamics of a quantum system distinct from the Schrödinger equation, yet completely equivalent as one can derive the path integral from the Schrödinger equation and vice-versa. The path integral was introduced by Richard Feynman in the late 1940s[1] and Feynman actually credits Paul Dirac with the idea, from an observation Dirac made in a paper from the 1930s.[2] This formulation is also similar to Huygen's principle that describes the propagation of coherent wave

---

[1] R. P. Feynman, "Space-time approach to nonrelativistic quantum mechanics," *Rev. Mod. Phys.* **20**, 367–387 (1948).
[2] P. A. M. Dirac, "The Lagrangian in quantum mechanics," *Phys. Z. Sowjetunion* **3**, 64–72 (1933).

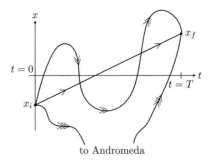

Fig. 11.3   Three example trajectories from initial point x_i to final point x_f.

fronts in optics, water, or other media. Huygen's principle states that from any point, a wave propagates outward as a spherical front and the observed wave is the linear combination, that is sum, of all the spherical fronts from all relevant points.

Before deriving the path integral, I want to make one more comment about the path of a quantum mechanical particle. These example paths I drew were both continuous and smooth (having continuous derivatives). The path of a quantum mechanical particle had better be continuous, otherwise a particle could travel from anywhere to anywhere instantaneously (i.e., teleport).[3] Another way to express continuity is that if you made measurements of the position of a particle in rapid succession, the particle had better not move far from one measurement to the next. However, there is no such logical requirement of smoothness; that is purely aesthetics. In the space of continuous functions, those functions that are also smooth are measure-zero: if you were just drawing continuous functions out of a hat you would never pull out a smooth function. That is to say, the trajectory of a quantum mechanical particle is continuous but not smooth. Perhaps a familiar example of a continuous, non-smooth function is the Koch snowflake (see Exercise 7.6 of Chap. 7) or the value of the stock market as a function of time.

11.2 Derivation of the Path Integral

With these preliminary comments out of the way, let's derive the path integral.[4] In the previous section, we had divided up the time from $t = 0$ to $t = T$ into N steps. At each of those time steps, we can inset a complete set of position eigenstates. For instance, at time $t_n = n\Delta t$, we have the completeness relation

---

[3] Actually, it has been explicitly proven that some non-relativistic quantum mechanical systems have a finite upper limit to the wavefunction's group velocity. This is known as the Lieb–Robinson bound and is quite surprising because in non-relativistic mechanics, either classical or quantum, the speed of light makes no appearance. See E. H. Lieb and D. W. Robinson, "The finite group velocity of quantum spin systems," *Commun. Math. Phys.* **28**, 251–257 (1972).

[4] This derivation follows closely Appendix A of J. Polchinski, *String Theory*, vol. 1 Cambridge University Press (1998).

$$\mathbb{I} = \int dx_n \, |x_n, t_n\rangle \langle x_n, t_n| \,, \tag{11.7}$$

where the integral over positions x_n represents the continuous sum over eigenstates. The integral extends over all positions $x_n \in (-\infty, \infty)$, but we suppress the bounds of integration for compactness. Inserting the identity operator at every time in our transition amplitude, we have

$$\langle x_f, t = T | x_i, t = 0 \rangle \equiv \langle x_N, t_N | x_0, t_0 \rangle \tag{11.8}$$

$$= \int dx_1 \, dx_2 \cdots dx_{N-1} \, \langle x_N, t_N | x_{N-1}, t_{N-1} \rangle \cdots$$

$$\times \langle x_2, t_2 | x_1, t_1 \rangle \langle x_1, t_1 | x_0, t_0 \rangle$$

$$= \int dx_1 \, dx_2 \cdots dx_{N-1} \prod_{i=0}^{N-1} \langle x_{i+1}, t_{i+1} | x_i, t_i \rangle \,.$$

Note that the inner products in the integrand are still evaluated between states separated in time by Δt. We can evaluate them at the same time by introduction of the time-translation operator

$$\hat{U}_H(\Delta t) = e^{-i\frac{\Delta t \hat{H}}{\hbar}} \,. \tag{11.9}$$

That is, for the inner product between states at time t_n and $t_n + \Delta t$, we can write

$$\langle x_{n+1}, t_{n+1} | x_n, t_n \rangle = \langle x_{n+1}, t_{n+1} | e^{-i\frac{\Delta t \hat{H}}{\hbar}} | x_n, t_n \rangle \tag{11.10}$$

$$\equiv \langle x_{n+1} | e^{-i\frac{\Delta t \hat{H}}{\hbar}} | x_n \rangle \,,$$

where we can drop the time in the states as the inner product involves states evaluated at the same time. Additionally, we will sandwich a complete set of momentum eigenstates in this inner product, where we note that

$$\mathbb{I} = \int dp_n \, |p_n\rangle \langle p_n| \,, \tag{11.11}$$

and we have

$$\langle x_{n+1} | e^{-i\frac{\Delta t \hat{H}}{\hbar}} | x_n \rangle = \int dp_n \, \langle x_{n+1} | p_n \rangle \langle p_n | e^{-i\frac{\Delta t \hat{H}}{\hbar}} | x_n \rangle \,. \tag{11.12}$$

As with positions, the momentum integral extends over all values $p_n \in (-\infty, \infty)$, but we suppress the bounds for compactness.

Now, it's straightforward to interpret the Hamiltonian sandwiched between the position and momentum eigenstates. These eigenstates just force a particular basis in which to express the Hamiltonian: the mixed position/momentum basis which is actually the way that the Hamiltonian is expressed classically. We are always free to re-order operators using their commutation relation, so if we push all position operators right and momentum operators left, then

$$\langle p_n | \hat{H}(\hat{p}, \hat{x}) | x_n \rangle = H(p_n, x_n) \langle p_n | x_n \rangle \,, \tag{11.13}$$

where now $H(p_n, x_n)$ is just a *function* of momentum p_n and position x_n, and not an operator.

With these manipulations, our product of position states becomes

$$\langle x_{n+1}|e^{-i\frac{\Delta t \hat{H}}{\hbar}}|x_n\rangle = \int dp_n \, \langle x_{n+1}|p_n\rangle\langle p_n|e^{-i\frac{\Delta t \hat{H}}{\hbar}}|x_n\rangle \tag{11.14}$$

$$= \int dp_n \, e^{-i\frac{\Delta t H(p_n,x_n)}{\hbar}} \langle x_{n+1}|p_n\rangle\langle p_n|x_n\rangle \, .$$

Continuing, we can evaluate these inner products of position and momentum eigenstates. Let's focus on $\langle x_{n+1}|p_n\rangle$ for a second. Reading from right to left, this is a momentum eigenstate $|p_n\rangle$ expressed in the position basis with coordinate x_{n+1}. We know what this is: it's just an imaginary exponential

$$\langle x_{n+1}|p_n\rangle \propto e^{i\frac{p_n x_{n+1}}{\hbar}} \, . \tag{11.15}$$

Correspondingly

$$\langle p_n|x_n\rangle \propto e^{-i\frac{p_n x_n}{\hbar}} \tag{11.16}$$

and so

$$\langle x_{n+1}|p_n\rangle\langle p_n|x_n\rangle \propto e^{\frac{i}{\hbar}p_n(x_{n+1}-x_n)} \, . \tag{11.17}$$

Now, I have only written that these inner products are proportional to the imaginary exponentials, not equal to. We fix the proportionality constant by demanding a normalization of the integral over momentum p_n:

$$\int_{-\infty}^{\infty} dp_n \, \langle x_{n+1}|p_n\rangle\langle p_n|x_n\rangle \propto \int_{-\infty}^{\infty} dp_n \, e^{\frac{i}{\hbar}p_n(x_{n+1}-x_n)} = 2\pi\hbar\,\delta(x_n - x_{n+1}), \tag{11.18}$$

where the final equality follows from identities of Fourier transforms that we had discussed in Chap. 2. If we want the result to exclusively be a δ-function with coefficient 1, then we define

$$\langle x_{n+1}|p_n\rangle\langle p_n|x_n\rangle = \frac{1}{2\pi\hbar} e^{\frac{i}{\hbar}p_n(x_{n+1}-x_n)} \, . \tag{11.19}$$

We can put this all together, first just evaluating the inner product over a single time step:

$$\langle x_{n+1},t_{n+1}|x_n,t_n\rangle = \langle x_{n+1}|e^{-i\frac{\Delta t \hat{H}}{\hbar}}|x_n\rangle \tag{11.20}$$

$$= \int dp_n \, \langle x_{n+1}|p_n\rangle\langle p_n|e^{-i\frac{\Delta t \hat{H}}{\hbar}}|x_n\rangle$$

$$= \int dp_n \, e^{-i\frac{\Delta t H(p_n,x_n)}{\hbar}} \langle x_{n+1}|p_n\rangle\langle p_n|x_n\rangle$$

$$= \int \frac{dp_n}{2\pi\hbar} \exp\left[\frac{i}{\hbar}\left(p_n(x_{n+1}-x_n) - H(p_n,x_n)\Delta t\right)\right] \, ,$$

as $\Delta t \to 0$. Plugging this into our expression for the transition amplitude, we then have

$$\langle x_f, t=T|x_i, t=0\rangle = \int \frac{dp_1}{2\pi\hbar} dx_1 \frac{dp_2}{2\pi\hbar} dx_2 \cdots \frac{dp_{N-1}}{2\pi\hbar} dx_{N-1} \tag{11.21}$$

$$\times \exp\left[\frac{i}{\hbar}\sum_{n=1}^{N-1}\left(p_n(x_{n+1}-x_n) - H(p_n,x_n)\Delta t\right)\right] \, .$$

As you might have expected, whenever we break some interval into N parts, we want to take the $N \to \infty$ limit, in which things (hopefully!) simplify. Let's first look at the factor in the exponent:

$$\sum_{n=1}^{N-1} \left(p_n(x_{n+1} - x_n) - H(p_n, x_n)\Delta t \right) . \tag{11.22}$$

To the first term, we can multiply and divide by Δt, which places an overall Δt in the sum:

$$\sum_{n=1}^{N-1} \left(p_n(x_{n+1} - x_n) - H(p_n, x_n)\Delta t \right) \tag{11.23}$$

$$= \sum_{n=1}^{N-1} \Delta t \left(p_n \frac{x_{n+1} - x_n}{\Delta t} - H(p_n, x_n) \right) .$$

Recall that $\Delta t = T/N$, so as $N \to \infty$, $\Delta t \to 0$. Correspondingly, $x_n = x(t_n) = x(n\Delta t)$ and so the difference $x_{n+1} - x_n$ is just the difference of positions of the particle at sequential times. As $\Delta t \to 0$ ($N \to \infty$), this becomes the time derivative of position evaluated at time t_n:

$$\lim_{\Delta t \to 0} \frac{x_{n+1} - x_n}{\Delta t} = \dot{x}(t_n) . \tag{11.24}$$

Further, the sum over time steps transmogrifies into an integral, that is a continuous sum

$$\lim_{\Delta t \to 0} \sum_{n=1}^{N-1} \Delta t \left(p_n \frac{x_{n+1} - x_n}{\Delta t} - H(p_n, x_n) \right) = \int_0^T dt \, (p\dot{x} - H(p,x)) , \tag{11.25}$$

where x and p are implicit functions of time t.

With this result, the transition amplitude can then be expressed as

$$\boxed{\langle x_f, t = T | x_i, t = 0 \rangle = \int \prod_{n=1}^{\infty} \left[\frac{dp_n}{2\pi\hbar} dx_n \right] \exp\left[\frac{i}{\hbar} \int_0^T dt \, (p\dot{x} - H(p,x)) \right] .} \tag{11.26}$$

This is called the **Hamiltonian path integral**. We will massage it a bit more shortly, but I want to point out the somewhat awkward infinity of integrals that we have to do:

$$\int \prod_{n=1}^{\infty} \left[\frac{dp_n}{2\pi\hbar} dx_n \right] . \tag{11.27}$$

Recall that at every time step $0 < t_n < T$, we have to allow for the possibility of the particle having any position x_n and any momentum p_n. As time is continuous, this means that we have to do a continuous number of integrals! This may sound like a bad trade-off, like we're getting swindled in our deal for manipulating the transition amplitude. However, in many cases this "infinity of integrals" can be re-interpreted and evaluated rather straightforwardly.[5]

---

[5] This is in some sense an example of the "no free lunch theorem," in that reformulation of a problem may seem to provide more solutions than other techniques, but in general just solves different problems, but

11.2.1 The Action Form of the Path Integral

Let's see an example of how to do half of these infinity of integrals now. In this book, when studying the quantum mechanics of a particle in space, we only considered those Hamiltonians for which the kinetic energy is quadratic in momentum:

$$H(p,x) = \frac{p^2}{2m} + V(x),$$ (11.28)

where $V(x)$ is a purely position-dependent potential. Note that this Hamiltonian is just a function of the phase-space variables p and x, and not an operator. If the form of this Hamiltonian is assumed, then the path integral is

$$\langle x_f, t = T | x_i, t = 0 \rangle$$ (11.29)

$$= \int \prod_{n=1}^{\infty} \left[\frac{dp_n}{2\pi\hbar} dx_n \right] \exp\left[\frac{i}{\hbar} \int_0^T dt \left(p\dot{x} - \frac{p^2}{2m} - V(x) \right) \right].$$

Next, we would like to perform all of the integrals over momentum p_n, as we know its explicit dependence in the exponent. The first thing we will do is to complete the square in the exponent. Note that

$$-\frac{p^2}{2m} + \dot{x}p = -\frac{1}{2m}(p^2 - 2m\dot{x}p) = -\frac{1}{2m}\left[(p - m\dot{x})^2 - m^2\dot{x}^2 \right]$$ (11.30)

$$= \frac{1}{2}m\dot{x}^2 - \frac{(p - m\dot{x})^2}{2m}.$$

Again, we re-emphasize that there are no operators here, so we can be careless with the order of multiplication. Then, the transition amplitude is

$$\langle x_f, t = T | x_i, t = 0 \rangle$$ (11.31)

$$= \int \prod_{n=1}^{\infty} \left[\frac{dp_n}{2\pi\hbar} dx_n \right] \exp\left[\frac{i}{\hbar} \int_0^T dt \left(\frac{1}{2}m\dot{x}^2 - \frac{(p - m\dot{x})^2}{2m} - V(x) \right) \right].$$

To completely integrate over momentum, we just have to evaluate

$$\int \prod_{n=1}^{\infty} \frac{dp_n}{2\pi\hbar} \exp\left[-\frac{i}{\hbar} \int_0^T dt \frac{(p - m\dot{x})^2}{2m} \right].$$ (11.32)

Note that each integral over momentum ranges over all real values:

$$\int_{-\infty}^{\infty} \frac{dp_n}{2\pi\hbar},$$ (11.33)

so we can make the change of variables $p_n \to p_n - m\dot{x}$, for which the Jacobian is 1 and the bounds of integration are unaffected. Thus, the momentum integral is equivalently

$$\int \prod_{n=1}^{\infty} \frac{dp_n}{2\pi\hbar} \exp\left[-\frac{i}{\hbar} \int_0^T dt \frac{(p - m\dot{x})^2}{2m} \right] = \int \prod_{n=1}^{\infty} \frac{dp_n}{2\pi\hbar} \exp\left[-\frac{i}{\hbar} \int_0^T dt \frac{p^2}{2m} \right].$$ (11.34)

fails at others. This folk theorem was introduced in D. H. Wolpert and W. G. Macready, "No free lunch theorems for optimization," *IEEE Trans. Evol. Comput.* **1**(1), 67–82 (1997).

To continue, note that $p^2 = p(t) \cdot p(t)$, momentum evaluated at time t, and the integral is a sum over all time $t \in [0, T]$. Thus, the integral in the exponent is just a dot product of momentum with itself at every time step t_n:

$$\int_0^T dt\, p^2 = \Delta t \sum_{n=1}^{\infty} p(t_n)^2 = \Delta t \sum_{n=1}^{\infty} p_n^2. \tag{11.35}$$

Then, with this identification, the integrals over momentum break up into a product of one-dimensional integrals:

$$\int \prod_{n=1}^{\infty} \frac{dp_n}{2\pi\hbar} \exp\left[-\frac{i}{\hbar} \int_0^T dt\, \frac{p^2}{2m} \right] = \int \prod_{n=1}^{\infty} \frac{dp_n}{2\pi\hbar} \exp\left[-\frac{i\Delta t}{2m\hbar} \sum_{n=1}^{\infty} p_n^2 \right] \tag{11.36}$$

$$= \lim_{N\to\infty} \left(\int_{-\infty}^{\infty} \frac{dp}{2\pi\hbar} \exp\left[-\frac{i\Delta t}{2m\hbar} p^2 \right] \right)^{N-1}.$$

The integral that remains is in the form of a Gaussian, and so we immediately know its value. The integral is

$$\int_{-\infty}^{\infty} \frac{dp}{2\pi\hbar} \exp\left[-\frac{i\Delta t}{2m\hbar} p^2 \right] = \sqrt{\frac{m}{2\pi i\hbar\Delta t}}, \tag{11.37}$$

which is just some (complex) value with units of inverse length.

We get this integral for each momentum p_n, so the transition amplitude can be expressed as

$$\langle x_f, t = T | x_i, t = 0 \rangle \tag{11.38}$$

$$= \int \prod_{n=1}^{\infty} \left[\sqrt{\frac{m}{2\pi i\hbar\Delta t}} \, dx_n \right] \exp\left[\frac{i}{\hbar} \int_0^T dt\, \left(\frac{1}{2}m\dot{x}^2 - V(x) \right) \right].$$

Going forward, we will just denote the remaining infinite integration measure over positions as

$$[dx] \equiv \prod_{n=1}^{\infty} \sqrt{\frac{m}{2\pi i\hbar\Delta t}} \, dx_n, \tag{11.39}$$

so the transition amplitude or *the* path integral Z is

$$\langle x_f, t = T | x_i, t = 0 \rangle \equiv Z = \int [dx]\, e^{\frac{i}{\hbar} \int_0^T dt\, \left(\frac{1}{2}m\dot{x}^2 - V(x) \right)}. \tag{11.40}$$

The factor in the exponent is the classical **action** of the quantum system of interest. The **Lagrangian** $L(t)$ is the difference of the kinetic and potential energies:

$$L(t) = \frac{1}{2}m\dot{x}^2 - V(x) \equiv L(x, \dot{x}), \tag{11.41}$$

and its time integral is the action $S[x(t)]$ for a given trajectory $x(t)$:

$$S[x(t)] = \int_0^T dt\, L(x, \dot{x}). \tag{11.42}$$

The action is a **functional** as its argument is a complete particle trajectory $x(t)$, and not just some independent variable. Finally, the most compact expression for the path integral is

$$\langle x_f, t = T | x_i, t = 0 \rangle = Z = \int [dx]\, e^{i\frac{S[x]}{\hbar}}. \tag{11.43}$$

Though this form explicitly requires a kinetic energy that is quadratic in momentum, this is essentially universally what people mean when they say "*the* path integral."

11.2.2 The Correspondence Principle Redux

Whole books can and have been written on the path integral formulation of quantum mechanics,[6] but here we'll just give you a flavor for its utility. We'll make some connections to other quantum aspects in the next section. I want to finish this section by making a connection to classical mechanics. The classical mechanics limit, as we have stated many times before, is the limit in which $\hbar \to 0$. Now, staring at the integrand of the path integral, you might think that as $\hbar \to 0$, the exponential factor wildly oscillates and so when integrated would evaluate to 0. Indeed, this would be the case for an arbitrary trajectory $x(t)$. The action $S[x]$ evaluated on an arbitrary trajectory $x_1(t)$ is, say, $S_1 \equiv S[x_1]$. A nearby trajectory, a path that differs by only a small distance in space at every time step, will in general have an action that differs significantly. Let's consider nearby trajectories $x_2(t)$, $x_3(t)$, $\ldots$, to $x_1(t)$ visualized as in Fig. 11.4a. The path integral sums up their exponentiated actions:

$$\langle x_f | x_i \rangle \sim e^{i\frac{S_1}{\hbar}} + e^{i\frac{S_2}{\hbar}} + e^{i\frac{S_3}{\hbar}} + \cdots, \tag{11.44}$$

and in general $S_1, S_2, S_3, \ldots$, will all be significantly different values of the action. Here, $e^{iS/\hbar}$ is a sinusoidal function, and summing up sinusoidal functions with essentially arbitrary phases produces 0, in the limit that $\hbar \to 0$. Thus, as $\hbar \to 0$, an arbitrary trajectory from x_i to x_f does not contribute to the path integral.

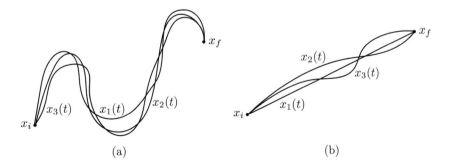

(a)   (b)

Fig. 11.4   (a) Illustration of three nearby trajectories from position x_i to x_f that are far from the classical trajectory. (b) Illustration of three nearby trajectories from position x_i to x_f that are also near the classical trajectory.

[6] R. P. Feynman and A. R. Hibbs, *Quantum Mechanics and Path Integrals* McGraw-Hill (1965).

However, nearby trajectories will contribute to the path integral if the value of the action varies sufficiently slowly. That is, trajectories $x_1, x_2, x_3, \ldots$ illustrated in Fig. 11.4b contribute to the path integral as

$$\langle x_f | x_i \rangle \sim e^{i \frac{S_1}{\hbar}} + e^{i \frac{S_2}{\hbar}} + e^{i \frac{S_3}{\hbar}} + \cdots . \tag{11.45}$$

Assuming that the values of the actions are all extremely close to one another:

$$S[x_1] \sim S[x_2] \sim S[x_3] \sim \cdots , \tag{11.46}$$

then the contribution to the path integral is non-zero. As we assume that the trajectories x_1 and x_2 (say) are close to one another, we can write

$$x_2(t) = x_1(t) + \epsilon(t), \tag{11.47}$$

for some small difference path, where $|\epsilon(t)| \ll |x_1(t)|$.

Now, demanding that the values of the actions differ only slightly, we take the difference

$$S[x_2] - S[x_1] = S[x_1 + \epsilon] - S[x_1] = S[x_1] + \epsilon \frac{\delta S[x]}{\delta x} + \mathcal{O}(\epsilon^2) - S[x_1] \tag{11.48}$$

$$= \epsilon \frac{\delta S[x]}{\delta x} + \cdots ,$$

where we have Taylor expanded in the small trajectory $\epsilon(t)$ and $\delta S[x]/\delta x$ represents the functional derivative of the action with respect to the path $x(t)$. Thus, if a trajectory is to contribute to the path integral for $\hbar \to 0$, then that path must be a trajectory for which the action is **stationary** or extremized:

$$\frac{\delta S[x]}{\delta x} = \frac{\delta}{\delta x} \int_0^T dt\, L(x, \dot{x}) = 0 . \tag{11.49}$$

This, however, is just the **principle of least action** from classical mechanics, and it produces the **Euler–Lagrange equations of motion**. This is satisfied if the functional derivative annihilates the Lagrangian:

$$\frac{\delta}{\delta x} L(x, \dot{x}) = 0 = \lim_{\epsilon \to 0} \frac{L(x + \epsilon, \dot{x} + \dot{\epsilon}) - L(x, \dot{x})}{\epsilon} \tag{11.50}$$

$$= \frac{\partial L(x, \dot{x})}{\partial x} - \frac{d}{dt} \frac{\partial L(x, \dot{x})}{\partial \dot{x}} .$$

In the limit as $\hbar \to 0$, only classical paths (those for which $\delta S[x]/\delta x = 0$ and thus the Euler–Lagrange equations are satisfied) contribute to the path integral. If that isn't a beautiful manifestation of the correspondence principle, then I don't know what is.

11.3 Derivation of the Schrödinger Equation from the Path Integral

In this section, we will start from the path integral Z as our quantum mechanical axiom and construct the wavefunction $\psi(x, t)$ and its time-evolution equation, the

Schrödinger equation. Thus, we will demonstrate the equivalence of the path integral as a complete description of a quantum system and the Schrödinger equation (and vice-versa).

To do this, we start with an arbitrary quantum state $|\psi\rangle \in \mathcal{H}$, the Hilbert space of our system. Recall that the state evolved from time $t = 0$ to $t = T$ is just acted on by the exponentiated Hamiltonian:

$$|\psi(T)\rangle = e^{-i\frac{T\hat{H}}{\hbar}}|\psi(0)\rangle . \tag{11.51}$$

Further, the wavefunction of this state is its representation in the position basis. To project to the position basis, we just act with the position eigenstate bra $\langle x|$ on our state $|\psi\rangle$:

$$\langle x|\psi(t)\rangle = \psi(x,t) = \langle x|e^{-i\frac{t\hat{H}}{\hbar}}|\psi\rangle , \tag{11.52}$$

the wavefunction evaluated at time t in terms of the initial state $|\psi\rangle \equiv |\psi(0)\rangle$. Next, we insert a complete set of position eigenstates between our state $|\psi\rangle$ and the time-evolution operator, where

$$\mathbb{I} = \int_{-\infty}^{\infty} dx_0 \, |x_0\rangle\langle x_0| , \tag{11.53}$$

and so

$$\psi(x,t) = \int_{-\infty}^{\infty} dx_0 \, \langle x|e^{-i\frac{t\hat{H}}{\hbar}}|x_0\rangle\langle x_0|\psi\rangle . \tag{11.54}$$

Now we recognize $\langle x_0|\psi\rangle \equiv \psi(x_0)$ as the wavefunction at time $t = 0$ and then the transition amplitude

$$\langle x|e^{-i\frac{t\hat{H}}{\hbar}}|x_0\rangle \tag{11.55}$$

is just the path integral for transition from position eigenstate $|x_0\rangle$ to final position $|x\rangle$ over total time t. Thus, from our expression for the path integral, we can write

$$\psi(x,t) = \int_{x_0}^{x} [dx'] \, e^{i\frac{S[x']}{\hbar}} \, \psi(x_0) . \tag{11.56}$$

The integral here is a bit of a compact notation, where we mean that

$$\int_{x_0}^{x} [dx'] = \int \prod_{n=0}^{N-1} \sqrt{\frac{m}{2\pi i\hbar\Delta t}} \, dx_n = \int \prod_{n=0}^{N-1} \sqrt{\frac{Nm}{2\pi i\hbar t}} \, dx_n , \tag{11.57}$$

where $N \to \infty$ and $x_N = x$, the final, fixed position (which we don't integrate over). Note that this wavefunction is L^2-normalized because we include the factor from integrating over intermediate momentum in the position integration measure.

If this is really the wavefunction, we should be able to derive the Schrödinger equation from the path integral representation. Our procedure for doing this will be similar to our operator derivation of the Schrödinger equation. We first only consider a time $t = \epsilon$, as $\epsilon \to 0$:

$$\psi(x,\epsilon) = \int_{x_0}^{x} [dx'] \, e^{i\frac{S[x']}{\hbar}} \, \psi(x_0) . \tag{11.58}$$

For small enough ϵ, the particle can't take arbitrary paths from x_0 to x; this is to say that x_0 and x are near one another by the continuity of quantum mechanical trajectories. In this expression for the wavefunction, there is just a single integral over x_0, the position at $t = 0$, and so we have

$$\lim_{\epsilon \to 0} \psi(x, \epsilon) = \sqrt{\frac{m}{2\pi i \hbar \epsilon}} \int_{-\infty}^{\infty} dx_0 \, e^{\frac{i}{\hbar} \int_0^\epsilon dt \left[\frac{1}{2} m \dot{x}^2 - V(x)\right]} \psi(x_0). \tag{11.59}$$

Now let's move on to the classical action in the exponent. As $\epsilon \to 0$, the integral of the Lagrangian over time can be approximated by just the Lagrangian times ϵ, as it is assumed to not vary wildly over $t \in [0, \epsilon]$ (again, by continuity). Additionally, note that the particle's velocity over time ϵ is

$$\dot{x} \approx \frac{x(\epsilon) - x(0)}{\epsilon} = \frac{x - x_0}{\epsilon}, \tag{11.60}$$

by the assumption that $x(\epsilon) = x$ and $x(0) = x_0$, the final and initial positions. Using these observations, we have

$$\lim_{\epsilon \to 0} \int_0^\epsilon dt \left[\frac{1}{2} m \dot{x}^2 - V(x)\right] = \epsilon \left[\frac{m}{2} \frac{(x - x_0)^2}{\epsilon^2} - V(x)\right], \tag{11.61}$$

where, by continuity, we can just evaluate $V(x)$ at x, as we assume that x and x_0 differ by a distance $\mathcal{O}(\epsilon)$, and so their difference is higher order in ϵ. Thus, as $\epsilon \to 0$, the wavefunction in the path integral formulation becomes

$$\lim_{\epsilon \to 0} \psi(x, \epsilon) = \sqrt{\frac{m}{2\pi i \hbar \epsilon}} \int_{-\infty}^{\infty} dx_0 \, \exp\left(\frac{i\epsilon}{\hbar} \left[\frac{m}{2} \frac{(x - x_0)^2}{\epsilon^2} - V(x)\right]\right) \psi(x_0) \tag{11.62}$$

$$= \sqrt{\frac{m}{2\pi i \hbar \epsilon}} \, e^{-\frac{i\epsilon}{\hbar} V(x)} \int_{-\infty}^{\infty} dx_0 \, e^{\frac{im}{2\hbar} \frac{(x - x_0)^2}{\epsilon}} \psi(x_0),$$

where, to $\mathcal{O}(\epsilon^2)$ corrections in the exponent, we can just pull out the potential factor. We're also assuming that the potential is non-singular for any finite x, otherwise we might not be able to claim that $\epsilon V(x)$ is small.[7]

We need to tackle the integral that remains. The exponential factor $\exp\left[\frac{im}{2\hbar} \frac{(x - x_0)^2}{\epsilon}\right]$ becomes an extremely narrow spike in the limit that $\epsilon \to 0$, centered at $x_0 = x$. Further, its integral is finite and so, as $\epsilon \to 0$, is proportional to a δ-function. If it were a δ-function, then evaluating the integral that remains would be easy. With this motivation, we want to expand the Gaussian in a series of δ-functions and its derivatives, centered about $x_0 = x$. That is, we want to write

$$\lim_{\epsilon \to 0} e^{\frac{im}{2\hbar} \frac{(x - x_0)^2}{\epsilon}} = a_0 \delta(x - x_0) + a_1 \delta'(x - x_0) + a_2 \delta''(x - x_0) + \cdots, \tag{11.63}$$

---

[7] This restriction may be troubling, especially because the most realistic system we discussed in this book, the hydrogen atom, has a potential that diverges at the origin. Weaker assumptions can be made to derive the Schrödinger equation than we employ here, but their justification is in general significantly complicated. Nevertheless, the path integral for the hydrogen atom has been explicitly calculated; see I. H. Duru and H. Kleinert, "Solution of path integral for H atom," *Phys. Lett. B* **84**, 185–188 (1979).

where $a_0, a_1, a_2, \ldots$ are some coefficients we would like to determine. Now, the derivative of a δ-function looks scary, but using integration by parts it just acts on a function $f(x_0)$ as

$$\int dx_0\, \delta'(x - x_0)\, f(x_0) = -\int dx_0\, \delta(x - x_0)\, f'(x_0)\,, \tag{11.64}$$

and similar for higher-order derivatives. This demonstrates that the integral of a δ-function with any number $n > 0$ of derivatives on it is 0:

$$\int_{-\infty}^{\infty} dx_0\, \delta^{(n)}(x - x_0) = 0\,. \tag{11.65}$$

This immediately tells us that

$$a_0 = \int_{-\infty}^{\infty} dx_0\, e^{\frac{im}{2\hbar} \frac{(x-x_0)^2}{\epsilon}} = \sqrt{\frac{2\pi\hbar\epsilon}{-im}}\,, \tag{11.66}$$

just evaluating the integral of a Gaussian.

Next, for the a_1 coefficient, we note that

$$\int_{-\infty}^{\infty} dx_0\, (x - x_0)\, \delta(x - x_0) = \int_{-\infty}^{\infty} dx_0\, (x - x_0)\, \delta^{(n)}(x - x_0) = 0\,, \tag{11.67}$$

for $n > 1$, because more than one derivative will kill the prefactor function. Further, note that

$$\int_{-\infty}^{\infty} dx_0\, (x_0 - x)\, \delta'(x - x_0) = -\int_{-\infty}^{\infty} dx_0\, \delta(x - x_0)\, \frac{d}{dx_0}(x_0 - x) = 1\,, \tag{11.68}$$

and so the a_1 coefficient is

$$a_1 = \int_{-\infty}^{\infty} dx_0\, (x_0 - x)\, e^{\frac{im}{2\hbar} \frac{(x-x_0)^2}{\epsilon}} = 0\,, \tag{11.69}$$

as the Gaussian is symmetric about $x_0 = x$.

Continuing to the a_2 coefficient, we note that

$$\int_{-\infty}^{\infty} dx_0\, (x_0 - x)^2\, \delta''(x - x_0) = \int_{-\infty}^{\infty} dx_0\, 2(x_0 - x)\delta'(x - x_0) = 2\,, \tag{11.70}$$

using integration by parts twice. Then, the a_2 coefficient is

$$a_2 = \frac{1}{2} \int_{-\infty}^{\infty} dx_0\, (x_0 - x)^2\, e^{\frac{im}{2\hbar} \frac{(x-x_0)^2}{\epsilon}} = \sqrt{\frac{\pi}{2}} \left(-\frac{\hbar\epsilon}{im} \right)^{3/2}\,. \tag{11.71}$$

Thus, this singular expansion of the very narrow Gaussian yields

$$\lim_{\epsilon \to 0} e^{\frac{im}{2\hbar} \frac{(x-x_0)^2}{\epsilon}} = \sqrt{-\frac{2\pi\hbar\epsilon}{im}}\, \delta(x - x_0) + \sqrt{\frac{\pi}{2}} \left(-\frac{\hbar\epsilon}{im} \right)^{3/2} \delta''(x - x_0) + \cdots\,, \tag{11.72}$$

suppressing terms at higher order in ϵ that can be ignored in the $\epsilon \to 0$ limit.

Now this expression can enable us to easily evaluate the integral over x_0 that remains in the path integral formulation of the wavefunction. We have

$$\int_{-\infty}^{\infty} dx_0 \, e^{\frac{im}{2\hbar} \frac{(x-x_0)^2}{\epsilon}} \, \psi(x_0) \tag{11.73}$$

$$= \int_{-\infty}^{\infty} dx_0 \left[\sqrt{-\frac{2\pi\hbar\epsilon}{im}} \, \delta(x - x_0) + \sqrt{\frac{\pi}{2}} \left(-\frac{\hbar\epsilon}{im} \right)^{3/2} \delta''(x - x_0) + \cdots \right] \psi(x_0)$$

$$= \sqrt{-\frac{2\pi\hbar\epsilon}{im}} \, \psi(x) + \sqrt{\frac{\pi}{2}} \left(-\frac{\hbar\epsilon}{im} \right)^{3/2} \psi''(x) + \cdots .$$

Putting this together and Taylor expanding the expression thus far, we have

$$\psi(x, \epsilon) = \psi(x, 0) + \epsilon \left. \frac{d\psi(x,t)}{dt} \right|_{t=0} + \cdots \tag{11.74}$$

$$= \sqrt{\frac{m}{2\pi i \hbar\epsilon}} \, e^{-i\frac{\epsilon V(x)}{\hbar}} \int_{-\infty}^{\infty} dx_0 \, e^{\frac{im}{2\hbar} \frac{(x-x_0)^2}{\epsilon}} \, \psi(x_0)$$

$$= \sqrt{\frac{m}{2\pi i \hbar\epsilon}} \left(\sqrt{-\frac{2\pi\hbar\epsilon}{im}} \, \psi(x) + \sqrt{\frac{\pi}{2}} \left(-\frac{\hbar\epsilon}{im} \right)^{3/2} \psi''(x) \right.$$

$$\left. - \frac{i\epsilon}{\hbar} \sqrt{-\frac{2\pi\hbar\epsilon}{im}} V(x) \psi(x) + \cdots \right)$$

$$= \psi(x) - \frac{1}{i} \frac{\hbar}{2m} \epsilon \, \psi''(x) - \frac{i\epsilon}{\hbar} V(x) \psi(x) + \cdots .$$

Now, to derive the Schrödinger equation, we just match terms at each order in ϵ on both sides of this equation. The terms at $\mathcal{O}(\epsilon)$ for general time t require

$$\frac{d\psi(x,t)}{dt} = -\frac{1}{i} \frac{\hbar}{2m} \psi''(x,t) - \frac{i}{\hbar} V(x) \psi(x,t), \tag{11.75}$$

or, more familiarly:

$$i\hbar \frac{d\psi}{dt} = -\frac{\hbar^2}{2m} \psi'' + V\psi, \tag{11.76}$$

which is just the Schrödinger equation in position space!

Thus, we have shown that the Schrödinger equation implies the path integral and vice-versa, thus one can start with either formulation to analyze quantum dynamics. While the Schrödinger equation seems nicer for the questions we have asked in this book, the path integral is a much more natural starting point for **quantum field theory**, the harmonization of quantum mechanics and special relativity.

11.4 Calculating the Path Integral

We will present an explicit calculation of the path integral for the harmonic oscillator potential in the next section, but here, we'll describe a couple of different ways that one can generally go about calculating it. First, let's go back to the definition of the path integral as the position space transition amplitude:

$$\langle x_f, t = T | x_i, t = 0 \rangle = \langle x_f | e^{-i\frac{T\hat{H}}{\hbar}} | x_i \rangle. \tag{11.77}$$

We had proceeded in our derivation of the path integral to insert complete sets of position eigenstates at infinitesimally separated times, but there's something much more natural that we would have done if this were presented a few chapters ago. Call $|\psi_n\rangle$ the nth energy eigenstate of the Hamiltonian $\hat{H}$ and by completeness and orthonormality of the energy eigenstates we have

$$\mathbb{I} = \sum_{n=1}^{\infty} |\psi_n\rangle\langle\psi_n|. \tag{11.78}$$

Then, inserting this identity operator into the transition amplitude, we have

$$\langle x_f | e^{-i\frac{T\hat{H}}{\hbar}} | x_i \rangle = \sum_{n=1}^{\infty} \langle x_f | e^{-i\frac{T\hat{H}}{\hbar}} |\psi_n\rangle\langle\psi_n|x_i\rangle = \sum_{n=1}^{\infty} e^{-i\frac{TE_n}{\hbar}} \langle x_f|\psi_n\rangle\langle\psi_n|x_i\rangle \tag{11.79}$$

$$= \sum_{n=1}^{\infty} e^{-i\frac{TE_n}{\hbar}} \psi_n(x_f)\psi_n^*(x_i).$$

Here, E_n is the energy eigenvalue of the state $|\psi_n\rangle$ and $\psi_n(x)$ is the energy eigenstate wavefunction. Thus, knowing the energy eigenstate wavefunctions and energy eigenvalues we can directly calculate the path integral. Specifically, this transition amplitude is also called the **propagator** as it represents the complex probability amplitude for a quantum mechanical particle to travel or *propagate* from point x_i to point x_f in time T.

Another way to approach the path integral is the following. Let's parametrize the trajectory $x(t)$ as the classical path plus a quantum deviation:

$$x(t) = x_{\text{class}}(t) + y(t), \tag{11.80}$$

where $x_{\text{class}}(t)$ is the classical path; namely, the path for which the action is extremized:

$$\frac{\delta S[x_{\text{class}}]}{\delta x_{\text{class}}} = 0. \tag{11.81}$$

Further, assuming that the classical path has the same initial and final positions as an arbitrary path, we require that the fluctuation vanishes at the initial and final times:

$$y(0) = y(T) = 0. \tag{11.82}$$

With this parametrization, we can express the path integral as

$$Z = \int [dy] \, e^{i\frac{S[x_{\text{class}}+y]}{\hbar}}. \tag{11.83}$$

Note that we only integrate over the quantum fluctuation $y(t)$ because the classical path is fixed for a fixed initial and final position.

Now, while the quantum fluctuation is completely general, we can Taylor expand the action in $y(t)$ and isolate the effect of the classical action individually. We have

$$S[x_{\text{class}}(t)+y(t)] = S[x_{\text{class}}(t)] + \int_0^T dt\, y(t)\, \frac{\delta L(x_{\text{class}},\dot{x}_{\text{class}})}{\delta x_{\text{class}}} \tag{11.84}$$

$$+ \int_0^T dt\, \frac{y^2(t)}{2}\, \frac{\delta^2 L(x_{\text{class}},\dot{x}_{\text{class}})}{\delta x_{\text{class}}^2} + \cdots,$$

only explicitly writing terms through second order in $y(t)$. By assumption, $x_{\text{class}}(t)$ is the classical path and solves the Euler–Lagrange equations, so the term linear in $y(t)$ just vanishes. Then, the path integral can be expressed as

$$Z = e^{i\frac{S[x_{\text{class}}]}{\hbar}} \int [dy]\, \exp\left[\frac{i}{\hbar} \int_0^T dt\, \frac{y^2(t)}{2}\, \frac{\delta^2 L(x_{\text{class}},\dot{x}_{\text{class}})}{\delta x_{\text{class}}^2} + \cdots\right]. \tag{11.85}$$

We'll see how evaluation of the path integral in this form works in the next section.

Example 11.1   In the next section, we will explicitly calculate the path integral for the harmonic oscillator. This will be highly non-trivial, and its calculation will involve many moving parts. Here, however, we can construct the path integral for the infinite square well using the techniques discussed above.

Solution

For the infinite square well, we explicitly know the energy eigenvalues and the energy eigenstate wavefunctions, so the form of the path integral from Eq. (11.79) will be the way that we proceed here. Recall that the energy eigenvalues of the infinite square well are

$$E_n = \frac{n^2 \pi^2 \hbar^2}{2ma^2}, \tag{11.86}$$

for a well of width a. The energy eigenstate wavefunctions are

$$\psi_n(x) = \sqrt{\frac{2}{a}}\, \sin\frac{n\pi x}{a}, \tag{11.87}$$

where $n = 1,2,3,\ldots$. Therefore, the path integral for the infinite square well is

$$\langle x_f | e^{-i\frac{T\hat{H}}{\hbar}} | x_i \rangle = \sum_{n=1}^{\infty} e^{-i\frac{TE_n}{\hbar}}\, \psi_n(x_f)\, \psi_n^*(x_i) \tag{11.88}$$

$$= \frac{2}{a} \sum_{n=1}^{\infty} e^{-iT\frac{n^2\pi^2\hbar}{2ma^2}}\, \sin\frac{n\pi x_f}{a}\, \sin\frac{n\pi x_i}{a}.$$

Sine is a linear combination of complex exponentials, for example

$$\sin\phi = \frac{e^{i\phi} - e^{-i\phi}}{2i}, \tag{11.89}$$

and so in general this infinite sum takes the form

$$\sum_{n=1}^{\infty} e^{-iT\frac{n^2\pi^2\hbar}{2ma^2}}\, \sin\frac{n\pi x_f}{a}\, \sin\frac{n\pi x_i}{a} \simeq \sum_{n=1}^{\infty} e^{-i(\alpha n^2 + \beta n)}, \tag{11.90}$$

for some constants (independent of n) α and β. This sum cannot be expressed in terms of elementary functions; a closed form does exist, but in terms of the Jacobi theta function.

One reason for the complication of the path integral for the infinite square well is because the multiple paths that can contribute involve reflection off the boundaries of the well. This is illustrated in Fig. 11.5 on a space-time graph for an initial position x_i and a final position x_f in the well. Three representative trajectories are shown: $>$, which corresponds to the particle traveling directly from x_i to x_f; $\ll$, which bounces off the upper wall once before reaching x_f; and $\lll$, which bounces off both the lower and upper walls before reaching x_f. The particle could also have bounced three, four, or more times before reaching x_f, and all of these possible trajectories must be summed together. This sum is encoded in the result of Eq. (11.88) in a very non-obvious way.[8] When we calculate the path integral for the harmonic oscillator, no reflections need to be included, which will make the analysis significantly simpler.

11.5 Example: The Path Integral of the Harmonic Oscillator

While the path integral has a profound and powerful physical interpretation for what quantum mechanics *is*, calculations with the path integral can be extremely challenging, not the least of which is due to the infinity of integrals that define the sum over paths. To demonstrate how the path integral can be calculated, in this section we will see how this is done for the quantum harmonic oscillator. This will be a very involved calculation with many steps and if you're interested just in the final result, you can skip to the end result, Eq. (11.137). Nevertheless, the calculation is illuminating and will be useful for comparison with the thermodynamic partition function that we will construct in the next chapter.

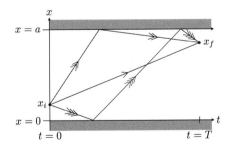

 Fig. 11.5   Illustration of three possible trajectories from initial point x_i to final point x_f in an infinite square well of width a.

[8] One can therefore also use the method of images, a technique familiar in electromagnetism for calculating potentials with conducting boundaries, to evaluate the path integral of the infinite square well. See M. Goodman, "Path integral solution to the infinite square well," *Am. J. Phys.* **49**, 843 (1981).

11.5.1 The Harmonic Oscillator Action

The path integral that we want to calculate is

$$Z = \int [dx] \, e^{i\frac{S[x]}{\hbar}} \, , \tag{11.91}$$

where $S[x]$ is the action of the harmonic oscillator whose classical Hamiltonian is

$$H(x,\dot{x}) = \frac{m}{2}\dot{x}^2 + \frac{m\omega^2}{2}x^2 \, . \tag{11.92}$$

The Lagrangian of the harmonic oscillator is just the difference of its kinetic and potential energies, so there's just a $-$ sign rather than a $+$ sign between terms:

$$L(x,\dot{x}) = \frac{m}{2}\dot{x}^2 - \frac{m\omega^2}{2}x^2 \, . \tag{11.93}$$

The action of the harmonic oscillator is just the time integral of the Lagrangian:

$$S[x] = \int_0^T dt \left(\frac{m}{2}\dot{x}^2 - \frac{m\omega^2}{2}x^2 \right) \, . \tag{11.94}$$

As mentioned in the previous section, we can extract the classical action by expressing the path as a fluctuation about the classical path. Because the Lagrangian for the harmonic oscillator is quadratic in the trajectory, the action very simply separates into

$$S[x] = S[x_{\text{class}}] + \int_0^T dt \left(\frac{m}{2}\dot{y}^2 - \frac{m\omega^2}{2}y^2 \right) \, , \tag{11.95}$$

where $y(t)$ is the quantum fluctuation and we'll leave the classical action $S[x_{\text{class}}]$ unevaluated here.

As the integration measure of the path integral is defined through a limiting procedure, where

$$[dy] = \lim_{N\to\infty} \sqrt{\frac{m}{2\pi i\hbar\Delta t}} \prod_{n=1}^{N-1} \sqrt{\frac{m}{2\pi i\hbar\Delta t}} \, dy_n \, , \tag{11.96}$$

with $y_n = x(n\Delta t)$ and $\Delta t = T/N$, it will be useful to express the action also as a limit. Note that there is one more normalization factor than integrals over the trajectory because the ends have 0 fluctuation, $y_0 = y_N = 0$, and there are a total of N momentum integrals that we had to do. The first step will be to use integration by parts to re-write the action a bit. Note that, by the product rule:

$$\dot{y}^2 = \frac{d}{dt}(y\dot{y}) - y\frac{d^2}{dt^2}y \, , \tag{11.97}$$

and so the action can be expressed as

$$S[x] = S[x_{\text{class}}] + \int_0^T dt \left(\frac{m}{2} \dot{y}^2 - \frac{m\omega^2}{2} y^2 \right) \tag{11.98}$$

$$= S[x_{\text{class}}] + \int_0^T dt \left(\frac{m}{2} \frac{d}{dt} (y\dot{y}) - y \frac{m}{2} \frac{d^2}{dt^2} y - \frac{m\omega^2}{2} y^2 \right)$$

$$= S[x_{\text{class}}] + \frac{m}{2} \left(y(T)\dot{y}(T) - y(0)\dot{y}(0) \right) - \int_0^T dt \left(y \frac{m}{2} \frac{d^2}{dt^2} y + \frac{m\omega^2}{2} y^2 \right).$$

By assumption, the quantum fluctuation of the particle at the initial and final points is 0, and so we can ignore the terms that explicitly depend on the boundary conditions:

$$S[x] = S[x_{\text{class}}] - \frac{m}{2} \int_0^T dt\, y \left(\frac{d^2}{dt^2} + \omega^2 \right) y. \tag{11.99}$$

Now, in this form, the object in parentheses is a linear operator sandwiched between positions evaluated at time t. The integral can be discretized and the action can be written in the form of a matrix element of this linear operator, for a given trajectory. That is

$$S[x] = S[x_{\text{class}}] - \lim_{\Delta t \to 0} \frac{m}{2} \Delta t \sum_{n=0}^N y_n \left(\frac{d^2}{dt^2} + \omega^2 \right) y_n, \tag{11.100}$$

where we have left the second derivative operator's action on this discretized sum implicit. Exactly like we did in Chap. 2, we can construct a matrix $\mathbb{A}$ that implements the action of the second derivative and multiplication by the angular velocity term. In particular, note that the second time derivative is the limit

$$\frac{d^2}{dt^2} y_n = \lim_{\Delta t \to 0} \frac{y(n\Delta t + \Delta t) + y(n\Delta t - \Delta t) - 2y(n\Delta t)}{\Delta t^2} \tag{11.101}$$

$$= \lim_{\Delta t \to 0} \frac{y_{n+1} + y_{n-1} - 2y_n}{\Delta t^2}.$$

This identification enables us to express the factor in parentheses as a matrix:

$$\left(\frac{d^2}{dt^2} + \omega^2 \right) \to \mathbb{A} \tag{11.102}$$

$$= \begin{pmatrix} -\frac{2}{\Delta t^2} + \omega^2 & \frac{1}{\Delta t^2} & 0 & 0 & 0 & \cdots \\ \frac{1}{\Delta t^2} & -\frac{2}{\Delta t^2} + \omega^2 & \frac{1}{\Delta t^2} & 0 & 0 & \cdots \\ 0 & \frac{1}{\Delta t^2} & -\frac{2}{\Delta t^2} + \omega^2 & \frac{1}{\Delta t^2} & 0 & \cdots \\ \vdots & \vdots & \vdots & \vdots & \vdots & \ddots \end{pmatrix}.$$

Then, if we denote the vector of positions at all times as

$$\vec{y}^{\mathsf{T}} = (y_0\ y_1\ y_2\ \cdots\ y_N), \tag{11.103}$$

the action can be expressed in vector form as

$$S[x] = S[x_{\text{class}}] - \lim_{\Delta t \to 0} \frac{m}{2} \Delta t\, \vec{y}^{\mathsf{T}} \mathbb{A} \vec{y}. \tag{11.104}$$

11.5.2 Diagonalizing the Path Integral

Now, with this form of the action, the path integral can be expressed as

$$Z = e^{i\frac{S[x_{\text{class}}]}{\hbar}} \lim_{\Delta t \to 0} \left(\frac{m}{2\pi i \hbar \Delta t}\right)^{\frac{N}{2}} \int_{-\infty}^{\infty} d\vec{y}\, e^{-\frac{im\Delta t}{2\hbar}\vec{y}^{\mathsf{T}}\mathbb{A}\vec{y}}. \tag{11.105}$$

Here, we have denoted the integration measure as

$$\int_{-\infty}^{\infty} d\vec{y} = \int_{-\infty}^{\infty} dy_1 \int_{-\infty}^{\infty} dy_1 \cdots \int_{-\infty}^{\infty} dy_{N-1}. \tag{11.106}$$

Our goal now will be to explicitly do these integrals and our procedure is the following. Note that the matrix $\mathbb{A}$ is Hermitian, as it is real and symmetric. Thus it is a normal matrix and so there exists an orthogonal matrix $\mathbb{O}$ that diagonalizes $\mathbb{A}$. That is, we can express the exponent in the path integral as

$$\frac{m\Delta t}{2\hbar}\vec{y}^{\mathsf{T}}\mathbb{A}\vec{y} = \frac{m\Delta t}{2\hbar}\vec{y}^{\mathsf{T}}\mathbb{O}^{\mathsf{T}}\mathbb{O}\mathbb{A}\mathbb{O}^{\mathsf{T}}\mathbb{O}\vec{y} \equiv \vec{u}^{\mathsf{T}}\mathbb{O}\mathbb{A}\mathbb{O}^{\mathsf{T}}\vec{u}, \tag{11.107}$$

where

$$\vec{u} = \sqrt{\frac{m\Delta t}{2\hbar}}\,\mathbb{O}\vec{y}. \tag{11.108}$$

Now the matrix $\mathbb{O}\mathbb{A}\mathbb{O}^{\mathsf{T}}$ is diagonal, so it is convenient to change integration variables from $\vec{y}$ to $\vec{u}$, where the integration measure becomes

$$d\vec{u} = \left|\det\left(\sqrt{\frac{m\Delta t}{2\hbar}}\,\mathbb{O}\right)\right| d\vec{y} = \left(\frac{m\Delta t}{2\hbar}\right)^{\frac{N-1}{2}} d\vec{y}, \tag{11.109}$$

by the assumption that $\mathbb{O}$ is an orthogonal matrix and so its determinant is ± 1. Then, the path integral can be expressed as

$$Z = e^{i\frac{S[x_{\text{class}}]}{\hbar}} \lim_{\Delta t \to 0} \sqrt{\frac{m}{2\pi i \hbar \Delta t}} \left(\frac{1}{i\pi\Delta t^2}\right)^{\frac{N-1}{2}} \int_{-\infty}^{\infty} d\vec{u}\, e^{-i\vec{u}^{\mathsf{T}}\mathbb{O}\mathbb{A}\mathbb{O}^{\mathsf{T}}\vec{u}} \tag{11.110}$$

$$= e^{i\frac{S[x_{\text{class}}]}{\hbar}} \lim_{\Delta t \to 0} \sqrt{\frac{m}{2\pi i \hbar \Delta t}} \left(\frac{1}{i\pi\Delta t^2}\right)^{\frac{N-1}{2}} \int_{-\infty}^{\infty} d\vec{u}\, e^{-i\sum_{n=1}^{N-1}\lambda_n u_n^2},$$

where λ_n is an eigenvalue of the matrix $\mathbb{A}$. Thus, we have successfully diagonalized the exponent of the path integral and so the integrals that remain transmogrify into a product of one-dimensional integrals:

$$\int_{-\infty}^{\infty} d\vec{u}\, e^{-i\sum_{n=1}^{N-1}\lambda_n u_n^2} = \prod_{n=1}^{N-1} \int_{-\infty}^{\infty} du_n\, e^{-i\lambda_n u_n^2}, \tag{11.111}$$

which are just a collection of Gaussian integrals. They can easily be evaluated using the result

$$\int_{-\infty}^{\infty} du\, e^{-i\lambda u^2} = \sqrt{\frac{\pi}{i\lambda}}. \tag{11.112}$$

Therefore, the integrals of the trajectories of the path integral can be evaluated to be

$$\int_{-\infty}^{\infty} d\vec{u}\, e^{-i\sum_{n=1}^{N-1}\lambda_n u_n^2} = \prod_{n=1}^{N-1} \int_{-\infty}^{\infty} du_n\, e^{-i\lambda_n u_n^2} = \prod_{n=1}^{N-1} \sqrt{\frac{\pi}{i\lambda_n}}. \tag{11.113}$$

Now, this result contains the product of eigenvalues of the matrix $\mathbb{A}$, which is thus just its determinant. That is

$$\prod_{n=1}^{N-1} \lambda_n = \det \mathbb{A}, \tag{11.114}$$

and so the path integral remarkably simplifies to

$$Z = e^{i\frac{S[x_{\text{class}}]}{\hbar}} \lim_{\Delta t \to 0} \sqrt{\frac{m}{2\pi i\hbar \Delta t}} \left(\frac{1}{i\pi \Delta t^2}\right)^{\frac{N-1}{2}} \prod_{n=1}^{N-1} \sqrt{\frac{\pi}{i\lambda_n}} \tag{11.115}$$

$$= e^{i\frac{S[x_{\text{class}}]}{\hbar}} \lim_{\Delta t \to 0} \sqrt{\frac{m}{2\pi i\hbar \Delta t}} \frac{(\det \mathbb{A})^{-1/2}}{(i\Delta t)^{N-1}}.$$

So, the evaluation of the path integral is reduced to simply determining the determinant of the matrix $\mathbb{A}$.

11.5.3 Evaluating the Determinant

The matrix $\mathbb{A}$ has a simple form that enables recursively solving for its determinant. To show this, we will denote $\mathbb{A}_n$ as the matrix $\mathbb{A}$ of Eq. (11.102) that is $n \times n$ dimensional. Then, the path integral we want to calculate is

$$Z = e^{i\frac{S[x_{\text{class}}]}{\hbar}} \lim_{\Delta t \to 0} \sqrt{\frac{m}{2\pi i\hbar \Delta t}} \frac{(\det \mathbb{A}_{N-1})^{-1/2}}{(i\Delta t)^{N-1}}, \tag{11.116}$$

where $N = T/\Delta t$. From the definition of matrix $\mathbb{A}$, we can rescale it to eliminate the overall Δt factor. Introducing matrix $\mathbb{B}$, where

$$\mathbb{B} = \begin{pmatrix} -2+\omega^2\Delta t^2 & 1 & 0 & 0 & 0 & \cdots \\ 1 & -2+\omega^2\Delta t^2 & 1 & 0 & 0 & \cdots \\ 0 & 1 & -2+\omega^2\Delta t^2 & 1 & 0 & \cdots \\ \vdots & \vdots & \vdots & \vdots & \vdots & \ddots \end{pmatrix}, \tag{11.117}$$

we have

$$\mathbb{A} = \frac{1}{\Delta t^2}\mathbb{B}. \tag{11.118}$$

With this rescaling, the path integral simplifies to

$$Z = e^{i\frac{S[x_{\text{class}}]}{\hbar}} \lim_{\Delta t \to 0} \sqrt{\frac{m}{2\pi i\hbar \Delta t}} \frac{(\det \mathbb{B}_{N-1})^{-1/2}}{i^{N-1}}. \tag{11.119}$$

Then, using Eq. (11.117) we see that its determinant satisfies the recursion relation

$$\det \mathbb{B}_n = \left(-2+\omega^2\Delta t^2\right)\det \mathbb{B}_{n-1} - \det \mathbb{B}_{n-2}. \tag{11.120}$$

This follows from starting to calculate the determinant of $\mathbb{B}_n$ from the two non-zero elements on the first row, then multiplying by the determinant of the appropriate minor of the matrix. This recursion relation can be conveniently expressed in a matrix relationship as

$$\begin{pmatrix} \det \mathbb{B}_n \\ \det \mathbb{B}_{n-1} \end{pmatrix} = \begin{pmatrix} -2+\omega^2\Delta t^2 & -1 \\ 1 & 0 \end{pmatrix} \begin{pmatrix} \det \mathbb{B}_{n-1} \\ \det \mathbb{B}_{n-2} \end{pmatrix}. \tag{11.121}$$

This can be recursed all the way down to the matrix acting on the vector of 1×1 and 2×2 matrices by repeated action of the mixing matrix as

$$\begin{pmatrix} \det \mathbb{B}_n \\ \det \mathbb{B}_{n-1} \end{pmatrix} = \begin{pmatrix} -2+\omega^2\Delta t^2 & -1 \\ 1 & 0 \end{pmatrix}^{n-2} \begin{pmatrix} \det \mathbb{B}_2 \\ \det \mathbb{B}_1 \end{pmatrix}. \tag{11.122}$$

From the form of $\mathbb{B}$ the initial determinants are simple:

$$\det \mathbb{B}_1 = -2+\omega^2\Delta t^2, \qquad \det \mathbb{B}_2 = (-2+\omega^2\Delta t^2)^2 - 1. \tag{11.123}$$

This matrix product can be evaluated in the following way. The matrix of interest has two eigenvalues; call them λ_1 and λ_2. As the trace of the matrix is the sum of the eigenvalues and the determinant is the product of the eigenvalues, we can express the matrix in the form

$$\begin{pmatrix} -2+\omega^2\Delta t^2 & -1 \\ 1 & 0 \end{pmatrix} = \begin{pmatrix} \lambda_1+\lambda_2 & -\lambda_1\lambda_2 \\ 1 & 0 \end{pmatrix}. \tag{11.124}$$

We can easily solve for λ_1 and λ_2, finding

$$\lambda_1 = -1 + \frac{\omega^2\Delta t^2}{2} + \omega\Delta t \sqrt{-1 + \frac{\omega^2\Delta t^2}{4}}, \tag{11.125}$$

$$\lambda_2 = -1 + \frac{\omega^2\Delta t^2}{2} - \omega\Delta t \sqrt{-1 + \frac{\omega^2\Delta t^2}{4}},$$

but we'll work with just λ_1 and λ_2 until the very end. Now, we would like to find the non-singular matrix $\mathbb{Q}$ that diagonalizes the matrix of interest; that is

$$\mathbb{Q}^{-1} \begin{pmatrix} \lambda_1 & 0 \\ 0 & \lambda_2 \end{pmatrix} \mathbb{Q} = \begin{pmatrix} \lambda_1+\lambda_2 & -\lambda_1\lambda_2 \\ 1 & 0 \end{pmatrix}. \tag{11.126}$$

If we had such a matrix $\mathbb{Q}$, then the product of the initial matrix with itself many times is simple:

$$\begin{pmatrix} \lambda_1+\lambda_2 & -\lambda_1\lambda_2 \\ 1 & 0 \end{pmatrix}^{n-2} \tag{11.127}$$

$$= \mathbb{Q}^{-1} \begin{pmatrix} \lambda_1 & 0 \\ 0 & \lambda_2 \end{pmatrix} \mathbb{Q}\mathbb{Q}^{-1} \begin{pmatrix} \lambda_1 & 0 \\ 0 & \lambda_2 \end{pmatrix} \mathbb{Q} \cdots \mathbb{Q}^{-1} \begin{pmatrix} \lambda_1 & 0 \\ 0 & \lambda_2 \end{pmatrix} \mathbb{Q}$$

$$= \mathbb{Q}^{-1} \begin{pmatrix} \lambda_1^{n-2} & 0 \\ 0 & \lambda_2^{n-2} \end{pmatrix} \mathbb{Q}.$$

There are many brute-force ways to determine $\mathbb{Q}$; here, we will just quote the result that one such $\mathbb{Q}$ is

$$\mathbb{Q} = \begin{pmatrix} -1 & \lambda_2 \\ -1 & \lambda_1 \end{pmatrix}, \qquad \mathbb{Q}^{-1} = \frac{1}{\lambda_2 - \lambda_1} \begin{pmatrix} \lambda_1 & -\lambda_2 \\ 1 & -1 \end{pmatrix}. \tag{11.128}$$

Using this result, we can easily evaluate the matrix product and find

$$\begin{pmatrix} \lambda_1 + \lambda_2 & -\lambda_1 \lambda_2 \\ 1 & 0 \end{pmatrix}^{n-2} = \mathbb{Q}^{-1} \begin{pmatrix} \lambda_1^{n-2} & 0 \\ 0 & \lambda_2^{n-2} \end{pmatrix} \mathbb{Q} \tag{11.129}$$

$$= \frac{1}{\lambda_2 - \lambda_1} \begin{pmatrix} \lambda_1 & -\lambda_2 \\ 1 & -1 \end{pmatrix} \begin{pmatrix} \lambda_1^{n-2} & 0 \\ 0 & \lambda_2^{n-2} \end{pmatrix} \begin{pmatrix} -1 & \lambda_2 \\ -1 & \lambda_1 \end{pmatrix}$$

$$= \frac{1}{\lambda_1 - \lambda_2} \begin{pmatrix} \lambda_1^{n-1} - \lambda_2^{n-1} & -(\lambda_1^{n-2} - \lambda_2^{n-2}) \\ \lambda_1^{n-2} - \lambda_2^{n-2} & -(\lambda_1^{n-3} - \lambda_2^{n-3}) \end{pmatrix}.$$

Now, we multiply this matrix on the initial vector of the recursion, as in Eq. (11.122). The result can be very compactly represented by noting that the initial determinants are just functions of the eigenvalues:

$$\det \mathbb{B}_1 = \lambda_1 + \lambda_2, \qquad \det \mathbb{B}_2 = (\lambda_1 + \lambda_2)^2 - 1. \tag{11.130}$$

Then, the linear equation for the determinants are

$$\begin{pmatrix} \det \mathbb{B}_n \\ \det \mathbb{B}_{n-1} \end{pmatrix} = \begin{pmatrix} -2 + \omega^2 \Delta t^2 & -1 \\ 1 & 0 \end{pmatrix}^{n-2} \begin{pmatrix} \det \mathbb{B}_2 \\ \det \mathbb{B}_1 \end{pmatrix} \tag{11.131}$$

$$= \frac{1}{\lambda_1 - \lambda_2} \begin{pmatrix} \lambda_1^{n-1} - \lambda_2^{n-1} & -(\lambda_1^{n-2} - \lambda_2^{n-2}) \\ \lambda_1^{n-2} - \lambda_2^{n-2} & -(\lambda_1^{n-3} - \lambda_2^{n-3}) \end{pmatrix} \begin{pmatrix} (\lambda_1 + \lambda_2)^2 - 1 \\ \lambda_1 + \lambda_2 \end{pmatrix}$$

$$= \frac{1}{\lambda_1 - \lambda_2} \begin{pmatrix} \lambda_1^{n+1} - \lambda_2^{n+1} \\ \lambda_1^n - \lambda_2^n \end{pmatrix}.$$

To get the final line, we used the relationship that $\lambda_1 \lambda_2 = 1$. In this form, we can immediately read off the necessary determinant.

11.5.4 Putting the Pieces Together

Now, with the determinant established, we need to take the appropriate limit to calculate the path integral for the harmonic oscillator. In particular, we want

$$\lim_{\Delta t \to 0} \det \mathbb{B}_{N-1} = \lim_{N \to \infty} \frac{\lambda_1^N - \lambda_2^N}{\lambda_1 - \lambda_2}, \tag{11.132}$$

where λ_1 and λ_2 are the eigenvalues established earlier. To first order in the $\Delta t \to 0$ or $N \to \infty$ limits, these eigenvalues are

$$\lambda_1 = -1 + i\omega \Delta t + \mathcal{O}(\Delta t^2) = -1 + \frac{i\omega T}{N} + \mathcal{O}(N^{-2}), \tag{11.133}$$

$$\lambda_2 = -1 - i\omega \Delta t + \mathcal{O}(\Delta t^2) = -1 - \frac{i\omega T}{N} + \mathcal{O}(N^{-2}).$$

Note that their difference, in the $N \to \infty$ limit, is

$$\lim_{N \to \infty} \lambda_1 - \lambda_2 = \frac{2i\omega T}{N} . \tag{11.134}$$

Further, the factors raised to the Nth power turn into exponentials:

$$\lim_{N \to \infty} \lambda_1^N = \lim_{N \to \infty} \left(-1 + \frac{i\omega T}{N} \right)^N = \lim_{N \to \infty} (-1)^N e^{-i\omega T} , \tag{11.135}$$

and similar for λ_2^N. Thus, in this limit, the determinant becomes

$$\lim_{\Delta t \to 0} \det \mathbb{B}_{N-1} = \lim_{N \to \infty} (-1)^N \frac{N}{\omega T} \frac{e^{-i\omega T} - e^{i\omega T}}{2i} \tag{11.136}$$

$$= \lim_{N \to \infty} i^{2(N-1)} N \frac{\sin(\omega T)}{\omega T} .$$

Finally, the path integral for the harmonic oscillator is[9]

$$Z = e^{i \frac{S[x_{\text{class}}]}{\hbar}} \lim_{N \to \infty} \sqrt{\frac{mN}{2\pi i \hbar T}} \frac{(\det \mathbb{B}_{N-1})^{-1/2}}{i^{N-1}} \tag{11.137}$$

$$= e^{i \frac{S[x_{\text{class}}]}{\hbar}} \sqrt{\frac{m}{2\pi i \hbar T}} \sqrt{\frac{\omega T}{\sin(\omega T)}} .$$

Indeed, as promised, actually calculating the path integral for even a simple system is extremely complicated, though the result is very simple. Nevertheless, it provides a profound interpretation of quantum mechanics and organizes problems in a very different way than the Schrödinger equation. You'll see some more examples and identify interesting properties of this harmonic oscillator path integral in the exercises.

Example 11.2   After calculating the path integral for the harmonic oscillator, you might need to take a breath or two. To relax, in this example, we will calculate the path integral for the free particle, using the same general techniques that were employed for the harmonic oscillator. That is, we will start from the form of Eq. (11.85), and calculate the necessary pieces for the free particle.

Solution

The first thing we will do is to calculate the classical action for the free particle. As a free particle, the potential energy is 0, and so the action is just the time integral of the kinetic energy:

$$S[x] = \int_0^T dt \, \frac{m}{2} \dot{x}^2 . \tag{11.138}$$

---

[9] As written, there is some ambiguity in the interpretation of "$\sqrt{i}$," because the square-root function has a branch cut in the complex plane. We won't concern ourselves with the ambiguity here, but it has been worked out; see N. S. Thornber and E. F. Taylor, "Propagator for the simple harmonic oscillator," *Am. J. Phys.* **66**, 1022–1024 (1998).

The Euler–Lagrange equation of motion of the free particle is just Newton's second law with no force:

$$m\ddot{x} = 0, \tag{11.139}$$

or the velocity of the particle is constant. The particle travels from x_i to x_f in time T, so the velocity is

$$\dot{x} = \frac{x_f - x_i}{T}. \tag{11.140}$$

Then, the classical action of the free particle evaluates to

$$S[x_{\text{class}}] = \int_0^T dt \, \frac{m}{2} \frac{(x_f - x_i)^2}{T^2} = \frac{m}{2} \frac{(x_f - x_i)^2}{T}. \tag{11.141}$$

The next thing to calculate from Eq. (11.85) would be the integrals over the different quantum deviation $y(t)$ from the classical trajectories. However, because the potential energy of the free particle is 0, this quantum deviation is rather trivial. There is no exponential phase factor to worry about because

$$\frac{\delta^2 L(x_{\text{class}}, \dot{x}_{\text{class}})}{\delta x_{\text{class}}^2} = 0, \tag{11.142}$$

and all higher functional derivatives vanish as well. So, the path integral for the free particle takes the simple form

$$Z = e^{i \frac{S[x_{\text{class}}]}{\hbar}} \int [dy]. \tag{11.143}$$

The path integral over y that remains just returns the normalization of the measure as established in Eq. (11.39), with total time $\Delta t = T$. Putting these pieces together, the free particle's path integral is

$$Z = e^{i \frac{m}{2} \frac{(x_f - x_i)^2}{\hbar T}} \sqrt{\frac{m}{2\pi i \hbar T}}. \tag{11.144}$$

There are a few things to note about this path integral that will be helpful for some of the exercises. First, this has a natural generalization to a free particle in multiple spatial dimensions. Instead of just squaring the difference between final and initial positions x_f and x_i, we need to take their vector difference, and then the dot product with itself:

$$(x_f - x_i)^2 \to (\vec{x}_f - \vec{x}_i) \cdot (\vec{x}_f - \vec{x}_i). \tag{11.145}$$

On the right, $\vec{x}_i$, for example, is the position vector of the initial location of the particle. Second, path integrals can be multiplied together if the particle passes through known intermediate points. For example, let's assume the particle starts at point x_i, then reaches point x_m after a time τ, then ends at point x_f after another time $T - \tau$. This is illustrated in Fig. 11.6. The path integral Z_1 for the the particle to reach point x_m from x_i is

$$Z_1 = e^{i \frac{m}{2} \frac{(x_m - x_i)^2}{\hbar \tau}} \sqrt{\frac{m}{2\pi i \hbar \tau}}. \tag{11.146}$$

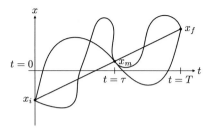

Fig. 11.6 Illustration of three paths from position x_i to position x_f in total time T, where all paths also pass through position x_m at intermediate time τ.

The path integral Z_2 for the particle to travel from x_m to x_f is

$$Z_2 = e^{i\frac{m}{2}\frac{(x_f - x_m)^2}{\hbar(T-\tau)}}\sqrt{\frac{m}{2\pi i\hbar(T-\tau)}}.$$ (11.147)

So, the path integral Z for the particle to travel from x_i to x_m and then to x_f is just their product:

$$Z = Z_1 Z_2 = e^{i\frac{m}{2\hbar}\left(\frac{(x_m - x_i)^2}{\tau} + \frac{(x_f - x_m)^2}{T-\tau}\right)}\frac{m}{2\pi i\hbar}\frac{1}{\sqrt{\tau(T-\tau)}}.$$ (11.148)

Exercises

11.1 The appearance of the classical Lagrangian and action in the path integral might be surprising given how we had constructed quantum mechanics throughout this book, but it was actually hidden in plain sight from the beginning. In this problem, we will show that we've seen the classical action many times before.

(a) A momentum eigenstate in position space can of course be expressed as

$$\langle x|p \rangle = e^{\frac{i}{\hbar}px},$$ (11.149)

for momentum p and position x. Show that the classical action for a free particle with momentum p and energy E that starts at position 0 at time $t = 0$ and ends at position x at time $t = T$ is

$$S_{\text{class}} = px - ET.$$ (11.150)

(b) For motion in two spatial dimensions, an eigenstate of the angular momentum operator in "angle" space can be expressed as

$$\langle \theta|L \rangle = e^{\frac{i}{\hbar}L\theta},$$ (11.151)

for angular momentum L and angle θ. Show that the classical action for a free particle with conserved angular momentum L and energy E that starts at angle 0 at time $t = 0$ and ends at angle θ at time $t = T$ is

$$S_{\text{class}} = L\theta - ET.\tag{11.152}$$

(c)   In Sec. 10.4, we introduced the WKB approximation to construct an estimate wavefunction by exploiting properties of the unitary translation operator when momentum is not conserved. We had demonstrated that the wavefunction $\psi(x)$ can be approximated from its known value at position x_0 as

$$\psi(x) \approx \exp\left[\frac{i}{\hbar}\int_{x_0}^{x} dx'\,\sqrt{2m\left(E - V(x')\right)}\right]\psi(x_0),\tag{11.153}$$

for some potential $V(x)$ and total energy E. Give this expression for the wavefunction time dependence, can you relate the factor in the exponent to the classical action for the particle in the potential? How is this expression related to the wavefunction constructed from the path integral, Eq. (11.56)?

11.2   In the evaluation of the path integral for the harmonic oscillator, we had left the classical action unevaluated. It's high time to fix that.

(a)   From the Lagrangian for the harmonic oscillator

$$L(x,\dot{x}) = \frac{m}{2}\dot{x}^2 - \frac{m\omega^2}{2}x^2,\tag{11.154}$$

show that its classical equation of motion, its Euler–Lagrange equation, is

$$m\ddot{x} + m\omega^2 x = 0.\tag{11.155}$$

(b)   Solve this differential equation for trajectory $x(t)$, with the boundary conditions that $x(t = 0) = x_i$ and $x(t = T) = x_f$, for initial and final positions x_i and x_f.

(c)   Now, with this trajectory $x(t)$, calculate the classical action

$$S[x] = \int_0^T dt\left(\frac{m}{2}\dot{x}^2 - \frac{m\omega^2}{2}x^2\right).\tag{11.156}$$

11.3   In evaluating the path integral of the harmonic oscillator, we had to perform a Gaussian integral for which the factor in the exponent was a Hermitian matrix $\mathbb{A}$ sandwiched between an N-dimensional real vector $\vec{x}$. We had shown that

$$\int_{-\infty}^{\infty} d\vec{x}\,e^{-\vec{x}^{\mathsf{T}}\mathbb{A}\vec{x}} = \pi^{N/2}\left(\det\mathbb{A}\right)^{-1/2}.\tag{11.157}$$

We then took the $N \to \infty$ limit to determine the path integral of the harmonic oscillator. However, we didn't have to discretize the time steps; we can directly evaluate the continuous infinity of position integrals directly. In that case we would have

$$\int_{-\infty}^{\infty} [dx]\,\exp\left[-\int_0^T dt\,x\left(\frac{d^2}{dt^2} + \omega^2\right)x\right] = \left(\det\frac{\frac{d^2}{dt^2} + \omega^2}{\pi}\right)^{-1/2}.\tag{11.158}$$

But what does the determinant of a differential operator mean?

(a) Determine the eigenvalues of the harmonic oscillator differential operator. That is, determine the values λ for a function of time $f_\lambda(t)$ such that

$$\left(\frac{d^2}{dt^2} + \omega^2\right) f_\lambda(t) = \lambda f_\lambda(t). \tag{11.159}$$

Make sure to enforce the boundary conditions that $f_\lambda(t = 0) = f_\lambda(t = T) = 0$.

(b) The determinant is the product of eigenvalues of an operator. Take all values λ established in the previous part and multiply them together. What is the determinant of this harmonic oscillator operator?

Hint: Leonard Euler first showed that the sine function can be expressed as an infinite product of its roots, analogous to a finite-order polynomial:

$$\sin x = x \prod_{n=1}^{\infty} \left(1 - \frac{x^2}{n^2 \pi^2}\right). \tag{11.160}$$

(c) How does this expression compare to the result we found in Sec. 11.5.4 from the limit of the discrete operator?

11.4 As emphasized in this chapter, calculating the path integral for even very simple quantum systems is highly non-trivial. In Example 11.2, we calculated the free-particle path integral one way, and here we will calculate it in two different ways.

(a) A zero-frequency harmonic oscillator is just a free particle. Take the $\omega \to 0$ limit of the harmonic oscillator path integral, Eq. (11.137), and show that it agrees with the result of Example 11.2.

(b) We can calculate this free-particle path integral in a different way. For a free particle, momentum eigenstates are also energy eigenstates, so we can also calculate the path integral using a procedure similar to that used in constructing Eq. (11.79). Generalize this equation for the case of continuous energy or momentum eigenstates and use it to evaluate the path integral for the free particle. Does it agree with your result in part (a)?

11.5 In this chapter, we only considered path integrals for one-dimensional systems, but in some cases it is straightforward to calculate the path integral for a system in three spatial dimensions. As an example of this, calculate the path integral for the three-dimensional harmonic oscillator, which has potential energy

$$U(\vec{r}) = \frac{m\omega^2}{2}\vec{r}\cdot\vec{r} = \frac{m\omega^2}{2}(x^2 + y^2 + z^2). \tag{11.161}$$

Don't forget that both the position and momentum of the particle in the potential have x, y, and z components.

11.6 In the calculation of the path integral, we of course have to sum over *all* possible paths between the initial and final positions. However, it can be useful to consider what contributions to the path integral look like for individual trajectories, to get a clearer idea of how the action varies for paths away from the classical trajectory. In this problem, we will focus on evaluation of the action

$$S[x] = \int_0^T dt \left(\frac{m}{2}\dot{x}^2 - U(x) \right),$$ (11.162)

where $x(t)$ is the trajectory that starts at x_i at $t = 0$ and ends at x_f at $t = T$, and $U(x)$ is the potential energy.

(a) Consider the system of a free particle which has trajectory $x(t)$ given by

$$x(t) = \begin{cases} x_i + \frac{2t}{T}(x' - x_i), & 0 < t < T/2, \\ x' + \frac{2}{T}\left(t - \frac{T}{2}\right)(x_f - x'), & T/2 < t < T. \end{cases}$$ (11.163)

Plot the value of the action corresponding to this trajectory as a function of the intermediate position x'.

(b) Consider the system of a free particle which has trajectory $x(t)$ given by

$$x(t) = x_i + \frac{t}{T}(x_f - x_i) + A \sin\left(\frac{\pi t}{T}\right).$$ (11.164)

Plot the value of the action corresponding to this trajectory as a function of the oscillation amplitude A.

(c) Consider the system of a free particle which has trajectory $x(t)$ given by

$$x(t) = x_i + \frac{t}{T}(x_f - x_i) + (x_f - x_i)\sin\left(\frac{n\pi t}{T}\right).$$ (11.165)

Plot the value of the action corresponding to this trajectory as a function of the frequency number $n \in \mathbb{Z}$, the integers.

(d) Repeat parts (a)–(c) for the harmonic oscillator with angular frequency ω. What do you notice that is different between the plots of the corresponding actions of the free particle and those of the harmonic oscillator?

11.7 Consider the familiar double-slit experiment: a quantum mechanical particle of mass m starts at some position to the left of the slits at time $t = 0$. As time evolves, the wavefunction of the particle passes through the slits and continues to a screen at which the particle is observed. This configuration is illustrated in Fig. 11.7. At $t = 0$, the particle is located a distance d to the left of the slits,

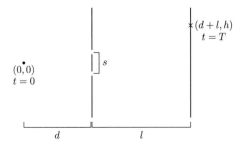

An illustration of the double-slit experiment, with the initial position of the particle left of the slits a distance d, and the viewing screen a distance l to the right of the slits.

directly in between them vertically. The slits are separated by a distance s. The screen is located a distance l to the right of the slits, and the particle is observed at time $t = T$ at a position $(d + l, h)$ with respect to the initial point. From the initial to final point the particle travels freely, under the influence of no potential.

(a) First, calculate the two path integrals corresponding to the two possible paths to the screen: through the upper or lower slit. Note that this is a two-dimensional problem.

(b) We do not measure which slit the particle passes through, so we need to sum these two path integrals as probability amplitudes. Do this, and determine the probability distribution for the particle to be detected at a location on the screen. Does this result seem familiar? Sketch a plot of the distribution you find as a function of the distance away from the center of the screen, h.

11.8 While we have focused on the Lagrangian and action for point particles in this chapter, it is possible to formulate the Schödinger equation itself through the principle of least action with a Lagrangian for the wavefunction $\psi(x,t)$. Consider the Lagrangian

$$\mathcal{L}(\psi, \psi', \dot{\psi}) = i\hbar \psi^* \dot{\psi} - \frac{\hbar^2}{2m} \psi^{*\prime} \psi' - V(x)\psi^* \psi. \tag{11.166}$$

Properly, this is a Lagrangian *density*, because its dimensions differ from a true Lagrangian by inverse length. Here, we have suppressed temporal and spatial dependence in the wavefunction, ψ^* is the complex conjugate of ψ, and $\dot{}$ and $'$ represent the temporal and spatial derivative, respectively. $V(x)$ is the potential, that only depends on spatial position x.

(a) Show that the Schrödinger equation follows from applying the principle of least action to this Lagrangian.
Hint: The wavefunction ψ and its complex conjugate ψ^* can be thought of as independent. Take the functional derivative of the Lagrangian with respect to ψ^* and set it equal to 0.

(b) Show that the complex conjugate of the Schrödinger equation follows from applying the principle of least action to this Lagrangian.

(c) The action that follows from this Lagrangian is

$$S[\psi] = \int_0^T dt \int_{-\infty}^{\infty} dx\, \mathcal{L}(\psi, \psi', \dot{\psi}). \tag{11.167}$$

Let's assume that ψ is an energy eigenstate wavefunction of the (arbitrary) quantum Hamiltonian. What is the action for such an eigenstate wavefunction?

12

The Density Matrix

Up until this point, we have essentially exclusively studied the quantum mechanics of individual particles. In this chapter, we will consider ensembles of quantum mechanical particles and attempt to understand their collective dynamics (or at least provide a brief introduction). By "ensemble," I mean that we are considering a quantum system with more than one particle in it. So, instead of, say, one spin-1/2 particle, we will consider a collection of N spin-1/2 particles, or N particles in the infinite square well, or something like that. With this set-up, we can immediately see that the quantum description of states on a Hilbert space is insufficient and incomplete for describing this collection of particles.

12.1 Description of a Quantum Ensemble

We have so far described a quantum system by a state on the Hilbert space $\mathcal{H}$, which can be represented by a sum over some set of basis states. Let's consider a state $|\psi\rangle \in \mathcal{H}$ and we will denote the orthonormal basis states as $|\chi_i\rangle$ so that

$$|\psi\rangle = \sum_{i=0}^{\infty} \beta_i |\chi_i\rangle, \tag{12.1}$$

and the β_i are some complex coefficients. Such a sum over the basis states is called a **coherent sum** because the phase of the coefficients is relevant in the sum. That is, a coherent sum can exhibit constructive or destructive interference. Indeed, when two waves pass through one another, the resulting wave is a superposition of the initial waves, in which the relative phases of the waves are important. If their phases differ by a multiple of 2π, the waves constructively interfere and if they differ by an odd multiple of π, they destructively interfere, for example.

Further, this $|\psi\rangle$ is a definite state on the Hilbert space. We happen to express it as a linear combination over some basis states, but that is merely a convenience: $|\psi\rangle$ is just the state $|\psi\rangle$, independent of basis. We had provided a physical interpretation for the β_i coefficients when $|\psi\rangle$ is expanded in some basis. $|\beta_i|^2$ represents the probability for an experiment to return the eigenvalue of the basis state $|\chi_i\rangle$. We defined this probability by considering N identical systems on which we perform the measurement. Then, as $N \to \infty$, the law of large numbers states that the number of measurements that produce the eigenvalue corresponding to state $|\chi_i\rangle$ is $|\beta_i|^2 N$. Note, however, that in this set-up as we have defined it so far, each of these N systems is isolated from the others, and we can unambiguously say that each system is state $|\psi\rangle$. Our "quantum system" is not N particles; we just have N quantum systems.

Now, let's consider a system of N particles. For concreteness and simplicity, we will consider N spin-1/2 particles for which each particle can be described by a linear combination of the two spin states. The term for a general two-state quantum system is **qubit**; that is, a quantum bit. With the formalism we have currently developed, we would want to write the quantum state of the system as some ket $|\psi\rangle$, which could be expressed as a coherent linear combination of eigenstates of $\hat{S}_z$, say

$$|\psi\rangle = \beta_\uparrow|\uparrow\rangle + \beta_\downarrow|\downarrow\rangle, \tag{12.2}$$

where $|\uparrow\rangle$ and $|\downarrow\rangle$ are the spin-up ($\hat{S}_z = +\frac{\hbar}{2}$) and spin-down ($\hat{S}_z = -\frac{\hbar}{2}$) eigenstates of $\hat{S}_z$, respectively. While this linear combination may seem to suggest an interpretation for the N particles in our system, we have to be careful. This state does *not* mean that a fraction $|\beta_\uparrow|^2$ spin-1/2 particles have spin-up and $|\beta_\downarrow|^2$ have spin-down. The state $|\psi\rangle$ is a definite state for spin and the expansion in eigenstates $|\uparrow\rangle$ and $|\downarrow\rangle$ just means that the definite spin points in a direction defined by the ratio of $\beta_\uparrow$ to $\beta_\downarrow$. We are representing the state of all N particles in our system by $|\psi\rangle$ and thus all N particles must be in state $|\psi\rangle$. That is, we say that these N particles are **polarized**, as all of their spins point in the same direction just like light when it passes through a polarizing filter. Further, because every particle is in the same state $|\psi\rangle$, it is therefore called a **pure state**. So, formally, the study of quantum mechanics of states in a Hilbert space is restricted to the study of the dynamics of pure states.

Why is this restrictive? Our N spin-1/2 particles clearly do not have to be polarized or all be in the same state. As an extreme example, we can consider an unpolarized collection of N spin-1/2 particles. Completely unpolarized means that $N/2$ particles have spin-up and $N/2$ have spin-down. Clearly in this case there is no way to write a single state on the Hilbert space $|\psi\rangle$ to represent this collection of particles. It is *not* correct to express the state of this unpolarized collection of particles as

$$|\psi\rangle = \frac{1}{\sqrt{2}}|\uparrow\rangle + \frac{1}{\sqrt{2}}|\downarrow\rangle, \tag{12.3}$$

as this actually describes a net spin pointed along the x-axis (it is an eigenstate of $\hat{S}_x$). However, for our unpolarized collection of particles, the net spin is 0: there are just as many up as down-spinning particles. So, this coherent sum of basis states is just the wrong way to express the quantum system. Instead, we must consider an **incoherent sum** of probabilities of the N particles to be in the spin-up or spin-down state. That is, each particle is in a definite eigenstate of $\hat{S}_z$, it's just that half have spin-up and the other half have spin-down. So, if we were to measure the spin of one particle at random, we would indeed find half the time spin-up and half the time spin-down. However, this would just be like putting your hand in a bag that contains an equal number of red and blue balls and pulling one out. The properties of the balls are fixed before you pull one out (there's no "collapsing" to an eigenstate); you happen to be ignorant of what ball you pull out because you couldn't see it before. We also call this unpolarized state of N particles **completely random** and it is very similar to classical probability.

Pure states and completely random states of an ensemble of N particles are ends of a spectrum of state configurations. In general, a state of N particles occupies a **mixed state**, which consists of some amount of polarization and some amount of randomness. More precisely, we can characterize a mixed state by a collection of pure states, which are present in the ensemble with some fraction. Let's call the collection of pure states present in the ensemble $\{|\psi_i\rangle\}_i$. As pure states, they are represented by an unambiguous ket on Hilbert space. Additionally, these pure states are present in the N-particle ensemble with fractions $\{w_i\}_i$. Note that these fractions/probabilities must sum to unity:

$$\sum_{i \text{ pure}} w_i = 1 \,. \tag{12.4}$$

Importantly, the pure states that compose our ensemble are not required to be an orthonormal basis, unlike what we discussed for specifying states on the Hilbert space.

In the following example, we will present a classical analysis of a system with two identical waves and you will study the analogous quantum system in Exercise 12.7.

Example 12.1   The **Hong–Ou–Mandel interferometer** is an experiment in which two identical photons are emitted from a laser toward separate mirrors that then reflect onto a **beam splitter** and can subsequently be detected.[1] When a photon hits the beam splitter, or half-silvered mirror, it has 50% probability of transmission and 50% probability of reflection. Two detectors are placed at the end of the experiment, and there are three possible observations: either one photon hits each detector, two photons hit one detector and no photons hit the other, or vice-versa. The experimental set-up is illustrated in Fig. 12.1. The beam splitter is the key device in this experiment and it has the important property that it can affect the phase of the photons that hit it. Recall that light that passes through an interface of higher to lower index of refraction does not change phase. However, light that is reflected from the interface of lower to higher index of refraction changes phase by π. While the analysis is a bit subtle, for a beam splitter, the phase difference between light that passes through and light that is reflected is π. Because light has 50% chance of transmission or reflection, we can say that the light that is reflected has a phase shift of $+\pi/2$ and the light that is transmitted has a phase shift of $-\pi/2$, for a difference of π.[2]

With this set-up, in this example, we will determine what would be detected in the two detectors *classically*. That is, assuming that light is just a classical electromagnetic wave, would you ever observe a hit in one detector and nothing in the other detector? Or, would you observe both detectors hit?

---

[1] C. K. Hong, Z. Y. Ou, and L. Mandel, "Measurement of subpicosecond time intervals between two photons by interference," *Phys. Rev. Lett.* **59**(18), 2044 (1987).

[2] V. Degiorgio, "Phase shift between the transmitted and the reflected optical fields of a semireflecting lossless mirror is $\pi/2$," *Am. J. Phys.* **48**, 81 (1980).

Solution

If we assume that light is a classical electromagnetic wave, we can just draw two waves emanating from the laser bound for the upper and lower mirror. Those waves then reflect and hit the beam splitter. Let's first consider the case where one wave is reflected off the beam splitter and the other passes through, as illustrated in Fig. 12.1. Up until the beam splitter, the two waves are always coherent, or in phase, because the light emitted by a laser is coherent, and the distance that both waves traveled to get to the beam splitter is identical. Because one wave reflects and the other transmits at the beam splitter, the two waves are then exactly out of phase by π in their travel to a detector. Otherwise identical waves that are out of phase by π superpose to a wave with zero amplitude, also known as nothing. Therefore, if both waves travel to a single detector, no signal is detected.

By contrast, if the two waves either both pass through or both reflect off the beam splitter, they travel to different detectors. Because the waves hit different detectors, they cannot interfere and would be observed as a signal in both detectors. Thus, classically, in the Hong–Ou–Mandel interferometer, you would only observe coincident hits in the two detectors, and never observe a single detector hit.

If one instead considers the photons to be identical quantum mechanical particles, the observation is very different. You will study this case in Exercise 12.7.

12.1.1 Ensemble Averages

Now, with this set-up, let's consider an observable $\hat{A}$ (i.e., a Hermitian operator) and calculate its expectation value or **ensemble average** on this system. For each pure state, we know the answer:

$$\langle \hat{A} \rangle_i \equiv \langle \psi_i | \hat{A} | \psi_i \rangle \,, \tag{12.5}$$

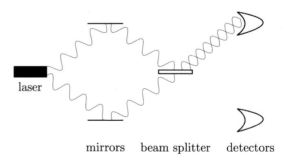

mirrors   beam splitter   detectors

Fig. 12.1   Simplified illustration of a Hong–Ou–Mandel interferometer in which two identical photons are emitted from the laser coincidentally, reflected off mirrors, have 50% probability of being reflected or transmitted off the beam splitter, and then are detected. Classical electromagnetic waves emitted from the laser are illustrated, and the wave reflected from the beam splitter shifts its phase by π.

and then each of these pure states are present with a fraction w_i, so we sum the $\langle A \rangle_i$ together weighted by w_i:

$$\langle \hat{A} \rangle = \sum_{i\text{ pure}} w_i \langle \hat{A} \rangle_i = \sum_{i\text{ pure}} w_i \langle \psi_i | \hat{A} | \psi_i \rangle. \tag{12.6}$$

To make a bit more sense of this, let's now introduce an orthonormal and complete basis on the Hilbert space in which the pure states live. Let's call this basis $\{|b_n\rangle\}_n$ and completeness means

$$\mathbb{I} = \sum_{n\text{ basis}} |b_n\rangle\langle b_n|, \tag{12.7}$$

where the sum runs over all basis states. Inserting this into the ensemble average for $\hat{A}$, we have

$$\langle \hat{A} \rangle = \sum_{i\text{ pure}} \sum_{n\text{ basis}} \sum_{m\text{ basis}} w_i \langle \psi_i | b_n \rangle \langle b_n | \hat{A} | b_m \rangle \langle b_m | \psi_i \rangle \tag{12.8}$$

$$= \sum_{n\text{ basis}} \sum_{m\text{ basis}} \left(\sum_{i\text{ pure}} w_i \langle b_m | \psi_i \rangle \langle \psi_i | b_n \rangle \right) \langle b_n | \hat{A} | b_m \rangle.$$

Going way back into the linear algebra discussed at the beginning of this book, we immediately recognize $\langle b_n | \hat{A} | b_m \rangle$ as the nth row, mth column matrix element of $\hat{A}$ in the basis $\{|b_n\rangle\}_n$: A_{nm}. Further, with this same interpretation, we identify

$$\rho_{mn} \equiv \sum_{i\text{ pure}} w_i \langle b_m | \psi_i \rangle \langle \psi_i | b_n \rangle \tag{12.9}$$

as the mth row, nth column of a matrix we will denote as ρ. That is, we can express the ensemble average of $\hat{A}$ as

$$\langle \hat{A} \rangle = \sum_{n\text{ basis}} \sum_{m\text{ basis}} \rho_{mn} A_{nm} = \text{tr}(\rho \hat{A}), \tag{12.10}$$

where tr denotes the trace, or sum of diagonal elements of the matrix $\rho \hat{A}$. Here, ρ is the **density matrix**[3] defined as

$$\rho = \sum_{i\text{ pure}} w_i |\psi_i\rangle\langle\psi_i|. \tag{12.11}$$

Thus, this density matrix is central for defining the most general state of a collection of quantum particles.

12.1.2 Properties of the Density Matrix

From its simple definition, we are able to immediately enumerate a number of interesting properties of the density matrix. First, it is Hermitian:

$$\rho^\dagger = \sum_{i\text{ pure}} w_i \left(|\psi_i\rangle\langle\psi_i| \right)^\dagger = \sum_{i\text{ pure}} w_i |\psi_i\rangle\langle\psi_i| = \rho, \tag{12.12}$$

[3] J. von Neumann, "Wahrscheinlichkeitstheoretischer Aufbau der Quantenmechanik," *Nachr. Ges. Wiss. Gott., Math.-Phys. Kl.* **1927**, 245–272 (1927); L. Landau, "Das dämpfungsproblem in der wellenmechanik," *Z. Phys.* **45**(5–6), 430–441 (1927).

as w_i is a real number on $[0, 1]$. Next, the trace of the density matrix is 1:

$$\text{tr}\rho = \sum_{n \text{ basis}} \sum_{i \text{ pure}} w_i \langle b_n | \psi_i \rangle \langle \psi_i | b_n \rangle = \sum_{i \text{ pure}} w_i \langle \psi_i | \psi_i \rangle = 1, \qquad (12.13)$$

where we have used the completeness of the basis states $\{|b_n\rangle\}_n$. Here $\text{tr}\rho = 1$ is the mixed-state formulation of probability conservation.

From the density matrix exclusively, we can define or determine if the system is in a pure state. Let's square ρ:

$$\rho^2 = \sum_{i,j \text{ pure}} w_i w_j |\psi_i\rangle \langle \psi_i | \psi_j \rangle \langle \psi_j| = \sum_{i,j \text{ pure}} (w_i w_j \langle \psi_i | \psi_j \rangle) |\psi_i\rangle \langle \psi_j|. \qquad (12.14)$$

Now, for a general mixed state, this is just some other matrix and in particular, note that the quantity in parentheses has an absolute value less than or equal to 1. It can only equal 1 if $w_i = w_j = 1$ and $\langle \psi_i | \psi_j \rangle = 1$, as $0 \le w_i \le 1$. However, because the sum of the probabilities w_i is 1, there must be only one non-zero w_i, or the entire system is in the same pure state. Thus, if one $w_i = 1$, we find

$$\rho^2 = |\psi_i\rangle \langle \psi_i| = \rho. \qquad (12.15)$$

So, an ensemble is in a pure state if and only if $\rho^2 = \rho$ or $\text{tr}\rho^2 = 1$. For this reason, $\text{tr}\rho^2$ is also called the **purity** of the ensemble.

Finally, let's quickly derive the time evolution of the density matrix, using the Schrödinger equation. Recall that the pure states evolve in time with the Schrödinger equation:

$$i\hbar \frac{d|\psi_i\rangle}{dt} = \hat{H}|\psi_i\rangle, \qquad (12.16)$$

and so the time derivative of the density matrix is

$$\frac{d\rho}{dt} = \sum_{i \text{ pure}} w_i \frac{d}{dt}(|\psi_i\rangle \langle \psi_i|) = \sum_{i \text{ pure}} w_i \left[\left(\frac{d}{dt}|\psi_i\rangle\right) \langle \psi_i| + |\psi_i\rangle \frac{d}{dt}\langle \psi_i| \right] \qquad (12.17)$$

$$= \sum_{i \text{ pure}} w_i \left(\frac{\hat{H}}{i\hbar} |\psi_i\rangle \langle \psi_i| - |\psi_i\rangle \langle \psi_i| \frac{\hat{H}}{i\hbar} \right)$$

$$= \frac{1}{i\hbar}[\hat{H}, \rho],$$

where we note that

$$\left(\frac{d|\psi_i\rangle}{dt}\right)^\dagger = \frac{d}{dt}\langle \psi_i| = \left(\frac{\hat{H}}{i\hbar}|\psi_i\rangle\right)^\dagger = -\langle \psi_i| \frac{\hat{H}}{i\hbar}. \qquad (12.18)$$

Thus

$$\boxed{i\hbar \frac{d\rho}{dt} = [\hat{H}, \rho],} \qquad (12.19)$$

which will be useful in what we do in the following sections.

12.2 Entropy

We have established that a system of particles is only in a pure state if the density matrix is **idempotent**; that is, the square of the density matrix is equal to itself $\rho^2 = \rho$. Can we quantify the extent to which a state is mixed? In a particular basis, we are always able to write the density matrix in diagonal form, where the w_i fractions occupy the diagonal entries, and 0 everywhere else:

$$\rho^{(\text{diag})} = \begin{pmatrix} w_1 & 0 & 0 & \cdots & 0 \\ 0 & w_2 & 0 & \cdots & 0 \\ 0 & 0 & w_3 & \cdots & 0 \\ \vdots & \vdots & \vdots & \ddots & \vdots \\ 0 & 0 & 0 & \cdots & w_N \end{pmatrix}. \tag{12.20}$$

The trace of a matrix is basis-independent, and so this diagonal form helps us analyze ρ^2. When in this diagonal form, ρ^2 is

$$\left(\rho^{(\text{diag})}\right)^2 = \begin{pmatrix} w_1^2 & 0 & 0 & \cdots & 0 \\ 0 & w_2^2 & 0 & \cdots & 0 \\ 0 & 0 & w_3^2 & \cdots & 0 \\ \vdots & \vdots & \vdots & \ddots & \vdots \\ 0 & 0 & 0 & \cdots & w_N^2 \end{pmatrix}, \tag{12.21}$$

and so its trace is

$$\text{tr}\rho^2 = \sum_{i \text{ pure}} w_i^2 = \sum_{i \text{ pure}} w_i^2 + \sum_{i<j \text{ pure}} 2w_i w_j - \sum_{i<j \text{ pure}} 2w_i w_j \tag{12.22}$$

$$= \left(\sum_{i \text{ pure}} w_i\right)^2 - \sum_{i<j \text{ pure}} 2w_i w_j = 1 - \sum_{i<j \text{ pure}} 2w_i w_j.$$

In these manipulations, we just added and subtracted a quantity that enabled us to complete the square. This demonstrates that $\text{tr}\rho^2 \leq 1 = \text{tr}\rho$, as all the w_i fractions are non-negative. Thus, a single number that quantifies the degree to which an ensemble is not in a pure state can be defined through the trace of ρ^2. Let's call this measure $S^{(2)}$:

$$S^{(2)} \equiv -\log\text{tr}\rho^2. \tag{12.23}$$

By the calculation above, this quantity is non-negative, $S^{(2)} \geq 0$, and is only 0 if the system is in a pure state.

Remember that if the system is in a pure state, every one of the quantum particles is in the same state $|\psi_i\rangle$. If the states are represented by spin, then we would say that the system is polarized. Another way to say this is that if your system is in a pure state, then you know exactly what state each individual particle occupies. If the system is not in a pure state then there exists a non-zero uncertainty in the state an individual particle occupies. Thus, $S^{(2)}$ is a measure of your knowledge of the states of individual particles, given information about the ensemble. If $S^{(2)} = 0$, then information about the ensemble is sufficient to have complete, perfect knowledge of all individual particle states. On the other hand, if $S^{(2)} > 0$, you only have incomplete knowledge of individual particle states.

Nevertheless, $S^{(2)}$ is just one possible measure for this information. We can define a one-parameter family of measures $S^{(\alpha)}$ where

$$S^{(\alpha)} = \frac{\log \operatorname{tr} \rho^{\alpha}}{1 - \alpha}, \tag{12.24}$$

where α is a positive real number. $S^{(\alpha)}$ satisfies the same properties that $S^{(2)}$ does: $S^{(\alpha)}$ is 0 if and only if the system is in a pure state, and $S^{(\alpha)} > 0$ otherwise. This quantity is called the **Rényi entropy**[4] with parameter α. Some values of α are particularly interesting. We have already discussed $\alpha = 2$, which exploits the idempotency of the density matrix for a pure state. As $\alpha \to \infty$, the Rényi entropy is set by the pure state with the largest fraction in the ensemble:

$$\lim_{\alpha \to \infty} S^{(\alpha)} = \lim_{\alpha \to \infty} \frac{\log \operatorname{tr} \rho^{\alpha}}{1 - \alpha} = -\log w_{\max}, \tag{12.25}$$

where $w_{\max}$ is the largest eigenvalue of the density matrix.

However, the most interesting and important value of α is as $\alpha \to 1$. Note that

$$\lim_{\alpha \to 1} \frac{\log \operatorname{tr} \rho^{\alpha}}{1 - \alpha} = \lim_{\alpha \to 1} \frac{\log \operatorname{tr} \left[\rho \rho^{\alpha - 1} \right]}{1 - \alpha} = \lim_{\alpha \to 1} \frac{\log \operatorname{tr} \left[\rho \, e^{(\alpha - 1) \log \rho} \right]}{1 - \alpha} \tag{12.26}$$

$$= \lim_{\alpha \to 1} \frac{\log \operatorname{tr} \left[\rho (1 - (1 - \alpha) \log \rho) \right]}{1 - \alpha} = -\operatorname{tr} (\rho \log \rho).$$

The $\alpha \to 1$ entropy is called the **von Neumann entropy**[5] S_{vN} and is defined to be

$$S_{\mathrm{vN}} \equiv -\operatorname{tr}(\rho \log \rho). \tag{12.27}$$

If the $\log \rho$ scares you, don't fret: we can always work with ρ in diagonal form or use the Taylor expansion of the logarithm:

$$\log \rho = \log \left(\mathbb{I} - (\mathbb{I} - \rho) \right) = -(\mathbb{I} - \rho) - \frac{(\mathbb{I} - \rho)^2}{2} + \cdots. \tag{12.28}$$

This von Neumann entropy is non-negative by construction and only vanishes for a pure state in which $\rho = 1$, so $\log \rho = 0$. A completely random state is a density matrix for which all diagonal entries are equal; that is, for a system in which every particle occupies a distinct pure state. For such a system, your knowledge about the ensemble provides the least possible useful information about each individual particle's state,

[4] A. Rényi, "On measures of entropy and information," in *Proceedings of the Fourth Berkeley Symposium on Mathematical Statistics and Probability*, Volume 1: Contributions to the Theory of Statistics, The Regents of the University of California (1961).

[5] J. von Neumann, "Thermodynamik quantenmechanischer gesamtheiten," *Nachr. Ges. Wiss. Gott., Math.-Phys. Kl.* **1927** 273–291 (1927); J. von Neumann, *Mathematical Foundations of Quantum Mechanics*, new edition, Princeton University Press (2018).

and so S_{vN} should be maximized. For a system of N particles, the completely random density matrix (when diagonalized) takes the form

$$\rho_{\text{random}} = \begin{pmatrix} \frac{1}{N} & 0 & 0 & \cdots & 0 \\ 0 & \frac{1}{N} & 0 & \cdots & 0 \\ 0 & 0 & \frac{1}{N} & \cdots & 0 \\ \vdots & \vdots & \vdots & \ddots & \vdots \\ 0 & 0 & 0 & \cdots & \frac{1}{N} \end{pmatrix}, \tag{12.29}$$

and so the von Neumann entropy of such a state would be

$$S_{vN} = -\sum_{i=1}^{N} \frac{1}{N} \log \frac{1}{N} = N \left(\frac{1}{N} \log N \right) = \log N. \tag{12.30}$$

Thus, in general, $0 \leq S_{vN} \leq \log N$ for an ensemble of N particles.

12.3 Some Properties of Entropy

Entropy is likely a quantity that you have some familiarity with, at least in the context of thermodynamics. We will touch on the connection to thermodynamics at the end of this chapter, but the von Neumann entropy that we have introduced thus far is really just a measure of the information content of a quantum ensemble. As such, the von Neumann entropy satisfies a number of properties and we will survey a few of them here.

12.3.1 Independent Systems

What makes the von Neumann entropy particularly nice and why it is *the* quantum mechanical entropy is due to its properties when two or more ensembles are combined. Consider two quantum ensembles, A and B, with density matrices ρ_A and ρ_B, respectively. If A and B are initially independent, but then combined to form a larger ensemble $A \oplus B$, the resulting density matrix is the **tensor product** of ρ_A and ρ_B:

$$\rho_{A \oplus B} = \rho_A \otimes \rho_B. \tag{12.31}$$

This can be proven easily when ρ_A and ρ_B are written in diagonal form. The tensor product means that each entry in ρ_A is multiplied by the entire matrix ρ_B. That is, the tensor product generalizes the outer product of two vectors to two matrices. The way to interpret this is the following. Given a state of system A with probability w_i, the state of B can be anything, by their assumed independence. We can draw a picture in Fig. 12.2 to illustrate this. In this figure, we have represented the ensembles as the boxes, and individual pure states by dots. Given the pure state i with probability w_i in A, the probabilities for i in A and j in B are just the product of the individual probabilities: $P(i \wedge j) = w_i v_j$, by the assumed independence. Now, the total probability to just be in

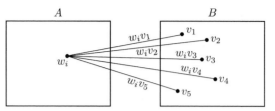

Fig. 12.2 Two independent systems A and B and the probabilities w_i and v_j, respectively, of measuring the pure states i and j in the two systems. As independent systems, their joint probability is just the product of the appropriate w_is and v_js.

state i in A is still w_i, which we can find by summing over the probabilities to be in any state of B:

$$P(i \in A) = \sum_{j \in B} P(i \wedge j) = \sum_{j \in B} w_i v_j = w_i \sum_{j \in B} v_j = w_i, \qquad (12.32)$$

as we have assumed that $\mathrm{tr}\rho_A = \mathrm{tr}\rho_B = 1$.

With this observation, we can then calculate the total entropy of the combined system:

$$S_{\mathrm{vN}}^{A \oplus B} = \mathrm{tr}(\rho_A \otimes \rho_b \log \rho_A \otimes \rho_B) = \mathrm{tr}\left[\rho_A \otimes \rho_B \log \rho_A + \rho_A \otimes \rho_B \log \rho_B\right] \qquad (12.33)$$
$$= \mathrm{tr}(\rho_A \log \rho_A) + \mathrm{tr}(\rho_B \log \rho_B) = S_{\mathrm{vN}}^A + S_{vN}^B.$$

By the manipulations above, the logarithm of a tensor product still expands out like you are familiar with for logarithms of products. That is, the von Neumann entropy is additive for independent systems. This also demonstrates that you can never gain information by adding a new, independent system. Because $S_{\mathrm{vN}} \geq 0$, we have

$$S_{\mathrm{vN}}^{A \oplus B} \geq \max(S_{\mathrm{vN}}^A, S_{\mathrm{vN}}^B). \qquad (12.34)$$

12.3.2 Subadditivity

A very powerful property of the von Neumann entropy, that does not exist for general Rényi entropies, is **subadditivity**. The property of subadditivity is the following. Consider two systems A and B, with no assumption of their independence. We will call the system formed from the union of A and B, system AB. Systems A, B, and AB have corresponding density matrices ρ_A, ρ_B, and ρ_{AB}, but we do not assume that ρ_{AB} is just the tensor product of ρ_A and ρ_B. Nevertheless, the density matrices for A and B individually can be found by tracing or summing over part of the density matrix ρ_{AB}. That is

$$\mathrm{tr}_B \rho_{AB} = \sum_{B \text{ states } |\psi_B\rangle} \langle \psi_B | \rho_{AB} | \psi_B \rangle = \rho_A, \qquad (12.35)$$

for example. This action of tracing or summing over a subset of the total system is referred to as the **partial trace**. Subadditivity is then the inequality that the von Neumann entropy of the combined system AB is less than or equal to the sum of entropies of A and B individually:

$$S_{vN}^{AB} = -\text{tr}_{A,B}\rho_{AB}\log\rho_{AB} \leq S_{vN}^{A} + S_{vN}^{B} = -\text{tr}_A\rho_A\log\rho_A - \text{tr}_B\rho_B\log\rho_B, \qquad (12.36)$$

where I've also denoted what system is being traced over in each term. While we won't provide a general proof of subadditivity here,[6] in the example below, we will construct an explicit counterexample for subadditivity of Rényi entropy with $\alpha \to \infty$ and we will see it in practice in a later section.

Of the Rényi entropies, only the von Neumann entropy enjoys the property of subadditivity. Its interpretation is similar to that of the triangle inequality, the simple statement that the sum of the length of two sides of a triangle is always at least the length of the third side. This is also related to the Cauchy–Schwarz inequality we had used in Chap. 4 to derive the uncertainty principle. We had seen above that if systems A and B are independent, then the inequality is saturated, and recall that smaller entropy means that information about the ensemble provides more information about each individual element. Thus, subadditivity means that a combined system has more information about its elements than the individual sum of its parts; or, that the correlations between the subsystems contain valuable information. We'll study this correlation and its interpretation in detail later. The von Neumann entropy satisfies a number of other very interesting relations that general Rényi entopies do not, that directly demonstrate its prominence as *the* quantum entropy.

Example 12.2 Let's see if we can make some more sense of the statement of subadditivity, and in particular, demonstrate that some particular Rényi entropies violate it. In this example, we will explicitly construct a counterexample for Rényi entropy with parameter $\alpha \to \infty$, and in Exercise 12.1, you will work to generalize this result. Let's consider a system of two spin-1/2 particles, call them 1 and 2, and their density matrix is diagonal and can be expressed as

$$\rho_{12} = \frac{1}{2}|\uparrow_1\uparrow_2\rangle\langle\uparrow_1\uparrow_2| + \frac{1}{4}|\uparrow_1\downarrow_2\rangle\langle\uparrow_1\downarrow_2| + \frac{1}{4}|\downarrow_1\uparrow_2\rangle\langle\downarrow_1\uparrow_2|, \qquad (12.37)$$

where the subscripts denote the specific particle's spin. What does subadditivity look like for this density matrix?

Solution

The first thing we need to do is to calculate the partial traces of this density matrix. The particle 1 and particle 2 reduced-density matrices are

$$\rho_1 = \text{tr}_2\rho_{12} = \langle\uparrow_2|\rho_{12}|\uparrow_2\rangle + \langle\downarrow_2|\rho_{12}|\downarrow_2\rangle = \frac{3}{4}|\uparrow_1\rangle\langle\uparrow_1| + \frac{1}{4}|\downarrow_1\rangle\langle\downarrow_1|,$$
$$\rho_2 = \text{tr}_1\rho_{12} = \langle\uparrow_1|\rho_{12}|\uparrow_1\rangle + \langle\downarrow_1|\rho_{12}|\downarrow_1\rangle = \frac{3}{4}|\uparrow_2\rangle\langle\uparrow_2| + \frac{1}{4}|\downarrow_2\rangle\langle\downarrow_2|. \qquad (12.38)$$

To evaluate the inner products in these partial traces, only the bras and kets of particle 1 combine together into inner products and the bras and kets of particle 2 similarly. No bra of particle 1 is taken with a ket of particle 2, for example. Further, the up- and

---

[6] Subadditivty of von Neumann entropy was first proved in H. Araki and E. H. Lieb, "Entropy inequalities," *Commun. Math. Phys.* **18**, 160–170 (1970).

down-spin states are orthogonal, $\langle \uparrow | \downarrow \rangle = 0$, which simplifies the expressions for the reduced-density matrices.

Let's first verify that subadditivity of the von Neumann entropy is satisfied by these density matrices. Because the density matrices are diagonal, it is trivial to read off their eigenvalues to input to the formulas for the entropies. The required entropies are

$$S_{\text{vN}}^{12} = -\frac{1}{2}\log\frac{1}{2} - \frac{1}{4}\log\frac{1}{4} - \frac{1}{4}\log\frac{1}{4} = \frac{3}{2}\log 2, \qquad (12.39)$$

$$S_{\text{vN}}^{1} = S_{\text{vN}}^{2} = -\frac{3}{4}\log\frac{3}{4} - \frac{1}{4}\log\frac{1}{4} = 2\log 2 - \frac{3}{4}\log 3. $$

Then, the statement of subadditivity of von Neumann entropy for this density matrix is

$$S_{\text{vN}}^{12} \leq S_{\text{vN}}^{1} + S_{\text{vN}}^{2} \quad \rightarrow \quad \frac{3}{2}\log 2 \approx 1.0397 \leq 4\log 2 - \frac{3}{2}\log 3 \approx 1.1247, \qquad (12.40)$$

which is indeed true.

Let's now consider the $\alpha \to \infty$ Rényi entropy, which had the definition from Eq. (12.25) of

$$S^{(\alpha \to \infty)} = -\log w_{\text{max}}, \qquad (12.41)$$

where w_{max} is the largest eigenvalue of the density matrix. Now, the relevant entropies with this set of density matrices are

$$S^{(\alpha \to \infty),12} = -\log\frac{1}{2} = \log 2, \qquad (12.42)$$

$$S^{(\alpha \to \infty),1} = S^{(\alpha \to \infty),2} = -\log\frac{3}{4} = \log\frac{4}{3}. $$

Subadditivity of this entropy would require

$$S^{(\alpha \to \infty),12} \leq S^{(\alpha \to \infty),1} + S^{(\alpha \to \infty),2} \quad \rightarrow \quad \log 2 \approx 0.6932 \leq 2\log\frac{4}{3} \approx 0.5754, \quad (12.43)$$

which is clearly false. So at least Rényi entropy with $\alpha \to \infty$ does not satisfy subadditivity. In fact, because this inequality is violated by nearly 20%, this density matrix will be a counterexample for subadditivity of Rényi entropies for a wide range of parameter α. You will study this in Exercise 12.1.

12.3.3 Time Dependence

It's also interesting to ask about the time derivative of the entropy. To evaluate this, we will work with the Rényi entropy as it is a bit easier to work with, but the $\alpha \to 1$ limit can be taken to find the time derivative of the von Neumann entropy. The time derivative is

$$\frac{dS^{(\alpha)}}{dt} = \frac{d}{dt}\frac{\log \text{tr}\rho^{\alpha}}{1-\alpha} = \frac{1}{1-\alpha}\frac{1}{\text{tr}\rho^{\alpha}}\text{tr}\frac{d\rho^{\alpha}}{dt} \qquad (12.44)$$

$$= \frac{\alpha}{1-\alpha}\frac{1}{\text{tr}\rho^{\alpha}}\text{tr}\left(\rho^{\alpha-1}\frac{1}{i\hbar}[\hat{H},\rho]\right),$$

where we used the chain rule for differentiation. We can further massage the trace as

$$\frac{dS^{(\alpha)}}{dt} = \frac{\alpha}{1-\alpha} \frac{1}{i\hbar \, \mathrm{tr}\rho^\alpha} \mathrm{tr}\left(\rho^{\alpha-1}\hat{H}\rho - \rho^{\alpha-1}\rho\hat{H}\right), \tag{12.45}$$

now we use the cyclicity of the trace:

$$\mathrm{tr}(\mathbb{ABC}) = \mathrm{tr}(\mathbb{CAB}), \tag{12.46}$$

for three matrices $\mathbb{A}, \mathbb{B}, \mathbb{C}$. This follows from just writing the trace as a sum over explicit matrix elements:

$$\mathrm{tr}(\mathbb{ABC}) = \sum_{i,j,k} \mathbb{A}_{ij}\mathbb{B}_{jk}\mathbb{C}_{ki} = \sum_{i,j,k} \mathbb{C}_{ki}\mathbb{A}_{ij}\mathbb{B}_{jk} \tag{12.47}$$

$$= \mathrm{tr}(\mathbb{CAB}).$$

Thus

$$\mathrm{tr}(\rho^{\alpha-1}\hat{H}\rho) = \mathrm{tr}(\rho^\alpha\hat{H}), \tag{12.48}$$

and so

$$\frac{dS^{(\alpha)}}{dt} = \frac{\alpha}{1-\alpha} \frac{1}{i\hbar \, \mathrm{tr}\rho^\alpha} \mathrm{tr}\left(\rho^\alpha\hat{H} - \rho^\alpha\hat{H}\right) = 0, \tag{12.49}$$

or, the entropy is constant in time. Now, you might be familiar with the second law of thermodynamics which states that the entropy of a closed system can never decrease. This, at least, is not inconsistent with that, but demonstrating that entropy can increase requires significantly more sophistication and analysis of mixed states and open quantum systems.

12.3.4 Invariance Under Unitary Transformation

Nevertheless, we can demonstrate a related property: that a pure state can never become a mixed state, assuming unitary time evolution. In terms of the density matrix at time $t = 0$, $\rho(0)$, the density matrix at time t is just

$$\rho(t) = \sum_{i \text{ pure}} w_i |\psi_i(t)\rangle\langle\psi_i(t)| = \sum_{i \text{ pure}} w_i e^{-i\frac{t\hat{H}}{\hbar}} |\psi_i(0)\rangle\langle\psi_i(0)| e^{i\frac{t\hat{H}}{\hbar}} \tag{12.50}$$

$$= e^{-i\frac{t\hat{H}}{\hbar}} \rho(0) e^{i\frac{t\hat{H}}{\hbar}},$$

assuming that the pure states evolve according to the Schrödinger equation. Now, let's calculate $\rho(t)^2$: if this is $\rho(t)$, then the system is in a pure state at time t:

$$\rho(t)^2 = e^{-i\frac{t\hat{H}}{\hbar}} \rho(0) e^{i\frac{t\hat{H}}{\hbar}} e^{-i\frac{t\hat{H}}{\hbar}} \rho(0) e^{i\frac{t\hat{H}}{\hbar}} \tag{12.51}$$

$$= e^{-i\frac{t\hat{H}}{\hbar}} \rho(0)^2 e^{i\frac{t\hat{H}}{\hbar}}.$$

If the system is initially in a pure state at $t = 0$, then $\rho(0)^2 = \rho(0)$, and so we further find

$$\rho(t)^2 = e^{-i\frac{t\hat{H}}{\hbar}} \rho(0)^2 e^{i\frac{t\hat{H}}{\hbar}} = e^{-i\frac{t\hat{H}}{\hbar}} \rho(0) e^{i\frac{t\hat{H}}{\hbar}} = \rho(t), \tag{12.52}$$

and thus pure states always evolve into pure states.[7]

We can prove even more general invariance properties of entropy. Consider an arbitrary unitary matrix $\mathbb{U}$ that transforms the density matrix:

$$\rho' = \mathbb{U}\rho\mathbb{U}^\dagger. \tag{12.53}$$

The Rényi entropy for the new density matrix is

$$S^{(\alpha)}(\rho') = \frac{\log \operatorname{tr}\rho'^\alpha}{1-\alpha} = \frac{\log \operatorname{tr}(\mathbb{U}\rho^\alpha\mathbb{U}^\dagger)}{1-\alpha} = \frac{\log \operatorname{tr}\rho^\alpha}{1-\alpha} = S^{(\alpha)}(\rho), \tag{12.54}$$

the Rényi entropy of the original density matrix. To establish this, we have used cyclicity of the trace and

$$\underbrace{\mathbb{U}\rho\mathbb{U}^\dagger\mathbb{U}\rho\mathbb{U}^\dagger\cdots\mathbb{U}\rho\mathbb{U}^\dagger}_{\alpha \text{ times}} = \mathbb{U}\rho^\alpha\mathbb{U}^\dagger, \tag{12.55}$$

by unitarity of $\mathbb{U}$. That is, under unitary transformations of the density matrix, the entropy is unaffected. No unitary transformation can change the information content of a quantum system.

Let's now consider an example which can illuminate properties of the density matrix and entropies that we have discussed in generality.

Example 12.3 Let's go back to our spin-1/2 friends and construct the density matrix for a collection of spin-1/2 particles. Let's say that we have a partially polarized beam for which a fraction of $3/4$ are polarized in the $+z$ direction and $1/4$ of the sample are polarized along the $+x$ direction. Our density matrix of this system is

$$\rho = \frac{3}{4}|\uparrow_z\rangle\langle\uparrow_z| + \frac{1}{4}|\uparrow_x\rangle\langle\uparrow_x|. \tag{12.56}$$

It's easy enough to calculate the von Neumann entropy of this density matrix, as it is already in diagonal form. We find

$$S_{\text{vN}} = -\operatorname{tr}(\rho\log\rho) = -\frac{3}{4}\log\frac{3}{4} - \frac{1}{4}\log\frac{1}{4} \simeq 0.562. \tag{12.57}$$

We'll calculate the entropy later in a different way, so it's important to emphasize this result. The von Neumann entropy does not assume anything about the relationship between the pure states that compose the density matrix; in fact, it assumes that there is no relationship between them whatsoever. In the case at hand, however, we know additional information that we did not tell our von Neumann entropy: we also know that eigenstates of the spin operator $\hat{S}_x$ can be expressed in terms of eigenstates of $\hat{S}_z$.

---

[7] Assuming the Schrödinger equation, pure states always evolve into pure states. One could imagine that there is a more general time-evolution operator for the density matrix that allows pure states to evolve into mixed states. However, such an operator necessarily violates energy conservation or other fundamental principles of physics; see T. Banks, L. Susskind, and M. E. Peskin, "Difficulties for the evolution of pure states into mixed states," *Nucl. Phys. B* **244**, 125–134 (1984).

Let's use this additional information to re-express the density matrix exclusively in terms of eigenstates of $\hat{S}_z$. Recall that the spin-up state along the x-axis can be expressed as

$$|\uparrow_x\rangle = \frac{1}{\sqrt{2}}(|\uparrow_z\rangle + |\downarrow_z\rangle),$$ (12.58)

exclusively in terms of the eigenstates of $\hat{S}_z$. Using this relationship, the density matrix can then be written as

$$\rho = \frac{3}{4}|\uparrow_z\rangle\langle\uparrow_z| + \frac{1}{4}\frac{1}{\sqrt{2}}(|\uparrow_z\rangle + |\downarrow_z\rangle)\frac{1}{\sqrt{2}}(\langle\uparrow_z| + \langle\downarrow_z|)$$ (12.59)

$$= \frac{7}{8}|\uparrow\rangle\langle\uparrow| + \frac{1}{8}|\uparrow\rangle\langle\downarrow| + \frac{1}{8}|\downarrow\rangle\langle\uparrow| + \frac{1}{8}|\downarrow\rangle\langle\downarrow|.$$

Here, we have dropped the z subscript on the second line as all of these spin states are eigenstates of $\hat{S}_z$. Now we can write this as an explicit 2×2 matrix, just by reading off the coefficients of the terms in this representation of the density matrix:

$$\rho = \begin{pmatrix} \frac{7}{8} & \frac{1}{8} \\ \frac{1}{8} & \frac{1}{8} \end{pmatrix}.$$ (12.60)

Note indeed that $\text{tr}\rho = 1$. Additionally, we can easily calculate ρ^2 to be

$$\rho^2 = \begin{pmatrix} \frac{7}{8} & \frac{1}{8} \\ \frac{1}{8} & \frac{1}{8} \end{pmatrix}\begin{pmatrix} \frac{7}{8} & \frac{1}{8} \\ \frac{1}{8} & \frac{1}{8} \end{pmatrix} = \begin{pmatrix} \frac{25}{32} & \frac{1}{8} \\ \frac{1}{8} & \frac{1}{32} \end{pmatrix},$$ (12.61)

which is not ρ and so this indeed is not a pure state.

In this representation of the density matrix, let's calculate the von Neumann entropy. We need to calculate

$$S_{\text{vN}} = -\text{tr}(\rho \log \rho) = -\sum_{i \text{ pure}} \rho_{ii}^{(\text{diag})} \log \rho_{ii}^{(\text{diag})},$$ (12.62)

when the density matrix is expressed in diagonal form. When diagonal, the density matrix elements that are non-zero are just its eigenvalues; hence another way to express the von Neumann entropy is

$$S_{\text{vN}} = -\sum_i \lambda_i \log \lambda_i,$$ (12.63)

where the set $\{\lambda_i\}$ are the eigenvalues of the density matrix. We can find the eigenvalues with the old trick that $\det\rho = \lambda_1\lambda_2$ and $\text{tr}\rho = 1 = \lambda_1 + \lambda_2$. Using the trace relation, the determinant is $\det\rho = \lambda(1-\lambda)$, where λ is either eigenvalue. The determinant of the matrix at hand is

$$\det\rho = \begin{vmatrix} \frac{7}{8} & \frac{1}{8} \\ \frac{1}{8} & \frac{1}{8} \end{vmatrix} = \frac{3}{32},$$ (12.64)

so the eigenvalues satisfy

$$\lambda - \lambda^2 = \frac{3}{32}.$$ (12.65)

The eigenvalues are then

$$\lambda_1 = \frac{1}{2} + \frac{1}{2}\sqrt{\frac{5}{8}}, \qquad\qquad \lambda_2 = \frac{1}{2} - \frac{1}{2}\sqrt{\frac{5}{8}}. \qquad (12.66)$$

Using this result, we find

$$S_{vN} = -\lambda_1 \log \lambda_1 - \lambda_2 \log \lambda_2 \qquad\qquad (12.67)$$

$$= -\left(\frac{1}{2} + \frac{1}{2}\sqrt{\frac{5}{8}}\right) \log \left(\frac{1}{2} + \frac{1}{2}\sqrt{\frac{5}{8}}\right) - \left(\frac{1}{2} - \frac{1}{2}\sqrt{\frac{5}{8}}\right) \log \left(\frac{1}{2} - \frac{1}{2}\sqrt{\frac{5}{8}}\right)$$

$$\simeq 0.335.$$

Interestingly, this value of the entropy is less than what we calculated in Eq. (12.57), suggesting that this density matrix contains more information about the individual states than in its original representation. While it may seem like all we did was re-write the density matrix from expressing pure states as eigenstates of $\hat{S}_x$ and $\hat{S}_z$ to states purely as eigenstates of $\hat{S}_z$, we used additional information to do so. We needed to know that there is indeed a transformation that enables one to express eigenstates of $\hat{S}_x$ in terms of eigenstates of $\hat{S}_z$. Again, the von Neumann entropy only measures the information that you explicitly give it: in the original form of the density matrix, the eigenstates of $\hat{S}_x$ and $\hat{S}_z$ are assumed to be completely independent. Indeed, there was no information provided to the von Neumann entropy to suggest otherwise. When the density matrix is written in an orthonormal basis, we have used all information possible to eliminate dependences between the pure states.

12.4 Quantum Entanglement

The density matrix and von Neumann entropy enable a concrete, quantitative study of phenomena that can happen quantum mechanically but not classically. Perhaps one of the strangest and most profound is that of quantum **entanglement**, correlations between particles in an ensemble that are not possible classically. We'll just introduce this extremely rich topic here through study of a particular illustrative example and through it attempt to define and identify general features of entanglement.

Let's consider a system of two spin-1/2 particles; call them 1 and 2. We will construct the state vector on the Hilbert space of the two particles that represents perfect anti-correlation of the two spins. That is, if one spin is up, the other is down, and vice-versa, with equal probability. We denote this state by $|\psi\rangle$ and it can be expressed as

$$|\psi\rangle = \frac{1}{\sqrt{2}}(|\uparrow_1\rangle|\downarrow_2\rangle + |\downarrow_1\rangle|\uparrow_2\rangle). \qquad (12.68)$$

Here, the spins are defined with respect to the z-axis, subscripts denote the particular particle, and the product of kets is understood to be their tensor product. The factor of $1/\sqrt{2}$ ensures that the state is normalized, namely

$$\langle\psi|\psi\rangle = 1 = \frac{1}{\sqrt{2}}((\langle\uparrow_1|\langle\downarrow_2| + \langle\downarrow_1|\langle\uparrow_2|)\frac{1}{\sqrt{2}}(|\uparrow_1\rangle|\downarrow_2\rangle + |\downarrow_1\rangle|\uparrow_2\rangle)) \qquad (12.69)$$

$$= \frac{1}{2}((\langle\uparrow_1|\uparrow_1\rangle\langle\downarrow_2|\downarrow_2\rangle + \langle\uparrow_1|\downarrow_1\rangle\langle\downarrow_2|\uparrow_2\rangle + \langle\downarrow_1|\uparrow_1\rangle\langle\uparrow_2|\downarrow_2\rangle + \langle\downarrow_1|\downarrow_1\rangle\langle\uparrow_2|\uparrow_2\rangle)).$$

The density matrix of this configuration of particles is especially simple because $|\psi\rangle$ is a pure state on Hilbert space. Let's denote the density matrix of the combined 1 and 2 particle system ρ_{12} and we have

$$\rho_{12} = |\psi\rangle\langle\psi| \qquad (12.70)$$

$$= \frac{1}{2}(|\uparrow_1\rangle|\downarrow_2\rangle\langle\uparrow_1|\langle\downarrow_2| + |\uparrow_1\rangle|\downarrow_2\rangle\langle\downarrow_1|\langle\uparrow_2| + |\downarrow_1\rangle|\uparrow_2\rangle\langle\uparrow_1|\langle\downarrow_2| + |\downarrow_1\rangle|\uparrow_2\rangle\langle\downarrow_1|\langle\uparrow_2|).$$

In the second line, we have expanded out the density matrix in the basis of individual particle spins, which will be useful in what follows. Because $|\psi\rangle$ is normalized and a pure state, we know that $\text{tr}\rho_{12} = \text{tr}\rho_{12}^2$ and so the von Neumann entropy vanishes:

$$S_{vN}^{12} = 0. \qquad (12.71)$$

We can also test the property of subadditivity. From this pure state, the density matrix of just particle 1, for example, is found by tracing over particle 2:

$$\rho_1 = \text{tr}_2\rho_{12} = \langle\uparrow_2|\rho_{12}|\uparrow_2\rangle + \langle\downarrow_2|\rho_{12}|\downarrow_2\rangle \qquad (12.72)$$

$$= \frac{1}{2}(|\uparrow_1\rangle\langle\uparrow_1| + |\downarrow_1\rangle\langle\downarrow_1|).$$

The density matrix for particle 2 is found similarly and takes a similar form:

$$\rho_2 = \text{tr}_1\rho_{12} = \frac{1}{2}(|\uparrow_2\rangle\langle\uparrow_2| + |\downarrow_2\rangle\langle\downarrow_2|). \qquad (12.73)$$

Note that the density matrices for both particles 1 and 2 individually actually describe a mixed state, and further that state is completely random! Thus, the von Neumann entropies of particles 1 and 2 individually are

$$S_{vN}^1 = S_{vN}^2 = \log 2. \qquad (12.74)$$

This equality of the von Neumann entropies for partial systems whose union is a pure state is true in general. That is, for any pure state AB, the entropies of its mutually exclusive partial traced systems A and B are equal:

$$S_{vN}^{AB} = 0, \qquad S_{vN}^A = S_{vN}^B. \qquad (12.75)$$

Then indeed subadditivity holds:

$$S_{vN}^{12} = 0 \leq S_{vN}^1 + S_{vN}^2 = 2\log 2. \qquad (12.76)$$

12.4.1 Purification

The observation that the von Neumann entropies of mutually exclusive partial-traced systems from a pure state are equal is a specific example of a more general feature of the partial trace. Consider an arbitrary system described by a density matrix ρ, which can be expressed as

$$\rho = \sum_{n \text{ basis}} w_n |b_n\rangle\langle b_n|. \tag{12.77}$$

Here, $\{|b_n\rangle\}_n$ is an orthonormal basis on Hilbert space and w_n is the probability of the state $|b_n\rangle$. Note that this density matrix is diagonal and its eigenvalues are the w_ns. In general, this is not a pure state, but it can always be expressed as the partial trace of some pure state, called the **purification** of ρ.[8] We will demonstrate this by explicitly constructing such a pure state.

Consider the pure state $|\psi\rangle$ formed from the tensor product of two orthonormal bases on the Hilbert space, $\{|b_n\rangle\}_n$ and $\{|c_n\rangle\}_n$:

$$|\psi\rangle = \sum_{n \text{ basis}} \sqrt{w_n} |b_n\rangle|c_n\rangle. \tag{12.78}$$

Note that the w_n are the eigenvalues of the density matrix of the original mixed state from Eq. (12.77). Now, let's form its density matrix and take the partial trace over the $\{|c_n\rangle\}_n$ basis states:

$$\text{tr}_c|\psi\rangle\langle\psi| = \text{tr}_c\left(\sum_{n \text{ basis}} \sqrt{w_n}|b_n\rangle|c_n\rangle\right)\left(\sum_{m \text{ basis}} \sqrt{w_m}\langle b_m|\langle c_m|\right) \tag{12.79}$$

$$= \sum_{n \text{ basis}}\sum_{m \text{ basis}} \sqrt{w_n w_m}|b_n\rangle\langle b_m|\langle c_m|c_n\rangle$$

$$= \sum_{n \text{ basis}}\sum_{m \text{ basis}} \sqrt{w_n w_m}|b_n\rangle\langle b_m|\delta_{nm}$$

$$= \sum_{n \text{ basis}} w_n|b_n\rangle\langle b_n| = \rho.$$

Note that this construction is completely general, and so every mixed state has a purification. Also, the purification is not unique, because the orthonormal basis $\{|c_n\rangle\}_n$ is only required to span the Hilbert space, but is otherwise unconstrained.

There are a number of interesting properties of the purification, but we'll focus on just a couple here. Importantly, the eigenvalue spectrums of the reduced-density matrices $\text{tr}_c|\psi\rangle\langle\psi|$ and $\text{tr}_b|\psi\rangle\langle\psi|$ are equal; indeed, note that

$$\text{tr}_b|\psi\rangle\langle\psi| = \sum_{n \text{ basis}} w_n|c_n\rangle\langle c_n|, \tag{12.80}$$

which has eigenvalues of $\{w_n\}_n$ by the orthonormality of the $\{|c_n\rangle\}_n$ basis. Because the trace of a matrix is the sum of its eigenvalues, the trace over any function f of these two partial traces is equal:

$$\text{tr}_b f\left(\text{tr}_c|\psi\rangle\langle\psi|\right) = \text{tr}_c f\left(\text{tr}_b|\psi\rangle\langle\psi|\right). \tag{12.81}$$

We can apply this result to the von Neumann entropy of the pure state with density matrix ρ_{12} and partial traces $\text{tr}_2\rho_{12} = \rho_1$ and $\text{tr}_1\rho_{12} = \rho_2$. The von Neumann entropy

---

[8] H. Araki and E. H. Lieb, "Entropy inequalities," *Commun. Math. Phys.* **18**, 160–170 (1970).

of ρ_1 is

$$
\begin{aligned}
S_{\text{vN}}^1 &= -\text{tr}_1\left(\rho_1\log\rho_1\right) = -\text{tr}_1\left[\text{tr}_2(\rho_{12})\log\text{tr}_2(\rho_{12})\right] = -\text{tr}_2\left[\text{tr}_1(\rho_{12})\log\text{tr}_1(\rho_{12})\right] \\
&= S_{\text{vN}}^2,
\end{aligned}
\tag{12.82}
$$

as we observed in the example above, but now proved in generality.

12.4.2 Entropy of Entanglement

This property that a pure state can transform into a mixed state when a particle is traced over is a manifestation of entanglement. The action of "tracing over" a particle is equivalent to ignorance of that particle. That is, on the state $|\psi\rangle$, if we simply ignore particle 2, then the spin of particle 1 is completely random; it has 50% probability of being either up or down. However, once you know the spin of particle 1, then the spin of particle 2 is fixed. This is precisely what the vanishing of the von Neumann entropy S_{vN}^{12} means. There is significant information in the correlation of the particles in the state $|\psi\rangle$ on Hilbert space, thus the entropy is small. Further, just knowing the fact that the spins of the two particles are perfectly anti-correlated means that we can essentially uniquely write down the pure state that describes such a system. Hence why the entropy vanishes. By ignoring one particle and tracing over it, we lose that correlation information, and an individual particle instead has a completely random spin. The entropy after tracing over a subset of the initial state is referred to as the **entropy of entanglement**. In this case, the entropy of entanglement is $\log 2$; or, the information contained in the correlation of the spins of the two particles is $\log 2$.

Note that the entropy of entanglement only vanishes when the state after performing the partial trace is also a pure state. The only way in which this can happen is if the state of the composite system is **separable**, or the state is just the tensor product of pure states of its constituents. For example, the state $|\chi\rangle$ is separable, where

$$
|\chi\rangle = |\uparrow_1\rangle|\downarrow_2\rangle.
\tag{12.83}
$$

Separable states exhibit no correlations between their constituent systems, therefore there is no information in their correlation.

While it may seem like entanglement, as we have defined thus far, is purely a quantum phenomenon, we have to be careful. For example, let's consider the following system of two spins that exhibits exclusively classical correlations. The system of interest now is a state of two particles, 1 and 2, in which their spins are opposite. This system is analogous to having two balls in a bag, one of which is red and the other is blue. The identity of the balls is always well-defined, you just don't know which is ball 1 and which is ball 2 before you draw one out. Note that this "classical" uncertainty is very different than the pure quantum state $|\psi\rangle$ studied earlier. In the state $|\psi\rangle$, the identity of the spin of particle 1 was undefined and unknowable before you make a measurement.

In this case, in which you have a spin-up and a spin-down but you just don't know *a priori* which is particle 1 and which is particle 2, the density matrix is

$$\rho_{\text{class}} = \frac{1}{2}(|\uparrow_1\rangle|\downarrow_2\rangle\langle\uparrow_1|\langle\downarrow_2| + |\downarrow_1\rangle|\uparrow_2\rangle\langle\downarrow_1|\langle\uparrow_2|). \tag{12.84}$$

This density matrix does not describe a pure state on Hilbert space, and actually describes a completely random system, given knowledge that there is one spin-up and one spin-down particle. As such, just knowing the anti-correlation between the spins is not sufficient to have perfect knowledge of the state of the system. In particular, the von Neumann entropy of this system is

$$S_{\text{vN}}(\rho_{\text{class}}) = \log 2, \tag{12.85}$$

as there are two pure states with equal probability in the density matrix. We can correspondingly determine the entropy of the reduced system, when we trace over particle 1 or particle 2. For example, the density matrix for particle 1 exclusively by tracing over particle 2 is

$$\rho_{1,\text{class}} = \text{tr}_2\,\rho_{\text{class}} = \langle\uparrow_2|\rho_{\text{class}}|\uparrow_2\rangle + \langle\downarrow_2|\rho_{\text{class}}|\downarrow_2\rangle \tag{12.86}$$

$$= \frac{1}{2}(|\uparrow_1\rangle\langle\uparrow_1| + |\downarrow_1\rangle\langle\downarrow_1|),$$

which is the same reduced-density matrix that we found from tracing over particle 2 in the pure state $|\psi\rangle$. That also means that the von Neumann entropy of this reduced-density matrix is still $\log 2$:

$$S_{\text{vN}}(\rho_{1,\text{class}}) = \log 2, \tag{12.87}$$

and there was no information gained or lost in ignoring particle 2.

In this case, we would still say that the entropy of entanglement is $\log 2$, as that is the entropy after performing the partial trace of the density matrix. While this happens to be the same as what we found for the partial trace of the pure state $|\psi\rangle$, there is a real sense in which this entropy is a measure of a different form of correlation and information about the state. Quantum entanglement, as exhibited in the state $|\psi\rangle$, is a manifestation of a correlation between subsystems of a larger ensemble for which the identity of the subsystems is undefined until a measurement is performed. For quantum-entangled subsystems, all that is known about them is their correlation on the Hilbert space, and not their individual identities. Classically entangled particles, or, perhaps more properly, classically correlated particles have a fixed and well-defined identity before any measurement is performed; you just happen to be ignorant of it. This subtle yet profound distinction between quantum and classical correlation has far-reaching consequences for interpreting and using quantum mechanics in our macroscopic world. In this book, we can only scratch the surface of this rich subject, but many excellent references exist for further study.[9]

---

[9] See, e.g., J. S. Bell, *Speakable and Unspeakable in Quantum Mechanics: Collected Papers on Quantum Philosophy*, Cambridge University Press (2004); J. Preskill, Lecture Notes for Physics 219/Computer Science 219 Quantum Computation, http://theory.caltech.edu/~preskill/ph219 (accessed November 9, 2020); R. Horodecki, P. Horodecki, M. Horodecki and K. Horodecki, "Quantum entanglement," *Rev. Mod. Phys.* **81**, 865–942 (2009) [arXiv:quant-ph/0702225 [quant-ph]].

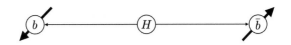

Fig. 12.3 Illustration of the decay of the spin-0 Higgs boson H to the spin-1/2 bottom b and anti-bottom $\bar{b}$ quarks. If the Higgs boson has zero momentum, then the quarks must be produced from the decay with back-to-back momentum. To conserve angular momentum, the spins of the quarks, illustrated by the thick arrows, must be pointed in opposite directions.

12.4.3 Observing Entanglement and Bell's Inequalities

One aspect of quantum entanglement that we will make a bit more precise in this book is a method for its unambiguous observation. As discussed above, the entropy of entanglement is a measure of the quantum correlations between particles, but the same entropy of a subsystem after partial tracing can also arise from purely classical correlations. So, on its own, entropy of entanglement is not an unambiguous observation of entangled particles because one also needs to know that the total system was initially in a pure state.

Entanglement of Spin from Particle Decay

In this section, we will re-evaluate the distinctions between the density matrix of classically correlated particles versus quantum-entangled particles and work to construct an observable that can unambiguously establish the presence of quantum entanglement. To start, let's first define the systems we are working with. Our system of interest will be two spin-1/2 particles produced from the decay of an unstable particle. As a concrete, physical example, we will consider the decay of a Higgs boson denoted as H to a bottom quark b and an anti-bottom quark $\bar{b}$, a process that has been conclusively observed in high-energy particle collision experiments.[10] The Higgs boson is a spin-0 particle and has exclusively a single angular momentum state. Bottom quarks and anti-bottom quarks are a type of spin-1/2 particle and can therefore exist in one of two spin states: either spin-up or spin-down with respect to a fixed axis. This decay system is illustrated in Fig. 12.3.

There are of course many ways in which the bottom and anti-bottom quarks are correlated through this decay. First, if the Higgs boson is at rest with zero momentum, then the sum of the momenta of the bottom and anti-bottom quark must be 0. Or, the momenta of the bottom and anti-bottom quarks lie along a line. Here, however, we will focus on how their spins can be correlated or entangled, and we will consider two possibilities and establish inequalities that must be satisfied classically, but can be broken quantum mechanically. As the Higgs is a spin-0 particle, to conserve angular

---

[10] M. Aaboud *et al.* [ATLAS], "Observation of $H \to b\bar{b}$ decays and VH production with the ATLAS detector," *Phys. Lett. B* **786**, 59–86 (2018) [arXiv:1808.08238 [hep-ex]]; A. M. Sirunyan *et al.* [CMS], "Observation of Higgs boson decay to bottom quarks," *Phys. Rev. Lett.* **121**(12), 121801 (2018) [arXiv:1808.08242 [hep-ex]].

momentum the bottom and anti-bottom quarks must have opposite spin along the z-axis, say, but that can be manifest in different ways. The spins could be entangled and described by a pure state $|\psi\rangle$ on Hilbert space. We can write $|\psi\rangle$ as

$$|\psi\rangle = \frac{1}{\sqrt{2}}|\uparrow_b\downarrow_{\bar{b}}\rangle + \frac{1}{\sqrt{2}}|\downarrow_b\uparrow_{\bar{b}}\rangle\,, \tag{12.88}$$

and the corresponding density matrix of this state would be

$$\rho_{\text{quant}} = |\psi\rangle\langle\psi| \tag{12.89}$$

$$= \frac{1}{2}\left(|\uparrow_b\downarrow_{\bar{b}}\rangle\langle\uparrow_b\downarrow_{\bar{b}}| + |\downarrow_b\uparrow_{\bar{b}}\rangle\langle\downarrow_b\uparrow_{\bar{b}}| + |\downarrow_b\uparrow_{\bar{b}}\rangle\langle\uparrow_b\downarrow_{\bar{b}}| + |\uparrow_b\downarrow_{\bar{b}}\rangle\langle\downarrow_b\uparrow_{\bar{b}}|\right)\,.$$

By contrast, the spins could just be anti-correlated classically. What this means is the following. If the measurements of the spins of bottom and anti-bottom quarks are performed along the same axis, then if the bottom quark is measured to have spin-up, the anti-bottom quark must be measured to have spin-down. By contrast, if the axis along which the spin of the anti-bottom quark is measured is different than for the bottom quark, the relative probability of measuring opposite spins is controlled by the relative angle between the axes. These probabilities can be calculated in the following way. Without loss of generality, we can say that we measure the spin of the bottom quark about the x-axis and find that it is spin-up. Thus, in the basis of the z-component of spin, the state of the bottom quark is

$$|\uparrow_{b,x}\rangle = \frac{1}{\sqrt{2}}|\uparrow_b\rangle + \frac{1}{\sqrt{2}}|\downarrow_b\rangle\,. \tag{12.90}$$

If we also measure the spin of the anti-bottom quark about the x-axis, we must find the opposite spin, by conservation of angular momentum:

$$|\downarrow_{\bar{b},x}\rangle = \frac{1}{\sqrt{2}}|\uparrow_{\bar{b}}\rangle - \frac{1}{\sqrt{2}}|\downarrow_{\bar{b}}\rangle\,. \tag{12.91}$$

Now, consider a spin measurement axis for the anti-bottom quark that is an angle χ with respect to the x-axis. To determine the relative fractions of spin-up and spin-down measurements with respect to this new axis, we can just rotate the state of Eq. (12.91) about, say, the z-axis. Recall from Eq. (8.57) that the unitary operator $\hat{U}(\chi)$ that rotates an angle χ about the z-axis can be expressed as

$$\hat{U}(\chi) = e^{i\frac{\chi}{2}}|\uparrow_{\bar{b}}\rangle\langle\uparrow_{\bar{b}}| + e^{-i\frac{\chi}{2}}|\downarrow_{\bar{b}}\rangle\langle\downarrow_{\bar{b}}|\,. \tag{12.92}$$

Acting this rotation on the initial anti-bottom quark spin state, we have

$$\hat{U}(\chi)|\downarrow_{\bar{b},x}\rangle = \frac{e^{i\frac{\chi}{2}}}{\sqrt{2}}|\uparrow_{\bar{b}}\rangle - \frac{e^{-i\frac{\chi}{2}}}{\sqrt{2}}|\downarrow_{\bar{b}}\rangle = i\sin\frac{\chi}{2}|\uparrow_{\bar{b},x}\rangle + \cos\frac{\chi}{2}|\downarrow_{\bar{b},x}\rangle\,. \tag{12.93}$$

Therefore, the probabilities of measuring spin-up and spin-down of the anti-bottom quark along an axis that is an angle χ from the axis along which the bottom quark was measured to be spin-up are $\sin^2(\chi/2)$ and $\cos^2(\chi/2)$, respectively.

The exact same analysis can be applied if the bottom quark were measured to be spin-down, and we assume that there is equal probability for it to be spin-up or spin-down. Therefore, the classical density matrix for the bottom quark and anti-bottom quark spins with measurement axes that are at a relative angle of χ with respect to each other is

$$\rho_{\text{class}} = \frac{\cos^2 \frac{\chi}{2}}{2} \left(|\uparrow_b \downarrow_{\bar{b}}\rangle \langle \uparrow_b \downarrow_{\bar{b}}| + |\downarrow_b \uparrow_{\bar{b}}\rangle \langle \downarrow_b \uparrow_{\bar{b}}| \right) \tag{12.94}$$

$$+ \frac{\sin^2 \frac{\chi}{2}}{2} \left(|\uparrow_b \uparrow_{\bar{b}}\rangle \langle \uparrow_b \uparrow_{\bar{b}}| + |\downarrow_b \downarrow_{\bar{b}}\rangle \langle \downarrow_b \downarrow_{\bar{b}}| \right).$$

To avoid an overabundance of subscripts, we have not identified the axis along which measurements are performed on the individual spins. However, we must keep that in mind when we calculate expectation values of operators on this state.

Spin Measurement Operators

Working with this reduced data of the quarks produced from the decay, the only measurements we can perform are of their spins projected along particular axes. In general, an axis in three spatial dimensions can be defined by its location on a sphere. Given the spin operators about the Cartesian axes $\hat{S}_x$, $\hat{S}_y$, and $\hat{S}_z$, we can express the spin operator $\hat{S}(\theta, \phi)$ about a general axis defined by the polar and azimuthal angles θ and ϕ on the sphere, where

$$\hat{S}(\theta, \phi) = \sin \theta \cos \phi \, \hat{S}_x + \sin \theta \sin \phi \, \hat{S}_y + \cos \theta \, \hat{S}_z \tag{12.95}$$

$$= \frac{\hbar}{2} \begin{pmatrix} \cos \theta & e^{-i\phi} \sin \theta \\ e^{i\phi} \sin \theta & -\cos \theta \end{pmatrix}.$$

In the second line, we expressed the spin operator as a 2×2 matrix in the basis of spins along the z-axis, the eigenstates of the $\hat{S}_z$ operator.

One important thing to note about this general spin operator is that its eigenvalues are identical to $\hat{S}_x$, $\hat{S}_y$, and $\hat{S}_z$ individually. The trace of $\hat{S}(\theta, \phi)$ is 0 and its determinant is

$$\det \hat{S}(\theta, \phi) = \frac{\hbar^2}{4} \left(-\cos^2 \theta - e^{-i\phi} e^{i\phi} \sin^2 \theta \right) = -\frac{\hbar^2}{4}, \tag{12.96}$$

and therefore its eigenvalues are $\lambda = \pm \hbar/2$. With this result, we can establish a robust bound on the expectation values of $\hat{S}(\theta, \phi)$ on any state. From the results of Sec. 10.3, where we used the power method to produce states that were more parallel to eigenstates with the largest eigenvalue of the Hamiltonian, the expectation value of $\hat{S}(\theta, \phi)$ is bounded from above by its largest eigenvalue $\lambda_{\max}$:

$$|\langle \hat{S}(\theta, \phi) \rangle| \leq |\lambda_{\max}| = \frac{\hbar}{2}. \tag{12.97}$$

Additionally, note that the square of the operator $\hat{S}(\theta, \phi)$ is proportional to the identity matrix $\mathbb{I}$:

$$\hat{S}(\theta, \phi)^2 = \frac{\hbar^2}{4} \mathbb{I}. \tag{12.98}$$

With this set-up, we can now begin to identify an observable that exclusively measures quantum entanglement. First, we want this observable to be *quantum*, and be expressed in terms of quantities that do not exist classically. For example, commutators of observables as Hermitian operators on Hilbert space do not exist classically. So, we can consider measuring the commutator of two general spin operators measured about different axes on our state of interest. Measuring the commutator of two operators just means that we make measurements consecutively in different orders, and then take their difference. So, we will measure the expectation value of the observable

$$\hat{S}(\theta,\phi)\hat{S}(\theta',\phi') - \hat{S}(\theta',\phi')\hat{S}(\theta,\phi) = \left[\hat{S}(\theta,\phi),\hat{S}(\theta',\phi')\right], \qquad (12.99)$$

where (θ,ϕ) and (θ',ϕ') are two sets of angles on the sphere.

Further, from the results established above, we can establish an upper bound on the expectation value of this commutator on any state. From Eq. (12.97), the upper bound on the expectation value of the product of two spin operators is just $\hbar^2/4$:

$$\left|\langle\hat{S}(\theta,\phi)\hat{S}(\theta',\phi')\rangle\right| \leq \frac{\hbar^2}{4}. \qquad (12.100)$$

The commutator is a linear combination of the product of two spin operators, so it is bounded from above by twice this value:

$$\left|\langle\left[\hat{S}(\theta,\phi),\hat{S}(\theta',\phi')\right]\rangle\right| \leq \frac{\hbar^2}{2}. \qquad (12.101)$$

Second, we want to establish non-zero entanglement between the spins of the bottom and anti-bottom quarks from decay of the Higgs. Their correlation or entanglement can be quantified by multiplying the measurement of the commutator of spins for the bottom quark by that of the anti-bottom quark. That is, we will measure the product of commutators

$$\left[\hat{S}_b(\theta,\phi),\hat{S}_b(\theta',\phi')\right]\left[\hat{S}_{\bar{b}}(\theta'',\phi''),\hat{S}_{\bar{b}}(\theta''',\phi''')\right] \equiv [\hat{S}_b,\hat{S}_b'][\hat{S}_{\bar{b}},\hat{S}_{\bar{b}}'], \qquad (12.102)$$

where the subscript denotes the quark on which the measurement is performed and we now choose four different axes to perform measurements, denoted by different numbers of primes ' on the angles on the sphere. Measurements on the bottom quark and anti-bottom quark are performed completely separately and so on the right, we have just introduced a compact notation for representation of all sets of angles of the measurement axes.

The quantity we will consider is therefore

$$\langle[\hat{S}_b,\hat{S}_b'][\hat{S}_{\bar{b}},\hat{S}_{\bar{b}}']\rangle = \text{tr}\left(\rho[\hat{S}_b,\hat{S}_b'][\hat{S}_{\bar{b}},\hat{S}_{\bar{b}}']\right), \qquad (12.103)$$

on some density matrix ρ. By our arguments above, this should take very different values on classically correlated or quantum-entangled states. The range that this expectation value of the product of commutators can take follows from the results of Eq. (12.101):

$$0 \leq \left|\langle[\hat{S}_b,\hat{S}_b'][\hat{S}_{\bar{b}},\hat{S}_{\bar{b}}']\rangle\right| \leq \frac{\hbar^4}{4}. \qquad (12.104)$$

We would like to choose spin measurement axes such that the entire range of this operator can be accessed and determine where in this range classically correlated or quantum-entangled states lie.

Actually, there is more simplification we can do. By shifting the product of the commutators, it can be re-expressed as the square of a Hermitian operator. We will define the operator $\hat{C}^2$ as

$$\hat{C}^2 \equiv \frac{\hbar^4}{4} - [\hat{S}_b, \hat{S}'_b][\hat{S}_{\bar{b}}, \hat{S}'_{\bar{b}}]. \tag{12.105}$$

By the bounds established above, all expectation values of this operator are non-negative, $\langle C^2 \rangle \geq 0$, so calling it a square makes some sense thus far. Further, by the bound established in Eq. (12.104), the largest possible expectation value is $|\langle C^2 \rangle| \leq \hbar^4/2$. Being careful with the order of multiplication of non-commuting operators, you can verify that the operator $\hat{C}$ that squares to the expression in Eq. (12.105) is

$$\hat{C} = \hat{S}_b(\hat{S}_{\bar{b}} + \hat{S}'_{\bar{b}}) + \hat{S}'_b(\hat{S}_{\bar{b}} - \hat{S}'_{\bar{b}}). \tag{12.106}$$

Operators that act on different particles commute (i.e., $[\hat{S}_b, \hat{S}_{\bar{b}}] = 0$) and we have used the fact that the square of any of the spin operators in this expression is proportional to the identity, from Eq. (12.98). It is this operator whose expectation values we will measure on the classical and quantum states and establish bounds on

$$\langle \hat{C} \rangle \equiv \mathrm{tr}(\rho\,\hat{C}). \tag{12.107}$$

By the bounds established on $\langle \hat{C}^2 \rangle$ above, it follows that

$$0 \leq |\langle \hat{C} \rangle| \leq \frac{\hbar^2}{\sqrt{2}}. \tag{12.108}$$

Evaluation of Expectation Values

We first measure the operator $\hat{C}$ on the classical state defined by the density matrix ρ_{class}. In our description of the classical state, we assumed that the direction of spin of the bottom quark, say, pointed in a completely random direction. Thus, the spin operator $\hat{S}_b$ takes the form

$$\hat{S}_b = \frac{\hbar}{2}\left(|\uparrow_b\rangle\langle\uparrow_b| - |\downarrow_b\rangle\langle\downarrow_b|\right), \tag{12.109}$$

for the spins defined along any axis. Therefore, the partial trace of this operator on the classical state is

$$\mathrm{tr}_b\left(\rho_{\mathrm{class}}\hat{S}_b\right) = \frac{\hbar}{4}\left(\sin^2\frac{\chi}{2} - \cos^2\frac{\chi}{2}\right)\left(|\uparrow_{\bar{b}}\rangle\langle\uparrow_{\bar{b}}| - |\downarrow_{\bar{b}}\rangle\langle\downarrow_{\bar{b}}|\right), \tag{12.110}$$

from the expression in Eq. (12.94). Then, because we next measure the spin of the antibottom quark along the axis that is an angle χ from the spin of the bottom quark, its spin operator takes an identical form:

$$\hat{S}_{\bar{b}} = \frac{\hbar}{2}\left(|\uparrow_{\bar{b}}\rangle\langle\uparrow_{\bar{b}}| - |\downarrow_{\bar{b}}\rangle\langle\downarrow_{\bar{b}}|\right). \tag{12.111}$$

The expectation value of the product of the spins of the bottom and anti-bottom quarks classically is therefore

$$\text{tr}\left(\rho_{\text{class}}\hat{S}_b\hat{S}_{\bar{b}}\right) = \text{tr}_{\bar{b}}\left(\text{tr}_b\left(\rho_{\text{class}}\hat{S}_b\right)\hat{S}_{\bar{b}}\right) = \frac{\hbar^2}{4}\left(\sin^2\frac{\chi}{2} - \cos^2\frac{\chi}{2}\right) \tag{12.112}$$

$$= -\frac{\hbar^2}{4}\cos\chi.$$

We will use this result for evaluation of the expectation value of the operator $\hat{C}$ with specific choices for the spin measurement axes. There are many ways to choose the spin operators to define $\hat{C}$, and what we do here will be the following. We take the operators to be

$$\hat{S}_b = \hat{S}_x, \qquad\qquad\qquad \hat{S}_b' = \hat{S}_y, \tag{12.113}$$

$$\hat{S}_{\bar{b}} = \sin\phi\,\hat{S}_x + \cos\phi\,\hat{S}_y, \qquad \hat{S}_{\bar{b}}' = \cos\phi\,\hat{S}_x - \sin\phi\,\hat{S}_y, \tag{12.114}$$

for some angle ϕ. Note that the two different measurement axes on the bottom quark are perpendicular to each other, and the two measurement axes on the anti-bottom quark are also perpendicular. We assume that the bottom and anti-bottom quarks' momenta lie along the z-axis, and we perform spin measurements in the x–y plane. The experimental set-up of this choice of spin axis measurements is illustrated in Fig. 12.4, where we place polarizing filters, say, that the quarks pass through to establish their projection of spin along the axis of the polarization.

With these choices of measurement axes, it is then straightforward to evaluate the expectation value on the classical state. With the result of Eq. (12.112), the expectation value on the classical state can be represented as

$$\text{tr}(\rho_{\text{class}}\hat{C}) = -\frac{\hbar^2}{4}\left(\cos\chi_{b\bar{b}} + \cos\chi_{b\bar{b}'} + \cos\chi_{b'\bar{b}} - \cos\chi_{b'\bar{b}'}\right), \tag{12.115}$$

where, for example, $\chi_{b\bar{b}}$ is the angle between the spin operators $\hat{S}_b$ and $\hat{S}_{\bar{b}}$, and similar for the other angles. With our choices of measurement axes in Eqs. (12.113) and (12.114), these angles can be immediately read off:

$$\cos\chi_{b\bar{b}} = \sin\phi, \qquad \cos\chi_{b\bar{b}'} = \cos\phi, \tag{12.116}$$

$$\cos\chi_{b'\bar{b}} = \cos\phi, \qquad \cos\chi_{b'\bar{b}'} = -\sin\phi.$$

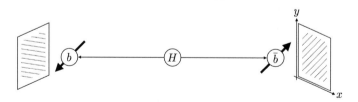

Fig. 12.4 An illustration of the measurements that we choose to make to establish entanglement. The spins of the bottom b and anti-bottom $\bar{b}$ quarks are measured along a direction in the x–y plane, perpendicular to the direction of their momenta. To do this, the quarks are passed through polarizing filters, represented by the sheets with dashed lines on the left and right.

Then, the classical expectation value of the operator $\hat{C}$ is

$$\text{tr}(\rho_{\text{class}}\hat{C}) = -\frac{\hbar^2}{2}\cos\phi. \tag{12.117}$$

Fascinatingly, we have demonstrated that the expectation value of $\hat{C}$ on a classical state, *any* classical state, is strictly bounded from above:

$$|\text{tr}(\rho_{\text{class}}\hat{C})| \leq \frac{\hbar^2}{2}. \tag{12.118}$$

What makes this especially intriguing is that this upper bound is stronger than the general result we had established in Eq. (12.108). Thus, a classical state of spins cannot explore the complete range of possible values of $\hat{C}$.

So, if we can show that quantum-entangled states can exhibit expectation values of the operator $\hat{C}$ that violate the bound of Eq. (12.118), we will have established an unambiguous experimental verification of entanglement. This expectation value of $\hat{C}$ on the pure state of quantum-entangled spins defined by state $|\psi\rangle$ of Eq. (12.88) is

$$\langle\psi|\hat{C}|\psi\rangle = \text{tr}(\rho_{\text{quant}}\hat{C}). \tag{12.119}$$

Note that the expectation value of $\hat{S}_x$ or $\hat{S}_y$ in the basis of spins along the z-axis is only non-zero if there is a spin-flip. For example

$$\langle\downarrow|\hat{S}_x|\uparrow\rangle = \langle\downarrow|\frac{\hbar}{2}(|\uparrow\rangle\langle\downarrow| + |\downarrow\rangle\langle\uparrow|)|\uparrow\rangle = \frac{\hbar}{2}. \tag{12.120}$$

So, the expectation value of $\hat{C}$ on the entangled state can be expressed as

$$\langle\psi|\hat{C}|\psi\rangle = \frac{1}{2}\langle\downarrow_b\uparrow_{\bar{b}}|\hat{S}_b(\hat{S}_{\bar{b}} + \hat{S}'_{\bar{b}}) + \hat{S}'_b(\hat{S}_{\bar{b}} - \hat{S}'_{\bar{b}})|\uparrow_b\downarrow_{\bar{b}}\rangle \tag{12.121}$$

$$+ \frac{1}{2}\langle\uparrow_b\downarrow_{\bar{b}}|\hat{S}_b(\hat{S}_{\bar{b}} + \hat{S}'_{\bar{b}}) + \hat{S}'_b(\hat{S}_{\bar{b}} - \hat{S}'_{\bar{b}})|\downarrow_b\uparrow_{\bar{b}}\rangle$$

$$= \frac{\hbar}{4}\langle\uparrow_{\bar{b}}|(\hat{S}_{\bar{b}} + \hat{S}'_{\bar{b}}) + i(\hat{S}_{\bar{b}} - \hat{S}'_{\bar{b}})|\downarrow_{\bar{b}}\rangle + \frac{\hbar}{4}\langle\downarrow_{\bar{b}}|(\hat{S}_{\bar{b}} + \hat{S}'_{\bar{b}}) - i(\hat{S}_{\bar{b}} - \hat{S}'_{\bar{b}})|\uparrow_{\bar{b}}\rangle.$$

Here, we have just explicitly evaluated the expectation values on the spin state of the bottom quark.

In the expectation values that remain, there are some simplifications we can perform. Given the particular operators $\hat{S}_{\bar{b}}$ and $\hat{S}'_{\bar{b}}$ that we chose, the combination of operators can be expressed in a more convenient form. For example, from the explicit expression in Eq. (12.114), note that

$$(\hat{S}_{\bar{b}} + \hat{S}'_{\bar{b}}) + i(\hat{S}_{\bar{b}} - \hat{S}'_{\bar{b}}) = -i(1+i)e^{i\phi}(\hat{S}_x + i\hat{S}_y) = -i(1+i)e^{i\phi}\hat{S}_+, \tag{12.122}$$

where $\hat{S}_+$ is the spin-raising operator. Its expectation value is

$$\langle\uparrow_{\bar{b}}|(\hat{S}_{\bar{b}} + \hat{S}'_{\bar{b}}) + i(\hat{S}_{\bar{b}} - \hat{S}'_{\bar{b}})|\downarrow_{\bar{b}}\rangle = -i(1+i)e^{i\phi}\langle\uparrow_{\bar{b}}|\hat{S}_+|\downarrow_{\bar{b}}\rangle \tag{12.123}$$

$$= -i(1+i)e^{i\phi}\hbar.$$

The other expectation value in the evaluation of the expectation value of $\hat{C}$ is the complex conjugate of this. Therefore, we find that

$$\langle\psi|\hat{C}|\psi\rangle = \frac{\hbar^2}{4}\left((1-i)e^{i\phi}+(1+i)e^{-i\phi}\right) = \frac{\hbar^2}{\sqrt{2}}\cos(\phi-\pi/4). \qquad (12.124)$$

For generic ϕ, this is non-zero and further saturates the bound of Eq. (12.108) when $\phi = \pi/4$.

Bell's Inequalities

We have thus established a concrete experimental procedure to conclusively observe entanglement of quark spins from this decay of the Higgs boson. We established that, for a generic classical state, its expectation value is strictly bounded from above as

$$\left|\langle\hat{C}\rangle\right|_{\text{class}} = \left|\langle\hat{S}_b(\hat{S}_{\bar{b}}+\hat{S}'_{\bar{b}})+\hat{S}'_b(\hat{S}_{\bar{b}}-\hat{S}'_{\bar{b}})\rangle\right|_{\text{class}} \leq \frac{\hbar^2}{2}. \qquad (12.125)$$

We explicitly demonstrated that we could violate this bound by a quantum-entangled state of particles in Eq. (12.124). Therefore, it is impossible for quantum entanglement to be a classical phenomenon. This celebrated result is called **Bell's inequality** after John Bell who established it.[11] This particular form of the inequality of Eq. (12.125) is due to John Clauser, Michael Horne, Abner Shimony, and Richard Holt who constructed simple concrete examples shortly after Bell's paper.[12]

Violations of Bell's inequalities are unambiguous signatures of entanglement, but the converse is not true. Numerous classes of states can be constructed that exhibit quantum entanglement but can be shown to satisfy Bell's inequalities.[13] You will study a very simple example of an entangled state that can be made to satisfy Bell's inequalities in Exercise 12.9.

While we introduced the analysis of this section from the perspective of desiring to observe entanglement, the historical motivation for Bell's inequalities and later developments was quite different. Quantum mechanics clearly challenges our notions of what it means to observe something, and the possibility that widely separated particles, like our bottom and anti-bottom quarks, could influence one another through entanglement seemingly doomed it as a possible theory of Nature. From the principles of special relativity, information cannot travel faster than the speed of light, but it would seem like entangled particles can communicate instantaneously, regardless of their separation. An entangled pair of particles seem to exactly "know" each other's state through their correlations after measurements are performed. One of the earliest and most famous criticisms of this feature of quantum mechanics is due to Albert Einstein, Boris Podolsky, and Nathan Rosen (EPR).[14] This is sometimes flippantly

---

[11] J. S. Bell, "On the Einstein–Podolsky–Rosen paradox," *Phys. Phys. Fiz.* **1**, 195–200 (1964).

[12] J. F. Clauser, M. A. Horne, A. Shimony, and R. A. Holt, "Proposed experiment to test local hidden variable theories," *Phys. Rev. Lett.* **23**, 880–884 (1969). Other weaker inequalities were established later in B. S. Cirelson, "Quantum generalizations of Bell's inequality," *Lett. Math. Phys.* **4**, 93–100 (1980).

[13] See, e.g., R. F. Werner, "Quantum states with Einstein–Podolsky–Rosen correlations admitting a hidden-variable model," *Phys. Rev. A* **40**, 4277–4281 (1989).

[14] A. Einstein, B. Podolsky, and N. Rosen, "Can quantum mechanical description of physical reality be considered complete?," *Phys. Rev.* **47**, 777–780 (1935).

referred to as quantum mechanics exhibiting "spooky action at a distance," a term attributed to Einstein.[15]

This and other criticisms of quantum mechanics inspired work on constructing classical theories of Nature that could explain all the data that validated quantum mechanics. In general, these theories required introducing quantities that were completely inaccessible to experiment, so-called "hidden variables," to explain the probabilistic nature inherent in quantum mechanics. Bell published his paper about 30 years after the EPR criticism, conclusively demonstrating that there were experiments that could be performed to test the validity of these hidden variable theories. The first experiments to test violations of Bell's inequalities were performed in the early 1970s,[16] but conclusive results were first obtained by Alain Aspect and collaborators in the early 1980s.[17] These experiments, and subsequent improvements, demonstrated that our universe cannot, at a fundamental level, be explained by classical mechanics. We will present more on what quantum mechanics implies for reality in Chap. 13.

12.5 Quantum Thermodynamics and the Partition Function

With a name like "von Neumann entropy," this should suggest a relationship to the entropy of thermodynamics, and not exclusively the "information entropy" for which we have interpreted it thus far. Let's see if we can formulate the notion of **thermal equilibrium** in this quantum statistical language.[18] From this formulation, we can establish a deep connection between the partition function and the path integral, which provides another interpretation as to what the path integral is.

12.5.1 Thermal Equilibrium

If a system is in thermal equilibrium, the most important feature of that system is that it is time-independent. Individual components or particles of the system might still have non-trivial dynamics (like the air molecules in a room), but in aggregate all of those little, microscopic fluctuations average out to stasis in the large ensemble. Hence the density matrix of such a system is time-independent:

---

[15] During the 1927 Solvay conference, Einstein himself was reported to have used the adjective "peculiar" rather than "spooky." In this context, Einstein was troubled by the more general principle of superposition in quantum mechanics, exhibited by the double-slit experiment, for example. See D. Howard, "Revisiting the Einstein–Bohr dialogue," *Iyyun: Jerusalem Phil. Q.*, 57–90 (2007).

[16] S. J. Freedman and J. F. Clauser, "Experimental test of local hidden-variable theories," *Phys. Rev. Lett.* **28**, 938–941 (1972).

[17] A. Aspect, P. Grangier, and G. Roger, "Experimental tests of realistic local theories via Bell's theorem," *Phys. Rev. Lett.* **47**, 460–6443 (1981); "Experimental realization of Einstein–Podolsky–Rosen–Bohm Gedankenexperiment: A new violation of Bell's inequalities," *Phys. Rev. Lett.* **49**, 91–97 (1982); A. Aspect, J. Dalibard, and G. Roger, "Experimental test of Bell's inequalities using time varying analyzers," *Phys. Rev. Lett.* **49**, 1804–1807 (1982).

[18] This section closely follows the presentation in J. J. Sakurai, *Modern Quantum Mechanics*, revised edition, Addison-Wesley (1993).

$$\frac{d\rho}{dt} = 0, \tag{12.126}$$

or ρ commutes with the Hamiltonian, $[\hat{H}, \rho] = 0$.

Further, as the density matrix is constant, the total energy of the system (the internal energy) is also constant. With N particles in the system, the average energy per particle is

$$\frac{\langle\hat{H}\rangle}{N} = \frac{1}{N}\mathrm{tr}(\rho\hat{H}) = \frac{1}{N}\sum_{i\,\mathrm{pure}}\rho_{ii}E_i = \mathrm{constant}. \tag{12.127}$$

Note that because $[\hat{H}, \rho] = 0$, we can choose a basis in which the density matrix and Hamiltonian are simultaneously diagonalized, hence the simple sum over eigenenergies E_i weighted by probability $\rho_{ii} = w_i$.

The final quality we enforce in thermal equilibrium is more subtle. We will assume that entropy is maximized for a given total energy in equilibrium. As some justification for this, recall that thermal equilibrium means that no net heat can be transferred from the system to the environment; otherwise, it wouldn't be in equilibrium. Entropy is a measure of the energy transferred per unit temperature, namely

$$dS = \frac{dQ}{T}, \tag{12.128}$$

where dQ is the work done by the system and T is its temperature. Thus, if $dQ = 0$ in thermal equilibrium, then so too must $dS = 0$. Thus entropy is extremized in equilibrium, and the second law of thermodynamics identifies that extremum to be a maximum.

So, in thermal equilibrium we have

(a) $\frac{d\rho}{dt} = 0$,
(b) $\sum_i w_i E_i = \mathrm{constant}$,
(c) $dS = 0$, and
(d) $\mathrm{tr}\rho = \sum_i w_i = 1$.

With these constraints, we would like to determine the entries of the diagonalized density matrix (its eigenvalues in an arbitrary basis), w_i. We will do this with the method of Lagrange multipliers. First, we identify all of our constraints which are required:

(a) $\sum_i w_i E_i = \mathrm{constant}$ implies that $\sum_i \delta w_i E_i = 0$, where δw_i is the variation of the density matrix element.
(b) $dS_{\mathrm{vN}} = 0 = -\delta\sum_i w_i \log w_i = -\sum_i \delta w_i(\log w_i + 1)$, by the product rule.
(c) $\mathrm{tr}\rho = \sum_i w_i = \mathrm{constant}$ implies that $\sum_i \delta w_i = 0$.

We next sum together the constraints, weighted by Lagrange multipliers β and γ:

$$\sum_i \delta w_i \left[(\log w_i + 1) + \beta E_i + 1\cdot\gamma\right] = 0. \tag{12.129}$$

$$\underbrace{}_{\delta S_{\mathrm{vN}}=0}\quad\underbrace{}_{\delta\langle\hat{H}\rangle=0}\quad\underbrace{}_{\delta\mathrm{tr}\rho=0}$$

Below each term we have written the variation it corresponds to. For this to vanish for an arbitrary variation δw_i, each term in the sum must individually vanish:

$$\log w_i + 1 + \beta E_i + \gamma = 0, \tag{12.130}$$

or

$$w_i = e^{-\gamma-1}e^{-\beta E_i}. \tag{12.131}$$

Using the normalization of the trace of ρ, we can solve for γ:

$$\mathrm{tr}\rho = 1 = \sum_i w_i = e^{-\gamma-1}\sum_i e^{-\beta E_i}, \tag{12.132}$$

or

$$e^{\gamma+1} = \sum_i e^{-\beta E_i} \equiv Z. \tag{12.133}$$

You might recognize this as the thermodynamic **partition function** and the elements of the density matrix as the **canonical ensemble**:

$$w_i = \frac{e^{-\beta E_i}}{\sum_i e^{-\beta E_i}} = \frac{e^{-\beta E_i}}{Z}. \tag{12.134}$$

Note that the sum extends over *all* energy eigenstates, and not just distinct energy eigenvalues. The individual exponential factor in the numerator, $e^{-\beta E_i}$, is called the **Boltzmann factor**. Everything we can possibly ask about our system is encoded in the partition function, and in thermodynamics, the Lagrange multiplier β is identified with inverse temperature $\beta = 1/k_B T$, where k_B is the Boltzmann constant.

12.5.2 Relationship to the Path Integral

Now, you can go to town with the partition function, but I want to connect it to the topic of the previous chapter: the path integral. First, note that the partition function is compactly expressed in a basis-independent way as

$$Z = \mathrm{tr}\left(e^{-\beta \hat{H}}\right), \tag{12.135}$$

where $\hat{H}$ is the Hamiltonian. Recall that the path integral is

$$Z = \int [dx]\, e^{i\frac{S[x]}{\hbar}}. \tag{12.136}$$

Though I have used the same letter, Z, to denote them, the partition function and path integral are not the same, right? I mean, the path integral has an imaginary number and the action, not the Hamiltonian, in the exponent.

Let's look at this exponent more carefully. The exponent in the path integral is

$$\frac{i}{\hbar}S[x] = \frac{i}{\hbar}\int dt\left[\frac{m}{2}\dot{x}^2 - V(x)\right]. \tag{12.137}$$

Let's see if we can massage the path integral into a form like that of the partition function. I'm going to do something really crazy: let's complexify time t, introducing a new coordinate $\tau \equiv it$, a procedure referred to as a Wick rotation.[19] Thus, the integration

[19] G. C. Wick, "Properties of Bethe–Salpeter wave functions," *Phys. Rev.* **96**, 1124–1134 (1954).

measure is $dt = -i d\tau$. Further, we have to account for the time derivative in the kinetic energy. Note that

$$\dot{x} = \frac{dx}{dt} = i\frac{dx}{d\tau},$$ (12.138)

so the kinetic energy is modified to

$$\frac{m}{2}\dot{x}^2 \rightarrow -\frac{m}{2}\left(\frac{dx}{d\tau}\right)^2 \equiv -\frac{m}{2}\dot{x}^2,$$ (12.139)

where $\dot{x}$ now means the derivative of position x with respect to τ.

With these identifications, the factor in the exponent becomes

$$\frac{i}{\hbar}S[x] = \frac{i}{\hbar}\int(-i)d\tau\left[-\frac{m}{2}\dot{x}^2 - V(x)\right] = -\frac{1}{\hbar}\int d\tau H(x,\dot{x}),$$ (12.140)

where $H(x,\dot{x})$ is just the classical Hamiltonian. The path integral with imaginary time is

$$Z(\tau = it) = \int[dx]\,e^{-\frac{1}{\hbar}\int d\tau H(x,\dot{x})}.$$ (12.141)

Further, by completeness of the eigenstates of the Hamiltonian, the imaginary time path integral can be re-expressed in terms of eigenvalues of the Hamiltonian. The path integral sums over every possible state/trajectory between two positions, but this sum in position space can be reorganized into a sum over all energy eigenstates, as the position and energy bases are both complete. Hence the sum over paths is also just a sum, or trace, over energy eigenstates:

$$Z(\tau = it) = \int[dx]\,e^{-\frac{1}{\hbar}\int dt\, H(x,\dot{x})} = \mathrm{tr}\left(e^{-\frac{\tilde{T}\hat{H}}{\hbar}}\right),$$ (12.142)

where $\hat{H}$ is now the quantum Hamiltonian operator and $\tilde{T}$ is now the total elapsed imaginary time. Comparing to the expression for the partition function from Eq. (12.135), the complex-time path integral is precisely the thermodynamic partition function, with the parameter $\beta = \tilde{T}/\hbar$, where $\hbar$ is Planck's reduced constant. Hence "time" is inversely related to temperature, in the thermodynamic representation.

Example 12.4 As we explicitly calculated the path integral of the harmonic oscillator in the previous chapter, it is useful to calculate the partition function for the harmonic oscillator to compare the explicit expressions. So what is this?

Solution

The partition function is again

$$Z = \mathrm{tr}(e^{-\beta\hat{H}}) = \sum_n e^{-\beta E_n},$$ (12.143)

where E_n is the nth-energy eigenvalue of the harmonic oscillator. From our analysis in Chap. 6, the energy eigenvalues are

$$E_n = \hbar\omega\left(n + \frac{1}{2}\right),$$ (12.144)

where $n = 0, 1, 2, \ldots$. Inserting this into the expression for the partition function, we find

$$Z = \sum_{n=0}^{\infty} e^{-\beta\hbar\omega\left(n+\frac{1}{2}\right)} = e^{-\frac{\beta\hbar\omega}{2}} \sum_{n=0}^{\infty} e^{-\beta\hbar\omega n} = \frac{e^{-\frac{\beta\hbar\omega}{2}}}{1 - e^{-\beta\hbar\omega}}, \tag{12.145}$$

where on the right we note that the sum is just a geometric series.

Now, this can be further manipulated into a form that appears similar to that which we found for the harmonic oscillator path integral. Note that we have

$$\frac{e^{-\frac{\beta\hbar\omega}{2}}}{1 - e^{-\beta\hbar\omega}} = \frac{1}{e^{\frac{\beta\hbar\omega}{2}} - e^{-\frac{\beta\hbar\omega}{2}}} = \frac{1}{2\sinh\left(\frac{\beta\hbar\omega}{2}\right)}, \tag{12.146}$$

using the definition of hyperbolic sine. Now, if we identify the parameter $\beta = \tilde{T}/\hbar$ with imaginary time $\tilde{T}$, we have

$$Z = \frac{1}{2\sinh\left(\frac{\omega\tilde{T}}{2}\right)}, \tag{12.147}$$

which is very similar to the expression for the path integral of Eq. (11.137), up to the square root and hyperbolic sine, rather than trigonometric sine. You'll identify the precise map between the result of the partition function and the path integral in the exercises.

Exercises

12.1 We had stated in this chapter that the general Rényi entropies do not satisfy subadditivity; only the von Neumann entropy does. In Example 12.2, we constructed an explicit counterexample for the Rényi entropy with the parameter $\alpha \to \infty$. In this exercise, we will study subadditivity of Rényi entropy more generally. Recall that the statement of subadditivity for a general entropy measure S

$$S^{AB} \leq S^A + S^B, \tag{12.148}$$

where AB is the combined system and A and B are its mutually exclusive subsystems.

(a) Consider two independent systems A and B and the system of their union AB. Does such a configuration of systems satisfy subadditivity for the Rényi entropy? It might help to consider particular values of Rényi parameter α, like $\alpha = 0$ or $\alpha \to \infty$.

(b) The density matrix constructed as a counterexample to subadditivity to the $\alpha \to \infty$ Rényi entropy in Example 12.2 is actually more powerful than presented there. Determine the range of the Rényi entropy parameter α such that that density matrix is a counterexample to subadditivity. Another way to ask this question is to determine the value of Rényi parameter α such that the density matrix of Eq. (12.37) renders the property of subadditivity an equality.

(c) Can you modify the counterexample density matrix of Eq. (12.37) such that subadditivity is violated for all Rényi entropies with $\alpha > 1$?

(d) What about for $0 < \alpha < 1$? Can you construct a counterexample density matrix that demonstrates that all Rényi entropies, except at the special value of $\alpha = 1$, violate subadditivity?

12.2 The von Neumann entropy satisfies a further inequality among three systems called *strong subadditivity*.[20] As with subadditivity, strong subadditivity is only a property of the von Neumann entropy, and not of the general Rényi entropies. Consider three systems A, B, and C and let AB, BC, and ABC be their combined systems, appropriately. The statement of strong subadditivity is that

$$S_{vN}^{ABC} + S_{vN}^{B} \leq S_{vN}^{AB} + S_{vN}^{BC}. \tag{12.149}$$

(a) Show that strong subadditivity holds if the systems A, B, and C are all independent.

(b) Show that strong subadditivity holds if the combined state ABC is pure. *Hint*: If ABC is pure, then the von Neumann entropies of systems A and BC, say, are equal.

(c) Consider the pure state $|\psi\rangle$ of three spins, where one spin is up and two are down:

$$|\psi\rangle = \frac{1}{\sqrt{3}} \left(|\uparrow_1\rangle|\downarrow_2\rangle|\downarrow_3\rangle + |\downarrow_1\rangle|\uparrow_2\rangle|\downarrow_3\rangle + |\downarrow_1\rangle|\downarrow_2\rangle|\uparrow_3\rangle \right). \tag{12.150}$$

Show that this state is normalized and construct the density matrices of the systems of spins 12, 23, and 2 alone.

(d) Now show that this state $|\psi\rangle$ satisfies strong subadditivity. What is the inequality that you find?

12.3 In our analysis of hydrogen in Chap. 9, we discussed the period of recombination, the time in the history of the universe at which protons and electrons became bound and formed neutral hydrogen. We were able to provide an estimate for the temperature at which recombination occurred according to the ground-state energy of hydrogen. However, this estimate was very coarse because we neglected thermodynamic effects. We now have a description of such effects through the partition function, so we can provide a better estimate here.

(a) Estimate the temperature at which 50% of hydrogen in the universe is in its ground state. To do this, you need to construct the partition function for hydrogen, which consists of a sum over *all* energy eigenstates. We can estimate the partition function here by simply considering just the first three energy eigenvalues of bound hydrogen. With this assumption, what is this recombination temperature?
Hint: Don't forget about degeneracy.

[20] J. Kiefer, "Optimum experimental designs," *J. Roy. Stat. Soc. B* **21**(2), 272–304 (1959); E. H. Lieb and M. B. Ruskai, "Proof of the strong subadditivity of quantum-mechanical entropy," *J. Math. Phys.* **14**, 1938–1941 (1973).

(b) The partition function that you would naïvely write down for hydrogen has a number of problems, not the least of which is that it is infinite. This happens because the higher bound states of hydrogen have energies that are closer and closer to 0, and so have Boltzmann factor 1, and there are an infinite number of them. Further, if the kinetic energy of the electron is sufficiently large, then it is not bound into hydrogen, and such scattering states also need to be accounted for in the partition function.

A much more complete partition function for hydrogen addressing these shortcomings is the **Planck–Larkin partition function.**[21] It is defined to be

$$Z_{\text{P-L}} = \sum_{n=1}^{\infty} n^2 \left(e^{-\beta \frac{E_0}{n^2}} - 1 + \beta \frac{E_0}{n^2} \right), \qquad (12.151)$$

where E_0 is the ground-state energy of hydrogen, n labels the energy eigenvalue, and $\beta = 1/k_B T$, proportional to the inverse temperature. While this partition function still has some problems,[22] it is at least not explicitly divergent. Actually, using the ratio test, show that this is indeed true, that the Planck–Larkin partition function converges.

(c) Additionally, show that the Planck–Larkin partition function reduces to what we expect at low temperature, as $T \to 0$.

(d) What does the Planck–Larkin partition function reduce to in the high-temperature limit, $T \to \infty$? Does this make sense?

12.4 We had established an intriguing relationship between the path integral of the previous chapter and the partition function here through "complexification" of the time coordinate. In this problem, we will make this relationship a bit more precise and attempt to understand what complexification of time actually accomplishes.

(a) Show that up to a normalization factor, the path integral Z_{path} is related to the partition function Z_{part} via: $Z_{\text{path}} Z_{\text{path}}^{\dagger} = Z_{\text{part}}$. For this to be true, what must the relationship between path integral time and partition function temperature be? What would we describe the trajectory of the particle to be in the path integral given this relationship?

(b) Now, use the position space energy eigenbasis representation of the path integral of Eq. (11.79) and the result of part (a) to derive the partition function from the path integral.

(c) Show that these relationships hold for the harmonic oscillator, with path integral given in Eq. (11.137) and partition function in Eq. (12.147).

12.5 We had mentioned that from the partition function Z, all possible thermodynamical quantities can be determined. In this problem, we will use the harmonic oscillator's partition function of Eq. (12.147) to do some calculations.

[21] M. Planck, "Zur Quantenstatistik des Bohrschen Atommodells," *Ann. Phys.* **380**(23), 673–684 (1924); A. I. Larkin, "Thermodynamic functions of a low-temperature plasma," *J. Exptl. Theoret. Phys. (U.S.S.R.)* **38**, 1896–1898 (1960), *Sov. Phys. JETP* **11**(6), 1363–1364 (1960).

[22] C. A. Rouse, "Comments on the Planck-Larkin partition function," *Astrophys. J.* **272**, 377–379 (1983).

(a)   Show that the average energy per particle $\langle E \rangle$ of a system in thermal equilibrium can be determined by

$$\langle E \rangle = -\frac{d \log Z}{d\beta} . \qquad (12.152)$$

What is the average energy of a particle in the harmonic oscillator?

(b)   Show that the entropy S of a system in thermal equilibrium can be expressed as

$$S = -\sum_n p_n \log p_n = \beta \langle E \rangle + \log Z . \qquad (12.153)$$

What is the entropy of the harmonic oscillator? Does the entropy in the temperature to zero limit make sense physically?

(c)   What if we used the path integral in these expressions instead of the partition function? What would β be in that case? Do these expressions make sense? Test it out for a free particle.

12.6   An extremely intriguing feature of quantum entanglement is a property that has been called the "monogamy of entanglement": a particle can only be maximally entangled with precisely one other particle.[23] In this problem, we will see this monogamy explicitly in systems of two and three spin-1/2 particles.

(a)   Consider two spin-1/2 particles, call them 1 and 2. Determine the pure states that exhibit maximum entanglement; that is, the states that have entropy of entanglement equal to $\log 2$, as we found in Eq. (12.74).

(b)   Is it possible for a partial trace of the density matrix of two spin-1/2 particles to be a pure state? Show that the only way that a pure state density matrix of two spin-1/2 particles reduces to a pure state of a single particle by tracing out the other particle is if the state was initially separable.

(c)   Now, consider a system of three spin-1/2 particles, call them 1, 2, and 3. Consider the most general density matrix of the three particles, ρ_{123}. Show that it is impossible for the reduced-density matrices ρ_{12} and ρ_{23} to simultaneously describe a pure state with maximum entropy of entanglement. Assume that ρ_{12} describes a maximally entangled pure state. What was the initial state of the three particles?

12.7   In Example 12.1, we introduced the Hong–Ou–Mandel interferometer and presented an analysis of thinking about the photons produced by the laser as classical electromagnetic waves. In this exercise, we will instead consider the photons as indistinguishable, quantum mechanical particles, to a rather surprising conclusion.

(a)   Let's analyze this experiment quantum mechanically, with individual, identical photons emitted from the laser. Let's call the state $|20\rangle$ when two photons hit the upper detector and none on the lower detector, the state $|02\rangle$ the opposite, and the state $|11\rangle$ when one photon hits each detector.

[23] M. Koashi and A. Winter, "Monogamy of quantum entanglement and other correlations," *Phys. Rev. A* **69**(2), 022309 (2004).

For the state $|11\rangle$, draw the possible photon trajectories from the laser to the detectors. Quantum mechanically, we have to sum together the probability amplitudes of all possible trajectories to determine the net probability amplitude of the observed state $|11\rangle$. What do you find? Do you ever detect the state $|11\rangle$ quantum mechanically? Don't forget to include the relative change in phase between the two photons at the beam splitter.

(b) Now, draw the photon trajectories for the observed states $|20\rangle$ and $|02\rangle$. Are these observed quantum mechanically?

12.8 Let's consider the harmonic oscillator immersed in a heat bath of temperature T. For any Hermitian operator $\hat{A}$, we can define its thermal average denoted as $\langle\langle\hat{A}\rangle\rangle$ through the Boltzmann factors and partition function:

$$\langle\langle\hat{A}\rangle\rangle = \frac{1}{Z}\sum_{n=0}^{\infty} e^{-\beta E_n}\langle\psi_n|\hat{A}|\psi_n\rangle, \qquad (12.154)$$

where E_n is the nth-energy eigenvalue and $|\psi_n\rangle$ is its corresponding eigenstate. In this problem, we'll consider how a thermal system interfaces with the Heisenberg uncertainty principle.

(a) First, determine the expectation value of the Hamiltonian $\hat{H}$, the mean value of the energy of this thermal harmonic oscillator.
     Hint: You can use the result of Eq. (12.152) in Exercise 12.5.
(b) Show that the expectation values of the position and momentum in this thermal harmonic oscillator are 0: $\langle\langle\hat{x}\rangle\rangle = \langle\langle\hat{p}\rangle\rangle = 0$.
(c) Now, determine the thermal averages of the squared position and momentum, $\langle\langle\hat{x}^2\rangle\rangle$, $\langle\langle\hat{p}^2\rangle\rangle$. What is the Heisenberg uncertainty principle for the thermal harmonic oscillator?
(d) Consider a *classical* harmonic oscillator with the energy you found in part (a). What is the probability distribution for the position x in the well? What about for the momentum p? What is the product of the classical variances of the position and momentum and how does it compare to the quantum mechanical result?

12.9 In Sec. 12.4.3, we motivated and introduced Bell's inequalities for unambiguous observation of quantum entanglement. There we demonstrated that a classical, random density matrix of two spins satisfied inequalities that could be violated by a pure, entangled quantum state. In this exercise, we will explore the extent to which the converse is true. Do there exist entangled states that do not violate Bell's inequalities?

(a) Let's first tie up a loose end from the analysis in Sec. 12.4.3. From the definition of the operator $\hat{C}$ in Eq. (12.106), show that it squares to the expression for $\hat{C}^2$ in Eq. (12.105). Don't forget to use the fact that each of the individual spin operators squares to an operator that is proportional to the identity.

(b) Consider the pure state $|\psi\rangle$ of the spins of a bottom b and anti-bottom $\bar{b}$ quark, where

$$|\psi\rangle = \sqrt{w}|\uparrow_b\downarrow_{\bar{b}}\rangle + \sqrt{1-w}|\downarrow_b\uparrow_{\bar{b}}\rangle. \qquad (12.155)$$

Here, $0 \leq w \leq 1$ represents the fraction of the state in which the bottom quark has spin-up and the anti-bottom quark has spin-down along the z-axis. If either $w = 0$ or $w = 1$, this pure state reduces to a separable state, exhibiting no entanglement. Therefore, in these limits, this separable state would satisfy Bell's inequality of Eq. (12.125). However, for all $0 < w < 1$, the pure state exhibits entanglement, so there must be a restricted range of w values that can possibly violate Bell's inequality. Determine this range of values for w, using the spin operators defined in Sec. 12.4.3.

(c) With this result, are there entangled states that do not violate Bell's inequality?

13

Why Quantum Mechanics?

The mathematical and physical construction of quantum mechanics is undeniably beautiful, but as hinted in the Introduction, *why* the universe should ultimately be quantum mechanical is mysterious. Of course, empirical science can never answer the question of "Why?" definitively, but only establish the rules that govern Nature through experiment. Nevertheless, there were several points in our discussion of the motivation for the Hilbert space, the Born rule, or the Dirac–von Neumann axioms that seemed to be completely inexplicable and potentially inconsistent with the guiding principles we used. In this chapter, we survey a few of these points from an introductory, modern perspective. Quantum mechanics works, makes precise predictions, and agrees with experiment, but what quantum mechanics *is* is still very much an open question.

13.1 The Hilbert Space as a Complex Vector Space

The formulation of quantum mechanics as a description of the evolution of linear operators on a state vector space, the Hilbert space, might seem like the simplest thing possible that you can imagine. Simplicity, as a guiding principle, is highly constraining, as mentioned several times throughout this book. Additionally, the Hilbert space is a complex vector space, and so the operators and state vectors are complex number-valued. This may not seem like the simplest possibility for the structure of the Hilbert space; indeed it is at least not unique. The Hilbert space could have been a real vector space, for example, and that may seem simpler if for no other reason than the fact that you are likely less familiar with complex numbers. However, there is a sense in which complex numbers are "simpler" than the reals. As one example, all roots of any polynomial with complex coefficients are still complex numbers, while there are polynomials with real coefficients that have no real roots, like $x^2 + 1$. As the characteristic equation for finding eigenvalues of a matrix is a polynomial, this feature that complex-valued polynomials are closed makes the description of such complex-valued matrices simpler.

Still, this possible explanation that a complex Hilbert space is simpler than a real Hilbert space leaves a lot to be desired. Indeed, we can just consider what the Hilbert space and therefore quantum mechanics would look like with a different number space. This exercise was first studied in the 1930s, along with the formalization of the mathematics of quantum mechanics.[1] In particular, one can generalize quantum mechanics to a real, complex, or **quaternionic** Hilbert space. Quaternions are like a generalization of the complex numbers that describe four non-commuting dimensions, identified by William Rowan Hamilton during a walk across the Broom Bridge in Dublin, Ireland, in the mid-nineteenth century.[2] In fact, it was discovered that these three possible expressions for the Hilbert space can be identified with one another, with appropriate additional structure.[3] This observation reduces the importance of the complex numbers in quantum mechanics on the one hand, but on the other, perhaps just pushes the problem a bit further down the road.

You are unlikely familiar with the quaternions, and this claim that complex and quaternionic Hilbert spaces could be identical may seem improbable, but can be shown in a simple example.

Example 13.1 The quaternions are three quantities i, j, and k such that, along with the real number 1, they form the basis of a four-dimensional space over the real numbers. The objects i, j, and k have non-trivial multiplication rules amongst each other, with

$$i^2 = j^2 = k^2 = ijk = -1 \qquad (13.1)$$

and anti-commutativity of

$$ij = -ji = k, \qquad jk = -kj = i, \qquad ki = -ik = j. \qquad (13.2)$$

Then, a quaternionic-valued Hilbert space would consist of those state vectors that are acted on by linear operators whose matrix elements are quaternions. If we just ignore the quaternions j and k, they would seem to just reduce to the complex numbers, which might suggest that the complex-valued Hilbert space can be described as a subset of the quaternionic-valued Hilbert space. However, if we can show that there is a map from the complex-valued Hilbert space to the quaternionic Hilbert space, this will actually demonstrate that they are isomorphic, with certain restrictions.

Just working with complex numbers, have we seen anything like the multiplicative relationships of i, j, and k before? Obviously, no complex numbers can satisfy these multiplication rules, but complex-valued matrices can. Indeed, recall the Pauli matrices, multiplied by i or $-i$, where

$$i\sigma_1 = \begin{pmatrix} 0 & i \\ i & 0 \end{pmatrix}, \qquad -i\sigma_2 = \begin{pmatrix} 0 & -1 \\ 1 & 0 \end{pmatrix}, \qquad i\sigma_3 = \begin{pmatrix} i & 0 \\ 0 & -i \end{pmatrix}. \qquad (13.3)$$

When we first introduced the Pauli matrices, we showed that they each squared to the identity matrix, so including the factor of i, these matrices square to $-\mathbb{I}$, like the quaternions. Further, we also showed the commutation relation

---

[1] P. Jordan, J. von Neumann, and E. P. Wigner, "On an algebraic generalization of the quantum mechanical formalism," *Ann. Math.* **35**, 29–64 (1934).

[2] W. R. Hamilton, "On quaternions; or on a new system of imaginaries in algebra," *Lond. Edinb. Dublin Philos. Mag. J. Sci.* **25**(169), 489–495 (1844).

[3] F. J. Dyson, "The threefold way. Algebraic structure of symmetry groups and ensembles in quantum mechanics," *J. Math. Phys.* **3**, 1199 (1962); for a modern perspective, see J. C. Baez, "Division algebras and quantum theory," *Found. Phys.* **42**, 819–855 (2012) [arXiv:1101.5690 [quant-ph]].

$$[\sigma_1, \sigma_2] = 2i\sigma_3, \tag{13.4}$$

and so multiplying on both sides by $1 = (i)(-i)$, we have

$$[i\sigma_1, -i\sigma_2] = 2i\sigma_3. \tag{13.5}$$

This is exactly the commutation relation that the quaternions satisfy, if we identify $i \equiv i\sigma_1$, $j \equiv -i\sigma_2$, $k \equiv i\sigma_3$. Apologies for overloading "i" to be both the imaginary number and a quaternion; both notations are standard. Finally, with this identification, we note that

$$(i\sigma_1)(-i\sigma_2)(i\sigma_3) = -\mathbb{I}, \tag{13.6}$$

again, exactly the same multiplication rule as the quaternions.

This demonstrates that the complex-valued Hilbert space on which the Pauli matrices act is isomorphic to a quaternionic-valued Hilbert space. We showed that the ket of a spin-1/2 particle was rotated by the Pauli matrices; therefore, the Hilbert space of a spin-1/2 particle *is* the quaternionic Hilbert space. The spin-1/2 particle is of course not a completely generic quantum system, but restricting the generic infinite-dimensional complex-valued Hilbert space to the two-complex dimensions spin-1/2 system is identical to the one-quaternionic dimensional Hilbert space. Quaternions have appeared elsewhere in physics historically. Interestingly, the quaternions were identified prior to the construction of modern vector notation and James Clerk Maxwell formulated electromagnetism with quaternions prior to Oliver Heaviside and others who wrote down what we now identify as the four Maxwell's equations.[4]

13.2 The Measurement Problem

In our motivation and formulation of quantum mechanics, we had identified Hermitian operators that act on states in the Hilbert space as special. Hermitian operators are the generalization of real numbers to arbitrary dimensional complex-valued operators and their eigenvalues correspond to the outcome of experimental measurement through the Born rule. At least, this is the hypothesis we were led to, in order to connect quantum mechanics to reality. These hypotheses, however, say nothing about what actually happens at the moment of measurement, and this is a problem.

Before making any measurements, a quantum system is in a state $|\psi\rangle$, and its time evolution is governed by the Schrödinger equation. This time evolution is unitary, ensuring that $|\psi\rangle$ remains normalized. Now, let's consider the process of making a measurement corresponding to a Hermitian operator $\hat{A}$. For now, we will assume that $\hat{A}$ has discrete and distinct eigenvalues, so we can expand $|\psi\rangle$ in the basis of the eigenstates of $\hat{A}$:

---

[4] J. C. Maxwell, *A Treatise on Electricity and Magnetism*, vol. 1, Clarendon Press (1873).

$$|\psi\rangle = \sum_n \beta_n |\chi_n\rangle, \tag{13.7}$$

where $|\chi_n\rangle$ is the nth eigenstate of $\hat{A}$ corresponding to eigenvalue λ_n and β_n is some complex number. By normalization of $|\psi\rangle$ we have that

$$\langle\psi|\psi\rangle = 1 = \sum_n |\beta_n|^2, \tag{13.8}$$

assuming that distinct eigenstates of $\hat{A}$ are orthogonal. So, in our framework we have developed, making this measurement on state $|\psi\rangle$ returns the eigenvalue λ_n with probability $|\beta_n|^2$. After the measurement is made, however, we know precisely the state of the system: if we measure eigenvalue λ_n, then the system must be in the state $|\chi_n\rangle$. That is, if we made many measurements in quick succession, they would all return the same value λ_n. Colloquially, this is sometimes referred to as the state $|\psi\rangle$ "collapsing to an eigenstate."

The problem with this is that unitarity or probability conservation has apparently been lost. Initially, before the measurement, the state $|\psi\rangle$ is normalized, but this normalization requires that the state has non-zero components along several eigenstates of $\hat{A}$ (assuming it's not an eigenstate itself). Once the measurement is made and the value λ_n is returned, it would seem like that process eliminated a probability of $1 - |\beta_n|^2$; that is, the probability of being not in state $|\chi_n\rangle$ vanished. This naïve physical interpretation of what happens in a measurement is referred to as the **measurement problem**, and suggests that the very act of *doing* science violates the principles of quantum mechanics.[5]

The measurement problem is more acute when considering measuring an operator with continuous eigenvalues. For instance, consider measuring the position x of a particle in state $|\psi\rangle$. The wavefunction $\psi(x) = \langle x|\psi\rangle$ describes the probability density for measuring any position x. Once the measurement is performed and position x is returned, we then know that the state of the system is the corresponding eigenstate of the position operator $\hat{x}$, $|x\rangle$. As we have discussed, eigenstates of position are not normalizable, and so $|x\rangle$ is not a state in the Hilbert space. This further suggests that the act of measurement removes a state from the Hilbert space, again apparently violating the conservation of probability in quantum mechanics.

From one perspective, this dissonance between the axioms of quantum mechanics and their naïve empirical application may seem deeply disturbing and demand that there be a deeper explanation. From this perspective, there are a number of potential philosophical resolutions that are referred to as interpretations of quantum mechanics.[6] The oldest such interpretation is called the **Copenhagen interpretation**, and is essentially the framework in which the measurement problem was introduced above. In the Copenhagen interpretation, physical systems simply do not have definite properties

---

[5] The measurement problem is nearly as old as quantum mechanics itself. To critique this interpretation of quantum indeterminacy, Erwin Schrödinger devised his now-famous cat paradox; see E. Schrödinger, "Die gegenwartige Situation in der Quantenmechanik," *Naturwiss.* **23**, 807–812 (1935).

[6] For a recent review of interpretations of quantum mechanics and their philosophical implications, see M. A. Schlosshauer, "Decoherence, the measurement problem, and interpretations of quantum mechanics," *Rev. Mod. Phys.* **76**, 1267–1305 (2004) [arXiv:quant-ph/0312059 [quant-ph]].

until a measurement is performed. This appears to be at odds with unitary time evolution in quantum mechanics, as already discussed; however, there is no measurement that we can perform to directly observe this non-unitary behavior by the assumption of the Copenhagen interpretation. Another possible resolution is the **Many-Worlds interpretation** of quantum mechanics.[7] The Many-Worlds interpretation assumes that time evolution and measurement is always unitary, but that the measurements we perform simply represent one outcome (or set of outcomes) that are realized in our universe. Time evolution is unitary if all of the "many worlds" that correspond to *all* possible outcomes of experiment are accounted for. Again, we can only explore our universe through experiment, so there is no possible way to observe these other worlds.

13.3 Decoherence, or the Quantum-to-Classical Transition

In parallel with the measurement problem there is another issue that we have been relatively cavalier about throughout this book. Our everyday experience of the world is of course classical, but we believe that fundamentally the universe is quantum mechanical and so there must be a connection between these two descriptions. We had identified the correspondence principle as a small $\hbar$ limit of quantum mechanics, in which classical mechanics "emerges." However, this correspondence principle was just the statement of a formal mathematical limit and not how the universe works. Further, the small $\hbar$ limit has many manifestations; one of which is the high-energy limit of quantum mechanics. Just taking a quantum state to have large energy is not a classical limit: in the infinite square well, for example, there is a wavefunction for the 10 quadrillionth energy eigenstate on Hilbert space that evolves according to the Schrödinger equation. Classically, of course, there is no such thing as a wavefunction or a Hilbert space, so this high-energy limit is still completely quantum mechanical.

A problem, if not the main problem, with the way that we discussed quantum systems throughout this book was that we assumed that the entire universe was described by the infinite square well potential, or the harmonic oscillator, or that there was a single hydrogen atom in the universe and nothing else. While it is of course necessary to simplify your system or imagine it perfectly isolated from anything else, this is not realistic and can lead to serious confusions about how the classical emerges from the quantum. If we do have a quantum particle in an infinite square well, then that well is, say, in a jar in a lab somewhere, and perhaps there are some probes in the jar so we can measure the particle's energy. If the particle were completely and perfectly isolated from the rest of the universe there would be no way that we could perform a measurement on it! To understand the problem at hand we are inextricably led to include interactions or couplings of our quantum system with the surrounding, classical, environment.

---

[7] H. Everett, "Relative state formulation of quantum mechanics," *Rev. Mod. Phys.* **29**, 454–462 (1957).

To make the problem a bit more precise, let's remind ourselves about features of a quantum system that have no classical analogue. A pure quantum state lives on Hilbert space and its time evolution is governed by the Schrödinger equation. As we assume that energy is conserved, the Hamiltonian is Hermitian and so time evolution is unitary. That is, all that time evolution does is affect the phase of the quantum state, as it is normalized to unit magnitude for all time. In our discussion of quantum ensembles, we noted that if a system of particles can be described by a state on Hilbert space, then that system is coherent; or, the states of all particles have a common phase which evolves according to the Schrödinger equation. By contrast, classical particles live on phase space, not Hilbert space, and a particle's trajectory is completely observable, in contrast to a quantum state on Hilbert space. As such, there is no notion of a phase for a classical particle. In the language of the density matrix, an ensemble of classical particles would be a completely random state for which their probabilities are incoherent. In this interpretation, then, the mechanism of "classicalization" of a quantum state is **decoherence**,[8] by which a pure state's density matrix transforms into a mixed or ultimately a completely random density matrix.

We had stated in Sec. 12.3 that pure states always evolve into pure states, assuming the von Neumann equation for the time evolution of the density matrix. A central assumption of the von Neumann equation is that the quantum system is a **closed system**, or perfectly isolated from the rest of the universe. As mentioned earlier, this can be a useful place from which to perform thought experiments, but is not realistic. A realistic quantum system always has some interaction with the environment, and is therefore an **open system**. The notion of open and closed systems might be familiar from a course on thermodynamics, in which we might consider a system immersed in a heat bath. From the perspective of the system, its energy is not conserved because energy can be freely transferred to and from the bath. An open quantum system is similar, and because it can transfer energy to and from its environment, the Hamiltonian of the system of interest is not Hermitian or from the system's perspective energy is not conserved. So a realistic, open quantum system cannot evolve in time according to the von Neumann equation; we must find the more general form of time evolution.

That more general time-evolution equation of the density matrix ρ with Hermitian Hamiltonian $\hat{H}$ is called the **Lindblad master equation**[9] and it takes the form

$$i\hbar \frac{d\rho}{dt} = [\hat{H}, \rho] + i\hbar \sum_{j=1}^{N^2-1} \gamma_j \left(\hat{L}_j \rho \hat{L}_j^\dagger - \frac{1}{2} \left\{ \hat{L}_j \hat{L}_j^\dagger, \rho \right\} \right). \qquad (13.9)$$

---

[8] For reviews on decoherence, see D. Giulini, C. Kiefer, E. Joos, J. Kupsch, I. O. Stamatescu and H. D. Zeh, *Decoherence and the Appearance of a Classical World in Quantum Theory*, Springer (2003); W. H. Zurek, "Decoherence, einselection, and the quantum origins of the classical," *Rev. Mod. Phys.* **75**, 715–775 (2003) [arXiv:quant-ph/0105127 [quant-ph]]; M. A. Schlosshauer, *Decoherence: and the Quantum-to-Classical Transition*, Springer Science & Business Media (2007).

[9] G. Lindblad, "On the generators of quantum dynamical semigroups," *Commun. Math. Phys.* **48**, 119 (1976); V. Gorini, A. Kossakowski, and E. C. G. Sudarshan, "Completely positive dynamical semigroups of N level systems," *J. Math. Phys.* **17**, 821 (1976). For a review of the Lindblad master equation and multiple derivations, see D. Manzano, "A short introduction to the Lindblad master equation," *AIP Adv.* **10**, 025106 (2020) [arXiv:1906.04478].

Here, γ_j are non-negative constants and the operators $\hat{L}_j$ called *jump operators* are traceless, $\text{tr}\hat{L}_j = 0$. The sum runs over a basis of all traceless operators on the N-dimensional Hilbert space, of which there are $N^2 - 1$ such operators. The total dimensionality of the space of operators on an N-dimensional Hilbert space is N^2, and the one operator not included in the sum is proportional to the identity operator. The curly bracket of two operators $\{\hat{A}, \hat{B}\} \equiv \hat{A}\hat{B} + \hat{B}\hat{A}$ is called the anticommutator.

The Lindblad master equation satisfies a number of interesting properties for describing the dynamics of open quantum systems. First and foremost, the Lindblad master equation still preserves probability or the trace of the density matrix if the jump operators are Hermitian. That is

$$i\hbar \frac{d}{dt}\text{tr}\rho = \text{tr}[\hat{H}, \rho] + i\hbar \sum_{j=1}^{N^2-1} \gamma_j \text{tr}\left(\hat{L}_j \rho \hat{L}_j^\dagger - \frac{1}{2}\left\{\hat{L}_j \hat{L}_j^\dagger, \rho\right\}\right) \tag{13.10}$$

$$= i\hbar \sum_{j=1}^{N^2-1} \gamma_j \text{tr}\left(\hat{L}_j \rho \hat{L}_j^\dagger - \hat{L}_j \hat{L}_j^\dagger \rho\right) = 0.$$

Here we used the cyclicity of the trace in the second and third equalities and noted that if the jump operators are Hermitian, then $\hat{L}_j = \hat{L}_j^\dagger$. While we won't do it here, it can also be shown that the time derivative of $\text{tr}\rho^2$ is non-positive in the Lindblad master equation:

$$\frac{d}{dt}\text{tr}\rho^2 \leq 0. \tag{13.11}$$

These identities enable us to define a useful effective entropy from the density matrix that satisfies the second law of thermodynamics. Consider the **linear entropy** S_L:

$$S_L = \text{tr}\left(\rho - \rho^2\right) = 1 - \text{tr}\rho^2. \tag{13.12}$$

In the previous chapter, we showed that $\text{tr}\rho^2 \leq 1$ and probability conservation requires that $\text{tr}\rho = 1$, so this is necessarily non-negative, $S_L \geq 0$. Further, $S_L = 0$ if and only if the system is a pure state for which $\rho^2 = \rho$. For a general mixed state, $S_L > 0$. While not the von Neumann entropy, the linear entropy can nevertheless be very useful for understanding a quantum system's dynamics. For example, by the property stated above of the Lindblad master equation, the linear entropy satisfies a second law of thermodynamics, where

$$\frac{d}{dt}S_L = -\frac{d}{dt}\text{tr}\rho^2 \geq 0. \tag{13.13}$$

The dynamics of general open quantum systems can therefore provide insight into decoherence through properties of the Lindblad master equation. An intriguing, concrete connection between the quantum and the classical is through coherent states. We introduced coherent states in Chap. 6 as quantum states that evolve according to classical equations of motion in the harmonic oscillator. Further, coherent states are localized in space like classical particles. Additionally, it has been shown that the evolution of coherent states with the Lindblad master equation generates the minimal linear entropy of any quantum state.[10] This perhaps points to why coherent states and classical particles share so much in common: time evolution of an open quantum system with the Lindblad master equation "washes out" all other states. Hopefully this and other hints can shed light on decoherence or the measurement problem and provide a deeper understanding of reality and answer what quantum mechanics *is*.

---

[10] W. H. Zurek, S. Habib, and J. P. Paz, "Coherent states via decoherence," *Phys. Rev. Lett.* **70**, 1187 (1993).

Mathematics Review

In this appendix, we present a concise review of the mathematics that are used throughout this textbook to understand quantum mechanics.

A.1 Linear Algebra

Quantum mechanics is fundamentally linear, and so linear algebra is the language to describe it. Here, we will review properties of matrices in particular, but it is easily extended to general linear operators. First, a matrix $\mathbb{M}$ is linear if it satisfies the two properties, where $\vec{u}$ and $\vec{v}$ are vectors:

(a) $\mathbb{M}(\vec{u} + \vec{v}) = \mathbb{M}\vec{u} + \mathbb{M}\vec{v}$, and
(b) $\mathbb{M}(a\vec{v}) = a\mathbb{M}\vec{v}$, where a is just a number.

For a complex-valued matrix, its elements can be determined in a particular basis in the following way. Let $\{\vec{v}_i\}_i$ be an orthonormal and complete basis on the space of interest. Orthonormality means that the vectors are normalized to unity and distinct vectors do not overlap under the inner product:

$$\vec{v}_i^\dagger \vec{v}_j = \delta_{ij}, \tag{A.1}$$

where $\dagger$ is the transpose-conjugation operation. Completeness means that the sum over the outer product of the vectors returns the identity matrix on the space:

$$\sum_i \vec{v}_i \vec{v}_i^\dagger = \mathbb{I}. \tag{A.2}$$

Then, the (i, j)th element of the matrix $\mathbb{M}$ in this basis is

$$M_{ij} = \vec{v}_i^\dagger \mathbb{M} \vec{v}_j. \tag{A.3}$$

There exists a special orthonormal basis $\{\vec{t}_i\}_i$ in which $\mathbb{M}$ is rendered diagonal, for which

$$M_{ij} = \vec{t}_i^\dagger \mathbb{M} \vec{t}_j = 0, \tag{A.4}$$

if $i \neq j$. The vector $\vec{t}_i$ is an eigenvector of $\mathbb{M}$ and the corresponding entry on the diagonal of $\mathbb{M}$ is its eigenvalue λ_i:

$$M_{ii} = \vec{t}_i^\dagger \mathbb{M} \vec{t}_i \equiv \lambda_i \,. \tag{A.5}$$

The existence of such an orthonormal basis is only possible if the matrix $\mathbb{M}$ is normal and commutes with its transpose-conjugate or Hermitian conjugate $\mathbb{M}^\dagger$:

$$\mathbb{M}\mathbb{M}^\dagger = \mathbb{M}^\dagger\mathbb{M} \,. \tag{A.6}$$

This result, that there exists an orthonormal basis in which a normal matrix $\mathbb{M}$ is rendered diagonal (and vice-versa), is called the spectral theorem. Examples of normal matrices include Hermitian matrices for which $\mathbb{M}^\dagger = \mathbb{M}$ and unitary matrices for which $\mathbb{M}\mathbb{M}^\dagger = \mathbb{I}$. Eigenvalues of a Hermitian matrix are exclusively real-valued, while eigenvalues of a unitary matrix lie on the unit circle in the complex plane.

The collection of eigenvalues $\{\lambda_i\}_i$ of a (normal) matrix $\mathbb{M}$ are basis-independent. Aspects of this basis-independence are represented in the determinant and trace of the matrix. The determinant of a matrix is the product of eigenvalues of that matrix:

$$\det(\mathbb{M}) = \prod_i \lambda_i \,. \tag{A.7}$$

To demonstrate its basis-independence, we can transform $\mathbb{M}$ to another basis through action of a unitary matrix $\mathbb{U}$:

$$\mathbb{M} \to \mathbb{U}\mathbb{M}\mathbb{U}^\dagger \,. \tag{A.8}$$

The determinant of this transformed matrix is

$$\det(\mathbb{U}\mathbb{M}\mathbb{U}^\dagger) = \det(\mathbb{U}^\dagger)\det(\mathbb{M})\det(\mathbb{U}) = \det(\mathbb{M})\det(\mathbb{U}\mathbb{U}^\dagger) = \det(\mathbb{M})\det(\mathbb{I}) \tag{A.9}$$
$$= \det(\mathbb{M}) \,,$$

where we used that the determinant of a product of matrices is the product of determinants of the matrices individually and that the determinant of the identity matrix is 1. The trace of a matrix is the sum of its diagonal elements or equivalently the sum of its eigenvalues:

$$\mathrm{tr}(\mathbb{M}) = \sum_i M_{ii} = \sum_i \lambda_i \,. \tag{A.10}$$

To demonstrate that this is basis independent, we again act with a unitary matrix and take the trace:

$$\mathrm{tr}(\mathbb{U}\mathbb{M}\mathbb{U}^\dagger) = \sum_{i,j,k} U_{ij} M_{jk} U_{ik}^* = \sum_{i,j,k} U_{ik}^* U_{ij} M_{jk} = \mathrm{tr}(\mathbb{U}^\dagger\mathbb{U}\mathbb{M}) = \mathrm{tr}(\mathbb{I}\mathbb{M}) \tag{A.11}$$
$$= \mathrm{tr}(\mathbb{M}) \,.$$

In this manipulation, we have exploited the cyclicity of the trace and recall that the elements of $\mathbb{U}^\dagger$ are the complex conjugate and transpose of $\mathbb{U}$.

A.2 Probability

The axioms of probability are

(a) The probability P of an event E is a non-negative real number:

$$P(E) \geq 0, \qquad P(E) \in \mathbb{R}. \tag{A.12}$$

(b) The total probability of any possible event occurring is unity:

$$\sum_E P(E) = 1. \tag{A.13}$$

(c) The probability P_{tot} of a collection of mutually exclusive events $\{E_i\}$ occurring is just the sum of their individual probabilities:

$$\sum_i P(E_i) = P_{\text{tot}}. \tag{A.14}$$

For a continuous random variable $x \in (-\infty, \infty)$, its probability distribution $p(x)$ integrates to 1 on its domain:

$$1 = \int_{-\infty}^{\infty} dx \, p(x). \tag{A.15}$$

The probability distribution $p(x)$ is also necessarily non-negative, $p(x) \geq 0$ for all x. This enables the interpretation of

$$P(a,b) \equiv \int_a^b dx \, p(x) \tag{A.16}$$

as the probability for $x \in [a,b]$, and $P(a,b) \geq 0$.

Other important quantities that can be defined from the probability distribution are its mean and variance. The mean, or expectation value, $\langle x \rangle$ is defined to be

$$\langle x \rangle = \int_{-\infty}^{\infty} dx \, x \, p(x). \tag{A.17}$$

Its interpretation is as the value of random variable x that is expected upon sampling the distribution $p(x)$ numerous times. The variance σ^2 is a measure of the spread of the distribution $p(x)$ about its mean and is defined to be

$$\sigma^2 = \langle (x - \langle x \rangle)^2 \rangle = \int_{-\infty}^{\infty} dx (x - \langle x \rangle)^2 \, p(x) \tag{A.18}$$

$$= \int_{-\infty}^{\infty} dx \left(x^2 - 2x\langle x \rangle + \langle x \rangle^2 \right) p(x)$$

$$= \int_{-\infty}^{\infty} dx \left(x^2 - \langle x \rangle^2 \right) p(x) = \langle x^2 \rangle - \langle x \rangle^2.$$

The standard deviation σ is the square root of the variance and its interpretation is that most of the integral of $p(x)$ is contained within one σ standard deviation of the mean; that is, for $x \in [\langle x \rangle - \sigma, \langle x \rangle + \sigma]$. The standard deviation and variance vanish if and only if there is a unique value of x which has non-zero probability.

A.3 Integration

At numerous points in this book we have encountered integrals of the Gaussian type, for which the integrand is quadratic in the integration variable:

$$I(a) = \int_{-\infty}^{\infty} dx\, e^{-ax^2}. \tag{A.19}$$

To evaluate this integral, we will actually consider its square, $I(a)^2$. We then have

$$I(a)^2 = \left(\int_{-\infty}^{\infty} dx\, e^{-ax^2} \right)^2 = \left(\int_{-\infty}^{\infty} dx\, e^{-ax^2} \right) \left(\int_{-\infty}^{\infty} dy\, e^{-ay^2} \right) \tag{A.20}$$

$$= \int_{-\infty}^{\infty} dx \int_{-\infty}^{\infty} dy\, e^{-a(x^2+y^2)}.$$

Note that the integrals over x and y are equivalent over the entire (x, y) plane, so we can re-express the integrals in polar coordinates (r, ϕ). The integration measure becomes

$$dx\,dy = r\,dr\,d\phi, \tag{A.21}$$

and the squared radius is

$$r^2 = x^2 + y^2. \tag{A.22}$$

The Gaussian integral then becomes

$$I(a)^2 = \int_0^{\infty} r\,dr \int_0^{2\pi} d\phi\, e^{-ar^2} = 2\pi \int_0^{\infty} dr\, r\, e^{-ar^2}. \tag{A.23}$$

Now, we make the change of variables $u = r^2$ and so $du = 2r\,dr$ and the integral becomes

$$I(a)^2 = \pi \int_0^{\infty} du\, e^{-au} = \frac{\pi}{a}. \tag{A.24}$$

Therefore the value of the Gaussian integral is

$$I(a) = \int_{-\infty}^{\infty} dx\, e^{-ax^2} = \sqrt{\frac{\pi}{a}}. \tag{A.25}$$

It is also straightforward to evaluate Gaussian integrals for which the integrand is also multiplied by some polynomial in the integration variable. Because the Gaussian is symmetric in $x \to -x$, including any odd power of x makes the integral vanish:

$$\int_{-\infty}^{\infty} dx\, x^{2n+1} e^{-ax^2} = 0, \tag{A.26}$$

for $n = 0, 1, 2, \ldots$. The integral with an even power of x can be evaluated with a trick. Consider the integral $I_{2n}(a)$, where

$$I_{2n}(a) = \int_{-\infty}^{\infty} dx\, x^{2n} e^{-ax^2}, \tag{A.27}$$

and $n = 0, 1, 2, \ldots$. We can pull down powers of x^2 by taking derivatives with respect to a:

$$I_{2n}(a) = \int_{-\infty}^{\infty} dx\, x^{2n} e^{-ax^2} = \int_{-\infty}^{\infty} dx\, (-1)^n \frac{d^n}{da^n} e^{-ax^2} \qquad (A.28)$$

$$= (-1)^n \frac{d^n}{da^n} I(a) = (-1)^n \frac{d^n}{da^n} \sqrt{\frac{\pi}{a}}.$$

Another important class of integrals are just those with an exponential integrand. Of course, we have

$$\int_0^{\infty} dx\, e^{-ax} = \frac{1}{a}. \qquad (A.29)$$

As with the Gaussian integrals, we can augment the integrand by a power of x with appropriate derivatives of a:

$$\int_0^{\infty} dx\, x^{n-1} e^{-ax} = (-1)^{n-1} \frac{d^{n-1}}{da^{n-1}} \frac{1}{a} = \frac{(n-1)!}{a^n}. \qquad (A.30)$$

Recall that the factorial $n!$ is defined to satisfy $n! = n \cdot (n-1)!$ and $0! = 1$. Therefore this integral satisfies the recursive relationship

$$\int_0^{\infty} dx\, x^n e^{-ax} = \frac{n}{a} \int_0^{\infty} dx\, x^{n-1} e^{-ax}. \qquad (A.31)$$

Now, if we generalize n from a natural number to a complex number, this integral is called the Euler Gamma function $\Gamma(n)$ and is defined to be

$$\Gamma(n) = \int_0^{\infty} dx\, x^{n-1} e^{-x}. \qquad (A.32)$$

The Gamma function satisfies the recursive relationship $\Gamma(n+1) = n\Gamma(n)$, just like the factorial, and $\Gamma(1) = 0! = 1$. The Gamma function is therefore a generalization of the factorial to continuous numbers. While this generalization is not unique, the presence of the Gamma function in physics is ubiquitous.

As an example of a non-integer value of the Gamma function, consider $\Gamma(1/2)$:

$$\Gamma(1/2) = \int_0^{\infty} dx\, x^{-1/2} e^{-x}. \qquad (A.33)$$

Let's make the change of variables $u = x^{1/2}$, so that $du = dx\, x^{-1/2}/2$:

$$\Gamma(1/2) = 2 \int_0^{\infty} du\, e^{-u^2} = \int_{-\infty}^{\infty} du\, e^{-u^2}. \qquad (A.34)$$

In the second equation, we have extended the bounds of integration to $(-\infty, \infty)$ because e^{-u^2} is an even function. Now this is in the form of a Gaussian that we had calculated earlier, and so $\Gamma(1/2) = \sqrt{\pi}$.

A.4 Fourier Transforms

For a function $f(x)$ defined on all $x \in (-\infty, \infty)$, its Fourier transform $\tilde{f}(p)$ is defined to be

$$\tilde{f}(p) = \int_{-\infty}^{\infty} \frac{dx}{\sqrt{2\pi}} f(x) e^{ipx} . \tag{A.35}$$

The action of the Fourier transform can be undone or inverted to transform $\tilde{f}(p)$ to $f(x)$ as

$$f(x) = \int_{-\infty}^{\infty} \frac{dp}{\sqrt{2\pi}} \tilde{f}(p) e^{-ipx} . \tag{A.36}$$

This can be established by simply inserting one expression into the other:

$$\tilde{f}(p) = \int_{-\infty}^{\infty} \frac{dx}{\sqrt{2\pi}} f(x) e^{ipx} = \int_{-\infty}^{\infty} \frac{dx}{\sqrt{2\pi}} \int_{-\infty}^{\infty} \frac{dp'}{\sqrt{2\pi}} \tilde{f}(p') e^{-ip'x} e^{ipx} \tag{A.37}$$

$$= \frac{1}{2\pi} \int_{-\infty}^{\infty} dp' \, \tilde{f}(p') \int_{-\infty}^{\infty} dx e^{ix(p-p')} = \frac{1}{2\pi} \int_{-\infty}^{\infty} dp' \, \tilde{f}(p') \, 2\pi \, \delta(p-p')$$

$$= \tilde{f}(p) .$$

To establish this identity, we used the definition of the Dirac δ-function $\delta(p - p')$ for which

$$2\pi \delta(p - p') = \int_{-\infty}^{\infty} dx \, e^{ix(p-p')} . \tag{A.38}$$

The Dirac δ-function integrates to unity:

$$\int_{a}^{b} dp \, \delta(p - p') = 1 , \tag{A.39}$$

for $p' \in [a, b]$. Also, its value is 0 if its argument is non-zero, with $\delta(p - p') = 0$, if $p \neq p'$.

The Fourier transform also maintains L^2-normalization of functions through a result called Plancherel's theorem. If $f(x)$ is L^2-normalized, then

$$\int_{-\infty}^{\infty} dx |f(x)|^2 = \int_{-\infty}^{\infty} dx f^*(x) f(x) < \infty . \tag{A.40}$$

Inserting the expression for $f(x)$ as a Fourier transform of $\tilde{f}(p)$, we have

$$\int_{-\infty}^{\infty} dx f^*(x) f(x) = \int_{-\infty}^{\infty} dx \int_{-\infty}^{\infty} \frac{dp}{\sqrt{2\pi}} \tilde{f}^*(p) e^{ipx} \int_{-\infty}^{\infty} \frac{dp'}{\sqrt{2\pi}} \tilde{f}(p') e^{-ip'x} \tag{A.41}$$

$$= \frac{1}{2\pi} \int_{-\infty}^{\infty} dp f^*(p) \int_{-\infty}^{\infty} dp' f(p') \int_{-\infty}^{\infty} dx e^{ix(p-p')}$$

$$= \frac{1}{2\pi} \int_{-\infty}^{\infty} dp f^*(p) \int_{-\infty}^{\infty} dp' f(p') \, 2\pi \delta(p - p') = \int_{-\infty}^{\infty} dp \tilde{f}^*(p) f(p)$$

$$= \int_{-\infty}^{\infty} dp |f(p)|^2 .$$

Poisson Brackets in Classical Mechanics

In the text, we formulated both the correspondence principle and canonical quantization with reference to Poisson brackets of functions on the phase space of some classical system. In particular, the correspondence principle is

$$\lim_{\hbar \to 0} \frac{[\cdot, \cdot]_{\text{quantum}}}{i\hbar} = \{\cdot, \cdot\}_{\text{classical}}, \tag{B.1}$$

where $\{\cdot, \cdot\}_{\text{classical}}$ is the classical Poisson bracket and $[\cdot, \cdot]_{\text{quantum}}$ is the quantum mechanical commutator. To study many quantum systems of interest, we can use the procedure of canonical quantization, by which quantum commutators are prescribed a value equal to the classical Poisson bracket, times $i\hbar$:

$$[\cdot, \cdot]_{\text{quantum}} = i\hbar \{\cdot, \cdot\}_{\text{classical}}. \tag{B.2}$$

This is a concrete procedure, but relies on knowledge of the classical Poisson brackets. In particular, depending on your exposure to formulations of classical mechanics, the Poisson bracket may be unfamiliar.[1]

Nevertheless, the Poisson bracket simply enables a distinct, yet equivalent, statement of quantities that may likely be more familiar from Newton's laws. For example, let's consider calculating the time derivative of a function $f(x, p)$ which is some function of the phase-space variables position x and momentum p. For simplicity, we will just work in one spatial dimension, but this whole framework naturally generalizes to arbitrary dimensions. For simplicity, we will also assume that $f(x, p)$ has no explicit time dependence, and so its total time derivative can be determined by the chain rule:

$$\frac{df(x, p)}{dt} = \frac{dx}{dt} \frac{\partial f(x, p)}{\partial x} + \frac{dp}{dt} \frac{\partial f(x, p)}{\partial p}, \tag{B.3}$$

where we have assumed that position and momentum are exclusively functions of time t. Now, let's use knowledge of classical kinematics and Newton's second law to re-write the derivatives of position and momentum. The time derivative of position is velocity v, or momentum divided by mass m:

$$\frac{dx}{dt} = v = \frac{p}{m}. \tag{B.4}$$

---

[1] For more information about Poisson brackets and the Hamiltonian formulation of classical mechanics, two good references are L. D. Landau and E. M. Lifshitz, *Mechanics: Volume 1*, Butterworth-Heinemann (1976) and H. Goldstein, C. Poole, and J. Safko, *Classical Mechanics*, Pearson (2002).

By Newton's second law, the time derivative of momentum is the force F:

$$\frac{dp}{dt} = F. \tag{B.5}$$

If we assume that F is a conservative force, then it can further be expressed as the spatial derivative of a potential energy $U(x)$:

$$\frac{dp}{dt} = F = -\frac{dU(x)}{dx}. \tag{B.6}$$

Then, the time derivative of the function $f(x, p)$ is

$$\frac{df(x, p)}{dt} = \frac{p}{m} \frac{\partial f(x, p)}{\partial x} - \frac{dU(x)}{dx} \frac{\partial f(x, p)}{\partial p}. \tag{B.7}$$

Continuing, we can define the Hamiltonian on phase space as the total energy functional:

$$H(x, p) = \frac{p^2}{2m} + U(x), \tag{B.8}$$

which is just the sum of kinetic and potential energies. In terms of the Hamiltonian, the velocity is just the partial derivative with respect to p:

$$\frac{\partial H(x, p)}{\partial p} = \frac{p}{m}. \tag{B.9}$$

Similarly, the partial derivative of the Hamiltonian with respect to position is the force:

$$\frac{\partial H(x, p)}{\partial x} = \frac{dU(x)}{dx}. \tag{B.10}$$

With these identifications, the time derivative of the function $f(x, p)$ is

$$\frac{df(x, p)}{dt} = \frac{p}{m} \frac{\partial f(x, p)}{\partial x} - \frac{dU(x)}{dx} \frac{\partial f(x, p)}{\partial p} \tag{B.11}$$

$$= \frac{\partial f(x, p)}{\partial x} \frac{\partial H(x, p)}{\partial p} - \frac{\partial f(x, p)}{\partial p} \frac{\partial H(x, p)}{\partial x} \equiv \{f, H\},$$

the classical Poisson bracket. Thus, this equation enables the interpretation that in classical mechanics, the Hamiltonian generates time translations.

The Poisson bracket can be generalized to be defined for any two functions on phase space $f(x, p)$ and $g(x, p)$, where

$$\{f, g\} = \frac{\partial f(x, p)}{\partial x} \frac{\partial g(x, p)}{\partial p} - \frac{\partial f(x, p)}{\partial p} \frac{\partial g(x, p)}{\partial x}. \tag{B.12}$$

The interpretation of this Poisson bracket is that it represents the transformation associated with g on the function f. For three functions on phase space, $f(x, p)$, $g(x, p)$, and $h(x, p)$, the Poisson bracket satisfies the Jacobi identity in which the sum over cyclic ordering of nested Poisson brackets is zero:

$$\{f, \{g, h\}\} + \{h, \{f, g\}\} + \{g, \{h, f\}\} = 0. \tag{B.13}$$

This can be proved by expanding out all Poisson brackets in terms of partial derivatives and noting that partial derivatives commute, that is

$$\frac{\partial}{\partial x}\frac{\partial}{\partial p}f(x,p) = \frac{\partial}{\partial p}\frac{\partial}{\partial x}f(x,p). \tag{B.14}$$

Existence of a Jacobi identity means that the Poisson brackets define an associative algebra of functions on phase space. The commutator of quantum operators also satisfies a Jacobi identity. Existence of a Jacobi identity ensures that when Hermitian operators are exponentiated, for example, the multiplication rule of the resulting unitary operators is associative.

C   Further Reading

Undergraduate

M. Beck, *Quantum Mechanics: Theory and Experiment*, Oxford University Press (2012).
Focuses around quantum properties of photons and provides undergraduate experiments to test the theory.

C. Cohen-Tannoudji, B. Diu, and F. Laloë, *Quantum Mechanics, Volume 1: Basic Concepts, Tools, and Applications*, 2nd ed., Wiley-VCH (2020).
Essentially encyclopedic introduction to quantum mechanics that introduces foundational concepts before principles of quantum mechanics.

D. J. Griffiths and D. F. Schroeter, *Introduction to Quantum Mechanics*, 3rd ed., Cambridge University Press (2018).
A classic, pragmatic introduction to quantum mechanics starting with the Schrödinger equation.

J. S. Townsend, *A Modern Approach to Quantum Mechanics*, University Science Books (2000).
Begins from spin and two-state systems to build up the modern, linear algebra formulation of quantum mechanics.

Graduate

L. D. Landau and E. M. Lifshitz, *Quantum Mechanics: Non-Relativistic Theory*, vol. 3, Elsevier (2013).
The classic reference in Landau and Lifshitz's course on theoretical physics.

J. J. Sakurai, *Modern Quantum Mechanics*, revised edition, Addison-Wesley (1993).
The extended, revised version of Sakurai's notes that were to become a book before his untimely death in 1982. Essentially unrivaled in its presentation and development of the theory of quantum mechanics.

R. Shankar, *Principles of Quantum Mechanics*, Springer (1994).
A standard graduate-course text, contains extensive discussion of path integrals.

M. Tinkham, *Group Theory and Quantum Mechanics*, Courier Corporation (2003).
Introduction to group theory and its application to quantum mechanics.

S. Weinberg, *Lectures on Quantum Mechanics*, Cambridge University Press (2015).
A collection of a wide range of topics not often presented in other books.

Lecture Notes

D. Tong, Lectures on Topics in Quantum Mechanics at University of Cambridge, `http://www.damtp.cam.ac.uk/user/tong/topicsinqm.html` (accessed December 3, 2020).
An advanced quantum mechanics course focusing on atomic physics, quantum foundations, and scattering theory.

N. Wheeler, Quantum Mechanics at Reed College, `https://www.reed.edu/physics/faculty/wheeler/documents/index.html` (accessed December 3, 2020).
An idiosyncratic, almost stream-of-consciousness collection of lectures on foundational ideas in quantum mechanics.

Glossary

A number of the quantum mechanical terms that are defined and used in the main text. We collect them here for ease of identification.

Abelian: the property of a group if all of its elements commute with one another

action: the time integral of the Lagrangian, the difference of kinetic and potential energies; extremization of the action over trajectories produces the classical equations of motion of a system

analytic function: a function whose Taylor expansion converges on a finite-sized domain; an analytic function of a complex variable z that is independent of its complex conjugate z^*

angular frequency: the rate at which the argument of a sinusoidal wave or other oscillation passes through one radian

angular momentum: a vector that quantifies the rotational motion of an object or system about a defined axis

annihilate: the action of an operator on a state that removes it from the Hilbert space, typically by transforming the state to a non-normalizable state

Argand diagram: the representation of a set of numbers on a circle in the complex plane; the circle in the complex plane with radius 1 centered at 0 on the real axis and i on the imaginary axis and corresponds to allowed eigenvalues of the interaction matrix

argument (of a complex number): the angle in the complex plane between the positive real axis and the line that extends from the origin to the complex number of interest

automorphism: a one-to-one and onto map of a space or set to itself that preserves the structure of the space

Baker–Campbell–Hausdorff formula: the rule for multiplication of exponentiated matrices that do not commute; their product can be represented by a series of nested commutators

beam splitter: an experimental device like a half-silvered mirror which reflects and transmits light with 50% probability each

Bell's inequality: a strict inequality on expectation values of multiparticle states that must be satisfied classically, but can be violated by quantum entangled states

Bohr magneton: the classical, non-relativistic magnetic moment of the electron, equal to the fundamental electric charge times $\hbar$ divided by twice the electron's mass

Bohr radius: the characteristic radius of the electron orbiting the proton in atomic hydrogen and often denoted by a_0, where $a_0 = 5.3 \times 10^{-11}$ m

Bohr–Sommerfeld quantization: the approximation to quantum mechanics that the spatial integral of momentum between the classical turning points of a potential is equal to an integer multiple of $\pi\hbar$

Boltzmann factor: the exponential of the negative energy of a state divided by the temperature that governs the relative probability of that state to be occupied in a system in thermal equilibrium

Born rule: the interpretation that the state vector of a quantum system represents a complex-valued probability amplitude

bound state: a state on the Hilbert space of a quantum system that has compact support; namely, it has 0 probability to be an arbitrary distance from the coordinate origin

bra: the transpose-conjugate of the state ket vector; also called a dual vector

canonical commutation relation: the commutation relation for the position $\hat{x}$ and momentum operators $\hat{p}$ for which $[\hat{x}, \hat{p}] = i\hbar$

canonical ensemble: the statistical description of a system in thermal equilibrium with a heat bath

canonical quantization: the formal procedure of rendering a classical system quantum by promoting all functions on phase space to be operators on Hilbert space and all Poisson brackets to commutation relations divided by $i\hbar$

Cartan subalgebra: the maximal subset of a Lie algebra that consists of mutually commuting operators

Casimir: the number that uniquely quantifies the specific representation of a Lie algebra; the quadratic Casimir is the sum of the squares of the elements of the Lie algebra

Casimir effect: an attractive or repulsive force between two parallel, electrically neutral plates held a very small distance apart that is a consequence of the quantum nature of electromagnetism

Cauchy–Schwarz inequality: for a vector space equipped with an inner product, the squared magnitude of the inner product of two vectors is less than or equal to the product of the squared magnitudes of the vectors

central potential: a potential energy that is strictly a function of the distance between the individual objects of the physical system

charge: a quantity whose value is conserved or constant in time; a generalization of electric charge to other quantities whose net value can neither be created nor destroyed

closed: a set equipped with a multiplication rule is closed if the product of any two elements in the set is also in the set

closed system: a system that is perfectly isolated from the rest of the universe or is the entire universe itself

coherent state: an eigenstate of the annihilation operator that saturates the Heisenberg uncertainty principle

coherent sum: a sum of complex-valued terms for which both the magnitude and argument of the complex number are important

commutative: a multiplication rule for which the order does not matter

commutator: the difference between the product of two linear operators taken in opposite orders

compact support: the property of a function in which the region over which it has non-zero and not exceedingly small value is finite; practically, the integral of the function over its entire domain is finite

complete: the property of a vector space if any linear combination of vectors in the space is also in the space

completeness relation: the relationship satisfied by a complete, orthonormal basis of a vector space where the sum over the all outer products of a basis vector with itself is the identify matrix

completely random: equal probability for any possible outcome of an experiment

Compton wavelength: the longest quantum wavelength of a particle of mass m; the quantum wavelength of a particle of mass m at rest

conjugacy class: a subset of a set for which an identification is imposed between different elements

conservation law: a non-trivial statement of the time-independence of a quantity

Copenhagen interpretation: the interpretation of quantum mechanics that states that physical systems do not have definite properties until a measurement is performed

correspondence principle: the formal $\hbar \to 0$ limit in which the predictions of quantum mechanics reduce to those of classical mechanics

cosmic microwave background: ubiquitous remnant electromagnetic radiation in the universe from the time of recombination

de Broglie wavelength: the characteristic quantum wavelength defined by an object's momentum

decoherence: the process by which a non-isolated quantum system becomes classical through weak interactions with the environment

degeneracy: the property of multiple, distinct states in the Hilbert space to have the same eigenvalue of an operator; typically refers to orthogonal states that have the same energy eigenvalue

density matrix: the matrix that describes the state of an ensemble of quantum particles through their probability densities

diagonalization: the process of determining the eigenvalues of a linear operator; when written in diagonal form, a linear operator's eigenvalues are its only non-zero entries

dilation: the operation that scales positions by a positive real value

Dirac δ-function: formally a distribution or generalized function that is zero almost everywhere but infinite where its argument vanishes and has finite integral over any domain

Dirac notation: a compact notation introduced by Paul Dirac for describing quantum states, their inner and outer products, and the action on them by linear operators

Dirac–von Neumann axioms of quantum mechanics: the three fundamental hypotheses from which the framework of quantum mechanics is derived: (1) observables are Hermitian operators on Hilbert space; (2) the state of a system is represented by a unit normalized state vector on Hilbert space; (3) expectation values of Hermitian operators are found by taking inner products with state vectors

dispersion relation: the functional dependence of angular velocity on wavenumber; the functional dependence of kinetic energy on momentum

Dyson series: the form of exponentiation for non-commuting Hermitian operators into a unitary operator

Ehrenfest's theorem: the general statement that expectation values of quantum mechanical operators obey the classical equations of motion by the correspondence principle

eigenvalue: a particular value associated to a matrix that corresponds to the scaling that the matrix acts on vectors

eigenvector: a particular vector associated with a matrix for which action by that matrix is a rescaling

electronvolt: the amount of energy gained by an electron in an electric potential of 1 V

energy: a quantity that represents a system's ability to perform a task; conservation of energy is associated with time translation invariance by Noether's theorem

ensemble average: an average or mean taken over a system composed of many individual elements or particles

entanglement: quantum correlation of a collection of particles that encodes information about their relationship to one another in a state vector on Hilbert space

entropy: a quantity that measures the information of individual particles in a composite system exclusively from knowledge of the whole system itself

entropy of entanglement: the entropy of a composite system after tracing out a subset of the system

Euler–Lagrange equations of motion: the equations of motion of a classical system derived from the principle of least action

expectation value: the mean or average value of a quantity when measured on a collection of independent and identical systems

exponentiation: the process of composing arbitrary infinitesimal transformations to produce a finite transformation; often, this is accomplished through making the infinitesimal transformation the argument of the exponential function

factorization: the process of decomposing a sum of terms into a product of terms, each of which is simpler than the original sum

fine structure constant: denoted as α, the dimensionless, pure number that quantifies the strength of electromagnetism; its value is approximately $1/137$

Fourier conjugate variables: two independent variables that are the arguments of a function and its Fourier transform, respectively

Fourier transform: the procedure of expressing an arbitrary function as a linear combination of sinusoidal functions

fractal: a set which exhibits non-smooth structure at any non-zero resolution scale

free particle: a particle which has no forces acting on it; equivalently, a particle that is under the influence of a potential whose value is independent of position and time

functional: a function whose argument is itself a complete function of an independent variable

Gamma function: $\Gamma(x)$, a continuous generalization of the factorial, satisfies $\Gamma(x+1) = x\Gamma(x)$ and $\Gamma(1) = 1$

Gaussian function: also called a bell curve or normal distribution, the exponential function with a quadratic function of the independent variable in the exponent

generalized uncertainty principle: the product of the variances of two Hermitian operators on any state in the Hilbert space is bounded from below by the absolute square of their commutator

g-factor: a dimensionless number that quantifies the size of a particle's quantum gyromagnetic ratio with respect to its classical value

ground state: the lowest-energy state of a system

group: a set of elements equipped with a bilinear multiplication rule that satisfies the properties of closure, existence of an identity, existence of an inverse, and associativity

group velocity: the velocity of the center-of-probability of a wave packet corresponding to the physical velocity of a particle; mathematically, the derivative of the energy or angular velocity with respect to the momentum or wavenumber, respectively

gyromagnetic ratio: the constant of proportionality between an electrically charged particle's angular momentum and its magnetic dipole moment

Hamiltonian: the energy functional of a classical mechanics system; the Hermitian energy or time-translation operator of a quantum system

Hamiltonian path integral: the expression of the transition amplitude from an initial position and time to a final position and time expressed as a sum over all possible trajectories and momenta at intermediate times, weighted by the imaginary exponential of the Legendre transformation of the time integral of the Hamiltonian

harmonic oscillator: a system for which the force acting on a particle is of Hooke's law form: restoring and linearly proportional to displacement; a system whose potential energy is quadratic in displacement

Hausdorff dimension: a generalized definition of dimension that allows for non-integer values based on scaling properties of lengths as a function of a resolution parameter

Heaviside theta function: a function which is 0 if its argument is negative and 1 if its argument is positive

Heisenberg uncertainty principle: the statement that the product of the standard deviations of position and momentum on any state in the Hilbert space is bounded from below by $\hbar/2$

Hermitian conjugate: the matrix found by transposition of rows and columns and complex conjugation of all elements of another matrix

Hermitian matrix: a matrix which equals its Hermitian conjugate

hidden symmetry: a transformation on a system that leaves its physical description unchanged, but is not obvious or designed in the form of the description of the system

Hilbert space: the space of physical states in quantum mechanics; the ray space of all normalizable, complex-valued vectors

Hong–Ou–Mandel interferometer: an interferometry experiment in which the quantum properties of a system of two identical photons exhibit exactly the opposite phenomena expected of classical electromagnetic waves

Hopf fibration: coordinates on the three-sphere that map it to the product of a two-sphere and a one-sphere: $S^3 = S^1 \times S^2$

idempotent: a quantity or object whose square is equal to itself

incoherent sum: the sum of complex-valued terms for which only the magnitude of the complex number is important

infinite square well: the quantum system for which its potential is 0 in the finite domain of the well and infinite outside; classically, corresponds to a box with perfectly rigid walls

inner product: the rule for multiplication of two vectors on a particular space that produces a scalar

in state: an initial state in a quantum scattering process; a free-particle state defined asymptotically far in the past

integrable system: a system for which there are sufficient conserved quantities to uniquely determine the time evolution of a system on phase space from given initial data

interaction matrix: a subcomponent of the S-matrix which completely describes non-trivial interaction in a scattering process

irreducible representation: the largest representation of a given dimension of a Lie algebra for which the action of its elements is closed

isomorphic: existence of a one-to-one and onto map between two sets that preserves the action of operations on the sets; colloquially, two sets are isomorphic if they are identical up to a renaming of elements

Jacobi identity: the identity of Poisson brackets or commutators that establishes their associativity; the sum of three nested Poisson brackets or commutators in cyclic order is 0

ket: a state vector on Hilbert space

Killing form: the normalization of basis elements or matrices of a Lie algebra; defined through the trace of the square of elements of the Lie algebra

Kronecker δ: a function of two integer arguments which is 1 if the arguments are equal and 0 otherwise

L^2-norm: the requirement that the integral over any domain of the squared magnitude of a function is finite and non-zero

ladder operator: a non-Hermitian linear operator whose action maps an eigenvector of a Hermitian operator to another eigenvector whose eigenvalue differs by a fixed, discrete value

Lagrangian: the difference of kinetic and potential energies for a system

Laplace–Runge–Lenz vector: a second conserved vector (along with angular momentum) in the central potential problem in three spatial dimensions with a potential that varies inversely with distance

Laurent series: a generalization of the Taylor series to describe functions at and near a point of non-analyticity; an infinite series formed from the linear combination of every integer power, positive, negative and zero, of a complex number z

Legendre polynomials: a set of polynomials in a real variable x that form an orthonormal and complete basis of functions on $x \in [-1, 1]$

Levi-Civita symbol: also the totally antisymmetric symbol, a three-index object that is 1 when its indices are in cyclic order, -1 when its indices are in anti-cyclic order, and 0 if any indices are repeated

Lie algebra: a vector space of linear operators additionally equipped with a Lie bracket or commutation relation that is linear in the operators

lifetime: the characteristic time an unstable particle exists before it decays; the time elapsed for the probability that an unstable particle still exists to be $1/e = 0.36787\ldots$

Lindblad master equation: the generalization of the von Neumann equation for time evolution of the density matrix to realistic, open quantum systems

linear entropy: an entropy that measures the degree to which a density matrix is not a pure state, defined by the difference of the trace of the density matrix and the trace of its square

linear operator: an object that maps vectors in an arbitrary space to one another and satisfies the conditions for linearity, namely the distributive property and commutativity with scalar multiplication

local operator: an operator whose action in position space is contained to a single point, or infinitesimal region about that point

localized: a function whose integral is dominated by a small region about a particular point

Lorentz transformation: a linear transformation on spatial and temporal coordinates that leaves the spacetime interval invariant in special relativity

Many-Worlds interpretation: the interpretation of quantum mechanics which asserts that time evolution is always unitary, but that our universe is just one of many possible realizations to the outcome of experiment

Maslov correction: the correction to the Bohr–Sommerfeld quantization that accounts for the $\pi/2$ change in phase of momentum from the region between the classical turning points of a potential to outside

matrix element: an individual entry at a particular row and column in a matrix

measurement problem: the apparent inconsistency between the probabilistic nature of quantum mechanics and unambiguous, classical outcomes of experiment

median: the point of a probability distribution at which half of the area is to the left and half of the area is to the right of that point

mixed state: the quantum state of an ensemble of particles that cannot be represented by a single state on the Hilbert space

momentum: a vector that quantifies the motion of an object or system through space; conservation of momentum is associated with spatial translation invariance by Noether's theorem

neutrinos: electrically neutral, very low mass, and in general extremely weakly interacting fundamental particles

Noether's theorem: the statement proved by Emmy Noether that continuous symmetries of a physical system described by a Lagrangian have corresponding conservation laws, and vice-versa

non-Abelian: a property of a group if its elements do not in general commute with one another

normal: a matrix or operator that commutes with its Hermitian conjugate

normal order: arranging the expression of an operator with commutation relations such that all raising/creation operators are to the left of all lowering/annihilation operators

number operator: the operator whose eigenvalues are non-negative integers and can be expressed as the product of the creation and annihilation operators of the harmonic oscillator; the operator that counts the number of the energy eigenstate that is occupied

observable: a quantity that can be the outcome of an experimental measurement of a quantum mechanical system

Occam's razor: the guiding principle of science that states when there are multiple descriptions of a physical system the simplest should be preferred

open system: a quantum system that has some interaction with or coupling to an external environment

optical theorem: the statement that the total probability for non-trivial interaction in a scattering process is equal to the amount of transparency lost to scattering

orthonormal: a set of vectors for which each vector has unit length and distinct vectors are orthogonal

out state: a final state in a quantum scattering process; a free-particle state defined asymptotically far in the future

outer product: the rule for multiplication of two vectors on a particular space that produces a matrix

partial trace: the action of tracing over a subset of a composite system's density matrix

partition function: the object that encodes the relative probabilities of energy states of a system in thermal equilibrium

path integral: the sum over all paths from a fixed initial position to a fixed final position weighted by an imaginary exponential with argument of the action of the path in units of $\hbar$

Pauli matrices: the three Hermitian 2×2 matrices that, along with the identity matrix, form a basis for all 2×2 Hermitian matrices; the matrices that form the two-dimensional representation of the $\mathfrak{su}(2)$ algebra

phase velocity: the speed at which a wave of fixed wavelength and frequency travels; mathematically, the ratio of the energy or angular velocity and the momentum or wavenumber, respectively

Plancherel's theorem: a function and its Fourier transform have the same L^2-normalization

Planck–Larkin partition function: an approximate partition function for hydrogen that accounts for the infinite number of bound states and incorporates scattering states of the electron

Planck's reduced constant $\hbar$: the fundamental quantum of action that sets all scales of a quantum mechanical system

plasma: an optically opaque fluid of electrically charged particles

Poisson bracket: the antisymmetric difference between the product of position and momentum derivatives of two functions on classical phase space; the Poisson bracket of a function on phase space and the Hamiltonian determines the time dependence of the function

polarized: the property that all individual particles in a quantum ensemble are in the same state on Hilbert space

pole: a point at which a function of a complex variable diverges

position basis: the basis of Hilbert space for which all states are expressed as linear combinations of eigenstates of the position operator; states on Hilbert space are expressed through the wavefunction

power method: the procedure for determining the eigenvector of a matrix corresponding to the largest in magnitude eigenvalue by repeated multiplication of the matrix on an arbitrary initial vector

precession: simple harmonic or circular motion of the angular momentum vector due to a perpendicularly applied torque

principle of least action: the classical equations of motion of a system follow from extremization of the action

propagator: the complex probability amplitude for a quantum mechanical particle to travel from one position to another in a fixed time

pure state: a state of an ensemble of quantum particles that can be represented by a single element of the Hilbert space

purification: the pure state and corresponding density matrix from which an arbitrary density matrix is its partial trace

purity: the trace of the square of the density matrix, $\operatorname{tr}\rho^2$, whose value is 1 if and only if the ensemble is in a pure state

quantum field theory: the union of special relativity and quantum mechanics that describes particles as the fundamental quanta of fluctuations of fields that permeate space and time

quantum number: an eigenvalue of an observable Hermitian operator that commutes with the Hamiltonian and is used to define a quantum state or system

quantum tunneling: the non-zero probability for a quantum particle to pass through a potential whose height is larger than the total energy of the particle

quark: a fundamental particle of which one bound state of quarks is the proton

quaternions: the four-dimensional space over the real numbers spanned by quantities i, j, k and the real number 1; i, j and k each square to -1 and satisfy $ijk = -1$ and $ij = -ji = k$, and all cyclic permutations thereof

qubit: a two-state quantum system; a "quantum bit"

ray space: a restricted vector space in which vectors $\vec{v}$ and $\vec{v}' = c\vec{v}$ for complex-valued $c \neq 0$ in the space are identified

recombination: the time in the history of the universe at which the temperature was low enough that the plasma of protons and electrons could combine into neutral hydrogen

reduced mass: the effective mass that orbits the center-of-mass in a two-body system

reflected: a wave scattered from a potential that travels back in the direction of where it originated

Rényi entropy: a one-parameter family of entropy measures; the limit in which the parameter goes to 1, Rényi entropy reduces to von Neumann entropy

representation: an explicit realization of a Lie algebra in which its elements are matrices

scalar: a quantity that is not affected by the action of rotation

scattering: the phenomena of an initial free quantum particle interacting with a localized potential and producing reflected and transmitted components of a final free quantum particle

Schrödinger equation: the time evolution equation for a state on Hilbert space

separable: a pure state of a composite system that is the tensor product of pure states of its constituents

simply connected: a space for which all closed paths in the space can be continuously deformed into a point

simultaneously diagonalizable: two matrices or operators that commute so that the same unitary transformation can render them both in diagonal form

S-matrix: the scattering matrix whose entries encode the probability amplitude for an initial state of fixed momentum to evolve into a final state of fixed momentum after interacting with a localized potential

SO(2): the group of 2×2 orthogonal matrices with unit determinant; the group of rotations in two dimensions

spectral theorem: the statement that every normal matrix or operator can be diagonalized by action of a unitary transformation and vice-versa

spectrum: the collection of eigenvalues of a matrix or operator

spin: intrinsic value of angular momentum of an object; the particular representation of the rotation group under which a particle transforms

spinor: an object that transforms under the two-dimensional representation of the rotation group SU(2)

spinor helicity: a formalism in which all quantities are expressed in terms of eigenstates of the z-component of spin-1/2 angular momentum operator

standard deviation: the square root of the variance of a probability distribution of a random variable; a measure of the domain of the random variable about its mean for which it is most likely to take its value

state: an abstract mathematical representation of a physical system

state space: the Hilbert space, the space in which the mathematical representation of physical states reside

stationary: those points of a function or functional at which its derivative vanishes

structure constants: the coefficients in the linear combination of elements of a Lie algebra to which its commutation relation equal

$\mathfrak{su}(2)$: the Lie algebra of rotations in three dimensions; the vector space formed from the three operators of angular momentum equipped with a cyclic commutation relation between the components

SU(2): the group of 2×2 unitary matrices with unit determinant; the group of rotations in three dimensions

subadditivity: the statement that the von Neumann entropy of a composite system is bounded from above by the sum of von Neumann entropies of its constituent systems

supersymmetry: a proposed symmetry algebra between half-integer spin (fermion) and integer spin (boson) fundamental particles that consists of both commutation and anti-commutation relations of its elements

surface of last scattering: the time in the history of the universe at which electrically neutral hydrogen formed and so the universe transformed from optically opaque to transparent

symmetry: an action performed on a system that leaves the physical description of the system unchanged

systematic improvability: an approximation procedure whose formal accuracy is well-defined and for which there is an algorithm for improving the accuracy

tensor product: a generalization of the outer product of vectors to matrices and higher-index objects; a product of matrices that produces an object for which each element is the product of an element of one matrix times the entire other matrix

test function: a formal, infinitely differentiable function that is used to test properties of distributions or operators

thermal equilibrium: a system of an ensemble of particles that is time-independent when averaged over all particles and whose total energy is characterized by a single quantity called temperature

transmitted: a wave scattered from a potential that travels away from the direction of where it originated

triangular matrix: a matrix with non-zero entries on the diagonal and above it (upper triangular) or on the diagonal and below (lower triangular)

trivial representation: a representation of a Lie algebra that trivially satisfies the commutation relations because all elements of the representation are 0

U(1): the unitary group of 1×1 matrices; the group that corresponds to the action of moving around the unit circle in the complex plane

unitary: a matrix whose product with its Hermitian conjugate is the identity matrix; an object of unit length, appropriately defined

variance: the difference of the mean square and the square mean of a random variable; a measure of the spread of a probability distribution about its mean

variational method: an approximation method in which one selects parameters of an ansatz quantum state by minimizing the expectation value of the Hamiltonian on that state

virial theorem: classically, proportionality between the kinetic and potential energies for particles in a system bound by a potential that is a power of relative distance; quantum mechanically, proportionality between the expectation values of the kinetic and potential energy operators for a bound energy eigenstate in a potential that is a power of the relative distance

von Neumann entropy: the negative trace of the product of the density matrix and its logarithm; the unique entropy measure that satisfies criteria of additivity, subadditivity, etc., and so has a direct information-theory interpretation

von Neumann equation: the equation that governs the time evolution of the density matrix of a closed system

wavefunction: the state vector of a quantum system expressed in position space

wavenumber: the inverse wavelength of a wave, multiplied by 2π; proportional to the momentum carried by the wave

wave packet: a linear combination of free-particle states that has compact support and is normalizable

Wigner's theorem: the statement that automorphisms of the Hilbert space are implemented by the action of unitary or anti-unitary operators

WKB approximation: an approximation to an energy eigenstate wavefunction of a potential through approximate application of the position-dependent unitary spatial translation operator

Zeeman effect: the splitting of energy eigenstates of different angular momentum in hydrogen or other elements due to the presence of a weak external magnetic field

zero-point energy: the minimal non-zero energy of a quantum system with no external energy source

Bibliography

Collected below are the in-text references presented in alphabetical order. This is formally superfluous, as all are provided as footnotes on the page where they are referenced. However, collecting them in one place can make finding a particular reference easier.

M. Aaboud *et al.* [ATLAS], "Observation of $H \to b\bar{b}$ decays and VH production with the ATLAS detector," *Phys. Lett. B* **786**, 59–86 (2018) [arXiv:1808.08238 [hep-ex]].

L. F. Abbott and M. B. Wise, "The dimension of a quantum mechanical path," *Am. J. Phys.* **49**, 37–39 (1981).

T. Andreescu and B. Enescu, *Mathematical Olympiad Treasures*, Springer Science & Business Media (2011).

H. Araki and E. H. Lieb, "Entropy inequalities," *Commun. Math. Phys.* **18**, 160–170 (1970).

A. Aspect, P. Grangier, and G. Roger, "Experimental tests of realistic local theories via Bell's theorem," *Phys. Rev. Lett.* **47**, 460–6443 (1981).

A. Aspect, P. Grangier, and G. Roger, "Experimental realization of Einstein–Podolsky–Rosen–Bohm Gedankenexperiment: A new violation of Bell's inequalities," *Phys. Rev. Lett.* **49**, 91–97 (1982).

A. Aspect, J. Dalibard, and G. Roger, "Experimental test of Bell's inequalities using time varying analyzers," *Phys. Rev. Lett.* **49**, 1804–1807 (1982).

J. C. Baez, "Division algebras and quantum theory," *Found. Phys.* **42**, 819–855 (2012) [arXiv:1101.5690 [quant-ph]].

H. F. Baker, "Further applications of matrix notation to integration problems," *Proc. Lond. Math. Soc.* **1**(1), 347–360 (1901).

M. Bander and C. Itzykson, "Group theory and the hydrogen atom," *Rev. Mod. Phys.* **38**, 330–345 (1966).

T. Banks, L. Susskind, and M. E. Peskin, "Difficulties for the evolution of pure states into mixed states," *Nucl. Phys. B* **244**, 125–134 (1984).

J. Bardeen, L. N. Cooper, and J. R. Schrieffer, "Theory of superconductivity," *Phys. Rev.* **108**, 1175–1204 (1957).

V. Bargmann, "Zur Theorie des Wasserstoffatoms," *Z. Phys.* **99**, 576–582 (1936).

J. S. Bell, "On the Einstein–Podolsky–Rosen paradox," *Phys. Phys. Fiz.* **1**, 195–200 (1964).

J. S. Bell, *Speakable and Unspeakable in Quantum Mechanics: Collected Papers on Quantum Philosophy*, Cambridge University Press (2004).

C. M. Bender and T. T. Wu, "Anharmonic oscillator," *Phys. Rev.* **184**, 1231–1260 (1969).

N. Bohr, "On the constitution of atoms and molecules," *Phil. Mag. Ser.* 6 **26**, 1–24 (1913).

N. Bohr, "On the constitution of atoms and molecules. 2. Systems containing only a single nucleus," *Phil. Mag. Ser. 6* **26**, 476 (1913).

M. Born and P. Jordan, "Zur Quantenmechanik," *Z. Phys.* **34**, 858–888 (1925).

M. Born, "Zur Quantenmechanik der Stoßvorgänge," *Z. Phys.* **37**(12), 863–867 (1926); reprinted and translated in J. A. Wheeler and W. H. Zurek (eds.), *Quantum Theory and Measurement*, Princeton University Press, 52 (1963).

L. Brillouin, "La mécanique ondulatoire de Schrödinger; une méthode générale de resolution par approximations successives," *Compt. Rend. Hebd. Seances Acad. Sci.* **183**(1), 24–26 (1926).

L. V. P. R. de Broglie, "Recherches sur la théorie des quanta," *Ann. Phys* **2**, 22–128 (1925).

J. E. Campbell, "On a law of combination of operators bearing on the theory of continuous transformation groups," *Proc. Lond. Math. Soc.* **1**(1), 381–390 (1896); **1**(1), 14–32 (1897).

H. B. G. Casimir, "Rotation of a rigid body in quantum mechanics," PhD thesis, Universiteit Leiden (1931).

H. B. G. Casimir, "On the attraction between two perfectly conducting plates," *Indag. Math.* **10**, 261–263 (1948).

H. B. G. Casimir and D. Polder, "The influence of retardation on the London–van der Waals forces," *Phys. Rev.* **73**, 360 (1948).

V. Chung, "Infrared divergence in quantum electrodynamics," *Phys. Rev.* **140**, B1110–B1122 (1965).

B. S. Cirelson, "Quantum generalizations of Bell's inequality," *Lett. Math. Phys.* **4**, 93–100 (1980).

J. F. Clauser, M. A. Horne, A. Shimony and R. A. Holt, "Proposed experiment to test local hidden variable theories," *Phys. Rev. Lett.* **23**, 880–884 (1969).

CODATA Internationally recommended 2018 values of the Fundamental Physical Constants, https://physics.nist.gov/cuu/Constants/index.html (accessed December 4, 2020).

A. H. Compton, "A quantum theory of the scattering of X-rays by light elements," *Phys. Rev.* **21**, 483–502 (1923).

F. Cooper, A. Khare, and U. Sukhatme, "Supersymmetry and quantum mechanics," *Phys. Rept.* **251**, 267–385 (1995).

V. Degiorgio, "Phase shift between the transmitted and the reflected optical fields of a semireflecting lossless mirror is $\pi/2$," *Am. J. Phys.* **48**, 81 (1980).

P. A. M. Dirac, "Quantum mechanics," PhD thesis, University of Cambridge (1926); http://purl.flvc.org/fsu/lib/digcoll/dirac/dirac-papers/353070

P. A. M. Dirac, "The Lagrangian in quantum mechanics," *Phys. Z. Sowjetunion* **3**, 64–72 (1933).

P. A. M. Dirac, "A new notation for quantum mechanics," *Math. Proc. Cam. Phil. Soc.* **35**(3), 416–418 (1939).

P. A. M. Dirac, *The Principles of Quantum Mechanics*, Oxford University Press, (1981).

I. H. Duru and H. Kleinert, "Solution of path integral for H atom," *Phys. Lett. B* **84**, 185–188 (1979).

E. B. Dynkin, "Calculation of the coefficients in the Campbell–Hausdorff formula," *Dokl. Akad. Nauk. SSSR (NS)*, **57**, 323–326 (1947).

F. J. Dyson, "The radiation theories of Tomonaga, Schwinger, and Feynman," *Phys. Rev.* **75**, 486–502 (1949).

F. J. Dyson, "The threefold way: Algebraic structure of symmetry groups and ensembles in quantum mechanics," *J. Math. Phys.* **3**, 1199 (1962).

P. Ehrenfest, "Bemerkung über die angenäherte Gültigkeit der klassischen Mechanik innerhalb der Quantenmechanik," *Z. Phys.* **45**(7–8), 455–457 (1927).

A. Einstein, B. Podolsky, and N. Rosen, "Can quantum mechanical description of physical reality be considered complete?," *Phys. Rev.* **47**, 777–780 (1935).

H. Everett, "Relative state formulation of quantum mechanics," *Rev. Mod. Phys.* **29**, 454–462 (1957).

R. P. Feynman, "Space-time approach to nonrelativistic quantum mechanics," *Rev. Mod. Phys.* **20**, 367–387 (1948).

R. P. Feynman and A. R. Hibbs, *Quantum Mechanics and Path Integrals*, McGraw-Hill (1965).

V. Fock, "Zur Theorie des Wasserstoffatoms," *Z. Phys.* **98**, 145–154 (1935).

S. J. Freedman and J. F. Clauser, "Experimental test of local hidden-variable theories," *Phys. Rev. Lett.* **28**, 938–941 (1972).

G. Gamow, "Zur Quantentheorie des Atomkernes," *Z. Phys.* **51**, 204–212 (1928).

D. Giulini, C. Kiefer, E. Joos, J. Kupsch, I. O. Stamatescu and H. D. Zeh, *Decoherence and the Appearance of a Classical World in Quantum Theory*, Springer (2003).

R. J. Glauber, "Coherent and incoherent states of the radiation field," *Phys. Rev.* **131**, 2766–2788 (1963).

H. Goldstein, "Prehistory of the 'Runge–Lenz' vector," *Am. J. Phys.* **43**, 737 (1975).

H. Goldstein, "More on the prehistory of the Laplace or Runge–Lenz vector," *Am. J. Phys.* **44**, 1123 (1976).

H. Goldstein, C. Poole and J. Safko, *Classical Mechanics*, Pearson (2002).

M. Goodman, "Path integral solution to the infinite square well," *Am. J. Phys.* **49**, 843 (1981).

V. Gorini, A. Kossakowski, and E. C. G. Sudarshan, "Completely positive dynamical semigroups of N level systems," *J. Math. Phys.* **17**, 821 (1976).

H. Grassmann, *Die Ausdehnungslehre: Vollständig und in strenger Form begründet*, T.C.F. Enslin (1862).

R. W. Gurney and E. U. Condon, "Wave mechanics and radioactive disintegration," *Nature* **122**(3073), 439–439 (1928).

R. W. Gurney and E. U. Condon, "Quantum mechanics and radioactive disintegration," *Phys. Rev.* **33**, 127–140 (1929).

W. R. Hamilton, "On quaternions; or on a new system of imaginaries in algebra," *Lond. Edinb. Dublin Philos. Mag. J. Sci.* **25**(169), 489–495 (1844).

F. Hausdorff, "Die symbolische Exponentialformel in der Gruppentheorie," *Leipz. Ber.* **58** 19–48 (1906).

F. Hausdorff, "Dimension und äußeres Maß," *Math. Ann.* **79**, 157–179 (1918).

W. Heisenberg, "A quantum-theoretical reinterpretation of kinematic and mechanical relations," *Z. Phys.* **33**, 879–893 (1925).

W. Heisenberg, "Uber den anschaulichen Inhalt der quantentheoretischen Kinematik und Mechanik," *Z. Phys.* **43**, 172–198 (1927).

W. Heisenberg, "The observable quantities in the theory of elementary particles," *Z. Phys.* **120**(513), 673 (1943).

C. K. Hong, Z. Y. Ou, and L. Mandel, "Measurement of subpicosecond time intervals between two photons by interference," *Phys. Rev. Lett.* **59**(18), 2044 (1987).

H. Hopf, "Über die Abbildungen der dreidimensionalen Sphäre auf die Kugelfläche," *Math. Ann.* **104**, 637–665 (1931).

R. A. Horn and C. R. Johnson, *Matrix Analysis*, 2nd ed., Cambridge University Press (2013).

R. Horodecki, P. Horodecki, M. Horodecki, and K. Horodecki, "Quantum entanglement," *Rev. Mod. Phys.* **81**, 865–942 (2009) [arXiv:quant-ph/0702225 [quant-ph]].

D. Howard, "Revisiting the Einstein–Bohr dialogue," *Iyyun: Jerusalem Phil. Q.* 57–90 (2007).

E. T. Jaynes, *Probability Theory: The Logic of Science*, Cambridge University Press (2003).

H. Jeffreys, "On certain approximate solutions of linear differential equations of the second order," *Proc. Lond. Math. Soc.* **2**(1), 428–436 (1925).

P. Jordan, J. von Neumann, and E. P. Wigner, "On an algebraic generalization of the quantum mechanical formalism," *Annals Math.* **35**, 29–64 (1934).

E. H. Kennard, "Zur Quantenmechanik einfacher Bewegungstypen," *Z. Phys.* **44**, 326–352 (1927).

J. Kiefer, "Optimum experimental designs," *J. Roy. Stat. Soc. B* **21**(2), 272–304 (1959).

J. R. Klauder and B-S. Skagerstam, *Coherent States: Applications in Physics and Mathematical Physics*, World Scientific (1985).

M. Koashi and A. Winter, "Monogamy of quantum entanglement and other correlations," *Phys. Rev. A* **69**(2), 022309 (2004).

H. von Koch, "Sur une courbe continue sans tangente, obtenue par une construction géométrique élémentaire," *Ark. Mat. Astron. Fys.* **1**, 681–702 (1904).

H. A. Kramers, "Wellenmechanik und halbzahlige Quantisierung," *Z. Phys.* **39**(10), 828–840 (1926).

P. P. Kulish and L. D. Faddeev, "Asymptotic conditions and infrared divergences in quantum electrodynamics," *Theor. Math. Phys.* **4**, 745 (1970).

L. Landau, "Das dämpfungsproblem in der wellenmechanik," *Z. Phys.* **45**(5–6), 430–441 (1927).

L. D. Landau and E. M. Lifshitz, *Mechanics: Volume 1*, Butterworth-Heinemann (1976).

A. I. Larkin, "Thermodynamic functions of a low-temperature plasma," *J. Exptl. Theoret. Phys. (U.S.S.R.)* **38**, 1896–1898 (1960); *Sov. Phys. JETP* **11**(6), 1363–1364 (1960).

A. J. Larkoski, *Elementary Particle Physics: An Intuitive Introduction*, Cambridge University Press (2019).

R. B. Laughlin, "Anomalous quantum Hall effect: An incompressible quantum fluid with fractionally charged excitations," *Phys. Rev. Lett.* **50**, 1395 (1983).

E. H. Lieb and D. W. Robinson, "The finite group velocity of quantum spin systems," *Commun. Math. Phys.* **28**, 251–257 (1972).

E. H. Lieb and M. B. Ruskai, "Proof of the strong subadditivity of quantum-mechanical entropy," *J. Math. Phys.* **14**, 1938–1941 (1973).

G. Lindblad, "On the generators of quantum dynamical semigroups," *Commun. Math. Phys.* **48**, 119 (1976).

B. B. Mandelbrot, "How long is the coast of Britain? Statistical self-similarity and fractional dimension," *Science* **156**(3775), 636–638 (1967).

D. Manzano, "A short introduction to the Lindblad master equation," *AIP Adv.* **10**, 025106 (2020) [arXiv:1906.04478].

V. P. Maslov, *Theory of Perturbations and Asymptotic Methods* (in Russian), Izv. MGU Moscow (1965). Translation into French 1972.

J. C. Maxwell, *A Treatise on Electricity and Magnetism*, vol. 1, Clarendon Press (1873).

A. McKellar, "Molecular lines from the lowest states of diatomic molecules composed of atoms probably present in interstellar space," *Publ. Domin. Astrophys. Obs. Vict.* **7**, 251 (1941).

N. D. Mermin, "What's wrong with this pillow?" *Phys. Today* **42**(4), 9 (1989).

N. D. Mermin, "Could Feynman have said this?" *Phys. Today* **57**(5), 10 (2004).

A. A. Michelson and E. W. Morley, "ART. XLVII.–On a method of making the wavelength of sodium light the actual and practical standard of length," *Am. J. Sci. (1880–1910)* **34**(204), 427 (1887).

J. von Neumann, "Wahrscheinlichkeitstheoretischer Aufbau der Quantenmechanik," *Nachr. Ges. Wiss. Gott., Math.-Phys. Kl.* **1927**, 245–272 (1927).

J. von Neumann, "Thermodynamik quantenmechanischer gesamtheiten," *Nachr. Ges. Wiss. Gott., Math.-Phys. Kl.* **1927**, 273–291 (1927).

J. von Neumann, "Allgemeine Eigenwerttheorie Hermitescher Funktionaloperatoren," *Math. Ann.* **102**, 49–131 (1930).

J. von Neumann, *Mathematical Foundations of Quantum Mechanics: New Edition*, Princeton University Press (2018).

E. Noether, "Invariant variation problems," *Gott. Nachr.* **1918**, 235–257 (1918).

W. Pauli, "Über das Wasserstoffspektrum vom Standpunkt der neuen Quantenmechanik," *Z. Phys.* **36**(5), 336–363 (1926).

A. A. Penzias and R. W. Wilson, "A measurement of excess antenna temperature at 4080-Mc/s," *Astrophys. J.* **142**, 419–421 (1965).

M. Planck, *Sitzungsber. Deutsche Akad. Wiss. Berlin, Math-Phys Tech. Kl* **5**, 440–480 (1899); *Ann. Phys.* **306** [1] (1900) 69–122.

M. Planck, "Zur Quantenstatistik des Bohrschen Atommodells," *Ann. Phys.* **380**(23), 673–684 (1924).

J. Polchinski, *String Theory*, vol. 1, Cambridge University Press (1998).

J. Prentis and B. Ty, "Matrix mechanics of the infinite square well and the equivalence proofs of Schrödinger and von Neumann," *Am. J. Phys.* **82**, 583 (2014).

J. Preskill, *Lecture Notes for Physics 219/Computer Science 219 Quantum Computation*, http://theory.caltech.edu/ preskill/ph219 (accessed November 9, 2020).

J. W. S. B. Rayleigh, *The Theory of Sound*, vol. 1. Macmillan & Co. (1877).

A. Rényi, "On measures of entropy and information," in *Proceedings of the Fourth Berkeley Symposium on Mathematical Statistics and Probability*, Volume 1: Contributions to the Theory of Statistics, The Regents of the University of California (1961).

S. Ross, *A First Course in Probability*, Pearson (2014).

C. A. Rouse, "Comments on the Planck–Larkin partition function," *Astrophys. J.* **272**, 377–379 (1983).

J. S. Russell, *Report on Waves*. Report of the 14th Meeting of the British Association for the Advancement of Science, York, September 1844 (1845).

M. A. Schlosshauer, "Decoherence, the measurement problem, and interpretations of quantum mechanics," *Rev. Mod. Phys.* **76**, 1267–1305 (2004) [arXiv:quant-ph/0312059 [quant-ph]].

M. A. Schlosshauer, *Decoherence: and the Quantum-to-Classical Transition*, Springer Science & Business Media (2007).

E. Schrödinger, "Quantisierung als Eigenwertproblem," *Ann. Phys* **384**(4), 361–376 (6), 489–527, **385**(13), 437–490, **386** (18) 109–139 (1926); "Der stetige Ubergang von der Mikro- zur Makromechanik," *Naturwiss.* **14**, 664–666 (1926).

E. Schrödinger, "An undulatory theory of the mechanics of atoms and molecules," *Phys. Rev.* **28**(6), 1049 (1926).

E. Schrödinger, "Die gegenwartige Situation in der Quantenmechanik," *Naturwiss.* **23**, 807–812 (1935).

K. Schwarzschild, "On the gravitational field of a mass point according to Einstein's theory," *Sitzungsber. Preuss. Akad. Wiss. Berlin (Math. Phys.)* **1916**, 189–196 (1916) [arXiv:physics/9905030 [physics]].

A. M. Sirunyan *et al.* [CMS], "Observation of Higgs boson decay to bottom quarks," *Phys. Rev. Lett.* **121**(12), 121801 (2018) [arXiv:1808.08242 [hep-ex]].

A. Sommerfeld, "Zur Quantentheorie der Spektrallinien," *Ann. Phys.* **356**(17), 1–94 (1916).

A. Sommerfeld, *Atombau und Spektrallinien*, Vieweg (1919).

M. J. Steele, *The Cauchy–Schwarz Master Class: An Introduction to the Art of Mathematical Inequalities*, Cambridge University Press (2004).

S. M. Stigler, "Stigler's law of eponymy," *Trans. N. Y. Acad. Sci.* **39**, 147–157 (1980).

E. C. G. Sudarshan, "Equivalence of semiclassical and quantum mechanical descriptions of statistical light beams," *Phys. Rev. Lett.* **10**, 277–279 (1963).

T. Sulejmanpasic and M. Ünsal, "Aspects of perturbation theory in quantum mechanics: The BenderWu Mathematica ® package," *Comput. Phys. Commun.* **228**, 273–289 (2018) [arXiv:1608.08256 [hep-th]].

N. S. Thornber and E. F. Taylor, "Propagator for the simple harmonic oscillator," *Am. J. Phys.* **66**, 1022–1024 (1998).

G. Wentzel, "Eine Verallgemeinerung der Quantenbedingungen für die Zwecke der Wellenmechanik," *Z. Phys.* **38**(6), 518–529 (1926).

R. F. Werner, "Quantum states with Einstein–Podolsky–Rosen correlations admitting a hidden-variable model," *Phys. Rev. A* **40**, 4277–4281 (1989).

H. Weyl, *The Theory of Groups and Quantum Mechanics*, Courier Corporation (1950).

J. A. Wheeler, "On the mathematical description of light nuclei by the method of resonating group structure," *Phys. Rev.* **52**, 1107–1122 (1937).

G. C. Wick, "Properties of Bethe–Salpeter wave functions," *Phys. Rev.* **96**, 1124–1134 (1954).

E. P. Wigner, *Gruppentheorie und ihre Anwendung auf die Quantenmechanik der Atomspektren*, Vieweg (1931).

D. H. Wolpert and W. G. Macready, "No free lunch theorems for optimization," *IEEE Trans. Evol. Comput.* **1**(1), 67–82 (1997).

P. Zeeman,"Over de invloed eener magnetisatie op den aard van het door een stof uitgezonden licht; Over de invloed eener magnetisatie op den aard van het door een stof uitgezonden licht; On the influence of magnetism on the nature of the light emitted by a substance," *Verslagen Meded. Afd. Natuurk. Kon. Akad. Wetensch.* **5**, 181 (1896).

P. Zeeman, "XXXII. On the influence of magnetism on the nature of the light emitted by a substance," *Lond. Edinb. Dublin Philos. Mag. J. Sci.* **43**(262), 226–239 (1897).

P. Zeeman, "The effect of magnetisation on the nature of light emitted by a substance," *Nature* **55**, 347 (1897).

W. H. Zurek, "Decoherence, einselection, and the quantum origins of the classical," *Rev. Mod. Phys.* **75**, 715–775 (2003) [arXiv:quant-ph/0105127 [quant-ph]].

W. H. Zurek, S. Habib, and J. P. Paz, "Coherent states via decoherence," *Phys. Rev. Lett.* **70**, 1187 (1993).

Index